MORPHOLOGY OF CRYSTALS: PART C
The Geometry of Crystal Growth

MATERIALS SCIENCE OF MINERALS AND ROCKS

MSMR Materials Science of Minerals and Rocks

MORPHOLOGY OF CRYSTALS

Edited by Ichiro Sunagawa

Part C
The Geometry of Crystal Growth

by J. van Suchtelen

Terra Scientific Publishing Company / Tokyo

KLUWER ACADEMIC PUBLISHERS
DORDRECHT / BOSTON / LONDON

Library of Congress Cataloging-in-Publication Data

Morphology of crystals.

(Materials science of minerals and rocks)
Includes bibliographies and index.
1. Crystallography. 2. Crystals--Growth.
I. Sunagawa, Ichiro. II. Series.
QD921.M758 1987 548 87-9571
ISBN 90-277-2507-1 (Part A)
ISBN 90-277-2508-x (Part B)
ISBN 0-7923-3592-9 (Part C)
ISBN 90-277-2510-1 (Set)

Published by Kluwer Academic Publishers,
P.O. Box 17, 3300 AA Dordrecht, The Netherlands
in co-publication with
Terra Scientific Publishing Company (TERRAPUB)
2003 Sansei Jiyugaoka Haimu, 5-27-19 Okusawa, Setagaya-ku, Tokyo 158, Japan

Distributors for Japan
Terra Scientific Publishing Company, 2003 Sansei Jiyugaoka Haimu, 5-27-19 Okusawa, Setagaya-ku, Tokyo 158, Japan

Distributors for the U.S.A. and Canada
Kluwer Academic Publishers,
101 Philip Drive, Assinippi Park, Norwell, MA 02061, U.S.A.

Kluwer Academic Publishers incorporates
the publishing programmes of
D. Reidel, Martinus Nijhoff, Dr W. Junk and MTP Press.

Distributors for all remaining countries
Kluwer Academic Publishers Group,
P.O. Box 322, 3300 AH Dordrecht, The Netherlands.

Printed in the Netherlands

The aim of all science is to understand the evolving, rather than to know the existing.

Dr. Nico van Suchtelen (1878-1949),
from his novel "Tat tvam asi" (Notebook of a crystal gazer), 1933 *)

*) *Dr. Nico van Suchtelen* (1878-1949), scientist and novelist, was the author's grandfather. His novel "Tat tvam asi" (1933) - Sanskrit for "That is you" - was given the subtitle "Notebook of a crystal gazer" and has an essayistic touch, discussing political and ethical topics in pre-WW II Europe in the framework of a novel. (the crystal, by the way, was a talisman pendulum)

CONTENTS

Preface

The manuscript in the drawer either rots or ripens.

Marie von Ebner-Eschenbach (1830-1916)

Years ago, working as a research scientist in the Philips Research Laboratories community, I was confronted with interface-instability and related crystal growth problems which were then a topic in several (different) research groups working on (different) crystals. The attempt to find a generalized attack strategy naturally suggested the search for mathematical similarities - which revealed that the "continuum description of crystal growth" was (as it still is) rather underdeveloped in comparison with the more traditional, "atomistic" versions of crystal growth theory. (the "Burton-Cabrera-Frank approach").

Since then I tried to work out some tools to cultivate this field. The present monograph reports results of this work. They are to be characterized as the development of a "toolbox" rather than as a set of worked-out applications, but it may be relevant to make these tools available here as a sort of interim report.

The project of preparing this monograph, I have to confess, had its ups and downs. The basis for chapters 2-8 was developed in my Philips Res. Lab. period (ending in 1983), and was originally intended for publication as a regular paper in J. Cr. Growth. Finding that the material was too massive for even a series of regular articles, I considered publication in the form of a monograph. In the meantime, I changed to Nijmegen University, where I worked on a technical project which left little time for writing. But I did discuss the "continuum approach" topic extensively with Bennema and Van der Eerden, whose comments and enthusiasm helped much to improve the manuscript. In 1986 I gladly accepted prof. Sunagawa's invitation to include my monograph as a closing chapter in his Morphology of Crystals series, although at that time I still had to rewrite most of it.

The idea of the network representation of growth rate functions gradually developed in the same period, and I tried to include this material in the monograph because I felt that the practical value of the rather abstract chapters 2-8 would be improved by the availability of analytic growth rate functions. As I had to do this as a private enterprise rather than as part of my regular job, this took (much) more time than anticipated - finally the topic grew out into a five-chapter Part II.

fication of eutectics, nucleation and whisker growth. Instead, most authors seem to focus here on the diffusion phenomenon in the (generally non-crystalline) parent phase from which the crystal grows. This, of course, cannot explain any shape characteristic which is related with the crystallographic orientation.

Two reasons may be responsible for this situation. In the first place, the availability of data is poor, both on orientation dependence of γ and on that of R. The determination of such data requires specialized experiments, and one may be tempted to use averaged values, especially when the fractional variation seems to be moderate. However, this criterion can be misleading. In the relevant formulae, $\mathbf{S}\gamma$ and $\mathbf{S}R$ take the place of γ and R for the orientation-independent case, $\mathbf{S}$ symbolizing the operator $(1 + \mathrm{d}^2/\mathrm{d}\varphi^2)$, where φ is an orientation-angle parameter. For a "smooth", almost-constant γ or R function (i.e., applying to low-α-factor surfaces), the influence of orientation dependence can be estimated, in the lower limit, by considering the function $(1 + \varepsilon \sin n\varphi)$ as a factor expressing this dependence. Then in $\mathbf{S}\gamma$ or $\mathbf{S}R$, we have the factor $\{1 - (n^2-1) \sin n\varphi\}$, i.e. the effect of the relative modulation by the fraction ε in the parameter itself is increased by the factor (n^2-1). For the "simplest" case of cubic symmetry, $n = 4$, so we underestimate the effect by a factor of 15 when we look only at the modulation in γ or R. For sixfold symmetry, the factor is 35, etc. Now this factor is related with curvature in one direction only; taking the other curvature into account brings in another factor of ~2. So the "underestimation factor" is 30 at least! For sixfold symmetry, the factor is 35 × 2, etc.

For less smooth orientation dependence (higher α-factor value), the functions become more cusped and correspondingly the $\mathbf{S}\gamma$ or $\mathbf{S}R$ will vary even stronger than for a sin $n\varphi$ modulation, due to higher harmonics - ultimately peaking to a δ function for a real cusp. Concluding, we can say that the importance of orientation dependence, when simply estimated from deviations from the average value, is underestimated by roughly two orders of magnitude or more.

In the second place, the familiar textbook discussions on equilibrium shape (Gibbs-Wulff plot) and on growth shape (kinematic wave theory) are restricted to the most simple case of constant growth affinity. In most practical situations, however, we have to do with nonequilibrium shapes and with an inhomogeneous affinity field. (as a further complication, the latter may have to be solved from a diffusion equation with boundary conditions defined at far-away surfaces). So, in order to tackle such problems, we need some sort of "equation of motion" (EOM) for a crystal surface which accounts for the dependence of growth rate on both orientation and affinity, plus the curvature-induced shift in the affinity due to (orientation- dependent) surface energy. CHERNOV (1962, 1963) has used such an equation (without capillarity corrections) in the discussion of growth forms and the stability of faceted crystals. MÜLLER-KRUMBHAAR c.s. (1977) have defined a "complete" equation for the two-dimensional case, including the capillarity term, and applied it to the shape of growth spirals.

In the more recent literature, a few more examples of continuum approaches to complex crystal-growth problems can be found, both via analytical and numerical treatment: e.g. on dendritic growth (e.g., BRUSH and SEKERKA (1989), TARABAEV, MASHIKHIN and ESIN (1991), SETHIAN and STRAIN (1992). Although OD γ and R are taken into account in these studies, the cases considered are limited to ad hoc postulated ε sin $n\varphi$ modulations. The lack of experimental possibilities for comparison between the simulated and the actual dendrite are a serious obstacle for further progress.

1.2 "LEVELS OF DESCRIPTION" OF TRANSPORT PHENOMENA

As indicated in 1.1, the scope of this book can be characterized as an attempt to develop a *continuum description of crystal growth* - or, more explicitly, to extend the existing continuum descriptions of transport phenomena (i.e., Navier-Stokes, diffusion, chemical reaction rate etc. equations) so as to include crystal growth phenomena. Mathematically, this implies that we have to complement the 3-D equations mentioned above with equations describing the transport across the 2-dimensional phase interfaces, i.e. with *boundary conditions* for the 3-D transport fields, depending on such parameters as orientation, curvature, temperature and composition.

In order to adstruct our view that such a continuum description has remained rather underdeveloped in comparison with the more traditional, atomistic versions of crystal growth theory, we start to point out that three levels of description can be discerned in transport theories in general: "atomistic", "continuum" and "reduced". Below, these levels are illustrated by the example of transport phenomena in gases and liquids:

1) The atomic/molecular level: the kinetic theory of gases and liquids. Transport phenomena such as diffusion and viscous flow are considered as the statistical outcome of collision kinetics between individual particles. Typical authors are Maxwell, Boltzmann, Enskog, Kirkwood etc. In parallel with the irreversible aspects (viscosity, diffusion, heat conduction) we have the equilibrium aspects as the outcome of "statistical mechanics": partition sum, definition of entropy, temperature etc. as based on many-particle equilibrium statistics.

2) The next level of description is that of continuous media. Transport processes are considered in terms of flux density vectors driven by potential gradients; relations between fluxes and potential gradients are governed by phenomenological coefficients (diffusion constant, viscosity, heat conduction coefficient).

At this level of description, the atomic structure of matter is irrelevant; rather, matter is considered as a continuum. We might say that the "magnification" of our observation is lowered so that the atomic structure can no longer be discerned. What comes into view by this same lowering of magnification, is the larger-area picture of transport processes as a map of flux-vector and potential "fields", guided by geometrical boundaries. These fields are mathematically described by equations or systems of equations like the Navier-Stokes, diffusion and reaction-rate equations, together with their boundary and continuity conditions.

This level of description is suitable to discuss transport phenomena on a technical or a biological scale, such as we have to deal with in combustion engines, chemical reactors, design of ships or airplanes and biological processes. The first level would be quite inadequate for this purpose. The connection between the levels is that the first level provides an understanding of the *phenomenological coefficients* (viscosity, conduction etc) which are implicit in the differential equations describing the transport fields.

In parallel with the *transport fields* again, we have at the second level a characterisation of the *equilibrium properties* of our continuous media as described by equilibrium thermodynamics. (with medium-specific "continuum" properties like free energy, specific heat etc.)

3) The third level of description comes into view when we reduce the magnification of the system once more so that the complete situation comes into view as a whole. Comparing different but similar systems, we can observe that their transport fields can be made to coincide provided that space and time dimensions are scaled by suitable factors. The similarity of the solutions (based on the similarity of the related differential equations) provides the possibility to describe the transport situation in terms of a considerably reduced set of parameters. At this level, the sight on details in the flux-vector and potential fields is lost and only overall or integrated fluxes and potential differences can be discerned.

This level of description - i.e. the technique of presenting overall results in terms of "dimensionless numbers" - is extremely useful in many physical and technical transport situations. In particular, the dimensionless numbers can be used to characterize a situation or to form the basis for a catalogue of all possible situations. Moreover, changes in microscopic behaviour of the physical system related with the onset of instabilities can often be expressed in terms of critical values of such dimensionless numbers (e.g., the laminar-turbulent flow transition at Re ~ 2000).

Of course, the three description levels sketched above do not stand apart - they are linked together by well-defined mathematical procedures.

Now we have clearly defined the three description levels of transport theory in general, we proceed to review the situation in the particular field of crystal growth theories. As the third ("reduced") level follows quite naturally from the second ("continuum"), we omit it from the review for reasons of compactness.

Atomistic and continuum descriptions of growth and equilibrium

The following framework illustrates the concept of the levels sketched in 1.2 but now applied to the special field of crystal growth. We will use it to define the scope of this monograph in more detail. The framework is represented below. The four distinctions which together lead to its structure are

1) "atomic-scale" models versus "continuum" models (vertical);
2) equilibrium versus nonequilibrium cases (horizontal);
3) unary case versus polynary case;
4) (as said in 1.2, we omit to mention the "reduced level" as it may be considered to be implicit in the continuum level)

<table>
<tr><th></th><th>EQUILIBRIUM</th><th>KINETICS</th></tr>
<tr><td rowspan="2">"ATOMIC SCALE"</td><td>Equilibrium thermodynamics of the surface; theory of roughening transition
Burton, Cabrera, Frank
Leamy, Gilmer, Jackson</td><td>Irreversible thermodynamics of the surface (ledge advance rate, birth-and-spread mech. etc)
Burton, Cabrera, Frank</td></tr>
<tr><td colspan="2">Temkin</td></tr>
<tr><td rowspan="2">"MACROSCOPIC"</td><td>Theory of equilibrium shape
Gibbs-Wulff
Herring</td><td>Kinematic-wave theory
Frank
Chernov</td></tr>
<tr><td colspan="2">generalized EOM
Müller-Krumbhaar
present monograph</td></tr>
</table>

Restriction to the unary case implies that we are only dealing with *geometrical* aspects of crystal growth. Extension to the aspect of chemical composition requires all discussions on the unary case to be embedded in the generalized polynary description. A rough sketch of such a generalization is given in ch. 8.5, but for the rest the subject is left aside in the present monograph. This restriction leaves us with the 4-block structure above.

The upper-left block "equilibrium thermodynamics of atomic-scale models" is concerned with the statistical thermodynamics of surfaces, including the theory of the roughening transition. Prototype examples are the papers by BURTON, CABRERA AND FRANK (1951) and by LEAMY, GILMER AND JACKSON (1975). The equilibrium (atomic-scale) structure of the interface is found together with the interface free energy by minimizing the expression for the free energy, obtained from the partition function. Due to the mathematical complexity, such calculations have been restricted to model systems obeying simplified statistics or interactions, such as solid-on-solid and two-dimensional Ising models. Moreover, most calculations have been focused on the structure of interfaces with fixed low-index orientations rather than of interfaces with variable orientation. (unfortunate for our purposes, because in our $\vec{\mathbf{Q}}$ and $\vec{\vec{\mathbf{S}}}$ operator formalism we need the first and second *derivatives with respect to the orientation variables*).

The lower-left block "equilibrium theory in continuum descriptions" refers to the classical theory of equilibrium shape (GIBBS 1876, HERRING 1951, 1952, CAHN AND HOFFMANN 1972). The point of departure here is the function $\gamma(\vec{n})$, i.e. the surface free energy as a function of the continuous orientation variable. In principle this function could be evaluated, for any atomic-scale model, as an end result of the upper-left block by rigorous application of the procedures of statistical thermodynamics. The most simplistic model of this kind is the broken-bond model, leading to a radial-plotted $\gamma(\vec{n})$ function built up of circles or spheres. ("raspberry" or "soapfroth" plot). Unfortunately, attempts to evaluate more realistic $\gamma(\vec{n})$ functions have been scarce. It seems to the author that this is a white area in the theory, especially with regard to the details of the $\gamma(\vec{n})$ function for large misorientations and, at near-facet orientations, for temperatures close to the roughening transition (smoothing of the cusp). In chs 9 - 12, we present a highly generalized approach to the construction of $\gamma(\vec{n})$ and $R(\vec{n})$ functions which in principle allows for the representation of such details.

The upper-right block refers to the atomistic models of "surface kinetics", i.e. of the potentially complex transport field involving adsorption, surface diffusion, ledge diffusion and kink-site incorporation. Additional complications due to interaction with impurities and to chemical steps (such as in CVD processes) are to be situated in the analogous polynary upper-right block. The prototype discussion of this "diffusion delta" of coupled and branching transport fluxes is that by BURTON, CABRERA and FRANK (1951). Note that the "supporting elements" of the diffusion delta, i.e. the ledges, kinks and interledge surfaces, are the equilibrium structures of the upper-left block (which explains the similarities between $\gamma(\vec{n})$ and $R(\vec{n})$ functions). A more direct approach, taking into account the effect of nonzero growth affinity on the surface structure itself, was given by TEMKIN (1964).

In the same block we have to situate discussions on the interactions between advancing ledges, oscillations of ledge trains etc.

The lower-right block, covering the macroscopic *"continuum"* description of kinetics, is essentially FRANK's (1958) kinematic-wave theory. Here the starting point is the $R(\vec{n})$ function. (Above, we argued that the surface mobility = R/A plays essentially a similar role as , e.g., the diffusion coefficient in Fick's law - with the complication that the R function depends on the orientation of the surface. This complication is the very basis for the wealth of crystal geometries and surface instabilities).

Just as in the $\gamma(\vec{n})$ case, the $R(\vec{n})$ function might in principle be derived from the atomistic theories of the upper-right block, but in practice the complications for higher-index orientations (interactions between neighbouring ledges) and for growth conditions near the roughening transition (simultaneous ledge-motion and birth-and-spread mechanisms) make such calculations unreliable. Just as in the left-hand blocks, the connection between upper and lower block, i.e. the determination of the $R(\vec{n},A)$ function for $\vec{n}$ variable over the unit sphere, remains a white area.

R and γ data and the equation of motion of a surface

In conclusion, in the discussion of OD effects, we have to distinguish two areas of interest. The first is that of the "data": the functions $\gamma(\vec{n})$ and $R(\vec{n},A)$ themselves. Either they should be measured by adequate experiments (see ch.13) or they should be calculated as the final result of the upper blocks. In chapters 1-8 of this monograph, this part of the problem is completely left out of consideration. In chs. 9 - 12, however, we present a generalized approach to the construction of R and γ functions. Although the results of this approach are not derived from the upper-right block, they represent a sort of "envelope" of solutions - i.e. they provide the means to express considerable detail in the form of analytical functions (which can be inserted in the equations of motion). The functions are parametrized, with parameters which can physically be assigned to phenomena like surface roughening, effects of adsorbed species, etc. The explicit correlation between these parameters on the one hand, and the details of the atomistic models on the other hand, remains an open - and intriguing - field of research for the moment.

On the other hand, the same parametrized functions provide a way to analyze results from, e.g., sphere-growth experiments with more subtlety, so as to derive information on surface roughening etc. via these parameters. (ch. 13).

The second area of interest is the rôle of these functions in an equation of motion and the consideration of the limitations and consequences of such equations. This, in short, is the subject of chapters 2-8 of our monograph, where the functions $\gamma(\vec{n})$ and $R(\vec{n},A)$ are considered as given data, defined as properties of a mathematical surface between continuous media. In the sketch of the framework, it is defined by the combination of the lower blocks, i.e. by the MÜLLER-KRUMBHAAR c.s. (1977) approach.

We emphasize that this subject belongs to the field of differential geometry rather than to "crystal growth theory" in the usual sense. The EOM, for instance, apply equally well to acoustic wavefronts in elastically anisotropic media or to Bloch walls in magnetic domain patterns, when the interpretation of R and γ functions is adapted accordingly.

We have to remark here that the formulation of an EOM (whichever version of them) is not much of an achievement in itself. The interesting points are rather lying in such complications which are *not* implicit in the EOM: i.e. in the consideration of the limitations of their applicability. Such limitations are, of course, inherent in the fact that crystal/parent phase interfaces are *not mathematical surfaces between continuous media at all.* In reality, neither the crystal nor the parent phase is a continuous medium and the interface cannot be said to have zero thickness at lattice-spacing resolution. We therefore have to be suspicious as to the applicability of the equations to cases where high curvatures are involved. In section 1.3 we will argue that these limits of applicability - in terms of curvature - can be pushed ahead somewhat further by considering both γ and R as curvature-dependent properties. However, even this trick does not work the whole way to "atomically" sharp edges.

1.3 THE FUNDAMENTAL RESTRICTION OF CONTINUUM DESCRIPTIONS

The limitations of a continuum description of crystal-growth physics, based on $R(\vec{n},A)$ and $\gamma(\vec{n})$ functions, become evident when we recognize that these functions are "delocalized", i.e. cannot be attributed to a single point of a mathematical surface but are rather to be considered as integrals of a distributed quantity which is spread out over a finite space element (size δ) surrounding the definition point. In the BCF formula for ledge advancement rate, for instance, the diffusion field over an area of at least the terrace width has to be taken into account. Obviously we are in trouble when the curvature of the surface is so high that the $\gamma(\varphi)$ and/or $R(\varphi)$ functions vary appreciably when φ varies over the range δ/r.

To a certain extent we can overcome this trouble, *still adhering to our continuum concept*, by considering $\gamma(\varphi)$ and $R(\varphi)$ as functions of curvature too. This approach has been worked out in some detail in ch. 2 for various physical situations. In general, the dependence on curvature takes the form of one or more terms containing powers of $1/r$. Of course, the curvature dependence complicates the EOM, but the amended EOM provides a more precise description of interface behaviour.*)

However, such an amendment really works "to a certain extent" only - the $1/r$ terms explode to a nonphysical singularity for $r \to 0$ and even for $r \simeq$ unit-cell size they are certainly not realistic. In the detailed balance statistics of equilibrium and kinetics of details in which only a few atoms are involved, energetic and kinetic effects cannot be separated and therefore the EOM (with separate terms for these effects) *loses its validity altogether. Basically this boils down to the point that the geometrical concept of curvature is in itself no longer applicable to such a (physical) surface.*

It is important to be aware of this restriction and consequently of the "status" of the EOM - either without or with curvature amendments - for small or "sharp" details in an interface structure. This is relevant, for instance, when we investigate the stability of a zigzag interface structure with negative-$\mathbf{S}\gamma$ and/or $\mathbf{S}R$ regions (ch. 4.2 and 6.3). Typically the situation is as follows:

1) The *unamended* EOM provides no clue for stability.

2) The *amended* EOM does: one or more of the $1/r$ terms result in a stabilized zigzag solution which may physically be considered as a solution with "solitons" (i.e., stabilized reentrant edges or other sharp features). If we are so lucky that this soliton solution can be worked out explicitly from the EOM (curvature corrections included), then we have a numerical result for, e.g., the vector velocity of the reentrant edge as well as for the orientation of the edges. But what is the value of such a result?

3) The value of this result is twofold: a) Quantitatively, the soliton properties (velocity, orientations) cannot be trusted. Most probably the

*) Also the capillary-shift term ΔA_c (see section 1.5 and eq. 1.2a), with its $1/r$ factor, can be considered as a "curvature correction" to an EOM in which surface energy effects were originally neglected. For this case too, the expression for ΔA_c (see eq. (3.10b) becomes suspect for r approaching the unit-cell size.

radius of the resulting solution is too small for the $1/r$ terms in the EOM to be applicable. b) However, we may consider the *existence* of the soliton solution qualitatively as a *hint that a stabilized solution exists*; only the exact data (velocity, orientations) cannot be extracted from the continuum EOM. We can obtain these exact data either by statistical calculation (Monte Carlo simulation on a suitable model) or by the actual growth experiment; i.e. use the crystal itself as the Ultimate Analog Computer.

In this monograph we will frequently refer to the non-applicability of the EOM to "too small" interface details as the "Fundamental Restriction" (FR) of the continuum approach. Associated with this FR is the suggestion that an eventual soliton solution of an EOM may at best be considered qualitatively as a hint to the existence of a stable-stationary structural detail, but certainly not quantitatively.

1.4 THE CRYSTAL, THE MODELS AND THE FUNCTIONS.

In this monograph we discuss the behaviour of crystal surfaces in terms of functions of a - supposed continuous - orientation variable.
By "functions" we mean in this section, very generally, functions for growth rate or surface free energy, or for whatever other physical parameter which might be useful to describe the behaviour of a surface. In fact, the crystals themselves - their structure and growth kinetics - are hardly spoken about, and the atomistically-detailed models, which might be used as a basis to *calculate* the functions, may seem to have escaped attention altogether.

As said elsewhere in this introduction, this choice of focus has been made deliberately - although the insight in relations between the details of the models and the parameters occurring in the functions is essential to make this monograph a *practical* tool. Yet it is instructive to discuss in some detail what is actually represented by such "functions".

To state things explicitly: consider two hypothetical crystal-growth scientists, both working on the topics described in this book, but differing in attitude and available equipment.

Scientist A lacks the insight that crystalline matter is built up of atomic or molecular units and really thinks that it is a continuous medium; although by some trick of nature its surface has orientation-dependent properties. Apart from that, he hates to think in terms of models, but he is a skilled experimentalist and equipped with a laboratory full of sophisticated contraptions to grow crystals and to measure their $\gamma(\vec{n})$ and $R(\vec{n})$ functions.

Scientist B, in contrast, has went through a solid education in crystallography; his desk is stuffed with models made of coloured balls, a powerful workstation with a megascreen and a mess of blocnotes, and he delights in computational physics.

For scientist A, the situation is enviably clear: there is only the crystal and the function (in singular), with nothing in between. (fig.1.1) The function, directly determined by experiments on the actual crystal, represents the property of the crystal surface "as it is" - basta. When scientist A reads this book (skipping chs. 9-12) and tries to verify the formulas, he will find (or so I hope) that they nicely conform with experiment. Sometimes, when working on very small crystals or very small details of big crystals, he may find that the equations do not work properly or rather that the "constants" in the functions seem to become size-dependent. (his situation then may somehow resemble that of the theoreticians who were intrigued by the black-body radiation law before the days of Planck).

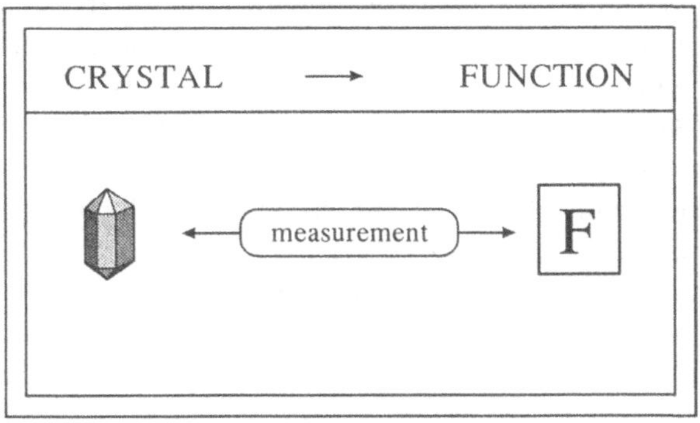

FIG. 1.1

For scientist B, life is more complicated. He is fully aware that a crystal is a complicated structure of nuclei of different species, surrounded by electron clouds and held together by interactions of all sorts. Phonons and eventual free electrons hurry around and interact with the surface. It is not easy to represent the actual crystal by a model to which "computational physics" - of whatever kind - can be applied. Although B has a library where he can find most data about interactions between atoms, molecules, nuclei and electrons, plus more or less attemptive algebraic representations for these interactions, it is quite difficult to work out a "model" which takes properly account of all the interactions and yet is compact enough to be digested by the computer in a reasonable time. Scientist B has to do with the actual crystal, *the model and* the function. There is a translation stage between the crystal and the model, and a calculus stage between the model and the function. (In some cases, the difficulties encountered in the calculus stage may induce B to choose an adapted model or even to discard a model altogether.) Anyhow, as a consequence of the possibility to make *choices* in the translation stage, he has to do with models and with functions, both in the plural, for one and the same crystal. In other words, all models and functions have the character of hypothetical or preliminary models and functions, until they have been verified by experiment. It is for the experimenter to find out which of the functions agrees best with reality - i.e. which of the models, by virtue or by coincidence, has the closest agreement with the real crystal.

It is important to be aware that the choices to be made in the translation stage are actually of two kinds. (fig. 1.2). The first kind of choice is for the interactions to be taken into account and for the analytical representation of these interactions; i.e. the basic assumptions of the model. Examples are: the number of neighbours in a crystal lattice to be accounted for; the choice for interaction potentials; the choice for an SOS model. Such choices are to be based on estimates of the relative importance of terms in equations. The functions which emerge from the models are functions in their own right - they need not correspond with reality, but they do correspond with the model, and as such it is quite legal to insert them into an EOM to find out how successfull the model is.

The second kind of choice is quite different. It emerges after the first choice has been made, but now our computational physicist is puzzling about *how the actual surface has to be imagined.*

This second kind of choice would not exist if crystalline matter were indeed continuous. A surface then could be mathematically smooth and differentiable and the definition of surface orientation would be straightforward. Not so for a real crystal. Looking at his coloured- ball models, our computational physicist soon realizes that "orientation" is a typically macroscopic concept which leaves some arbitrariness - or a “degree of freedom” - on a microscopic scale. *The second kind of choice is the choice within this degree of freedom.* *)

*) A typical example which will be familiar to most readers of this book is the degree of freedom which is “absorbed” in the phenomenon of surface roughening. The most simple and elementary version of a theoretical description may be that by JACKSON, UHLMANN and HUNT (1967) where the freedom is confined to a monolayer of adatoms.

As we will see in ch. 7, the different possibilities allowed by this freedom actually represent the elements of a statistical ensemble of situations, rather like the ensembles of statistical thermodynamics. Whereas the latter represent N-particle systems which can be represented in a 6N-dimensional phase space, the “particles” or units dealt with in the surface-geometry ensembles are rather adatoms (for roughened crystal surfaces), ledges (for faceted surfaces) or even macrosteps or terraces (= “condensates” of ledges). In ch. 7 we will propose the view that ledges and macrosteps can be considered as solitons in the adatom-density and the ledge-density fields, respectively. For a discussion of *their* statistical physics, we would have to represent such ensembles in an appropriate “phase space” which, as far as the author is aware, is still to be invented.

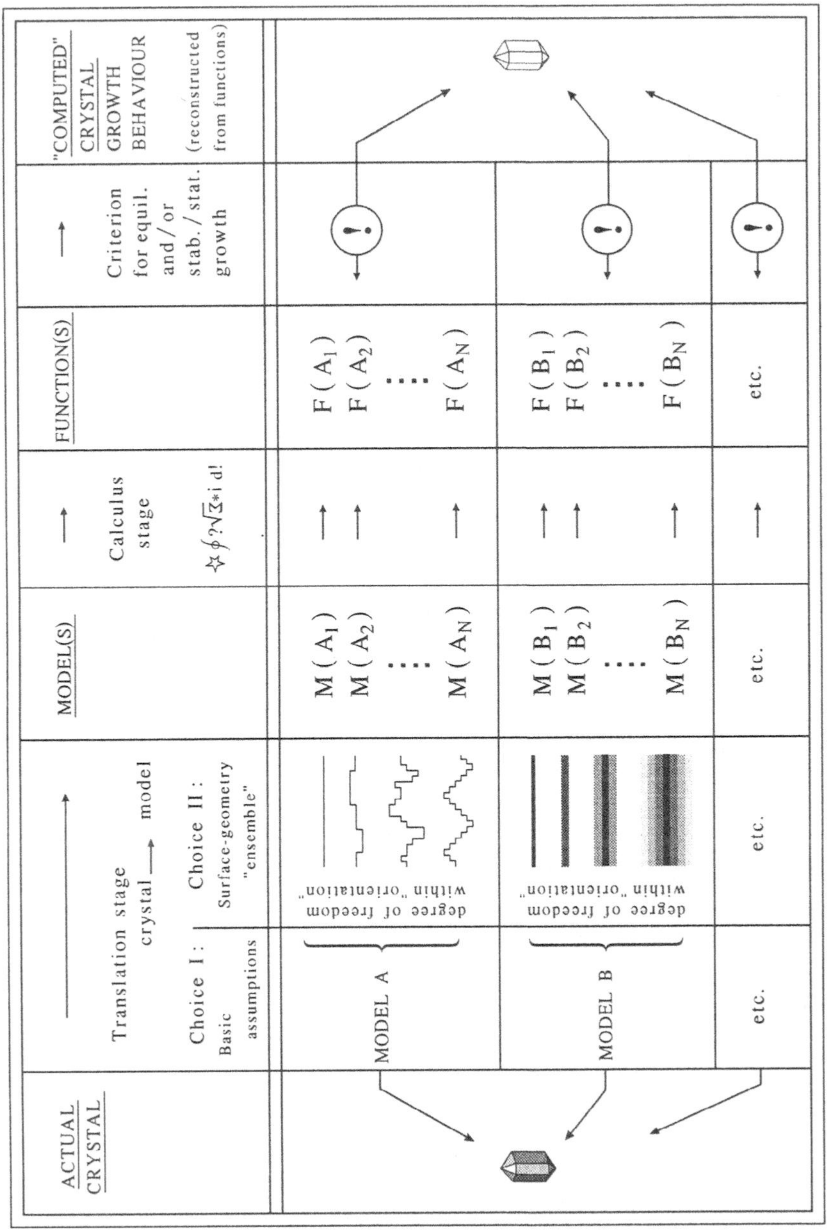
ACTUAL CRYSTAL
Translation stage
crystal → model
Choice I : Basic assumptions
Choice II : Surface-geometry "ensemble"
MODEL A
degree of freedom within "orientation"
MODEL B
degree of freedom within "orientation"
etc.
etc.
MODEL(S)
M (A1)
M (A2)
M (AN)
M (B1)
M (B2)
M (BN)
etc.
Calculus stage
FUNCTION(S)
F (A1)
F (A2)
F (AN)
F (B1)
F (B2)
F (BN)
etc.
Criterion for equil. and / or stab. / stat. growth
"COMPUTED" CRYSTAL GROWTH BEHAVIOUR
(reconstructed from functions)

FIG. 1.2

Different choices result in different models (i.e., differing in microscopic geometrical structure) and these, again, could yield different functions. Once these functions have been worked out, they can either be directly confronted with experiments to see which fits best to reality, or they can be selected by some criterion (such as minimum-free-energy for equilibrium situations, minimum entropy production for growth situations). The selected function corresponds with a selected model and thus with a particular choice of surface structure.

In the definition of orientation as a statistical average, an evasive difficulty is hidden which is related with the fact that averaging can be performed at different scales. To state that the orientation parameter is an average is not enough: it is necessary to specify the length scale over which the averaging is carried out, and what is the statistical deviation between different averages. (obviously, there is a minimum length over which to average for the orientation concept to make sense: e.g. at least two lattice units). The possibility to average over different length scales implies that we can have "models within models". An illustrative example is the discussion of functions $\Psi(\vec{n})$ in relation with an orientation distribution function $F(\vec{n})$, to be found in ch. 11. In principle this discussion is based on the idea that a crystal surface with macroscopic orientation $\vec{n}$ can have, on closer look, a wavy or corrugated structure of a "statistical" kind (fig. 11.2,c). Such a corrugation might be stabilized, for instance, by the entropy gain brought about by "absorbing" the corresponding degree of freedom.

When the Fourier spectrum of the surface structure cuts off below a length scale ≫ unit-cell size, we might for the sake of surface properties such as Ψ and R consider such properties as depending on a local orientation $\vec{n}_{\text{local}}$. This implies that the "macroscopic" orientation $\vec{n}_o$ is in fact a statistical average - fluctuations in $\vec{n}_{\text{local}}$ cannot be seen by lack of resolution. and "macroscopic" could actually be characterized as "lores" in computer-screen terminology. Following this line of thought, the wavy surface of fig. 11.2c then would have to be characterized as "medres" - there is still an averaging process needed to determine a local orientation $\vec{n}_{\text{local}}$ but this variable is fluctuating over the surface and its spread might be represented by a distribution function $F(\vec{n})$. *) Finally, in "hires" magnification we would see the surface structure at atomic scale and the concept of an orientation distribution - or its relation with surface properties - would cease to be applicable. Note that the latter remark hints at an intrinsic difficulty in the use of the "orientation" concept and in the definition of a distribution function $F(\vec{n})$ in particular. (what kind of surface

*) One particular and more specific definition of "medres" is the resolution in which individual atoms cannot be seen, but ledges (single-atom height steps) can be seen - the individual ledges being translatable into "orientation". We will work out this point in ch.7.

would correspond with a delta-spike distribution, for instance?)

In chapter 11 we will show in more detail that the tensor operator **S**, applied to an orientation-dependent surface quantity Ψ, has a relation with the $F(\vec{n})$ distribution function - whatever its meaning may be (eqs. 11.10b and 11.19b). This also illustrates the vagueness in the definition of $F(\vec{n})$.

For the time being, we want to highlight that there are as many functions as models. Some of the models may be physically realistic, while others are not. The same applies to the functions. Still, all such functions can be inserted into the equations of motion - when they relate to nonrealistic hypothetical models, they still describe the behaviour of those models. Indeed, the equations can *prove* that a model is unrealistic, by showing that the surface rapidly evolves towards another more stable form. An ultimate stabilized form of the surface will correspond with one particular choice of the model and of the surface geometry. This procedure is quite analogous to the ensemble procedures of statistical (equilibrium) thermodynamics.

In most of this book we will hardly be concerned with the question whether a function describing $\gamma(\vec{n})$ or $R(\vec{n})$ corresponds with a hypothetical but unrealistic case or with a realistic (equilibrium, stabilized) case. We might think of expressions for such functions as printed in colours and in black, respectively - to become relevant only as soon as we confront the end results of the equations with experiment. Still, we thought it appropriate to make the reader aware of the various "levels of hypothesis" involved in this computational-physics approach of crystal growth phenomena.

1.5 CHARACTERISATION OF THE THREE-DIMENSIONAL AFFINITY FIELD IN CRYSTAL GROWTH SITUATIONS

1.5.1 Nonequilibrium interface kinetics and the growth affinity.

In the generalized approach to crystal growth phenomena which is emphasized in this monograph, we avoid to distinguish between growth situations such as growth from the melt, from solution, from the vapour phase etc. Moreover, the comparison of growth kinetics for the same crystal but in different growth situations is interesting from a theoretical point of view. It is therefore essential to express the driving force for such different growth situations in identical units, rather than in case-specific units such as (absolute or relative) undercooling, supersaturation etc. Another (practical) advantage of the use of universal units is that comparison of experimental data from different sources is facilitated. (unfortunately, most authors are still to be converted to this habit).

We are thus led to use either the "growth affinity" or the "growth pressure" as the elementary driving force for crystal growth (VAN LEEUWEN (1979), SOHNEL and GARSIDE (1979), LUNDAGER MADSEN (1991)). They can be expressed in units of energy (e.g., kcal/mole) or of pressure, respectively.

First, we have to point out that there are two alternative and complementary points of view in the description of interface motion. Consider fig. 3.1 which represents a number of grains in a polycrystalline sample during a sintering process. The grain boundaries are schematically indicated at two different moments (drawn lines first, dotted later). So, for instance, the small grain numbered 5 is "dissolving" into the surrounding grains, etc.

The transport process at the 1/2 interface can alternatively be considered as a transport of the *atoms* from grain 1 to the *outside* or as a transport of the *interface between grains 1 and 2* to the *inside* of grain 1. The first description applies to bulk-phase material; consequently we can consider the affinity $A_{1/2} = \mu_1 - \mu_2$ as the driving force for the transport of grain-1 atoms from the inside to the outside. The complementary point of view is to ignore the nature of the bulk phases and to consider the interface *itself* as the moving entity. In the same process as above, the interface is moving to the inside of grain 1 and consequently we have to define the driving force as a force working on the interface (i.e., a pressure) pointing into grain 1. From now on we will refer to this pressure as the *growth pressure,* P_g. *)

In accordance, the affinity for this interface motion is identical with that for bulk-phase transport but the motion is in the opposite direction. We will call it the *growth affinity* (A_g). When one mole has crystallized, the amount of energy dissipated by the irreversible growth process is A_g. Obviously, in terms of growth pressure the energy dissipated by the crystallization of one mole is $V_c . P_g$, so that we have the identity

$$A_g = V_c \cdot P_g \quad (V_c \text{ is molar volume of the crystal phase}) \tag{1.1}$$

Growth affinity and growth pressure are alternative definitions of the driving force for the interface motion. (Note that we can formally use any suitable dimension for the forces and fluxes describing an irreversible process, provided the phenomenological coefficient is adapted correspondingly. E.g, we can express growth rate in cm/sec. or in mole/cm^2 sec, etc.)

This picture is analogous with the dissipative transport of a fluid through a porous membrane. On any surface element dO of the membrane,

*) Strictly taken, the growth pressure P_g (and consequently also the growth affinity, A_g) include also: a) a term accounting for the momentum transfer by the solidifying particles, when their velocities differ in the c and p phases; b) a term associated with eventual viscous effects at the interface. (see, e.g., CAROLI c.s. (1992). Both effects seem to be negligible for all practical purposes, except maybe in growth from the vapour phase at high growth rates.

we have a force equilibrium $\Delta p_{\text{bulk}} = p_{\text{surface}}$. The latter pressure results from all forces working on the membrane and includes a term resulting from its eventual curvature and surface "tension".

Growth affinity, planar-interface affinity and curvature-induced affinity shift

In equations, it can be convenient to split up the growth affinity or the growth pressure into components ΔA_{c} and $A_{|}$:

$$A_{\text{g}} = A_{|} + \Delta A_{\text{c}} \tag{1.2a}$$

$$P_{\text{g}} = P_{|} + \Delta P_{\text{c}} \tag{1.2b}$$

$A_{|}$ is, by definition, the growth affinity in agreement with a *planar* interface, i.e.

$$A_{|} \equiv \mu_p(T_{\text{trans}}, C_{\text{trans}}, p_{\text{trans}}) - \mu_c(T_{\text{cis}}, C_{\text{cis}}, p_{\text{trans}}) \tag{1.3}$$

where the suffix "cis" represents the values at the crystal side of the interface and "trans" those at the parent-phase side, at a (mathematically identical) position on the interface. As the interface presents no "Kapitza" barrier for heat flow (unless in exotic cases such as He crystals) we can also substitute $T_{\text{cis}} = T_{\text{trans}}$ in (1.3), so that *the concentration is the only different argument in the* μ_p *and* μ_c *terms for almost all practical purposes*:

$$A_{|} \equiv \mu_p(T_{\text{trans}}, C_{\text{trans}}, p_{\text{trans}}) - \mu_c(T_{\text{trans}}, C_{\text{cis}}, p_{\text{trans}}) \tag{1.4}$$

The term ΔA_{c} represents the shift in μ_p (and therefore also the change in A_{g}) due to the pressure difference (between c and p phases) induced by the *interface curvature*: (equations borrowed from ch. 3)

$$\Delta p_{\text{c}} = \overset{\Rightarrow}{\boldsymbol{K}} : \overset{\Rightarrow}{\mathbf{S}}\gamma \ ; \quad \Delta A_{\text{c}} = V_c \ \overset{\Rightarrow}{\boldsymbol{K}} : \overset{\Rightarrow}{\mathbf{S}}\gamma \tag{3.7a/b}$$

In a practical situation where an inhomogeneous T and/or C_{p} field around the crystal is imposed by external boundary conditions, $A_{|}$ can be found from the T and C values at the interface. (analytical expressions for μ_p and μ_c as a function of T and C have to be available). Still, $A_{|}$ is an "abstract" parameter, only relevant as a term in equations; the *physical* parameter is A_{g}.

Note: The "growth pressure" concept refers exclusively to the driving force for the *irreversible* (i.e., dissipative) motion of the crystal interface. By definition, the growth pressure is zero when equilibrium conditions at the interface are assumed. It is to be distinguished from any actual changes, e.g. in the external vapour pressure, which might bring about the nonequilibrium situation. In a unary solid-vapour system with a flat interface and no Marangoni force, a deviation Δp_{vap} of the vapour pressure from the equilibrium value brings about a growth affinity A_g and corresponding growth pressure $P_g = A_g / V_{cr.} (\gg \Delta p_{vap})$; the corresponding pressure balance equation is

$$P_g = (1/V_c)(\mu_{vap} - \mu_{sol}). \tag{1.5}$$

However, *macroscopically* there are no forces working on the interface. The pressure forces P_g and Δp_{vap} are in equilibrium in an "internal balance" because the dissipative motion of the interface (over a distance agreeing with one mole in the crystal) has to be compared with the transition of one mole in the vapour, induced by Δp_{vap}. The "lever" involved in the molecular-volume difference is given by the relation

$$P_g = \Delta p_{vap.} \frac{V_v - V_c}{V_c} \tag{1.6}$$

The nonequilibrium situation may equally be brought about by a deviation of another parameter (e.g. temperature, concentration) from an equilibrium value.

1.5.2 Nonlinear R(A) dependence and minimum rate of entropy production

In this monograph we adopt the interface-based point of view. Summarizing, we describe the (nonequilibrium, i.e. dissipative) crystal growth process (interface advance rate R) as driven by a growth affinity A_g :

$$R = A_g \cdot \chi \tag{1.7}$$

where the (perpendicular) advance rate R is measured in a reference frame which is attached to the crystal's bulk lattice. For the α/β interface, the growth affinity is ($\mu_\beta - \mu_\alpha$) plus eventual terms accounting for capillarity and Marangoni effects. (see 2.2 for the normal-direction convention). The proportionality factor χ may be called the "growth mobility". For a crystal, χ is in general orientation-dependent, even for crystals with roughened interfaces ($T > T_R$). It may also depend on A_g : this is essentially the case for $T \leqslant T_R$, for orientations close to facet orientations (even under common "technical" growth conditions). The consequence of the nonlinearity in the $R(A_g)$ relation is that the entropy production is not a quadratic function of A_g, and that the principle of *minimum rate of entropy production* (minimum REP), as a characteristic of stable-stationary dissipative states, is not necessarily applicable.

In this context it is relevant to notice (in a terminology borrowed from following chapters) that the implication of minimum-REP in the

$\vec{Q}$R shape (i.e., the shape of a crystal grown from a *small nucleus*) is not equivalent with a proof that *any arbitrary evolving shape* would obey this criterion. On the contrary, growth on a concave interface, according to kinematic-wave theory construction principles (ch. 5.2) agrees with *maximum* rate of entropy production. (in the $\vec{Q}$R shape, orientations with **SR** < 0 are automatically eliminated). If the minimum-REP principle would apply, orientations growing at **SR** < 0 orientations would break up immediately into a zigzag structure as described in ch. 6.3. Experimental evidence is that they can, but do not necessarily break up; they may even be remarkably stable. In 6.3 we argue that a nucleation barrier is to be expected. In other words: the "metastability" is the very consequence of the non-applicability of the minimum-REP principle for this case, as associated with nonlinear $R(A_g)$ behaviour.

The above is illustrated again by fig. 13.2. Both concave and convex *equilibrium shapes* obey the criterion of *minimum total free energy*, but convex and concave *growth shapes* obey minimum and maximum REP, respectively.

1.5.3 Interface growth affinity as part of the overall affinity

In this subsection we consider the overall 3-dimensional transport situation in which the crystal/parent phase interface is embedded. In general we have to do with a "technical" situation in which atoms or molecules travel from a source (parent phase) to a sink (bulk of crystal) under the influence of an overall driving force or affinity A_{total}, their flux being limited by one or more resistances or barriers acting in series. Such a process can be described by an overall equation which is essentially the combination of the continuity conditions for the flux- and the potential fields, plus the "constitutive relation" between flux and affinity (= potential drop) for each of the resistances. (notice that, in contrast with subsection 1.5.2, we formulate in terms of "bulk-species" quantities again).

Symbolically, we can represent such an equation as:

$$\underset{(1)}{A_{total}} - \underset{(2)}{\Delta A_{curv.}} = \underset{(3)}{A_{surf.kin.}} + \underset{(4)}{A_{bulk\,diff.}} \tag{1.8}$$

The first term represents the externally applied "total" affinity. Usually, the field cannot be prescribed in detail but is imposed by boundary conditions at some distance from the interface. (thermal and chemical sources and sinks).

The second term accounts for the shift in the affinity across the interface, induced by the interface curvature. Terms (3) and (4) are the dissipation terms: (3) is the affinity A_g which accounts for the "surface kinetics" (i.e., surface diffusion, ledge diffusion, adsorption-desorption kinetics and chemical reactions at the surface) and (4) that of the diffusion field in the *bulk* of the parent phase and of the crystal itself. Term (3) might be split up into sub-terms representing surface-diffusion, ledge-diffusion and the passage of single barriers, if appropriate. (terms (3) and (2) together are defined as the "interface affinity" $A_|$ in eq. (1.2a)).

Whereas (1) and (4) represent three-dimensional fields, (2) and (3) are defined at the interface only, i.e. have to be considered as boundary conditions for the diffusion field defined by (1) and (4), defined on the (moving) crystal interface. This implies that the equation with term (4) included is a moving-boundary ("Stefan-type") problem. The mathematics is considerably simplified when (4) can be neglected ("surface-controlled growth").

Special cases emerge when one or more of the terms (1) - (4) are zero or negligible. For instance:

(3) = (4) = 0	static equilibrium
(2) = (4) = 0	surface-controlled growth: "classical" kinematic wave theory
(2) = (3) = 0	bulk-diffusion controlled growth
(3) = 0	Mullins-Sekerka theory of interface instability
(2) = 0	growth of "big" crystals; neglects surface-energy
(1) = 0	theory of ageing and sintering kinetics
.... etc.	

Such a catalogue has been used as a guideline in the structure of this monograph. Further subdivisions can be made according to the variables occurring in terms (2) and (3). We will come back to this guideline in 1.7 where this structure is sketched.

1.5.4 The three-dimensional extension of the two-dimensional interface-affinity field

At this point it is relevant to call attention to the use of the expressions "affinity field" and "affinity gradient" for an interface, as contrasted with *potential* fields and gradients as applicable to the *species* in a bulk phase.

The terms in eq. (1.8) are really affinities - i.e. potential *differences* - which can be discerned as the driving force for parts of the transport process in series. But a 3-dimensional T or C field related with a bulk diffusion process (term (4) in (1.8)) is related with a "potential field" and with "potential gradients". The affinity associated with such a diffusion process is the difference in potential *at two different locations*, bridging the region where the potential drop, i.e. the dissipation, is appreciable. For bulk diffusion we have to do with a (3-dimensional) *potential field*, not an affinity field. (some authors define the affinity for a diffusion process as the potential *gradient* in a bulk phase).

The affinity associated with dissipation in an *infinitesimally thin barrier* - i.e., our model of an interface between continuous media, or term (3) in eq. (1.8) - is the difference in potential "cis" and "trans" the interface - i.e. *in two different phases*. These are mathematically *identical positions*. Accordingly it makes sense to speak of the *2-dimensional affinity field defined on the interface* (as a mathematical surface). This includes the use of a surface-tangential gradient ($\vec{\nabla}_s$) of this affinity.

In eq. (1.2a) we have discerned between the actual growth affinity A_g and the abstract interface parameter $A_|$. Both are defined on the interface. For an interface in equilibrium, $A_g \equiv 0$ so that also $\vec{\nabla}_s A_g = 0$. However, an interface can be in equilibrium in an inhomogeneous T and/or C field (e.g., the grain-boundary groove to be discussed in 3.3.3). This illustrates why it is practical to split up A_g according to (1.2a) : the $A_|$ corresponds with the values of the inhomogeneous (T,C) field at the interface, whereas ΔA_c represents the compensating curvature shift, so that the curvature can be calculated from it (cf. the procedure described in 3.3.3).

Although, strictly taken, A_g and $A_|$ are only "physically" defined on the interface, we will in this monograph also refer to the "interface affinity field", somewhat loosely, in the sense of a field defined in *3-dimensional* space. This may seem an improper use of the term, because in the bulk of the parent or crystal phase we have only to do with μ_p or μ_c, i.e. with *potential* fields.

Still, the 2-dimensional interface affinity field $A_|(\vec{s})$ can formally be embedded in a virtual, 3-dimensional interface-affinity field. We will start to illustrate this for an extreme and simple case: a unary melt/crystal interface in a unidirectional-growth set-up. We suppose that the solidifying melt is held in a thin cell which is pulled through a heater-cooler combination at a constant ("mechanical") growth rate R_{mech}. The thermal contact is so perfect that the temperature field can be considered as to be "imposed"; it is unaffected by the liberated heat of solidification. The s/l interface is planar. The situation is illustrated in fig. 1.3.

In a coordinate frame which is fixed to the temperature field, we can graphically represent curves for μ_p and μ_c (known as a function of T; therefore also of x : fig. 1.3a). In equilibrium ($R_{mech} = 0$, or $R_{mech} > 0$ but $\chi^{-1} = 0$), the interface settles at the melting-temperature isotherm (fig. 1.3a). When the interface (remaining planar) is displaced from that position (by

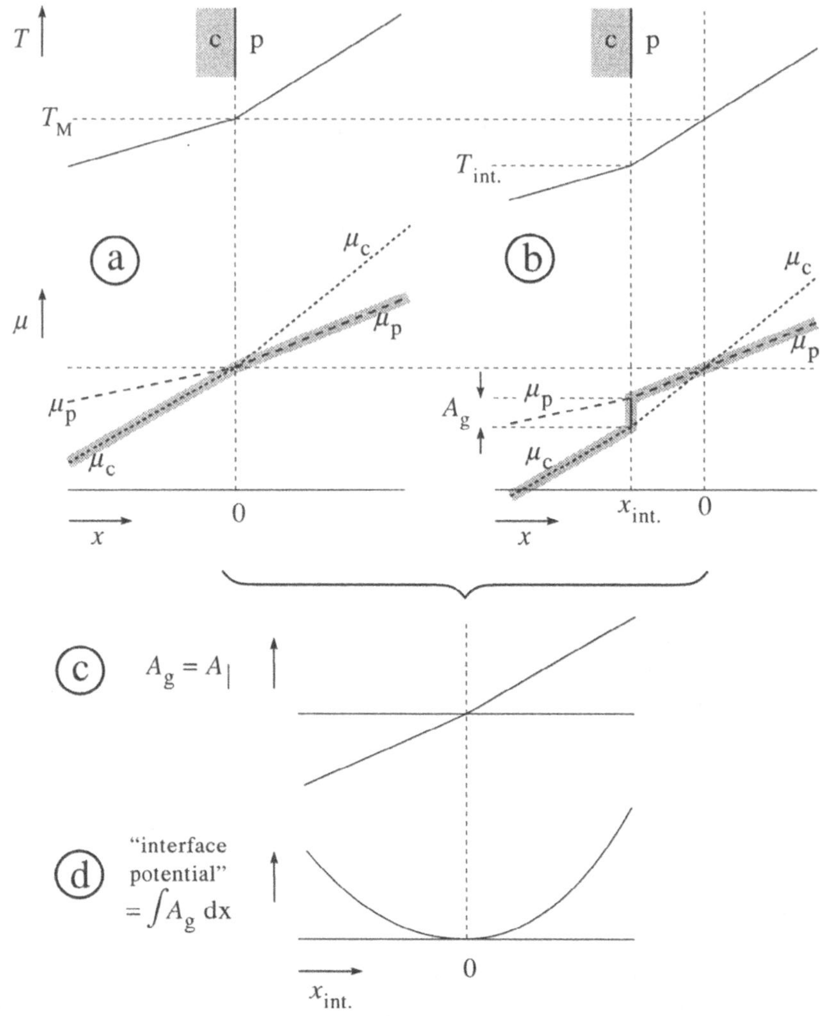

FIG. 1.3

any magic), it "feels" the growth affinity A_g ($= A_|$ for a planar interface) or the growth pressure P_g corresponding with the temperature at the new position. (fig. 1.3b)

Now our technical means to displace it are limited. At $R_{mech} = 0$, we would need an extra force working on the interface; a Marangoni force (cf. eq. (3.10)) might do such a trick, e.g. an interface bearing an electric charge could be displaced by an electric field. For $R_{mech} \neq 0$, we may perform a mental experiment by supposing that we can modify the interface mobility χ (in $R = \chi A$) by some external parameter. This will vary $A_|$ in proportion with χ^{-1} , and accordingly the interface position will shift.

Anyhow, *we can plot* $A_g = A_|$ *as a function of the distance from the* T_M *isotherm (fig. 1.3c):* in other words, we can consider this as an extension of the 2-dim. $A_|$ field as relevant on the actual interface, to a 3-dim. $A_|$ field for a virtual, displaced interface, in which the 2-dim $A_|$ field is embedded. This 3-D extension enables the definition of a *normal* gradient, $\vec{\nabla}_n A_|$.

> (by the way: in this example the interface is shown to be "elastically bound" to the T_M-isotherm position "x = 0". Ergo, we might also plot the *energy* or "potential" *of the displaced interface* as a function of x which is thus proportional to x^2; as a consequence of the potentials μ_c and μ_p of both phases varying *linearly*. (fig. 1.3d) This illustrates the jump in the rôles of bulk phases and interfaces in the tiered hierarchy of structural differentiation, to be discussed in ch. 7.)

This example illustrates that the 3-dim. (extension of the) interface affinity field represents how the interface will react to an infinitesimal shape perturbation, in a way which is also implicit in an instability analysis of the RUTTER-CHALMERS (1953) or the MULLINS-SEKERKA (1963, 1964) type. In such an analysis, we consider the interface in a stationary-growth situation. To find out whether the stationary situation is stable or unstable, we analyse the behaviour of a "perturbed" part of the interface which is shifted in the normal direction over a small distance δ: $\mathrm{d}\delta/\mathrm{d}t > 0$ corresponds with instability, $\mathrm{d}\delta/\mathrm{d}t < 0$ with stability. For the calculation of $\mathrm{d}\delta/\mathrm{d}t$, we have to calculate the T, C etc. fields in the bulk of the parent and crystal phases which correspond with the perturbation. (in a quasi-stationary approach). When capillarity effects are neglected (in agreement with the definition of the $A_|$ field), the relevant quantity for distinction between an accelerating and decelerating perturbation amplitude, i.e. $\mathrm{d}\delta/\mathrm{d}t > 0$ or < 0, is the *normal gradient* of the $A_|$ field as defined above. Note that this applies to a moving *non-dissipative* interface ($\chi^{-1} = 0$) as well as to a moving *dissipative* interface ($\chi^{-1} > 0$): the difference is only in the position of the interface: $A_|$ = 0 and > 0, respectively. Note also that in such a perturbation analysis we really have to do with the μ_p *and* the μ_c values at the perturbed interface, i.e. with the *affinity* ($\mu_p - \mu_c$) and not just with the potential μ_p .

Now in our pure-melt example the situation was simple because μ_p and μ_c could just be translated from the temperature map and were unrelated with the interface position. This is no longer true for a binary melt or a solution where the stationary C and/or T field is "attached" to the interface. In order to calculate the 3-D extended $A_|$ field for that situation, we have to work out the following procedure:

1) *Fixed data*: external boundary conditions: sources and sinks for C and T fields; R_{mech}. Consider planar interface.
2) *Variable*: interface χ (for really unary case, this is a single coefficient). $A_g = A_l$ varies together with χ.

Now find interface position x as a function of χ - or inversely, A_l as a function of x. (note that this $A_l(x)$ function is specific for a given R_{mech}; which it was not in the pure-melt, fixed-T field case).

Note: It is important to be aware that in a multi-component situation, the relevant parameter is $\Sigma_k(\widetilde{A}_l)_k$, i.e. the "geometrical" growth affinity as defined in (8.27).

We have to emphasize that we can only define the 3-D A_l field in this way, not a 3-D A_g field which depends on interface curvature. In interface-instability theory, $\vec{\nabla}_n A_l$ corresponds with the normal gradient of (effective liquidus temperature minus actual temperature) which is the critical parameter for C.S. (in the R-Ch and M-S procedures). The "intuitive" meaning of $\vec{\nabla}_n A_l$ is a squeezing or flattening pressure on the interface for $\vec{\nabla}_n A_l > 0$, or a destabilizing "negative" pressure for $\vec{\nabla}_n A_l < 0$. In instability analysis, the effect of this "pressure" has to be confronted with the "tension" effects of $\mathbf{S}\gamma$ (which can also be flattening for $\mathbf{S}\gamma > 0$, or destabilizing for $\mathbf{S}\gamma < 0$).

When reference is made to the (3D) "affinity field" in this monograph, it is in most cases actually concerned with its value A_l *at the interface.* Although for explicit calculation of this field, we have to work out the 3-D T and C fields (in both phases), the A_l field and the expressions in which it occurs can *in principle* be determined without ambiguity. Such applications can be related, for instance, with the shape of grain-boundary grooves, of the s/l interface of unidirectionally solidified eutectics and of cellular structures.

In a few cases we have used the 3-D affinity field concept in connection with the *normal* component of $\vec{\nabla}$A. (for instance, in equations (3.13/14), (4.18) and (6.22)). In all such cases we are discussing *instability,* and we use the 3D affinity field concept in an attempt to summarize the Rutter-Chalmers/Mullins-Sekerka procedure in a compact way. Also in this case we have to work out the T and C fields in both phases, but the variational procedure of the stability analysis now implies to do this for a *perturbed* interface (see above).

In view of the possible extra complications of such a stability analysis, the "status" of the equations mentioned above, and indeed of the quantity $\vec{n}\cdot\vec{\nabla}A$ itself, is somewhat more symbolic or hypothetical than the 2-D interface A_l field. This, of course, agrees with the fact that we have extended the use of the term "affinity" beyond its traditional meaning, where it refers to *bulk matter*. In defining "interface affinity" in eqs. (1.1-1.4), we have made it applicable to a *surface*.

1.6 ON "EQUATIONS OF MOTION" FOR SURFACES.

A mathematical formulation of the evolution of an arbitrarily shaped surface may in general be called an equation of motion for such a surface. However, we may choose from several alternative representations of such an equation of motion, differing both in explicitness and specific applicability. Such representations will be discussed in detail in ch. 5, but it seems appropriate to indicate the differences in this introduction.

The description of the motion of a line or of a surface involves a "degree of freedom" which is inherent in the *dimension* of this line or surface being > 0. For a point, motion can only be expressed in terms of its position and velocity vectors and consequently its EOM has to be formulated in terms of these vectors. But for a line or a surface, we have the choice of the *direction* in which the displacement is measured, as shown in figs

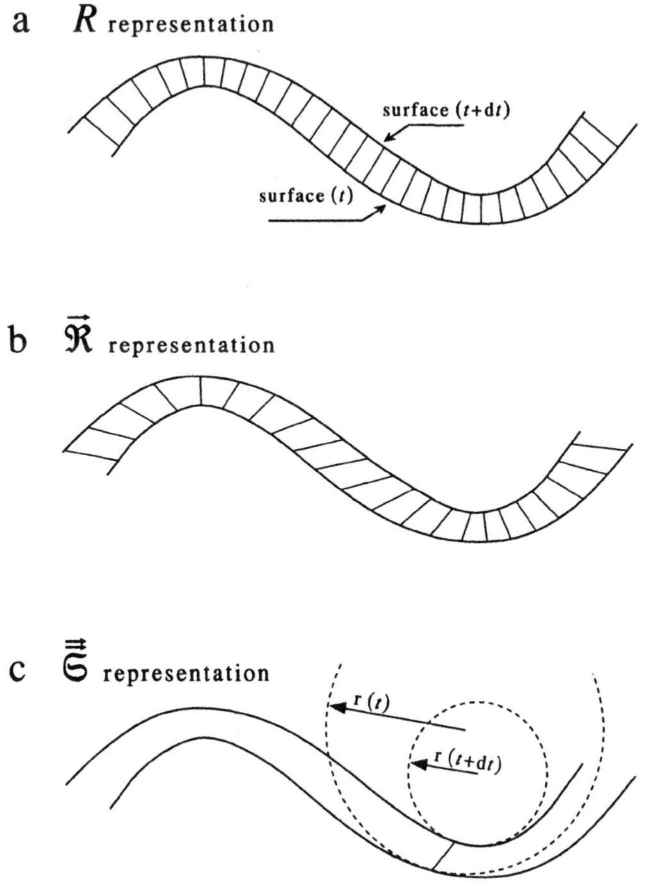

FIG. 1.4

1.4a and b. In agreement with this freedom, we can formulate different "equations of motion" for the evolution of the surface:

In the first place, displacement can be measured along normals (fig.1.4a). The equation of motion is then represented by a function $R(\vec{n},m)$ = perpendicular advance velocity as a function of orientation $\vec{n}$ and any scalar "wild card" parameter(s) m (for instance, affinity). Although such a function describes the evolution of any arbitrarily shaped interface unambiguously (apart from possible curvature effects and the properties of stabilized sharp corners) and hence can be considered as one possible version of an equation of motion in itself, we will in this monograph usually refer to this function as the *growth rate function*. The orientation of the interface, when followed along the normal direction, is not fixed: the *tilt rate* can be expressed by a derived equation (eq. (5.8 a or b)).

Alternatively we might measure the displacement along any other field of direction vectors - and again, in general, the interface orientation will tilt. One specific choice in the latter case is to measure the displacement along *kinematic wave trajectories;* i.e. loci of constant surface orientation. (fig. 1.4b) (by definition, the orientation of the interface when following these trajectories is *fixed*).*) The practical advantage of the latter description in the case of crystal surfaces is that many physical parameters of the surface depend on orientation (only) and hence are constant along these trajectories. The directed growth rate along the constant-orientation loci is a vector quantity which we will call the *vector velocity*, to be written as $\vec{R}$ or $\vec{\Re}$. This vector was introduced already by CHERNOV (1963) and by MÜLLER-KRUMBHAAR c.s. (1977). For the case of uniform m, it can be derived from the growth rate function $R(\vec{n},m)$ by application of the vector operator $\vec{Q} = \vec{n} + d/d\vec{n}$, as we will discuss in detail in ch. 5.

The vector velocity can also be used to describe the motion of lower-dimensional "objects" such as D-junctions, 3-junctions, 4-junctions and sharp edges and vertices. For such objects we will also use the $\vec{R}$ or $\vec{\Re}$ symbol, but supplement it with an appropriate suffix such as $\vec{R}_{D\text{-}j}$ or $\vec{R}_{3\text{-}j}$ etc. to discern it from an ordinary surface-element $\vec{R}$. Because such lower-dimensional elements are "alien" elements in the crystal surface, specific

*) This definition of kinematic wave trajectories leads to an indeterminacy in those cases where one or both curvatures of the surface are zero. This indeterminacy can be relieved by slightly refining the definition: superpose an infinitesimal corrugation on the flat surface so that the original definition makes sense again. This refined definition works for orientations $\vec{n}$ with finite $\vec{\vec{S}}R(\vec{n})$. If $\vec{\vec{S}}R$ is infinite (cusped extremum in $R(\vec{n})$) the kwt definition is still indeterminate, to be remedied only by removing the singularity from $\vec{\vec{S}}R$. The latter amounts to replacing the $\vec{\vec{S}}R$ δ-function of the cusped $R(\vec{n})$ by a finite-width distribution function. Taking the limit to obtain the δ-function again, we can maintain the kwt definition.

physical effects can be active in their kinetic behaviour so that their $\vec{R}_{3\text{-j}}$ etc vectors do not necessarily correspond with the $R(\vec{n})$ plot of the adjoining surface(s). In such cases they function mathematically as *boundary conditions (BC)* for the EOM of the surface, and we distinguish their $\vec{R}$ or $\vec{\mathfrak{R}}$ vectors by an asterisk (see discussion in ch. 6, 6.2 and 6.4). In this way a surface can effectively be "pulled" or "held back" by BC-providing elements such as a dislocation ending on a facet or by a grain boundary. In agreement with this rôle we introduced the concept of "velocity source" for those BC-producing elements which actively affect the evolution of a surface. The mathematical formulation of this concept is given in 6.4.

Finally, a third alternative for representing the evolution of a surface is by the *divergence of the vector velocities* of neighbouring surface elements, i.e. the rate of change of surface-curvature radius when followed along kinematic wave trajectories (fig. 1.4c). This is, for two orientational degrees of freedom, a 2 x 2 tensor quantity which we will write as $\vec{\vec{\mathfrak{S}}}$ and may call the "Spreading Rate Tensor" (expressing the fact that small-amplitude perturbations on the surface, considered as features on a map, expand or shrink in size according with this tensor; i.e. two points on the surface, differing in orientation by a small amount $\Delta\varphi$, "drift away" at a mutual velocity $\Delta\varphi.\vec{\vec{\mathfrak{S}}}$) So we can write

$$\vec{\vec{\mathfrak{S}}} \overset{\text{(def)}}{\equiv} \left(\frac{\mathrm{d}}{\mathrm{d}t}\right)_{\text{kwt}} \left(\vec{\vec{K}}^{-1}\right) = \left(\frac{\mathrm{d}}{\mathrm{d}t}\right)_{\text{kwt}} \left(\frac{\mathrm{d}\vec{s}}{\mathrm{d}\vec{n}}\right) \tag{1.9}$$

In fig. 1.5 we illustrate the basic relation between $\vec{\vec{\mathfrak{S}}}$ and $\vec{\mathfrak{R}}$. Snapshots of a curved surface have been taken at successive times t and $t+\Delta t$. A point $\vec{s}(t)$ on the first surface is connected with a point of identical orientation $\vec{n}(\vec{s})$ on the second surface by the vector $\Delta t.\vec{\mathfrak{R}}(\vec{s})$. Similarly for the nearby point $\vec{s}(t)+\Delta\vec{s}(t)$. Ergo

$$\vec{\vec{\mathfrak{S}}} \equiv \left(\frac{\mathrm{d}}{\mathrm{d}t}\right)_{\text{kwt}} \left(\frac{\mathrm{d}\vec{s}}{\mathrm{d}\vec{n}}\right) = \lim_{\Delta t\to 0} \lim_{\Delta\vec{n}\to 0} \frac{\Delta\vec{s}(t+\Delta t)-\Delta\vec{s}(t)}{\Delta\vec{n}.\Delta t} =$$

$$= \frac{\vec{\mathfrak{R}}(\vec{s})-\vec{\mathfrak{R}}(\vec{s}+\Delta\vec{n})}{\Delta\vec{n}} = \frac{\mathrm{d}\vec{s}}{\mathrm{d}\vec{n}} \frac{\mathrm{d}\vec{\mathfrak{R}}(\vec{s})}{\mathrm{d}\vec{s}} \quad \Rightarrow$$

$$\vec{\vec{\mathfrak{S}}} = \vec{\vec{K}}^{-1} \cdot \vec{\nabla}_{s} \vec{\mathfrak{R}} \tag{1.10}$$

Note that eq. (1.10) just expresses the *basic relation between the "gothic"* $\vec{\mathfrak{R}}$ *vector and* $\vec{\vec{\mathfrak{S}}}$ *tensor*, i.e. it is valid regardless of the eventual homogeneity of A or m fields. (it has nothing to do with A or m fields).

Gothic versus Roman symbols:

We have introduced the Gothic characters $\vec{\mathfrak{R}}$ and $\vec{\vec{\mathfrak{S}}}$ alongside with the expressions $\vec{\mathrm{Q}}R$ (also written as $\vec{R}$) and $\vec{\vec{\mathrm{S}}}R$ in order to highlight the difference in the variable space and the "character" of these symbols:

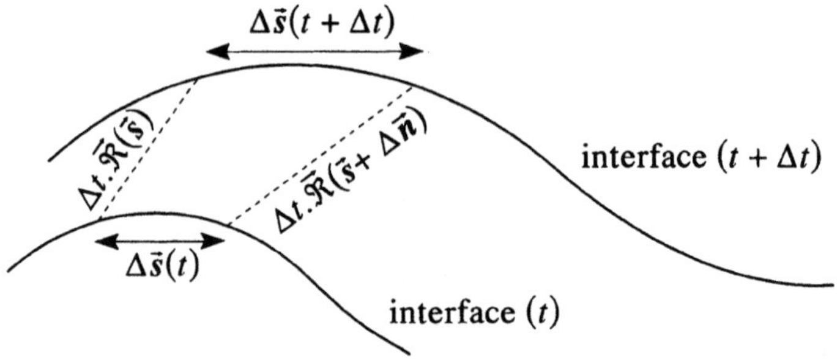

FIG. 1.5

Consider fig. 1.4a: only the Roman symbol R $(= R(\vec{n},m))$ is used there. For any part of the curved surface, even in an inhomogeneous m field so that that both $\vec{n}$ and m vary as a function of position, R can be found from the $\vec{n}$ vector coordinates and the m value corresponding with that particular position alone, *just because R is supposed to be known as an explicit function of* $\vec{n}$ *and m.* In other words, the information provided by the $R(\vec{n},m)$ function is sufficient to describe the evolution of any interface (any shape, any m field) - the interface parameters themselves (position, curvature, m gradient etc) are irrelevant, only the local orientation and m value are relevant.

Now consider fig. 1.4b. In the general case (inhomogeneous m field, and/or capillarity effects on effective affinity) the vector growth rate cannot be found from the $\vec{n}$ and m parameters only; in fact (as we will see in ch. 5) it is given by an equation which also involves the curvature of the surface and the local gradient of the m field. In other words, it can no longer be considered as a function of $\vec{n}$ and m alone but rather as a vector *which is specific for a certain point on a surface* (with its $\vec{n}$ and m, but also its $\vec{\vec{K}}$ and $\vec{\nabla}m$). The same applies (a fortiori) to the spreading-rate tensor $\vec{\vec{\mathfrak{S}}}$ (fig. 1.4c).

The Gothic symbols are used to represent explicitly that they refer to a physical surface. In contrast, the Roman symbols refer to a function of the variables $\vec{n}$ and m.

Note: The vector velocity can also be used to describe the motion of a *point* (e.g., a 4-junction) or a *line* (e.g., a 3-junction). In the case of a point or of the perpendicular-projected representation of a line, the reason for distinguishing between $\vec{\mathfrak{R}}$ and $\vec{R}$ symbols vanishes.

Although the three alternative "equations of motion" provided by explicit expressions for $R(\vec{n},m)$ and the derived expressions for the surface-based $\vec{\mathfrak{R}}$ and $\vec{\vec{\mathfrak{S}}}$ can be readily derived one from another, we have in this monograph used the somewhat looser convention of calling $R(\vec{n},m)$ a "growth rate function" and the derived explicit expressions for $\vec{\mathfrak{R}}$ and $\vec{\vec{\mathfrak{S}}}$ "equations of motion". This convention differs from that followed by the cited authors who have also called $R(\vec{n},m)$ *itself* an equation of motion. The motivation of our choice is based on the similarity with other equations involving coefficients. Consider diffusion: stated as Fick's law, the diffusion equation reads

$$\vec{j} = D\,\vec{\nabla}C$$

where C plays the rôle of a potential. D is the "diffusion constant" (scalar for isotropic, tensor for anisotropic media).

A similar "equation" for crystal growth - i.e., for the displacement rate of a crystal/parent phase interface - would read

$$R = \chi . A \tag{1.7}$$

where A ("affinity") = the potential difference across the interface, and χ is something like a *mobility*, relating transport rate with a potential difference, i.e. playing the similar rôle as D in Fick's law. The χ is not a tensor, but it is *orientation-dependent*, i.e. although it plays the rôle of a "phenomenological constant" such as D, it is in fact a function of orientation $\vec{n}$. Moreover, in most practical cases it is also a function of A, i.e. the relation between R and A is nonlinear and therefore it makes hardly sense to use equation (1.7) , with the coefficient $\chi(\vec{n},A)$ depending on $\vec{n}$ and A, instead of the function $R(\vec{n},A)$ itself.

Nevertheless, the above argument may help to adstruct the view that expressions for $\vec{\mathfrak{R}}$ and $\vec{\vec{\mathfrak{S}}}$ - which may contain several terms for more than one contributions to the total affinity, for instance - have a different, more equation-like character than an analytic expression for $R(\vec{n},A)$ itself, which really represents just a phenomenological entity which happens to depend on $\vec{n}$ and on A. Only in the "classical kinematic-wave theory limit", i.e. $\vec{\nabla}m = 0$ and neglect of capillarity effects, where $\vec{\mathfrak{R}} = \vec{\mathbf{Q}}R$ and $\vec{\vec{\mathfrak{S}}} = \vec{\vec{\mathbf{S}}}R$, also $\vec{\mathfrak{R}}$ and $\vec{\vec{\mathfrak{S}}}$ are functions of $\vec{n}$ and m on the same level as the $R(\vec{n},m)$ function. But in the context of the generalized treatment in this monograph, this is really a limit case.

Extension of $\vec{\mathfrak{R}}$ and $\vec{\vec{\mathfrak{S}}}$ definitions to 3D fields

In order to specify the evolution of an interface with given initial shape, we should be able to calculate A (or m) at any time at any position at the interface. Only in that case, the EOM can be considered as an explicit equation. Now this condition is an exception rather than a rule: only when an inhomogeneous A or m field is imposed by external boundary conditions and this field is not affected by the position and the motion of the interface, the A or m values at the interface can be found simply in a coordinate frame fixed to the external boundaries (an "*A* or *m* map")

The more common situation is that the *m* field is influenced by the motion of the interface. The *m* field obeys a diffusion equation (for heat, matter or both); the interface acts as a source or as a sink in the diffusion field, but moves itself as a result of the transport. ("Stefan-type" problem). Then the EOM cannot be considered as an explicit equation, but rather as a boundary condition to be fulfilled at the interface, specifying its thermodynamic and dissipative properties. (OD γ and R). For an explicit description of the interface evolution, the *combined* diffusion-field equations and the interface EOM have to be solved.

For an inhomogeneous field of a three-dimensional scalar variable m we can also define a vector-velocity field, in analogy with that for a curved interface, as the field of vector velocities of the iso-m surfaces (in general curved). So, moving with the interface along a kwt, we not only walk along a surface coordinate but we also feel a changing m:

$$(\mathrm{d}/\mathrm{d}t)_{\mathrm{kwt}}\, m = (\vec{\mathbf{Q}}R_{\mathrm{surf.}} - \vec{\mathbf{Q}}R_{\mathrm{iso\text{-}m}}) \cdot \vec{\nabla} m \tag{1.11}$$

This illustrates again that an explicit solution for the interface evolution cannot be complete without an explicit solution for the m field.

The $\vec{\vec{\mathfrak{S}}} = \vec{\vec{0}}$ *criterion for stationary structures*

A stationary structure is defined as one for which a coordinate system, fixed to the interface and moving with constant vector velocity, can be defined in which the structure is immobile. For the parts of the interface with nonzero curvature this implies:

$$\vec{\mathfrak{R}} = \text{constant}, \quad \text{or } \vec{\vec{\mathfrak{S}}} = \vec{\vec{0}} \tag{1.12}$$

for all points on the interface and for all times. (note that for growth in an inhomogeneous m field, this implies also: $\vec{\vec{\mathfrak{S}}}$ (m field) $= \vec{\vec{0}}$; i.e. the m field is stationary too with respect to the interface).

A stationary structure obeying $\vec{\vec{\mathfrak{S}}} = \vec{\vec{0}}$ can be either stable or unstable against perturbations. A perturbation (small deformation) of the surface can be defined by a function $\Delta\vec{x}(\vec{s})$, where $\vec{s}$ represents a surface coordinate on the unperturbed surface and $\Delta\vec{x}$ is the vector difference between $\vec{s}$ and that point on the perturbed interface which has the same orientation. (in analogy with the definition of vector velocity). The structure is asymptotically stable, i.e. the perturbation decays, when

$$\lim_{t \to \infty} \int \mathrm{d}t\, \Delta\vec{x}(t) \cdot \Delta\vec{\mathfrak{R}}(t) = 0 \tag{1.13}$$

where $\Delta\vec{\mathfrak{R}}$ is the difference between the $\vec{\mathfrak{R}}(\vec{s})$ vectors of the perturbed and the original surface: $\Delta\vec{\mathfrak{R}} = \vec{\mathfrak{R}}(\vec{s}) - \vec{\mathfrak{R}}(\vec{s}+\Delta\vec{x}(\vec{s}))$. (note that the simpler condition $\Delta\vec{x} \cdot \Delta\vec{\mathfrak{R}}(t) \rightarrow 0$ is too restrictive, as the perturbation may disappear by a decaying oscillation. In some instances, sustained oscillations seem to be stabilized, e.g. in the vicinity of a dendrite tip of low α-factor materials (LANGER and MüLLER-KRUMBHAAR 1978). This suggests orbital stability of kwt paths in the $(1/r\ ,\ \mathrm{d}^2m/\mathrm{d}x^2)$ space).

An obvious alternative expression for stability is

$$\lim_{t\to\infty} \vec{\vec{\mathfrak{S}}} = \vec{\vec{O}} \tag{1.14}$$

However, as a checking procedure the first expression seems to be more appropriate mathematically.

The expression $\vec{\vec{\mathfrak{S}}} = \vec{\vec{O}}$ is particularly appropriate for the definition of a stationary state because it does not specify the direction or magnitude of the vector velocity. Thus, it can be used to investigate if the condition for stationary growth can be fulfilled for a certain parametrized surface shape (e.g., a Fourier series). If so, both the stationary surface shape and the corresponding m field can be calculated, and $\vec{\mathfrak{R}}$ can be found.

This strategy is somehow complementary to the well-known strategies for the investigation of (in)stability, e.g. in the MULLINS-SEKERKA approach (1964). There a geometrically simple initial structure (moving plane, sphere, paraboloid etc.) is given a small perturbation, and by quasi-stationary analysis of the diffusion field it is determined whether the initial structure is stable or unstable. That is, the structure whose stability is investigated has to be specified a priori, and the analysis does not predict whether the structure, if unstable, will evolve into a new (this time stable) structure, let alone how this new structure might look like. In terms of stability theory: it describes the destabilisation of an originally stable operating point when some parameter exceeds a critical value, but does not predict the new operating point.

The strategy suggested by the $\vec{\vec{\mathfrak{S}}} = \vec{\vec{O}}$ condition starts from the other side. When a given initial interface can be shown to be unstable, it is argued (on intuitive or on experimental grounds) that a new "superstructure" will probably develop which is again stabilized against perturbations (a "dissipative structure" in the nomenclature of NICOLIS AND PRIGOGINE, (1977)). Some overall characteristics of the new superstructure can be predicted on symmetry arguments (e.g., macroscopic shape imposed by external boundary conditions). For the microscopic structure we can try a periodic structure (Fourier series) with adaptable period and shape. In order to be a possible stationary shape, the system has to obey:

a) the boundary equations of motion,
b) the diffusion equation (with velocity term),
c) the $\overset{\rightrightarrows}{\mathfrak{S}} = \overset{\rightrightarrows}{O}$ condition,
d) in order to be also a stable solution, it has to obey a stability criterion, e.g. eq. (1.13) or (1.14).

This procedure does not limit the class of stationary shapes to small-amplitude or small-angle structures. (apart from mathematical difficulties). For instance, the method could be applied to cellular growth, terrace formation in liquid-phase epitaxy, multi-level spiral growth (instability in step trains), eutectic growth, grain-boundary groove shape in directional solidification, dendrite tips and whisker growth.

An instance of a typical time-dependent problem is that of the evolution of dendritic shapes, which is notorious for its mathematical complexity, even without OD parameters. However, important aspects of this evolution may be brought back to a stable/unstable transition (ch. 8.1). The dendrite tip region may be considered as a $\overset{\rightrightarrows}{\mathfrak{S}} = \overset{\rightrightarrows}{O}$ region as suggested above. Moreover, the survival of sidebranches is dictated by a mechanism which seems comparable to the stabilisation of a periodic cellular array in unidirectional solidification, and finally the ageing of dendrite branches is an instance of Ostwald ripening in a highly unstable geometry, which is driven by differences in capillarity shift.

In the absence of OD parameters, the kinematic-wave description of interface evolution is an artefact, at least it has no advantages over the usual procedures where perpendicular velocities are used. When OD γ and R have to be introduced, however, the advantage is a more compact notation of the EOM.

1.7 STRUCTURE OF THIS BOOK.

The structure of this monograph is as follows. Apart from the introductory Ch. 1 and the summarizing Ch. 14, the bulk of the material is split up into
part I *(chapters 2-8)* covering the formal description of crystal growth in the continuum language, leaving the $R(\vec{n},A)$ and $\gamma(\vec{n})$ functions unspecified;
part II *(chapters 9-13)* dealing with the $R(\vec{n},A)$ and $\gamma(\vec{n})$ functions, in particular analytic representations of these functions.

More explicitly, in ch. 2 we introduce an operator formalism to facilitate the notation of equations related with orientation-dependent quantities. The vector operators $\vec{\mathsf{P}}$ (defined by $\vec{\mathsf{P}}\Psi \equiv \vec{n}\Psi$) and $\vec{\mathsf{Q}}$ (defined by $\vec{\mathsf{Q}}\Psi \equiv \vec{\nabla}\{\vec{r}\,\Psi(\vec{r}/|\vec{r}\,|)\}$ and the tensor operator $\overset{\rightrightarrows}{\mathsf{S}}$ (defined by

	Ch.1	*Introduction*	
part I: basic formulation	*Ch.2*	*Operators for OD parameters*	
		Equilibrium	Growth
Surfaces away from boundaries		*Ch.3*	*Ch.5*
Boundary conditions: junctions, edges and vertices		*Ch.4*	*Ch.6*
Comments and applications	*Ch.7*	*Family tree of structural differentiation*	
	Ch.8	*Miscellaneous*	
part II: $R(\vec{n})$ functions	*Ch.9*	*Introduction*	
Formal mathematical approach	*Ch.10*	*Assembling notation; nonlinear networks*	
	Ch.11	*Physical aspects*	
	Ch.12	*Construction toolbox for $R(\vec{n})$ functions*	
Experimental approach	*Ch.13*	*Sphere-growth experiments*	
	Ch.14	*Summary & conclusions*	

$\vec{\vec{S}}\Psi \equiv (d/d\vec{n})(\vec{Q}\Psi))$ can be applied to any scalar field Ψ defined on the interface and depending on the orientation variable $\vec{n}$ (Ψ is not specified a priori; in later chapters we substitute γ and R, but other substitutions are possible). The vector field $\vec{Q}\Psi(\vec{n})$ and the tensor field $\vec{\vec{S}}\Psi(\vec{n})$ have a geometrical meaning which, for the substitutions $\Psi = \gamma$ and $\Psi = R$, can be correlated with the mechanical and the kinetic properties of the interface, respectively. The surface defined by all vectors $\vec{Q}\Psi(\vec{n})$ is identical with the surface which is obtained by the well known Gibbs-Wulff construction from the "Ψ plot" or $\vec{P}\Psi(\vec{n})$ surface.

A special section of ch. 2 is devoted to the implications of the delocalisation of Ψ and its consequences in terms of curvature dependence.

In chapters 3 and 4 the operator formalism is applied to the $\gamma(\vec{n})$ function, implying that these chapters deal with *equilibrium* structures. In contrast, in chapters 5 and 6 it is applied to the $R(\vec{n})$ function and accordingly these chapters are about *growth* of structures. Another division can be made by distinguishing equations for interfaces "away from boundaries" (chapters 3 and 5) and those which are specific for boundaries and /or boundary conditions: i.e. junctions, edges and vertices. (chs 4 and 6).

Chapter 3 considers the case (3) = (4) = 0 (in equation (1.8); i.e. static equilibrium. The simplest case occurs for homogeneous $A_|$: then we have essentially the theory of equilibrium shapes. (the $\vec{\xi}$ vector introduced by CAHN AND HOFFMANN (1972) is identical with $\vec{\mathbf{Q}}\gamma$). In principle, the $\vec{\mathbf{Q}}\gamma$ vector can be considered as a force vector and the $\vec{\vec{\mathbf{S}}}\gamma$ tensor as a mechanical **S**tiffness, and so the formal application to the γ function seems to concern mainly the "mechanics" of interfaces. This can be translated into thermodynamics, as a pressure difference can be related with a chemical-potential difference by the Clausius-Clapeyron relation.

Complications can be induced either by an inhomogeneous $A_|$ field or by an inhomogeneous m field working on γ. Both result in a distortion of the equilibrium shape: in the first case it can be characterized as a capillarity distortion, in the second case as a Marangoni distortion.

Chapter 5 considers first the case (2) = (4) = 0 (in eq.(1.8) (growth shapes). Here we have to do with the $\vec{\mathbf{Q}}R$ vector or "vector velocity", already defined by CHERNOV (1963b) and parallel with the kinematic-wave trajectory introduced by FRANK (1958). In the most simple version ($A_|$= constant), the "classical" kinematic-wave theory is reproduced (5.2), and $\mathbf{S}R$ is identical with the **S**preading rate tensor. But the more interesting situations are those where the affinity field is inhomogeneous. (5.3). The appropriate EOM is then generated by extending the constant-A equation by terms accounting for the effects of the affinity gradient.

Once we have the EOM for an explicit A field, we also have the possibility of substituting the effective affinity = $A_| + \Delta A_c$ (i.e., the MüLLER-KRUMBHAAR growth rate function (1977)). The expression for ΔA_c is provided by ch.3. This results in an "exact" EOM, taking account of OD growth rate and surface free energy, i.e. capillarity effects: the condition (2) = 0 is thereby relieved. This "exact" EOM is discussed in 5.3.

In this EOM we can omit certain terms so as to obtain equations for special situations. For instance, by omitting the external driving force $A_|$ a we obtain the equations for ageing and sintering processes. By putting $\gamma = 0$, we describe the long-wavelength limit of shape evolution (only small features are affected by capillarity) and by putting $dR/dA = \infty$ we neglect the resistance of interface kinetics altogether. Above all, a major simplification is obtained by considering only one orientation variable instead of two ("cylindrical" surfaces).

From a geometrical point of view, there is an essential difference between the constant-$A_|$ and the inhomogeneous-$A_|$ cases. In the first, kinematic-wave trajectories are straight lines, and surface orientations which are not present in some initial shape cannot occur after $t=0$. (at $t=0$, they can eventually develop from non-stabilized sharp corners or vertices in the initial shape). In the second case, kinematic-wave trajectories can be curved and also they can bifurcate and be generated and annihilated; i.e. new surface orientations can be generated continuously. The latter is essential, for instance, in the description of dendritic growth and of surface instability in general.

Boundary conditions: junctions, edges and vertices

The BC rôle of *junctions* in the establishment of interface structures (in equilibrium as well as in motion) is obvious enough. (e.g. the "3-junction" = meeting line of three interfaces; "D-junction" = meeting point between a dislocation and an interface). That also *edges and vertices* (= meeting lines and points between facets on a crystal surface) can play a similar rôle may seem surprising.

The *mechanical* aspects of this BC rôle are discussed in ch. 4, as an essential complement to ch. 3 which deals only with interfaces away from such boundaries. Edges and vertices are related with negative eigenvalues of the $\vec{\vec{\mathbf{S}}}\gamma$ tensor. A planar interface at a $\mathbf{S}\gamma < 0$ orientation is unstable: the restabilized interface has a zigzag or pyramid structure, the period being related with $(\vec{\nabla}A \cdot \vec{n})$.

The *kinetic* aspects of the BC role are covered by ch. 6, complementary to ch. 5 in the same way as ch. 4 to ch. 3. One of the interesting features discussed in this chapter is the interaction between energetic effects as expressed by the BC's imposed by junctions etc. and the kinetics as expressed by the $R(\vec{n})$ functions of the interfaces involved. It turns out that the BC can induce a junction (or edge, or vertex) to act as a "velocity source" for one or more of the interfaces. This *velocity source concept* is worked out in detail in ch. 6.4. As an application of the sections on junction and edge behaviour, we discuss (in 6.5) the special case of shapes which consist *uniquely of planar areas* connected by junctions, edges and/or vertices ("multiplanar geometries"): their evolution can simply be described by the trajectories of these junctions etc. only.

Chapter 7 is devoted to interface instability or rather to the more general subject of morphological phase transitions in crystal/parent phase interfaces. (in fact, this is another extension of the analysis of stationary morphologies as stationary solutions of the equation of motion). Such transitions include the roughening transition, transition from a planar into a cellular, terraced or zigzag/pyramid morphology, and the formation of superterraces. Orientation dependence of the surface free energy and of the growth mobility are considered in combination as a possible mechanism which can destabilize a planar interface.

Being essentially related with critical phenomena, the subject as defined in this wider sense is to be approached by nonlinear stability analysis. The relevant equations of motion are nonlinear partial differential equations and the morphological phase transitions can be related with the marginal stability of soliton solutions of these equations.

Finally, ch. 8 is a collection of applications or conclusions from part I to practical situations, e.g. in dendritic growth, effects of a boundary layer in the parent phase and ageing kinetics. Also, we consider here the possibility of an extension of the purely geometrical (= unary) case discussed in this monograph, to a multi-component description.

Part II falls apart in a theoretical and an experimental section, after a general introductory ch. 9. Chapters 10, 11 and 12 form together one mega-chapter in which we introduce a new principle for the construction of analytic R and γ functions, based on the set of PBC vectors of the crystal. In the construction of a complete function describing all orientations, contributions of individual PBC's or PBC pairs (= facets) are "assembled". This assembling is a mathematical operation by which a number of functions is combined into a single one. The principle is explained in chapter 10. The operation has a parameter: the "assembling exponent". For γ functions, the assembling exponent is +1, implying that the "assembling" is just a summation and that the γ function plot, for $T = 0$, reduces to the well-known raspberry or soap- froth plot. *However, for R functions the assembling exponent is* $\leqslant -1$ -which may be the most elementary characteristic for the difference between R and γ functions.

The same assembling procedure is used, at another stage in the construction procedure, to describe roughening of facets. (chapter 11) For this purpose we introduce a roughening parameter, ε (typical for a facet or a facet family) which can range between 0 and 1. For $\varepsilon = 0$, cusps in R and γ plots are sharp and growth rate at the exact facet orientation is zero. For $\varepsilon = 1$, the surface is completely rough or liquid-like, and orientation dependence has vanished altogether. *The "roughened" function is obtained by assembling the original* ($\varepsilon = 0$) *OD function with the constant* ε *with an assembling exponent* = +2.

Physically, a relevant parameter for the representation of roughness is the value of $R/\mathsf{S}R$ taken at the facet orientation. (this value also ranges between 0 and 1 during the transition from faceted to rough). The relation with the ε parameter (for assembling exponent a = +2) is $R/\mathsf{S}R = \varepsilon^2$.

In a sense, the procedure of assembling an $R(\vec{n})$ function, in a number of steps, from discernible subelement functions or constant parameters, is equivalent to the formation of a (nonlinear) network. We will discuss the relation between the topology of this network and the physical aspects of the $R(\vec{n})$ end function extensively in ch. 11.4.

The principle of function construction by a sequence of assembling procedures (chapter 12) enables the definition of R and γ functions based on a) the set of relevant PBC's and b) a reduced number of parameters to account for ledge-advance velocity, roughening etc. These parameters are themselves functions of affinity, temperature etc. It is possible to construct such functions, depending on the purpose of the description, to a variable degree of detail. On one end of the scale, "prototype" functions can be defined which just represent the typical behaviour of one type of facet in a crystal of simple symmetry. On the other end, real crystals with complex symmetry and many facet "families" can be represented in considerable detail, even including orientation-dependent growth rate of the ledges on every individual facet.

The functions can be used for theoretical analysis (and computer simulation) of crystal growth situations in which orientation- dependent behaviour is important. (e.g., interface instability, dendritic growth). The parameter approach enables the definition of dimensionless parameters which are suitable to characterize stability bifurcations.

Chapter 13 is the experimental counterpart of chs. 10-12. Here we discuss possible refinements of sphere-growth experiments, again with an emphasis on methods to determine the roughening parameter. We show that this can be done without any thickness measurement. Apart from this, "orientation mapping" is introduced as an alternative for the conventional "thickness mapping" in sphere-growth experiments.

1.8 MAIN THEMES

To conclude this introductory chapter, let us summarize a few main themes:

1) One of the aims is to develop an "exact" equation of motion for crystal surfaces, to investigate the limitations of their applicability (resulting in necessary refinements) and to draw conclusions about the consequences of combined energetic and kinetic effects. This includes a critical discussion of the very foundations of such equations.

We emphasize that the neglect of curvature effects (not only the capillarity effect, but also other effects such as result from "delocalisation" as mentioned in 2.8) leads to results which conflict with actual physical behaviour. Proper amendment makes the equations nonlinear and can result in soliton-like behaviour of the solution, physically equivalent with zigzag or pyramid surface structures.

2) Another main theme is the analysis of stationary structures which are in agreement with both the kinetic and the energetic requirements of the EOM. This has led us to the formulation of the concept of the "velocity source". A well-known example is the intersection of a dislocation with a crystal facet (spiral growth) - others are junctions of three or four interfaces, but possibly also stabilized re-entrant edges. Whether or not a certain structural element acts as a velocity source depends on the solubility of the combined equations of stationary motion (see section 6.4).

3) We present a construction principle for analytic R and γ functions. Such functions are based on the relevant PBC vectors and contain an "economically" reduced number of parameters. The functions can be constructed to any required degree of accuracy and can be used for computer simulation of crystal growth and for (in)stability analysis. An important aspect of such functions is the possibility to describe the degree of roughening by a parameter, specific for each facet "family".

4) Finally, as another branch of the stationary-structure analysis, we discuss stationary cellular, zigzag, terrace- etc. structures at the interface of growing crystals. We conclude that many morphological (and chemical) instability phenomena at the interface of growing crystals can be interrelated as in a family tree of generalized phase transitions, being a small part of the universal family tree of structural differentiation of the physical world.

*What is **not** in this book*

As mentioned before, this monograph is a *toolbox* - with tools designed specifically to deal with OD surface properties. The tools have not actually been brought into use - e.g. for the explicit theoretical analysis of interface instability or dendritic growth. Such applications, covering a much wider field than the interface-related topics on their own, are quite beyond the scope of this book. Still it may be relevant to state explicitly what is needed for such applications above the material presented here - in other words, to sum up "what is *not* in this book".

In 1.5.3, eq. (1.8), we have sketched the complementary rôles of a) the bulk-diffusion fields and their "external" BC's = sources/sinks for heat and matter flux fields, and b) the interface BC's = "interface EOM" + continuity conditions. For the explicit statement of the general situation in terms of (coupled sets of) equations, we have to *adapt* the interface EOM in order to match it to the bulk-diffusion equations. (an "interfacing problem").

The nature of the bulk-diffusion equations, and in particular the parameters occurring in it, depend on the technical situation (growth from melt, from solution, eventually imposed Δp, $\vec{\nabla} T$ or $\vec{\nabla} C$, or growth rate such as in Bridgman or Czochralsky situations). Accordingly, the parameters occurring in the (bulk) diffusion equations are either p, or T, or C (for more species, $\tilde{C}$), or a combination of those. Again, the continuity conditions to be obeyed at the interface are concerned with heat transport (where the heat of crystallisation has also to be accounted for), with matter transport (related with segregation at the interface) or with both. Therefore, we have to use the same p, T or $\tilde{C}$ parameters in the interface EOM, in particular in the γ and R functions. In this way the equations become "case-specific".

However, in this book we have deliberately tried to follow a generalized approach, emphasizing physical concepts and principles related with OD surface properties and illustrating them at most by simple prototype examples. Apart from that, we thought it appropriate to neglect bulk-diffusion and thermodynamic aspects, because these have already been covered extensively in existing textbooks and review papers. (see below). In agreement with this generalized (but limited) point of view, we have expressed our equations and functions in terms of "wild-card" parameters A and/or m, rather than explicit and more case-specific parameters p, T or $\tilde{C}$ etc. In Part I this abstract formulation concerns the EOM themselves; in part II it concerns the parameters occurring in $R(\vec{n})$ and/or $\gamma(\vec{n})$ functions. This means that we are just presenting a "formal" or abstract formulation in this monograph, which is only concerned with the *geometry* of crystal growth. For practical application, a) this formulation has to be adapted to the relevant situation by making the wild-card parameters A or m explicit in terms of the actual parameters (p, T or $\tilde{C}$); b) the description has eventually to be supplemented by equations for diffusion fields or by thermodynamic relations - in short, by equations concerning those subjects which have been neglected in this book:

* thermodynamics
* (bulk) diffusion, eventually including polynary diffusion and combined diffusion of heat and matter + cross-relations (Soret effect etc.)
* chemical reactions in the bulk phase

This same list of neglected subjects explains why this monograph is *not* a suitable introductory textbook on crystal growth in general. For reference on these subjects, see e.g. PARKER (1970),OHARA and REID (1973), ROSENBERGER (1979), WILCOX (1976), CAROLI, CAROLI and ROULET (1992).

Chapter 2

OPERATORS FOR ORIENTATION-DEPENDENT PARAMETERS

2.1 INTRODUCTION

For a compact formulation of continuum surface-evolution equations, it is appropriate to introduce vector and tensor operators working on OD parameters. The present chapter is devoted to these mathematical tools. In order to label symbols as operators, we will use slightly oversized sanserif characters as a typographic distinction. (e.g. $\vec{\mathsf{P}}$, $\vec{\mathsf{Q}}$, S etc).

We assume that the reader is familiar with textbook discussions on the equilibrium form and the stationary growth form of crystals in connection with the $\gamma(\vec{n})$ and $R(\vec{n})$ functions. The connection between these forms and the "plot representation" of the γ and R functions is given by the "Gibbs-Wulff construction". (e.g., HERRING 1952). There is a unique correspondence between a point on the (γ or R) plot and a point on the (equilibrium or growth) crystal shape. Moreover, certain orientation ranges are unstable and do not occur on the crystal.

Usually this correspondence between plot and crystal shape is explained as a geometrical construction principle. This would be sufficient if we were only interested in the equilibrium and stationary-growth forms, respectively. But in this monograph, we want also to discuss nonequilibrium and nonstationary growing interfaces and their evolution, and we need a mathematical expression for this correspondence in the equation of motion for the surface (even although this EOM is not restricted to equilibrium or stationary shapes at all).

The correspondence between plot and (equilibrium/stationary growth) shape is a *relation between vectors;* i.e. between the normal vector $\vec{n}$ defining the point on the polar γ or R plot as the argument of the γ or R function, and the points on the equilibrium and growth shapes obtained from these plots by the GW construction (hereafter called the *GW shape*).

We can define this vector relation symbolically by a vector operator $\vec{\mathsf{Q}}$, working on an orientation-dependent scalar, in general indicated as $\Psi(\vec{n})$.

We may call it the *Gibbs-Wulff operator*. The vector function $\vec{\mathbf{Q}}\Psi(\vec{n})$ transforms the points on the unity sphere (i.e., the $\vec{n}$-variable space) into the points of the GW shape (which therefore may also be called the $\vec{\mathbf{Q}}\Psi$ shape). Note that it is a trivial consequence of the GW construction that the normal to the GW shape at the point $\vec{\mathbf{Q}}\Psi(\vec{n})$ is also $\vec{n}$ (see discussion in 2.1). As we will see below, $\vec{\mathbf{Q}}$ can be identified as $(\vec{n} + \mathrm{d}/\mathrm{d}\vec{n})$.

Although the correspondence between the $\vec{\mathbf{Q}}\gamma$ and $\vec{\mathbf{Q}}R$ shapes and equilibrium- resp. growth shapes suggests a direct relation between the $\vec{\mathbf{Q}}$ operator and actual surfaces, the $\vec{\mathbf{Q}}$ operator and the vector functions $\vec{\mathbf{Q}}\Psi(\vec{n})$ are defined without reference to any physical surface. Of course we may say that the differentiation $\mathrm{d}/\mathrm{d}\vec{n}$ in the operator corresponds to a variation of the "surface property" Ψ (γ or R) when we move over an actual (curved) surface or when we stay at a fixed point but tilt the surface. But such infinitesimal variations in γ and R are the very variations which occur when we have to calculate other surface properties than γ or R (but related with them) by a variational principle. To be specific, $\vec{\mathbf{Q}}\gamma(\vec{n})$ is related with a force vector exerted by an interface at orientation $\vec{n}$, and $\vec{\mathbf{Q}}R(\vec{n})$ with the kinematic wave trajectory followed by infinitesimal disturbances of the surface. These relations are worked out in detail in chapters 3 and 5, respectively.

Consequently, it seems logical to consider $\vec{\mathbf{Q}}\gamma$ and $\vec{\mathbf{Q}}R$ as new (vector) properties of the interface: also orientation-dependent, and derivable from the γ and R functions by application of the $\vec{\mathbf{Q}}$ operator.

The correspondence between $\vec{\mathbf{Q}}\gamma$ and $\vec{\mathbf{Q}}R$ shapes and the shape of actual crystal surfaces implies that also the *curvatures or the radii* of the $\vec{\mathbf{Q}}\gamma$ and $\vec{\mathbf{Q}}R$ shapes are important physical parameters. (the curvature of $\vec{\mathbf{Q}}\gamma$ is related with the Gibbs-Thomson effect, that of $\vec{\mathbf{Q}}R$ with the curvature evolution rate). For a compact notation of equations it is practical to introduce a tensor operator $\vec{\vec{\mathbf{S}}}$ which expresses the inverse of the curvature tensor, i.e. the radius-tensor, of $\vec{\mathbf{Q}}\Psi$ shapes directly: $\vec{\vec{\mathbf{S}}}\Psi = \mathrm{d}/\mathrm{d}\vec{n}\,(\vec{\mathbf{Q}}\Psi)$

2.2 NOTATIONS

The orientation variable is written as the normal vector $\vec{n}$. To describe variations of this normal vector around a fixed direction $\vec{n}_0$, it is convenient to use angular variables φ and ϑ. (some models of the crystal surface can be worked out into $\gamma(\vec{n})$ or $R(\vec{n})$ functions expressible as a power series of the deviation angles from a given low-index orientation). Note, however, that the use of such angular variables implies the definition of a basis set of reference directions on the unit sphere, for each $\vec{n}_0$. The coordinates of a given deviation from $\vec{n}_0$ with respect to different basis sets are related by a rotation transformation in the tangent plane at orientation $\vec{n}_0$.

Because the normal can be chosen in two opposite directions, we have to label the OD parameter with a suffix denoting the direction convention. For an interface between phases or grains α and β, we use the suffix α/β to indicate that the α phase is "behind" and the β phase "ahead" of the interface, looking in the direction of the normal $\vec{n}$. Note that in general

$$\Psi_{\alpha/\beta}(\vec{n}) \neq \Psi_{\beta/\alpha}(\vec{n})$$

as interface structures are not necessarily identical when the positions of the phases α and β are interchanged, both keeping their crystallographic orientation in space.

The symbols α and β can also refer to grains of the same crystal, but of different crystallographic orientation (α/β grain boundaries). In the course of this monograph, we will often have to do with the interface between a crystal and a *noncrystalline parent phase*. We will adopt the convention to choose the normal in the crystal/parent phase direction, and omit the direction suffix in this case.

Note also that for a crystal/noncrystalline (isotropic medium) interface, the value of the OD parameter depends uniquely on the interface normal relative to the crystallographic axes of the crystal: the Ψ plot is given and fixed with respect to the crystal. For an interface between two crystalline phases, however, the situation is more complicated: the shape of the plot depends also on the *relative orientations of both crystals*, i.e. an extra 3-dim. mutual-orientation parameter $\vec{n}_{rel.}$: $\Psi_{\alpha/\beta}(\vec{n}_{rel.}, \vec{n})$.

2.3 THE VECTOR OPERATORS $\vec{P}$ *AND* $\vec{Q}$

In the discussion to follow we will consider in parallel the three- and the two-dimensional or "cylindrical" case: the latter is concerned with an OD parameter depending on a single angular variable φ. (note that this φ variable, just as $\vec{n}$, is a vector on the unit circle. We may emphasize the vector character of the φ variable, when used as an orientation parameter, by actually writing it as a vector: $\vec{\varphi}$). Discussion of the single-variable case, being inherently simpler and more familiar, is often instructive as most physical effects can also be demonstrated for this case and their importance estimated to the correct order of magnitude.

The parameter $\Psi(\vec{n})$ is a scalar parameter defined on the unit sphere. Usually it has a dimension: in this paper we are mainly concerned with the parameters γ (energy per unit area), R (growth rate) or χ (growth mobility, i.e. growth rate per unit driving force). In mechanics, optics, acoustics and semiconductor physics we can find other examples: free energy of Bloch and ferroelectric-domain walls, velocities and mobilities of particles and waves, and surface-layer concentrations.

We can conveniently "plot" these parameters in a graphical representation by defining the *plotting operator* $\vec{\mathsf{P}}$ which is simply the multiplication with the normal vector:

$$\vec{\mathsf{P}}\Psi(\vec{n}) \equiv \vec{n}\,\Psi(\vec{n}) \tag{2.1}$$

The function $\Psi(\vec{n})$ is now represented by the three-dimensional surface $\vec{n}\,\Psi(\vec{n})$ (polar diagram) defined in a 3-dim. space where length has the same dimension as Ψ. (the two-dimensional plots which are often referred to in, e.g. demonstration of the Gibbs-Wulff construction are cross-sections of the 3-dimensional plots). The Gibbs-Wulff operator can now be defined by

$$\vec{\mathsf{Q}}\Psi(\vec{n}) \equiv \vec{n}\,\Psi(\vec{n}) + \frac{d}{d\vec{n}}\Psi(\vec{n}) = \vec{\nabla}(\vec{r}\,\Psi(\frac{\vec{r}}{|\vec{r}|})) \tag{2.2}$$

The operator $d/d\vec{n}$ is the *surface gradient* $\vec{\nabla}_s$ *on the unit sphere*; it is locally an only two-dimensional vector operator as $\vec{n}$ cannot vary in length: the component of $d\vec{n}$ perpendicular to the surface of the unit sphere is zero as a consequence of the normalisation restriction. So, $d\Psi/d\vec{n}$ is a vector in the tangent plane at orientation $\vec{n}$. (written out in coordinates: when $\vec{e}_2$, $\vec{e}_3$ are the ortho-normal basis set in the tangent plane at $\vec{n}$ $(= \vec{e}_1)$ and φ ,ϑ are the corresponding variables for $\vec{n}$, so that $d\vec{n} = \vec{e}_2\, d\varphi + \vec{e}_3\, d\vartheta$, then $d\Psi(\vec{n})/d\vec{n}$ is a direct vector notation for $\vec{e}_2\, d\Psi/d\varphi + \vec{e}_3\, d\Psi/d\vartheta$).

$\vec{\mathsf{Q}}\Psi(\vec{n})$ can be considered as a transformation working on the points of the plot $\vec{\mathsf{P}}\Psi(\vec{n})$. The geometrical aspects of this transformation are illustrated in Fig. 2.1 which shows a two-dimensional plot $\vec{\mathsf{P}}\Psi(\vec{n})$. The point of the plot corresponding with the orientation φ_o is A; in this point we have drawn the line perpendicular to the normal. The same has been done for a neighbouring direction φ_o+$d\varphi$. The intersection of the perpendicular lines, for infinitesimal $d\varphi$, is B. It is easily verified that the length $\overline{AB}$ is equal to $(d\Psi/d\varphi)_{\varphi o}$ (the triangles OAP and BA'P are similar, $AP = \Psi(\varphi_o)\, d\varphi$ and

$$\frac{A'P}{AP} = \frac{1}{\Psi}\,\frac{d\Psi}{d\varphi}.$$ So as OA is equivalent with the vector $\vec{n}\Psi(\vec{n})$ and

AB with $(d/d\vec{n})\ \Psi(\vec{n})$, (h.l. $(d/d\varphi)\ \Psi(\varphi)$), the vector OB is equivalent with $(\vec{n} + d/d\vec{n})\ \Psi(\vec{n}) = \vec{\mathsf{Q}}\Psi(\vec{n})$. The above can be generalized to the 3-dim. case by substituting $\vec{n}$ for φ. The vector $\vec{\mathsf{Q}}\Psi$ corresponding with the point A, i.e. the vector $\vec{\mathsf{P}}\Psi$, on the $\vec{\mathsf{P}}\Psi$ plot is now the vector OB where B is the point

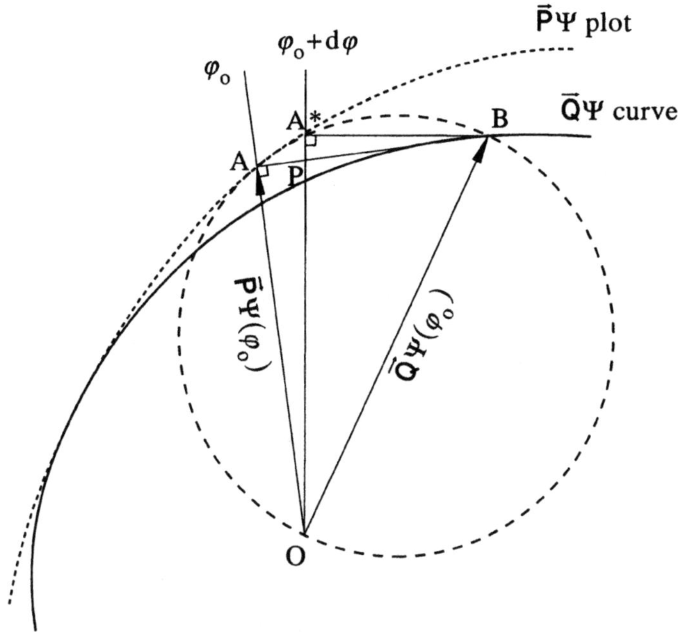

FIG. 2.1

on the sphere, passing through O and tangent to the $\vec{P}\Psi$ plot at A, diametrically opposite O. Fig. 2.1 can be considered as the construction in a cross-section through the 3-dim. plot, through O and one of the $\vec{e}_2$, $\vec{e}_3$ vectors, to find the contribution to the $d\Psi/d\vec{n}$ = AB vector from the $\vec{e}_2\, d\Psi/d\varphi$ or the $\vec{e}_3\, d\Psi/d\vartheta$ components.

The operator $\vec{Q}$ transforms the $\vec{P}\Psi(\vec{n})$ plot into another surface, which we will refer to as the $\vec{Q}\Psi$ surface. The two are identical only in the trivial case that $\Psi(\vec{n})$ is independent of $\vec{n}$. In the general case, $\vec{P}\Psi$ and $\vec{Q}\Psi$ coincide in the extrema of $\vec{P}\Psi$.

In chapters 3 and 5 we will see that the $\vec{Q}\Psi$ shape has a direct physical meaning, being similar to the shape of real objects such as crystals grown under special conditions or of wavefronts. In contrast, the shape of the $\vec{P}\Psi$ plot has no such meaning, except as a convenient means to represent the set of $\Psi(\vec{n})$ data in graphical form. It is therefore relevant to investigate the geometry of the $\vec{Q}\Psi$ shape in some detail.

2.4 *THE TENSOR OPERATOR* $\vec{\vec{\mathbf{S}}}$

In the following we focus on the geometry of the $\vec{\mathbf{Q}}\Psi$ surface. In the two-dimensional case where we have a single-parameter $\vec{\mathbf{P}}\Psi(\varphi)$ plot and $\vec{\mathbf{Q}}\Psi(\varphi)$ curve, the curvature $\boldsymbol{K}$ of the $\vec{\mathbf{Q}}\Psi$ curve, for orientation φ, is (ds following the curve $\vec{\mathbf{Q}}\Psi$):

$$\boldsymbol{K} = \frac{1}{r(\varphi)} = \left\{\frac{\mathrm{d}}{\mathrm{d}\varphi}\left(\varphi + \frac{\mathrm{d}}{\mathrm{d}\varphi}\right)\Psi(\varphi)\right\}^{-1} = \left\{\left(1 + \frac{\mathrm{d}^2}{\mathrm{d}\varphi^2}\right)\Psi(\varphi)\right\}^{-1} \tag{2.3}$$

(note that the normal to the $\vec{\mathbf{Q}}\Psi(\varphi)$ curve is just in the φ direction, and that $\vec{\varphi}\cdot \mathrm{d}\Psi/\mathrm{d}\vec{\varphi}$, being the interior product of two orthogonal vectors, equals zero).

So, the radius of the $\vec{\mathbf{Q}}\Psi$ curve, corresponding with orientation φ, is $(1+\mathrm{d}^2/\mathrm{d}\varphi^2)\Psi(\varphi)$.

This can also be seen from fig. 2.2. For neighbouring points A_o and A_1 on the $\vec{\mathbf{P}}\Psi$ plot, we have corresponding B_o and B_1 on the $\vec{\mathbf{Q}}\Psi$ curve. As

$$r(B_o) : B_oB_1 = OA_o : A_oP \quad \text{and} \quad A_oP = \Psi\mathrm{d}\varphi,$$

$$A_1B_1 = \left(\frac{\mathrm{d}\Psi}{\mathrm{d}\varphi}\right), \quad A_oB_o = \left(\frac{\mathrm{d}\Psi}{\mathrm{d}\varphi}\right) \quad \text{, we have}$$

$$B_oB_1 \simeq A_1B_1 + A_oP - A_oB_o = \mathrm{d}\varphi\frac{\mathrm{d}^2\Psi}{\mathrm{d}\varphi^2} + \mathrm{d}\varphi.\Psi$$

and so
$$r(B) = \Psi + \frac{\mathrm{d}^2\Psi}{\mathrm{d}\varphi^2}$$

In the following, we will use the symbol $\mathbf{S}$ ("stiffness", see 3.2) for the operator $(1 + \mathrm{d}^2/\mathrm{d}\varphi^2)$.

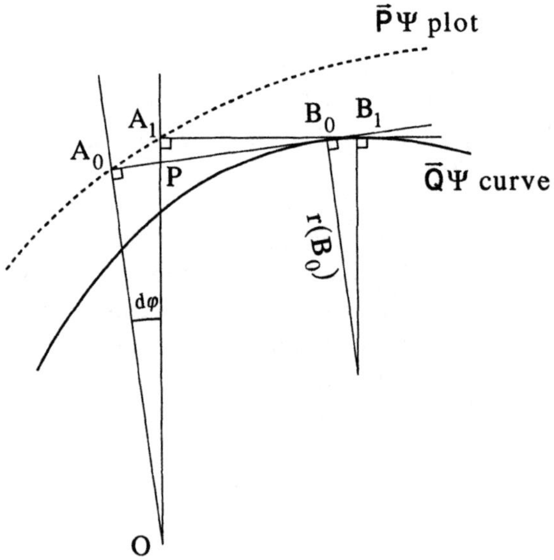

FIG. 2.2

This can be generalized to the 3-dim. case. The curvature ($\vec{\vec{K}}$) of a 2-dim. surface in 3-dim. space is a tensor quantity, usually referred to as the "second fundamental tensor" of the surface. As we will mostly be concerned with radii rather than with curvatures, we will preferably use the inverse of this second fundamental tensor. Therefore we define the tensor operator $\vec{\vec{S}}$:

$$\vec{\vec{K}}^{-1}\{\vec{Q}\Psi \text{ shape}\} \equiv \vec{\vec{S}}\Psi(\vec{n}) \equiv \frac{d}{d\vec{n}}\{\vec{Q}\Psi(\vec{n})\} = \frac{d}{d\vec{n}}(\vec{n} + \frac{d}{d\vec{n}})\Psi(\vec{n}) =$$

$$= \left(\vec{\vec{I}} + \frac{d^2}{d\vec{n}^2}\right)\Psi(\vec{n}) \tag{2.4}$$

giving, per definition, the shift along the $\vec{Q}\Psi$ surface corresponding with a specified change in orientation $d\vec{n}$. (here again, it is essential to realize that the normal of the $\vec{Q}\Psi(\vec{n})$ surface, in a point corresponding with $\vec{P}\Psi(\vec{n})$, is just $\vec{n}$. We write $\vec{\vec{I}}$ for the unity tensor, $\begin{pmatrix}1 & 0\\ 0 & 1\end{pmatrix}$ for all basis sets.

Choosing a basis set of reference directions in the tangent plane to the $\vec{Q}\Psi$ surface for which the tensor $\vec{\vec{S}}\Psi(\vec{n})$ takes the diagonal form (i.e., these reference directions coincide with the "principal directions" of the $\vec{Q}\Psi$ surface at that point), the tensor $\vec{\vec{S}}\Psi(\vec{n})$ is represented by the matrix

$$\vec{\vec{S}}\Psi(\vec{n}) = \begin{pmatrix} r_1 & 0 \\ 0 & r_2 \end{pmatrix} \tag{2.4a}$$

where $r_1 = \dfrac{d\vec{Q}\Psi}{d\varphi_p} = \Psi + \dfrac{d^2\Psi}{d\varphi_p^2}$

$r_2 = \dfrac{d\vec{Q}\Psi}{d\vartheta_p} = \Psi + \dfrac{d^2\Psi}{d\vartheta_p^2}$

φ_p and ϑ_p are the set of angular variables for $\vec{n}$ corresponding to the principal directions of the $\vec{Q}\Psi$ surface. For the same basis set,

$$\left(\vec{\vec{S}}\right)^{-1}\Psi(\vec{n}) = \begin{pmatrix} 1/r_1 & 0 \\ 0 & 1/r_2 \end{pmatrix} \tag{2.4b}$$

with the principal curvatures (eigenvalues of the second fundamental tensor) on the diagonal.

For an arbitrarily chosen direction, making angles ξ_1 and $\xi_2 = (\pi/2 - \xi_1)$ with the principal directions, the curvature $1/r_\xi$ of the $\vec{Q}\Psi$ surface can be found from Euler's formula:

$$\frac{1}{r_\xi} = \frac{1}{r_1}\cos^2\xi_1 + \frac{1}{r_2}\cos^2\xi_2 = \frac{1}{r_1}\cos^2\xi_1 + \frac{1}{r_2}\sin^2\xi_1 \tag{2.5}$$

2.5 DISTANCE ON THE $\vec{\mathbf{Q}}\Psi$ SURFACE

Eq. (2.4) can be integrated to find the distance between two points of the $\vec{\mathbf{Q}}\Psi$ surface with given orientations $\vec{n}_1$ and $\vec{n}_2$, as measured along the cross-section of the $\vec{\mathbf{Q}}\Psi$ surface and the plane defined by $\vec{n}_1$ and $\vec{n}_2$:

$$\int_{\vec{n}_1}^{\vec{n}_2} d\vec{s} = \int_{\vec{n}_1}^{\vec{n}_2} \frac{d\vec{s}}{d\vec{n}} \cdot d\vec{n} = \int_{\vec{n}_1}^{\vec{n}_2} \vec{\vec{\mathbf{S}}}\Psi(\vec{n}) \cdot d\vec{n} \tag{2.6}$$

($d\vec{n}$ to be kept in the plane through $\vec{n}_1$ and $\vec{n}_2$). Using φ for the $\vec{n}$ variable in the plane through $\vec{n}_1$ and $\vec{n}_2$ we have

$$\int_{\vec{n}_1}^{\vec{n}_2} d\vec{s} = \int_{\varphi_1}^{\varphi_2} \left(1 + \frac{d^2}{d\varphi^2}\right) \Psi(\varphi) d\varphi \tag{2.7}$$

An interesting consequence is that the total circumference of this cross-section, $\oint (1 + d^2/d\varphi^2)\, \Psi(\varphi) d\varphi$, is equal to $\oint \Psi(\varphi) d\varphi$, i.e. the surface of the *linear* graph of Ψ as a function of φ. This is because the second derivative of Ψ, being a 2π- periodic function of φ, gives no contribution to this integral. For those cases where $\mathbf{S}\Psi(\varphi)$ is $\geqslant 0$ for all φ, i.e. where the $\vec{\mathbf{Q}}\Psi$ curve has no self-intersections (see 2.7), this total length can be identified with the "circumference" of the closed $\vec{\mathbf{Q}}\Psi$ curve without complications. In contrast, when orientation ranges with $\mathbf{S}\Psi < 0$ occur - and consequently "ears" develop in the $\vec{\mathbf{Q}}\Psi$ curve by self-intersection - we have a complication which is discussed in 2.7.

Whereas $\oint \Psi(\varphi) d\varphi$ (= circumference of $\vec{\mathbf{Q}}\Psi$ curve) is the surface of the *linear* graph of Ψ as a function of φ, the *surfaces* of the *polar* plot $\vec{\mathbf{P}}\Psi$ and that of the $\vec{\mathbf{Q}}\Psi$ curve have also a simple algebraic expression:

$$\text{surface of (polar) } \vec{\mathbf{P}}\Psi \text{ plot} = \tfrac{1}{2}\oint \Psi^2(\varphi)\, d\varphi). \tag{2.8}$$

$$\text{surface of } \vec{\mathbf{Q}}\Psi \text{ plot} = \tfrac{1}{2}\oint \Psi(\varphi).\ \mathbf{S}\Psi(\varphi)\, d\varphi). \tag{2.9}$$

(in the case of a $\vec{\mathbf{Q}}\Psi$ plot with self-intersections, it is easily verified that in the "surface of the $\vec{\mathbf{Q}}\Psi$ plot", the surface of the ears has to be taken negative for eq. 2.9 to be valid).

2.6 PATHOLOGY OF THE $\vec{\mathbf{Q}}\Psi$ SURFACE: FACETS, EDGES, VERTICES, VIRTUAL REGIONS AND THE EFFECTIVE PLOT

We now discuss two types of complications:

1) the plot $\vec{\mathbf{P}}\Psi$ can have cusps, so that (for certain directions) $\vec{\mathbf{Q}}\Psi$ is discontinuous and the elements of $\vec{\vec{\mathbf{S}}}\Psi$ are singular; (section 2.6)

2) the $\vec{\mathbf{Q}}\Psi$ surface can show self-intersections and/or cusps. (section 2.7)

As to 1), consider first the 2-dim. case (fig. 2.3). For the cusp orientation φ_0, the (now scalar) function $\mathbf{S}\Psi$ is infinite. ($+\infty$ for an inward cusp as shown in fig. 2.3, $-\infty$ for a protruding cusp). Since $\mathbf{S}\Psi$ is the radius of curvature of $\vec{\mathbf{Q}}\Psi$, this implies that the $\vec{\mathbf{Q}}\Psi$ curve has a flat region corresponding with the cusp in the $\vec{\mathbf{P}}\Psi$ plot: this region is called a *facet*. We can obviously define $\mathbf{S}\Psi(\varphi_0)$ more explicitly in order to maintain the validity of eq. (2.7):

$$\mathbf{S}\Psi(\varphi_0) \equiv \left| \vec{\mathbf{Q}}\Psi(\varphi_0^+) - \vec{\mathbf{Q}}\Psi(\varphi_0^-) \right| . \delta(\varphi_0) \tag{2.10}$$

where δ is the Dirac delta function, and φ_0^+ and φ_0^- are the directions which immediately adjoin the cusp direction.

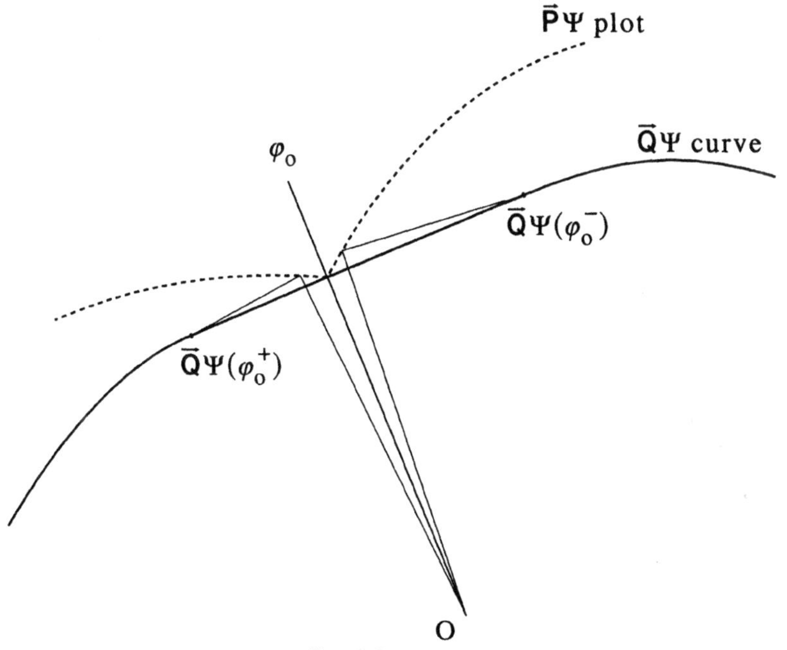

FIG. 2.3

For a "symmetric" cusp, we can write

$$\Psi(\varphi_0 + \varphi) = \Psi(\varphi_0)\,(1 + \tfrac{a}{2}|\varphi| \quad + \text{higher-order terms}) \tag{2.11}$$

so that the size of the corresponding facet is $\mathbf{S}\Psi(\varphi_0) = a\,\Psi(\varphi_0)\,.\,\delta(\varphi_0)$. The cusp angle is $2(\pi/2 - \arctan \tfrac{1}{2}a)$, both in the polar plot $\vec{\mathbf{P}}\Psi(\varphi)$ and in the cartesian plot of the reduced quantity $\Psi/\Psi(\varphi_0)$. The generalisation to asymmetric cusps is obvious.

When the cusp is smoothed to a finite radius, so that Ψ in the neighbourhood of the cusp orientation can be written as

$$\Psi(\varphi_o + \varphi) = \Psi(\varphi_o)\,(1 + \frac{b}{2}\varphi^2 + \ldots) \tag{2.12}$$

then the "facet" becomes curved: the radius is

$$\mathbf{S}\Psi(\varphi_o) = \Psi(\varphi_o)\,.\,(1 + b) \tag{2.13}$$

The radius in the cartesian plot of the reduced quantity $\Psi/\Psi(\varphi_o)$ as a function of φ is $r = b^{-1}$. In the polar plot $\vec{\mathbf{P}}\Psi(\varphi)$, the radius is $\Psi(\varphi_o).b^{-1}$.

The generalisation to the 3-dim. case is obvious. For a "smooth" plot, with continuous derivatives, we can expand Ψ in the vicinity of an orientation (φ_o,ϑ_o):

$$\begin{aligned}\Psi(\varphi_o+\varphi,\vartheta_o+\vartheta) = \Psi(\varphi_o,\vartheta_o)\,(\,1 &+ a\varphi + \frac{b}{2}\,\varphi^2 + \frac{c}{3}\,\varphi^3 + \ldots + \\ &+ k\vartheta + \frac{\ell}{2}\,\vartheta^2 + \frac{m}{3}\,\vartheta^3 + \ldots + \\ &+ p\varphi\vartheta + \frac{q}{2}\,\varphi^2\vartheta + \frac{r}{2}\,\varphi\vartheta^2 + \ldots)\end{aligned} \tag{2.14}$$

$$\vec{\mathbf{Q}}\Psi = \Psi(\varphi_o,\vartheta_o)(\,a + b\varphi + p\vartheta + \ldots + k + \ell\vartheta + p\varphi + \ldots) + \vec{n}\Psi \tag{2.15}$$

and then
$$\vec{\vec{\mathbf{S}}}\Psi(\varphi_o,\vartheta_o) = \Psi(\varphi_o,\vartheta_o)\begin{pmatrix} 1+b & p \\ p & 1+\ell \end{pmatrix} \tag{2.16}$$

When the basis set is chosen along the principal directions, the mixed terms in Ψ vanish and $\vec{\vec{\mathbf{S}}}\Psi$ is represented by a diagonal matrix.

When an edge is present, we have to add a term $e|\Psi|$ to the expression for Ψ. The first basis vector is now chosen perpendicular to the projection of the edge line. It is no longer possible in general to omit the mixed term, but often it will be zero for reasons of symmetry. The edge shows up in the $\vec{\vec{\mathbf{S}}}\Psi$ matrix as an extra term $e\delta(\varphi_o)$; i.e., the $\vec{\mathbf{Q}}\Psi$ surface is cylindrical.

$$\vec{\vec{\mathbf{S}}}\Psi(\varphi_o,\vartheta_o) = \Psi(\varphi_o,\vartheta_o)\begin{pmatrix} 1+b+e\delta(\varphi_o) & (p) \\ (p) & 1+\ell \end{pmatrix} \tag{2.16a}$$

For a vertex point where three or more edges meet, both radii are infinite. When the projections of the edges are perpendicular, the δ functions can be kept separated and we get $e\delta(\varphi_o)$, $f\,\delta(\vartheta_o)$ terms in the diagonal of the $\vec{\vec{\mathbf{S}}}\Psi$ tensor.

Note that $\mathbf{S}\Psi$ for the 2-dim. case, as well as the four tensor elements for the 3-dim. case, have no "vectorial" significance. They can, for the purpose of graphical illustration, be plotted by applying the $\vec{\mathbf{P}}$ operator, but the surfaces $\vec{\mathbf{P}}\mathbf{S}\Psi(\varphi)$ etc. obtained in this way have no physical interpretation. (the same is true for $\vec{\mathbf{P}}\Psi$ itself).

2.7 SELF-INTERSECTIONS OF THE $\vec{Q}\Psi$ SURFACE.

The $\vec{Q}\Psi$ surface can have self-intersections and, correspondingly, the volume bounded by it can be divided into separate regions. The situation is illustrated in fig. 2.4 for the 2-dim. case. We will see in chapter 3 that those parts of the $\vec{Q}\gamma$ surface which belong to the "disconnected" volume, i.e. disconnected from the central part containing the origin, correspond to physically unstable situations. It is therefore appropriate to investigate the topology of the $\vec{Q}\Psi$ surface.

Starting again with the 2-dim. case, we see from the tangent-circle construction of the $\vec{Q}\Psi$ curve that a selfintersection point corresponds with the situation where the tangent circle touches the $\vec{P}\Psi$ plot in two points simultaneously. The angle made by the two parts of the $\vec{Q}\Psi$ curve at the intersection point is the angle between the corresponding tangent-point normals of the $\vec{P}\Psi$ plot. We will call these normals the "edge orientations".

For the part of the $\vec{P}\Psi$ plot between the edge orientations, the corresponding $\vec{Q}\Psi$ curve belongs to the disconnected part. ("ears" of the $\vec{Q}\Psi$ curve). Particular features of this part are the negative cusps which occur when the tangent circle touches the $\vec{P}\Psi$ plot to second order (i.e., $S\Psi = 0$) and the part between these cusps where $S\Psi$ is negative.

It is of interest to consider how the $\vec{P}\Psi$ plot might be reshaped in order to eliminate the disconnected part of the $\vec{Q}\Psi$ curve altogether, while maintaining the rest. It is easily seen that the part of the $\vec{P}\Psi$ plot between the edge orientations then has to be replaced by the common tangent circle. For every point on this circle, the corresponding point of the $\vec{Q}\Psi$ curve is the same (the edge point, i.e. the point on the circle which is diametrically opposite the origin). Indeed, a circle - i.e., $\Psi(\varphi) = \Psi_0 \sin(\varphi-\varphi_0)$ - obeys the equation $S\Psi(\varphi) = 0$, in agreement with the sharp edge. In view of the physical interpretation given later (in particular for the γ function), we will call the range of orientations between the edge orientations the *"virtual range"* and the part of the $\vec{P}\Psi$ plot in this range the *"virtual part of the $\vec{P}\Psi$ plot"*. The plot which is obtained by replacing the virtual part(s) by tangent circle(s) will be referred to as the *"effective $\vec{P}\Psi$ plot"*. We will write the corresponding function as Ψ^*. The "inner" part of the $\vec{Q}\Psi$ curve, i.e. without the ears, will be called the "real $\vec{Q}\Psi$ curve" ($\equiv \vec{Q}\Psi^*$ curve).

The occurrence of negative-$S\Psi$ ranges has a consequence for the interpretation of eqs. (2.6) and (2.7) - and a fortiori for the identification of the total circumference of the $\vec{Q}\Psi$ curve with $\oint \Psi(\varphi)d\varphi$. We have to be aware that the left-hand side of the equations refers to the $\vec{Q}\Psi$ curve *including* the ears, not to the $\vec{Q}\Psi^*$ curve without the ears. (an actual equilibrium crystal corresponds with the $\vec{Q}\gamma$ curve, an actual - large enough - growing crystal with the $\vec{Q}R$ curve). *But the length integral over the ears is not zero.*

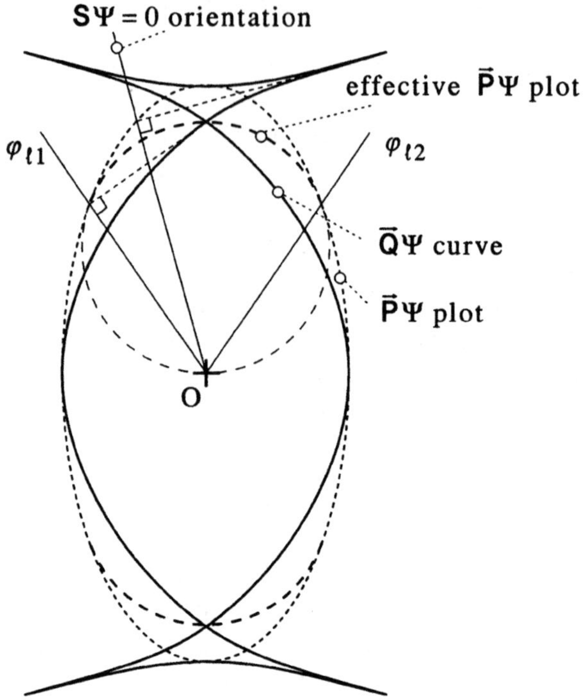

FIG. 2.4

Although the part between the cusps, with **S**Ψ negative, is subtracted from the rest, the "net ear length" is positive (for small $|\varphi_{t1}-\varphi_{t2}|$, second order in $\sin(|\varphi_{t1}-\varphi_{t2}|)$). This implies that eqs (2.6) or (2.7) are no longer correct when considered to apply to a path on a real crystal crossing an edge; i.e., when the range $(\varphi_1 - \varphi_2)$ contains a region with **S**Ψ<0. The validity of eqs (2.6) or (2.7) can be restored for that case by replacing $\Psi(\varphi)$ in the right-hand term by $\Psi^*(\varphi)$. By definition, for the Ψ^* function **S**Ψ^* is $\geqslant 0$ everywhere, but it can have **S**Ψ^* discontinuities. As a consequence of the replacement by tangent circles, $\Psi^*<\Psi$, so that in the total circumference $\oint \Psi^*(\varphi)\mathrm{d}\varphi < \oint \Psi(\varphi)\mathrm{d}\varphi$. (This is dramatically demonstrated by the $R(\varphi)$ plot of a perfect crystal with all facets below the roughening transition: the real $\vec{\mathbf{Q}}R$ curve then condenses into a point!) Note that for a "soapfroth-type" surface-energy plot, i.e. a plot consisting of circle sections so that **S**γ is $\geqslant 0$ (either 0 or a δ function), the condition is marginally satisfied so that the circumference of the (faceted) *equilibrium* crystal with such a γ function is still, in agreement with eq. (3.15), $= (V/\Delta A_c).\oint \gamma(\varphi)\mathrm{d}\varphi$.

The above can directly be extended to 3-dimensional plots. The points of the $\vec{\mathbf{Q}}\Psi$ surface can be found by the tangent-sphere construction. When the tangent sphere touches the $\vec{\mathbf{P}}\Psi$ plot in two points, the $\vec{\mathbf{Q}}\Psi$ surface has a (protruding) edge; when it does so in three or more points it has a vertex where three or more edges meet. An edge can end either in a vertex or in a "critical point" where the edge orientations coincide. All orientations where the tangent sphere cuts the $\vec{\mathbf{P}}\Psi$ plot in more than one other point are virtual, and the effective $\vec{\mathbf{P}}\Psi$ plot is obtained as the envelope of the tangent sphere as it "inflates" within the virtual plot. It consists of purely spherical parts (corresponding to vertex points) and of parts of anisotropic curvature.

2.8 CURVATURE DEPENDENCE OF SURFACE-RELATED QUANTITIES

A crystal-parent phase interface is not a mathematical surface: it has a certain thickness and "vagueness", inherent to its atomic structure. Consequently, surface-related physical quantities cannot be considered as to be strictly related with a mathematical point on such a surface: rather they are distributed according to some density function, which has to be integrated over a certain volume to obtain the "macroscopic" quantity. If we insist on using a mathematical surface as a basis for the description of the surface (as we do in this monograph), then we have to be aware that this integration procedure leads to a curvature dependence of Ψ for curvatures exceeding some critical value. As such dependence can have an effect on kinetic stability (ch. 6), we will discuss it here in some detail.

In this section we focus on the curvature dependence as such, including its implications for the $\vec{\mathbf{Q}}\Psi$ and $\vec{\vec{\mathbf{S}}}\Psi$ quantities, whereas its relations with the equation of motion and stability will be discussed in 6.3. For simplicity, we consider one curvature variable only, rather than the tensor curvature. We do this because the series expansion formula given below inherently falls short to describe really sharp edges and vertices, so that its practical value to predict their line energy and vector velocity is limited. Still, the expansion is instructive as it can be used to identify the coefficients of the successive orders in $1/r$ with "physical" parameters.

Our aim, then, is to obtain an expression for the coefficients $\Psi_1, \Psi_2, \ldots$ in an expansion for Ψ :

$$\Psi\left(\frac{1}{r}\right) = \Psi_0 + \frac{1}{r}\Psi_1 + \frac{1}{r^2}\Psi_2 + \ldots \tag{2.18}$$

We discuss the "delocalisation" of Ψ in the tangential and perpendicular directions separately, as their physical interpretation is somewhat different.

First the tangential contribution. Consider a curved interface as in fig. 2.5. The delocalisation implies that Ψ as measured at a particular point x_0 on the surface, is not exclusively determined by the orientation and eventual other field parameters at this point only, but by their values within a range Δx from this point. (For instance: when Ψ is growth rate of a surface

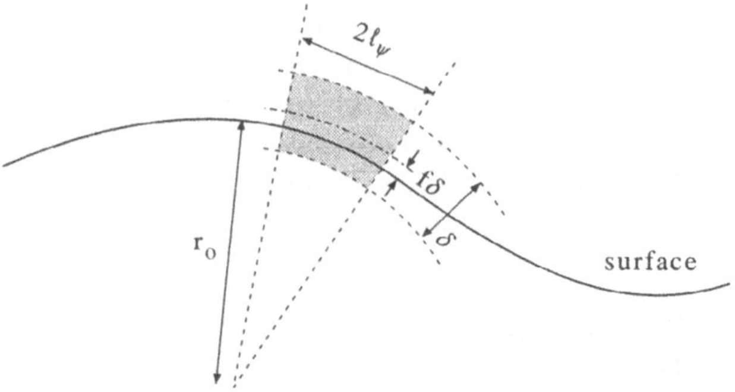

FIG. 2.5

at $T < T_R$, the atoms which are built in into the lattice at x_o have arrived at the surface at some average distance x_d away, and their transport rate can be influenced by the positions of nearby ledges (i.e., the "orientations" at $x_o + x_d$ and $x_o - x_d$) and by the A value there. When Ψ is surface free energy, the contribution by the interaction term of the ledge system depends on the distances of ledges at the right and at the left of x_o).

In other words, when Ψ ($= \Psi_{local}$ in (2.21)) varies very fast with x, we are led to define a correlation or contributive function $C_x(\Delta x)$ for Ψ:

$$\Psi(x_o) \equiv \int_{-\infty}^{\infty} C_x(\Delta x)\, \Psi_{local}(x_o+\Delta x)\, d(\Delta x) \tag{2.21}$$

with

$$\int_{-\infty}^{\infty} C_x(\Delta x)\, d(\Delta x) = 1 \tag{2.22}$$

The $\Psi(x_o)$ is the "macroscopically observable" Ψ. $C_x(\Delta x)$ is in general positive, but can be orientation-dependent and also asymmetric (i.e. $C_x(\Delta x) \neq C_x(-\Delta x)$), for instance in the case of the Schwoebel effect for ledge velocity. It extends to a certain range, roughly to be considered as a correlation length (ℓ_Ψ), but drops to zero for $x \gg \ell_\Psi$. These general features can be stated more or less intuitively; the exact nature of $C_x(\Delta x)$, of course, has to be determined by an averaging procedure on an atomistic expression for Ψ.

On a curved surface, the orientation variation within the range of the correlation function is of the order $d\varphi = (1/r) . \ell_\Psi$; i.e. the curvature dependence of Ψ can be considered as resulting from the averaging over the range $d\varphi$. The result depends on the details of the correlation function. As a first approximation we might try a symmetric rectangular correlation function; i.e. Ψ is simply averaged over the surface between $x - \ell_\Psi$ and $x + \ell_\Psi$:

$$\Psi(\varphi_o, r) = \frac{1}{2}\, \frac{r}{\ell_\Psi} \int_{-\ell_\Psi/r}^{\ell_\Psi/r} \Psi(\varphi_o + d\varphi) \frac{d\varphi}{\cos d\varphi} \tag{2.23}$$

The factor $\cos^{-1} d\varphi$ is required because we have to average over the *projected* surface area. Expanding $\Psi(\varphi)/\cos d\varphi$ around $\varphi = \varphi_o$, we find

$$\frac{\Psi(\varphi_o + d\varphi)}{\cos d\varphi} = \Psi(\varphi_o) + \Psi'(\varphi_o) + \tfrac{1}{2}\,\{\Psi(\varphi_o) + \Psi''(\varphi_o)\}\,(d\varphi)^2 + \ldots \tag{2.24}$$

Integrating, we have

$$\Psi(\varphi_o\,, r) = \Psi(\varphi_o\,, r=0) + \frac{\ell_\Psi^2}{6r^2}\,\mathbf{S}\Psi(\varphi_o) + \ldots \tag{2.25}$$

In other words, we have identified the "virial coefficient" Ψ_2 with $\frac{1}{6}\ell_\Psi^2\,\mathbf{S}\Psi$, where ℓ_Ψ is something as a correlation length of the Ψ-determining "action" on the surface. Of course, the concept of a simple correlation length is highly schematic. Higher orders in the expansion are required to express the details of the correlation function.

The cancellation of the first-order term is a consequence of the symmetry of our assumed correlation function. When it is not symmetric, it hardly makes sense to write down an explicit expression for the resulting first-order term as the asymmetry can be brought about in a $C_x(\Delta x)$ function in so many ways. Rather we will write the corresponding first-order coefficient symbolically as A_p, referring to the asymmetry in the parallel (p) direction.

Now we turn to the perpendicular or radial delocalisation of Ψ. (An instance for the γ function is an interface at $T > T_R$, where the thickness extends to several lattice spacings and γ can be considered as to be proportional with a squared density gradient, integrated over the thickness (CAHN and HILLIARD, 1958). An instance for the R function is the contribution to the growth resistance by the diffusion resistance in a diffusion boundary layer in the parent phase. For the latter case, it is rather the $\vec{\nabla}C$ field than the transport-flux field which has the additive characteristic in the radial direction (series resistance). Also, it is inherently asymmetric).

Repeating the above procedure for the radial direction, we have:

$$\Psi(z) \equiv \int_{-\infty}^{\infty} C_z(\Delta z)\,\Psi(z_o + \Delta z)\,d(\Delta z) \tag{2.26}$$

with

$$\int_{-\infty}^{\infty} C_z(\Delta z)\,d(\Delta z) = 1 \tag{2.27}$$

where z_o is the position of the mathematical surface chosen, and $C_z(\Delta z)$ the radial Ψ-contribution function. As in the C_x case, the function C_z has to be determined from a microscopic model. Here we will discuss two cases:

a) Ψ is a "volumetric" quantity: then $\Psi(z_o + \Delta z)$ is to be considered as a density at height Δz and $C_z(\Delta z)$ as the volume of a layer with thickness dz, at height Δz, per unit area (as measured at z_o):

$$C_z(\Delta z) = \left(1 + \frac{\Delta z}{r}\right) \tag{2.28}$$

b) Ψ is a potential difference, its contribution $C_z(\Delta z)\Psi(z_o+\Delta z)$ being given by a current density

$$j(z_o+\Delta z) = j(z_o)\,\frac{1}{1+\Delta z/r} \tag{2.29}$$

multiplied by a resistivity.

Taking again "rectangular" distributions of density (case a) and resistivity (case b), we have for the integrand in case a: $(1 + \Delta z/r)$ and in case b: $(1 + \Delta z/r)^{-1}$.

As in practice (boundary layer!) asymmetry is the rule rather than the exception, we allow for this by situating the layer boundaries at $+(\delta/2)(1+f)$ and $-(\delta/2)(1+f)$, respectively, so that a symmetric situation corresponds with $f = 0$ and a fully asymmetric situation with $f = 1$. Expanding the integrand for case b and integrating, we obtain for the thickness-delocalisation contribution, up to second order in δ/r:

$$\Psi(r) = \Psi(0)\,\{\,1 \pm \tfrac{1}{2}\,\delta f\,\tfrac{1}{r} + \tfrac{1}{12}\left(\tfrac{\delta}{r}\right)^2 (1 + 3f^2 + \ldots\ )\,\} \tag{2.30}$$

(+ for case a, − for case b; the term $\frac{1}{12}\left(\frac{\delta}{r}\right)^2 (1 + 3f^2 + \ldots\ )$ applies to case b only).

Summarizing, we can combine (2.23) and (2.28) into an expression for the curvature effect to second order in $1/r$, due to both the tangential and the radial delocalisation:

$$\Psi(\varphi_o, r) = \Psi(\varphi_o, 0) + \frac{1}{r}\left\{A_p \pm \frac{\delta f}{2}\right\}\Psi(\varphi_o, 0) + \\ + \frac{1}{r^2}\left\{\frac{\ell_\Psi^2}{6}\,\overline{\overline{\mathbf{S}}}\Psi(\varphi_o) + \frac{1}{12}\,\delta^2\,(1 + 3f^2)\,\Psi(\varphi_o)\right\} \tag{2.31}$$

(again, the term $\frac{1}{12}\left(\frac{\delta}{r}\right)^2 (1+3f^2 + ..)$ applies to case b only, and the + sign in the $\frac{1}{r}\,\frac{\delta f}{2}$ term applies to case a, the − sign to case b). Here A_p is a factor accounting for asymmetry in the tangential delocalisation, ℓ_Ψ a correlation length for this delocalisation, δ is the "thickness" of the interface or boundary layer, and f is an asymmetry variable for the radial delocalisation). Important conclusions are:

1) first-order effects in $1/r$ are related with asymmetry in the delocalisation, either in the tangential direction (e.g., Schwoebel effect) or in the radial direction (e.g., diffusion boundary layer).

2) The second-order term in $1/r$ contains a contribution proportional to $\mathbf{S}\Psi$, plus, in case b (potential-drop type) a thickness-related term which is nonzero even for symmetric position.

These results can be relevant for the stabilisation of reentrant edges, as discussed in 6.3.

Chapter 3:

MECHANICS OF INTERFACES AND ITS APPLICATION TO EQUILIBRIUM STRUCTURES

3.1 INTRODUCTION

In this chapter we apply the operator formalism developed in ch.2 to the description of the mechanics and thermodynamics of surfaces (with OD free energy). When the only parameter is the orientation, the result is equivalent to the discussion of this subject by HERRING (1951, 1952), by JOHNSON (1965), by HOFFMANN and CAHN (1972) or by NOZIERES (1992), apart from the slightly more compact operator notation.

The approach of this chapter is to discuss first the "mechanics" of interfaces in terms of forces. By relating forces with affinities, the translation into thermodynamics is obvious. The result of this discussion (section 3.2) is the basic relation (3.14) showing that the total driving force working on an interface can be written as a sum of three terms: a "bulk" term, a capillarity term and a "Marangoni" term. (The Marangoni force is felt by a surface with m-dependent free energy, in the direction of an m gradient, where m is some field parameter such as temperature, etc.)

Up to that point, the results apply as well to interfaces in motion and in rest. Interfaces in motion are discussed further in ch.5, and eq. (3.14) can be used there to be inserted into the driving-force parameter, in those cases where capillarity and/or Marangoni corrections are relevant. Interfaces in equilibrium are the specific subject of the rest of ch. 3, i.e. sections 3.2 and 3.3. As the condition for equilibrium is $A = 0$, the equation defining "interfaces in equilibrium" is just obtained by equating the sum of the three terms in (3.14) to zero. (the "classical" Wulff-Herring equilibrium shape is a special case corresponding with $\vec{\nabla} A = 0$ and $\vec{\nabla} m = 0$).

Boundary conditions related with equilibrium are a subject on their own, and chapter 4 is specially devoted to it as a complement to ch. 3. Equilibrium configurations at junctions are discussed there, and also complications arising from negative eigenvalues of the $\vec{\vec{\mathbf{S}}}\gamma$ tensor, i.e. zigzag or pyramid structures.

Detailed description of force equilibrium: line forces and surface forces.

In "complex" interface structures (like the polygrain or polycrystalline structure sketched in fig. 3.1), we have not only to do with interfaces ("2-junctions"), but also with 3- and 4-junctions (and eventually also D-junctions, i.e. junctions between interfaces and dislocations). The mechanics of such interface structures is identical with that of soap froth, apart from the possible orientation dependence of interface properties.

Equilibrium conditions for such complex structures consist of the *combined* equilibrium conditions for the interfaces and for the junctions. As 3-junctions are the boundaries of interfaces, the 3-junction equilibrium conditions can be considered to supply the boundary conditions for the interface-equilibrium equations.

Consequently, we will have to consider both the equilibrium equations for interfaces (2-junctions; the subject of ch.3) as well as those for 3-,4- and D-junctions (ch.4). An equilibrium equation for a 3-junction can be considered to represent a force equilibrium between three "line forces" (i.e., effective forces working on the junction line per unit length), whereas at an interface we have equilibrium of "surface forces" (i.e., effective forces per unit area, or pressures).

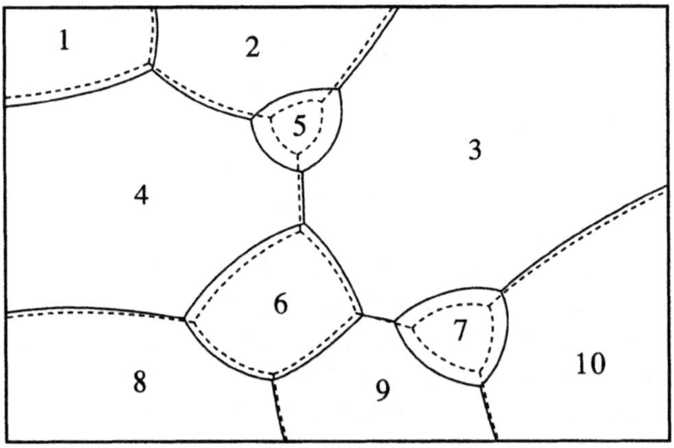

FIG. 3.1

Note 1: In discussing forces working on an interface, we have to be aware that the term "equilibrium" may be used in connection with two quite distinct situations:

1) In crystal growth literature it is common practice to define an "interface in equilibrium" or an "equilibrium shape" as one which is thermody-

namically in equilibrium, i.e. which is not moving because the growth affinity is zero all over the surface: $A_g = 0, \rightarrow \quad R \equiv 0$.

2) In a more general sense, we have an "equilibrium of surface forces or pressures" also in the case of a moving (i.e. dissipative) interface. The equilibrium concept then applies to the pressure balance at the interface; i.e. it forms the principle on which the equation is based:

$$P_g = \frac{\mu_1 - \mu_2}{V} + \Delta p_c + \Delta p_m \tag{3.1}$$

which is valid also for nonzero values of P_g .

In order to distinguish between the two "equilibria", we will use the term "equilibrium condition" for $A_g \equiv 0$ (in agreement with convention) and the term "pressure balance equation" for eq. (3.1). Even for the equilibrium condition $A_g \equiv 0$ we can make a further distinction, as will be argued in 3.3.

Note 2: For a "complex" interface structure *in motion*, we have to distinguish the "surface dissipation" by the interfaces and the "line dissipation" by the junctions. (moving edges can be dissipative too, and even 4-junctions, D-junctions and vertices can formally have a "point dissipation". Dissipative motion of an interface means that χ^{-1} in eq. (1.7) is nonzero; that of a 3-junction or an edge means that an extra force on the junction/edge is required to keep it moving, to be expressed by the extra term $\vec{F}_\ell$ in eq. (6.6) = force per unit length of the 3-junction etc.: $R_{\text{junc.}} = |\vec{F}_\ell| . \chi^*$ (where χ^* is a "line mobility"). This line force should be considered as independent of the growth pressures of the interfaces involved in the 3-junction or edge. Such a force is quite similar to the force which is required to move a dislocation through a lattice. In the case of edges and junctions, this force can be associated with nucleating activity, as will be explained in 6.2.2.1.

3.2 BASIC RELATIONS.

3.2.1 Line force.

The line force working on the edge of a surface element of an interface can be derived by considering the change in free energy resulting from an infinitesimal displacement of this edge. Under the appropriate conditions (see below, note 1) we have only to take the changes in surface free energy into account, and this makes the case identical with that of the mechanics of a soap film whose surface free energy is by some mechanism made dependent of orientation and/or a second parameter *m*.

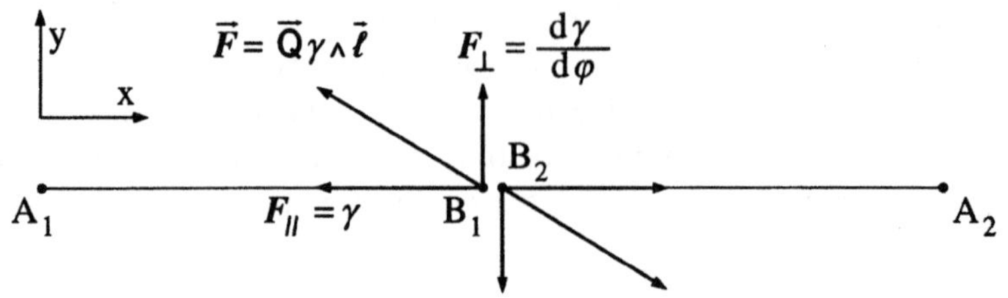

FIG. 3.2

Consider fig. 3.2. It represents a cross-section through a rectangular piece of soap film, framed by two horizontal wires A_1 and A_2 and two parallel plates perpendicular to the wires , outside the plane of the drawing but parallel to it. Let B_1 and B_2 be two free horizontal wires, also in the film, which fit closely between the plates but can move between them without friction (the distance between B_1 and B_2 is negligible against that between A_1 and A_2). When the soap film in the narrow rectangle between B_1 and B_2 is punctured, the intact film rectangles (between A_1, B_1 and between B_2, A_2) exert a force on the wires B_1 and B_2 which can be expressed as the gradient of the energy of the film (the gradient is defined here as $d/d\vec{s}$ where $d\vec{s}$ is the two-dimensional displacement vector of the horizontal wires B_1 or B_2).

In this way we have defined a model which is suitable to demonstrate the effect of orientation dependence on one orientation vaiable , i.e. the angle φ about the wire axis. Assuming γ to be continuously differentiable around the original orientation angle φ_o we can write

$$\gamma(\varphi) = \gamma(\varphi_o) + \frac{d\gamma}{d\varphi}\Delta\varphi\,; \qquad (\Delta\varphi = \varphi - \varphi_o) \tag{3.2}$$

The components of the force, per distance-unit between the end plates, in tangential (x) and perpendicular (y) direction, are

$$\left.\begin{aligned} F_{//} &= \frac{\partial}{\partial x}\{\,\gamma(\varphi).s\} = \gamma(\varphi_o) \\ F_\perp &= \frac{\partial}{\partial y}\{\,\gamma(\varphi).s\} = \frac{1}{A_iB_i}\,\frac{d}{d\varphi}\{\,\gamma(\varphi).s\} = \left(\frac{d\gamma}{d\varphi}\right)_{\varphi_o} \end{aligned}\right\} \tag{3.3}$$

In other words, we have to do with an equilibrium of forces which in general are not parallel with the surface. The tangential component (per unit length) is the "ordinary" γ , the perpendicular component is $d\gamma/d\varphi$.

These two components and their mutual orientation correspond with the vectors $\vec{n}\gamma(\vec{n})$ and $d/d\vec{n}\{\gamma(\vec{n})\}$, together giving the vector $\overline{Q}\gamma(\vec{n})$. (and identical with the $\vec{\xi}$ vector of HOFFMANN and CAHN (1972). Note, however, that the vector $\vec{Q}\gamma(\vec{n})$ is rotated by $\pi/2$ with respect to the actual force on the wires B_1 and B_2. Mathematically this is expressed by the relation

$$\vec{F} = \vec{Q}\gamma(\vec{n}) \wedge \vec{\ell} \tag{3.4}$$

(equation 15 in Cahn and Hoffmann's paper) where $\vec{F}$ is the force vector acting on an edge or border, h.l. the wires B_1 and B_2, with oriented length $\vec{\ell}$. In other words, a point on the curve or surface $\vec{Q}\gamma(\vec{n})$ can be associated with the force equilibrium in an interface with the same orientation as the $\vec{Q}\gamma(\vec{n})$ surface in that point. The $\vec{Q}\gamma$ surface is thus identified as a "force plot".

NOTES:

1) On the validity of the "soap-film" concept.

In the derivation of $F_\perp$, i.e. in considering a "vertical" displacement of the surface, we have tacitly dropped all bulk terms in the total free-energy change. This is permitted as a consequence of the surface-pressure balance condition. For a surface with zero curvature, $\Delta p_c = 0$ and then the compensation of the $p_g.\Delta V$ term by $(p_o + \Delta p_m)\Delta V$ is trivial because it applies identically along A_1B_1. For a curved surface, the compensation is somewhat less trivial because the (locally) vertical displacement of the B_1 wire (during which the curvature is kept constant) results in a first-order increase of the surface area A_1B_1 . (A_1 is no longer situated in the tangent plane in B_1 because of the curvature). However, it is easily verified that this extra first-order surface-energy increase is just compensated by the bulk term $\Delta p_c.\Delta V$ (the calculation, of course, is essentially the same as that given in 3.2.2 for the Δp_c expression).

Consequently, eq. (3.4) is valid regardless of curvature, Marangoni gradients, eventual A gradients, and growth. The line force depends only on the $\gamma(\vec{n})$ function for the local A_g and m values and not on the A and m gradients.

2) Two orientation variables.

Equation (3.4) has been derived above for one orientation variable, $\vec{Q}\gamma$ representing a vector in the plane perpendicular to $\vec{\ell}$. However, the three-dimensional vector $\vec{Q}\gamma$ (including a $d\gamma/d\vartheta$ vector in the $\vec{\ell}$ direction) gives exactly the same result. Therefore eq. (3.4), with the 3-dimensional (two orientational variables) $\vec{Q}\gamma$ vector, represents the general expression for the line force for arbitrary edge direction $\vec{\ell}$.

3.2.2 *Surface forces.*

3.2.2.1 *The curvature-induced surface force and growth-pressure shift.*

Curvature of an interface with nonzero free energy generates a pressure difference across it. This pressure difference is obtained directly by integrating the line force (given by eq. (3.4) over the edge contour of an infinitesimal (nonplanar) surface element and dividing by its area.

For the cylindrical case (one orientation variable) we have, per unit length perpendicular to the drawing plane (fig. 3.3):

$$\vec{n}\,\Delta p_{nn}\Delta s = \{\vec{\mathbf{Q}}\gamma(\vec{n}) - \vec{\mathbf{Q}}\gamma(\vec{n}+d\vec{n})\} \wedge \vec{\ell} =$$

$$= d\vec{n}\,\{\gamma(\vec{\varphi}) + d^2\gamma/d\varphi^2\} \wedge \vec{\ell} =$$

$$= \vec{n}\,d\varphi\,\mathbf{S}\gamma \tag{3.5a}$$

or, in other words, $\Delta p_{nn} = \mathbf{S}\gamma/r$. The identification of $\mathbf{S}$ as a "stiffness operator" is illustrated by this relation.

For the general case of curvature in two directions, we use a form of Green's theorem to transform the contour integral into a surface integral:

$$\oint \vec{a} \wedge d\vec{\ell} = \iint_{\text{surface}} (dO \wedge \vec{\nabla}) \wedge \vec{a} = \iint_{\text{surface}} dO\,(\vec{\nabla}_s \cdot \vec{a}\,) \tag{3.6}$$

for any vector field $\vec{a}$ defined on the surface with continuous derivatives to the surface cordinates. When applied to $\vec{\mathbf{Q}}\gamma$, we obtain for the total line force on the edge:

$$\vec{F}_{\text{surf.curv.}} = \oint \vec{\mathbf{Q}}\gamma \wedge d\vec{\ell} = \iint_{\text{surface}} dO(\vec{\nabla}_s.\vec{\mathbf{Q}}\gamma) = \iint_{\text{surface}} dO\,(\vec{\vec{K}} : \vec{\vec{\mathbf{S}}}\gamma) \tag{3.5b}$$

(for the last result, we have substituted eq. (2.19)). In other words, the curvature-induced force is exactly in the normal direction, i.e. it can be considered as resulting from a hydrostatic pressure difference

$$\Delta p_{\text{curv.}} = \vec{\vec{K}} : \vec{\vec{\mathbf{S}}}\gamma \tag{3.7a}$$

and consequently this pressure difference can be considered as a curvature-induced shift in the growth pressure, so that

$$\Delta A_{\text{curv.}} = V\,\vec{\vec{K}} : \vec{\vec{\mathbf{S}}}\gamma \tag{3.7b}$$

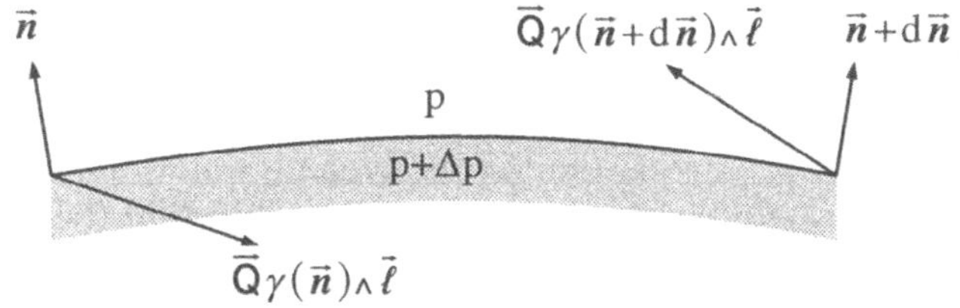

FIG. 3.3

3.2.2.2 The nonisostatic stress field in nonequilibrium-shaped crystals

In a nonequilibrium-shaped crystal, the internal stress field which results from the surface tension is not isostatic (e.g., in a cylindrical needle the axial stress is $2\gamma/r$, the radial stresses γ/r for OID γ). Only the perpendicular tensor element Δp_{nn} is determined simply by the local curvature according to eqs. (3.7); the *tangential* components are determined by the normal pressure distribution over the total surface of the body; their determination requires the solution of the distortion field in the whole body which agrees with these normal forces. This latter problem involves the elasticity properties of the crystal. Moreover, we assume that the enveloping phase is fluid so that tangential forces do not occur (for coherent solid-solid interfaces, the tangential force transferred from the enveloping phase has also to be taken into account).

The question of the correct expression for Δp to be used in the case of nonisostatic stress has been considered in detail by LI, ORIANI and DARKEN (1966), by LARCHE and CAHN (1973) and by NOZIERES (1992). Strictly taken, we have to make a distinction between "mobile" and "immobile" components in the stressed phase. Immobile components are those which can be identified with the atoms in the crystal lattice and which, by definition, do not diffuse; mobile components are, e.g., dissolved interstitials which readily diffuse and can be considered as a permeating gas, having an isostatic partial pressure. Clearly, in the one-component crystal system we are considering, we have to do with an immobile component. The chemical potential of such a component at the surface of a stressed solid can be written (according to Li c.s., eq. 3) as:

$$\mu_s = \mu_o + \overline{w} + p_\perp V \tag{3.8}$$

Here $\overline{w}$ is the strain energy per mole of the solid. Neglect of this term (which implies that the crystal is considered as completely rigid) reduces the equation to the form used above in the expression for ΔA_c. In practice, this neglect is justified except in highly exotic cases. This again illustrates that eq. (3.7) is of general validity and not restricted to equilibrium-shaped surfaces.

An excellent review of the thermodynamic relations between OD γ-related forces and thermodynamic potential *including* the strain terms is given by Nozières (see above).

3.2.2.3 *Curvature-induced affinity shift versus curvature-induced equilibrium shift.*

It is important to recognize the difference between
1) the curvature-induced shift ΔA_c in the growth affinity as such, unaffected by an equilibrium-restoring shift in temperature, pressure etc., and
2) the shift in thermodynamic potential (of both phases, $\Delta\mu_{eq.}$) which results when equilibrium is *restored* by such a shift. (the shift in equilibrium conditions is generally known as the *Gibbs-Thomson effect*).

For ΔA_c , expression (3.7b) is exact; for $\Delta\mu_{eq.}$ it is an approximation. The difference is illustrated by fig. 3.4.

Consider, for instance, a solid-vapour equilibrium at fixed temperature. (unary system). We have (fig. 3.4)

$$\Delta\mu_{\text{vapour}} = V_v(p_v - p_e) \;\; ; \;\; \Delta\mu_{\text{crystal}} = V_{cr}(p_{cr} - p_e) \tag{3.9}$$

(p_e = equilibrium pressure for planar, p_v = idem for curved interface). Solving ($p_v - p_e$), i.e. the shift in vapour pressure required to restore equilibrium, as a function of $(p_{cr} - p_v) = \Delta p_{\text{curv.}} = \vec{\vec{K}} : \vec{\vec{S}}\gamma$, we obtain

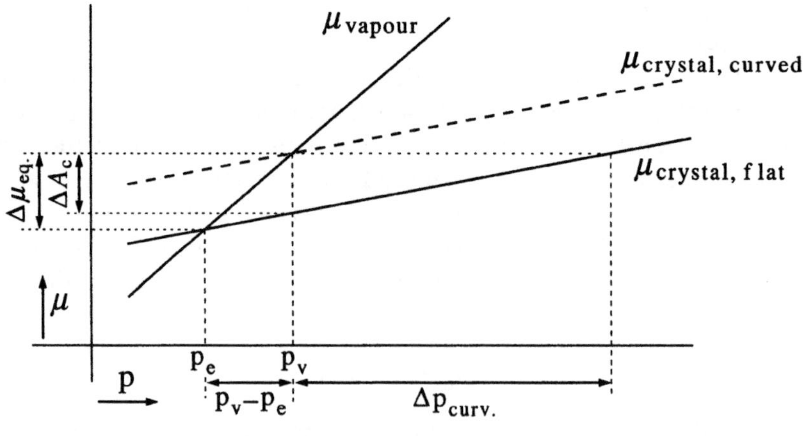

FIG. 3.4

$$p_v - p_e = \frac{V_{cr}}{V_v - V_{cr}} \cdot \Delta p_{curv.} \tag{3.10}$$

(which is actually identical with (1.6), but now specified for curvature-induced pressure), and the corresponding shift in thermodynamic potential is

$$\Delta\mu_{eq.} = \Delta p_{curv.} \frac{V_{cr}\, V_v}{V_v - V_{cr}} \tag{3.11}$$

i.e. (for $V_{cr} \ll V_v$) slightly more than the curvature-induced affinity shift itself.

3.2.2.4 The Marangoni force.

The Marangoni force is simply the gradient of the energy of a surface element resulting from a gradient in any parameter field, other than orientation, on which the surface free energy depends. So

$$\vec{F}_m = |\,\mathrm{dO}|\, \vec{\nabla} m\; \mathrm{d}\gamma/\mathrm{d}m \tag{3.12}$$

i.e. a force in the direction of the *m* gradient , for any surface orientation. So in general, the Marangoni force has both a normal and a tangential component (the latter is responsible for Marangoni convection in liquids).

The classical example of a Marangoni situation is generated by the temperature dependence of surface energy; this case is nonexistent in isothermal equilibrium interfaces, but it can play a role in the *A*-gradient cases discussed in 3.3.3. However, other examples do provide a nonzero Marangoni effect even in isothermal surfaces. For instance, the interface may carry a certain density of electric charge and so be pulled or pushed by an electric field, or a dipole layer by a field gradient.

In the "soap-film" terminology, the Marangoni effect can be thought of as analogous to the case of a "heavy" soap film (with a weight $\mathrm{d}\gamma/\mathrm{d}m$ per area unit) in a "gravitational field" $\vec{\nabla} m$.

Now the tangential component, although it executes a shear force on the bulk of the interface, does not influence the growth affinity. It is only the normal component which is part of the growth pressure:

$$\Delta p_m = \left(\frac{\mathrm{d}F_m}{\mathrm{dO}}\right)_{\text{normal}} = \frac{\mathrm{d}\gamma}{\mathrm{d}m}\,(\vec{n}\cdot\vec{\nabla} m) \tag{3.13}$$

3.2.2.5 *Explicit expression of the surface-force balance and the total growth affinity.*

We can now write down the expression for the growth pressure or the affinity explicitly, substituting eqs (3.7) and (3.14) into eq.(3.1):

$$P_g = \frac{A_|}{V_{\text{crystal}}} + (\vec{\vec{K}} : \vec{\vec{S}}\gamma) + (\vec{n} \cdot \vec{\nabla} m)\frac{d\gamma}{dm} \tag{3.14a}$$

$$A_g = A_| + V.(\vec{\vec{K}} : \vec{\vec{S}}\gamma) + V\frac{d\gamma}{dm}(\vec{n} \cdot \vec{\nabla} m) \tag{3.14b}$$

($A_|$ = affinity corresponding with T, C etc parameters at the interface, but for a planar interface)

Equations (3.14) form the starting point for two different branches of the continuum description of interface behaviour:

1) the "exact" equations of motion to be discussed in ch.5. (exact in the sense that capillarity effects are taken into account). Here, the "direction" of application of eqs. (3.14) is "shape→affinity": the (in general non-equilibrium) shape is included in the starting conditions. The effective affinity, found directly from eq. (3.14), is used to calculate the motion of the interface, in agreement with the $R(\vec{n},A)$ function and with the eventual diffusion processes in the bulk phases.

2) determination of the class of equilibrium shapes defined by the condition $A_g = 0$, to be the subject of section 3.3. In contrast with the above case, the "direction" of this application is "affinity ⟹ shape".

It should be stressed that from a mathematical point of view, the latter problem is quite different from the previous one. In the case of equilibrium shapes, we are interested in the shape of a crystal which is free except for certain imposed restraints. Such restraints can be of several kinds, e.g.:

a) given volume in a closed surface whose free energy is to be minimized;

b) (as extra): surface has to contain a given closed contour in space; cf. a wire frame suspending a soap bubble;

c) surface has to contain part of a given fixed surface (e.g., a solid substrate in heterogeneous nucleation);

d) an imposed inhomogeneous parameter field on which γ depends (e.g. a temperature gradient determining the shape and scale of a grain-boundary groove in a solid-melt interface).

In all such cases, we have to do with variational problems requiring an "equilibrium condition" (i.e., $A_g = 0$) to be satisfied everywhere at the interface. Such problems can be fairly difficult except in some special cases. The case of ($\vec{n}$ and m)-independent surface free energy is that of "soap film mechanics", and the corresponding equilibrium-shape problems are known as Plateau's problems in the mathematical literature. When the only imposed condition is that of minimum total surface free energy for a given enclosed volume , the solution is simple (a sphere for a soap bubble, the $\vec{\mathbf{Q}}\gamma$ shape for γ dependent on $\vec{n}$ only). But when a boundary condition is imposed, e.g. a given arbitrary contour has to lie in the surface, then the solution is not so easy except for certain highly symmetrical contours (cf. wire frames for soap films and bubbles). The reason is that the surface has in general two curvature parameters, whereas the equilibrium condition is expressed in terms of the combination $\vec{\vec{K}} : \vec{\vec{\mathbf{S}}}\gamma$. When the problem is quasi-one-dimensional, e.g. in the calculation of straight grain-boundary grooves or lamellar- eutectic solidification fronts in a constant-gradient field, the problem is relatively easier as it leads to a differential equation for one variable.

We will not try to discuss solution methods for the Plateau's problems in general, even apart from the special cases to be discussed below. The interested reader is referred to the monograph by F.J. ALMGREN on this subject (1966). In the following sections we just treat the "classical" cases of equilibrium shape (in absence of Marangoni terms) for homogeneous A ($\vec{\mathbf{Q}}\gamma$ shape) and constant-gradient A (grain-boundary groove shape), and we will touch upon the Marangoni effect only superficially by showing the difference with capillarity distortion in a grain-boundary groove (only in the OID case, as the most simple example).

3.3 *EQUILIBRIUM SHAPE OF INTERFACES*

Inserting the equilibrium condition $A_g \equiv 0$ into (3.14b) demonstrates that we still have many degrees of freedom within the class of "equilibrium shapes": only the *sum* of three terms has to be zero, and the imposed conditions may be such that both the bulk μ field and the m field (whatever the latter may be) are inhomogeneous. We may define a number of elementary cases by choosing some of the terms in eq. (3.14) and/or some of the gradients zero. Far from exhausting all possibilities, we have chosen the following cases to work out:

1) the "classical" equilibrium shape: i.e. shape of a "freely floating" crystal (without spatial boundary conditions) in a uniform affinity field and in absence of a Marangoni term (section 3.3.1).

2) In order to highlight the difference between Marangoni distortion and $\vec{\nabla} A$ -induced distortion, we work out the $\vec{\nabla} A \neq 0$, $\vec{\nabla} m \neq 0$ situation for the OID γ case only (section 3.3.2)

3) The $\vec{\nabla} A \neq 0$ situation for the OD γ case is of physical relevance for the shape of grain-boundary grooves and of crystals adherent to other phases (and for shape measurements on such geometries as used to determine $\gamma(\vec{n})$ functions experimentally). This case, under neglect of the Marangoni term, is discussed in section 3.3.3.

3.3.1 The equilibrium shape in a homogeneous affinity field.

A direct consequence of eq. (3.7) is that the equilibrium shape of a simply connected surface in a homogeneous (nonzero) affinity field is geometrically similar with the $\vec{\mathbf{Q}}\gamma$ shape. (i.e., the classical Gibbs-Wulff result). The "size factor" correlating the size of both shapes is

$$\frac{\vec{\mathbf{Q}}\gamma \text{ shape}}{\text{equil. crystal shape}} = \Delta p = \frac{\Delta A_c}{V} \qquad (3.15)$$

This result is obvious from the fact that equilibrium requires ΔA_c (and thus $\vec{\vec{K}} : \vec{\vec{\mathbf{S}}}\gamma$) to be constant over the surface. Since $\vec{\vec{K}}(\vec{\mathbf{Q}}\gamma)$ is the inverse of $\vec{\vec{\mathbf{S}}}\gamma$ by definition, indeed ΔA_c is constant all over the $\vec{\mathbf{Q}}\gamma$ shape.

The internal pressure of equilibrium-shape crystals is isostatic. As all the used relations keep their validity for cusped γ functions (by the δ -function extension of the definition of $\vec{\vec{\mathbf{S}}}$; see eq. (2.10)) there is no exception to this isostatic-pressure condition in the vicinity of facets and edges, a fact which is not always properly recognized.

Crystals which are not similar with the $\vec{\mathbf{Q}}\gamma$ shape feel an inhomogeneous affinity distribution, trying to restore them to the $\vec{\mathbf{Q}}\gamma$-similar shape. Only within very small distance from sharp corners, (i.e. of the order of the correlation length of the γ function (see 2.8), where the surface free energy becomes curvature-dependent, deviations from isostatic pressure would occur for the exact $\vec{\mathbf{Q}}\gamma$-similar shape so that deviations from this shape are generated. On such a scale, we have to go back to atomistic models. This is particularly true for the shape of very small nuclei, consisting of up to 10^2 - 10^3 atoms, which as a consequence can have equilibrium shapes different from the macroscopic equilibrium crystal and even different symmetries (BINDER and STAUFFER 1976, RASMUSSEN 1982). On this scale, the continuum concepts of phases and interfaces cease to be applicable - together with the identification possibility of $\vec{\mathbf{Q}}\gamma$ and equilibrium shape.

3.3.2 Marangoni effect versus Gibbs-Thomson effect

It is important to distinguish between two complicating effects when considering interfaces in inhomogeneous A and/or m fields. In this case the condition $A_g = 0$ can be met when the differences in $A_|$ are compensated by differences in ΔA_c (eq. (3.7b), = the second right-hand term in eq. (3.14) and/or the Marangoni term (third term in eq. (3.14)). However, "floating" closed surfaces are no longer possible; the interface has necessarily to be adhered to a fixed frame or boundary which can take up a net force. One-dimensional examples are the shape of a sessile drop in a gravitational field and the shape of a grain-boundary groove in an $A_|$ gradient as discussed by several authors (e.g. NASH and GLICKSMAN (1971), GLICKSMAN and VOLD (1969), BOLLING and TILLER (1960), SCHAEFER, GLICKSMAN and AYERS (1975), ARBEL and CAHN (1977)). All of these calculations assume OID surface energy; we do the same in the following.

Although the capillarity term and the Marangoni term are clearly discernible as 2nd and 3rd terms in eq. (3.14), it is quite possible that in practice the effects are coupled because both A and γ depend on an inhomogeneous field parameter such as temperature. In such a case, the constants a and b in eq. (3.18) are related; but this does not prevent us from distinguishing their mathematical rôle in this equation.

For the purpose of demonstrating these rôles we choose to omit all non-essential complications. So we are led to consider a crystal-parent phase interface with OID γ which depends linearly on some parameter m: $\gamma = \gamma_o + m\gamma_1$. We consider only one orientation variable (straight grain-boundary groove). The parameters m and/or A are supposed to vary linearly with z, the perpendicular distance from the (planar) $A = 0$ surface, so that we can write $m = \mathrm{a}z$, $A = \mathrm{b}z$. We may call a the Marangoni variable, b the Gibbs-Thomson gradient variable.
For a = b = 0, the equilibrium-shape equation is simply:

$$A_| = \frac{V\gamma_o}{r} = \text{constant} \Rightarrow r = \frac{V\gamma_o}{\text{const.}} \quad \text{(a circle)}$$

For the gradient case we can substitute

$$\frac{\mathrm{d}}{\mathrm{d}z} = \frac{1}{\sin\varphi}\frac{\mathrm{d}}{\mathrm{d}s} \tag{3.16}$$

and the equation reads $A_| - \Delta A_{\text{curv.}} = \mathrm{K}$ (const.); i.e.

$$\mathrm{b}z - \frac{V}{r}\left\{\gamma_o + \mathrm{a}\gamma_1\left(z + \frac{\mathrm{d}^2 z}{\mathrm{d}\varphi^2}\right)\right\} = \mathrm{K} \tag{3.17}$$

or

$$V\sin\varphi\left\{\gamma_o + \mathrm{a}\gamma_1\left(z + \frac{\mathrm{d}^2 z}{\mathrm{d}\varphi^2}\right)\right\} - (\mathrm{b}z - \mathrm{K})\frac{\mathrm{d}z}{\mathrm{d}\varphi} = 0 \tag{3.18}$$

Suitable boundary conditions have to imposed for the interface to be uniquely determined. For a grain-boundary groove between two infinitely large half-crystals, K = 0. Case a = 0 (the classical grain-boundary groove equation) is discussed, for instance, by NASH and GLICKSMAN (1971).

3.3.3 Interfaces in an affinity gradient: the orientation-dependent case.

We extend the discussion of the 2-dimensional case a = 0 of 3.3.2 to OD surface energy. This is relevant in the determination of grain-boundary groove shapes in an A gradient from given $\gamma(\vec{n})$ functions, or, inversely, for the determination of γ plots from a measured grain-boundary shape.

Starting point is eq. (3.18) for a = 0, b ≠ 0, but now OD surface energy. Also, we take K = 0, i.e. the grain-boundary groove width is ≪ the size of the grains themselves. This yields

$$\sin\varphi .\, \mathbf{S}\gamma = \frac{\mathrm{bz}}{V}\,\frac{\mathrm{dz}}{\mathrm{d}\varphi} = \frac{\mathrm{b}}{2V}\,\frac{\mathrm{d}(z^2)}{\mathrm{d}\varphi} \tag{3.19}$$

For convenience, we change the z variable to $\zeta = \sqrt{\mathrm{b}/2V}$. (ζ^2 has the dimension of surface energy). When the problem is stated such that an angle φ_0 is imposed for a certain z_0 (or φ_0), we have

$$\frac{\mathrm{b}}{2V}(z^2 - z_0^2) = \zeta^2 - \zeta_0^2 = \int_{\varphi_2}^{\varphi_1} \mathbf{S}\gamma(\varphi)\,\sin\varphi\,\mathrm{d}\varphi \tag{3.20}$$

When $\mathbf{S}\gamma = \gamma$ = constant, (OID case), an algebraic solution can be obtained by direct integration. This case has extensively been investigated, mainly in connection with experimental methods for surface-energy measurements (see cited literature above).

When $\mathbf{S}\gamma(\varphi)$ is not available as an explicit analytic expression, we are forced to consider graphical solution methods. Now the right-hand side of eq. (3.20) is the vertical (ζ direction) distance between the φ_0 -and the φ_1 -oriented points of the $\vec{\mathbf{Q}}\gamma$ curve. So we are led to use the $\vec{\mathbf{Q}}\gamma$ plot in combination with ruled ζ^2 paper, i.e. paper with parallel lines (numbered n) at a distance ~ n^2 from an origin line. A linear scale for ζ^2 (vertical) and ξ^2 (horizontal) in erg/cm^2 can be plotted in addition. Of course the size on which the plot is drawn should agree with these ζ^2 scales. (fig. 3.5a)

On this paper, the $\vec{\mathbf{Q}}\gamma$ plot or plots have to be translated to the specific (ζ, ξ) positions which agree with the boundary conditions of the problem. This representation will be referred to as the (ζ^2, ξ^2) shape. (fig. 3.5b) It can be transferred to the real (x,z) shape as follows: the z values can be read directly from the (ζ^2, ξ^2) drawing. (when the only problem is to find a groove depth, we can read the solution at once). Transfer of the interface shape to (x,z) paper involves a multiplication factor ~ $1/z$ for each segment between z and z+dz. In order to maintain the angle φ, we have to multiply the corresponding dξ steps with the same factor:

$$\frac{\mathrm{d}x}{\mathrm{d}(\xi^2)} = \frac{\mathrm{dz}}{\mathrm{d}(\zeta^2)} = \frac{V}{\mathrm{bz}} \tag{3.21}$$

The shape of any interface configuration in the A-gradient field can be constructed in this way with any desired accuracy, except in the vicinity of $z = 0$ (for $\vec{\mathbf{Q}}\gamma$ plots which are smoothly curved for $\varphi \simeq 0$). In the $z \to 0$ limit, x tends to $\pm\infty$ (fig. 3.5c)

So far, we have sketched the procedure for the construction of the groove slopes. The position z_0 and the orientation of the 3-junction line has

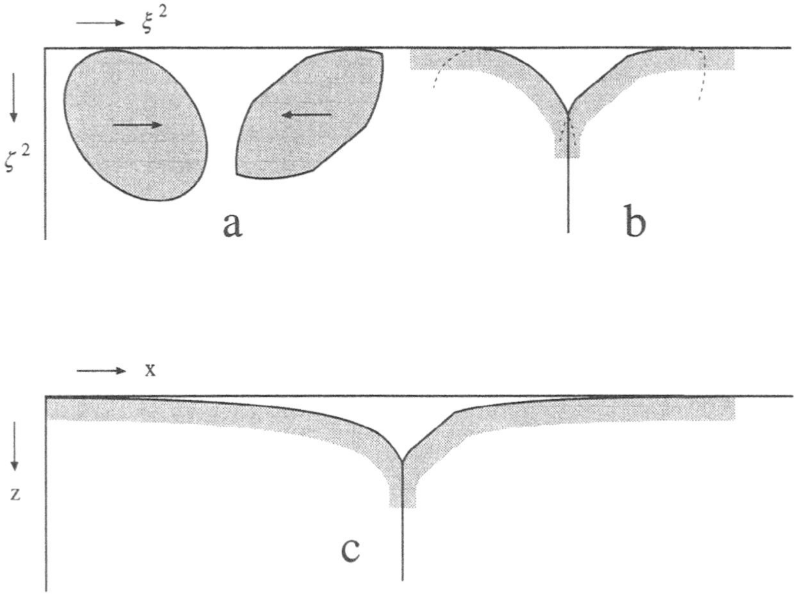

FIG. 3.5: Construction of grain-boundary groove shape for OD surface energies.

to be determined separately. Indicating the grains by α and β and the opposing parent phase by γ, we have to satisfy the following conditions:

1) Both the $\vec{Q}\gamma_{\alpha\gamma}$ and the $\vec{Q}\gamma_{\beta\gamma}$ plots have to touch the $\zeta = 0$ horizontal ($\varphi = 0$, $\zeta = 0$). So at the 3-junction, $\varphi_{\alpha\gamma}$ and $\varphi_{\beta\gamma}$ are tangents to the $Q\gamma$ plots at equal vertical distance from the top, and we can accordingly plot $\varphi_{\beta\gamma}$ as a function of $\varphi_{\alpha\gamma}$.

2) The second, and different, relation between $\varphi_{\beta\gamma}$ and $\varphi_{\alpha\gamma}$ to be satisfied is imposed by the 3-junction equilibrium, represented by a plot as discussed in 4.1.3, fig. 4.4.

We can find the solution for $\varphi_{\beta\alpha}$ and ζ at the 3-junction by superimposing both plots and finding the intersection of both $\varphi_{\beta\gamma}$ lines. Having thus obtained $\varphi_{\beta\alpha}$, we can proceed to draw the (ξ^2, ζ^2) shape (fig. 3.5a) and further construct the (x,y) shape (fig. 3.5b) *)

*) The simple graphical solution method described above can only be applied for a unidirectional gradient. In other cases we can still construct the interface shape from the $\vec{Q}\gamma$ plot, provided a map of the size-factor field is available. On a map with iso-affinity contours we can, starting from a given point and φ, draw sections of the $\vec{Q}\gamma$ plot multiplied with the appropriate size factor. By this procedure the interface can be "integrated" section by section.

Chapter 4

EQUILIBRIUM STRUCTURE OF JUNCTIONS, EDGES AND VERTICES

4.1 EQUILIBRIUM AT THE JUNCTIONS OF INTERFACES.

4.1.1 Introduction

In chapter 3 we have only considered interfaces as such. When more than two phases or grains are involved, we are confronted with:

a) junctions between three interfaces, at the meeting line of three grains*) (3-junctions).

b) junctions between six interfaces, at the meeting point of four grains (4-junctions).

(In this context an ordinary interface, i.e. a meeting plane between *two* grains, can be called a *2-junction*. In this section we will sometimes use this term in order to stress the similarity of equilibrium principles).

Meeting lines of more than three, or meeting points of more than four grains, although geometrically conceivable, are in general thermodynamically unstable, i.e. split up into combinations of 3- and 4-junctions, and will not be further considered.

In this section we discuss the geometry of 3- and 4-junctions in the immediate vicinity of the meeting line or point, respectively. This direct vicinity implies, in connection with the finiteness of pressure differences and consequently of interface curvature between adjoining grains, that the interfaces can be considered as planar and their intersections as straight lines in the region of interest. This means that we have essentiallly to do with equilibrium between three and six force vectors, respectively. Outside the immediate vicinity of junctions, curvature is relevant again for the interfaces on a larger scale, but that is the subject of chapter 3.

*) In this section, we use the term "grains" for shortness but tacitly include different phases. As far as the mechanics of interfaces is concerned, there is no difference in principle between grain boundaries and interphase interfaces.

In other words, when the determination of the equilibrium configuration of a whole structure involving many grains is given as a problem, then the equilibrium configurations of the 3- and 4-junctions are to be considered as boundary conditions to be obeyed at these junctions, necessary to specify the problem fully. The problem as a whole is a complex version of a "Plateau problem", and the position and orientation of the 3- and 4-junctions are part of the solution (see 3.2.2.5).

The force-equilibrium condition in a 3-junction can be written, as the obvious consequence of (3.4):

$$\left(\sum_{i=1}^{3} \vec{Q}\, \gamma_i \right) \wedge \vec{\ell} = 0 \tag{4.1a}$$

which for OID γ reduces to

$$\left(\sum_{i=1}^{3} \vec{n}_i\, \gamma_i \right) \wedge \vec{\ell} = 0 \tag{4.1b}$$

($\vec{\ell}$ = unit vector along the 3-junction axis. Of course, the vector product with $\vec{\ell}$ can be omitted - we merely write the equation in this form in order to let the force direction agree with physical intuition). The geometrical significance of conditions (4.1) agrees with the construction of a triangle with given side lengths: the "Neumann triangle" of forces. For the OID case, the angles between the interfaces equal sums of angles in the triangle: that between interfaces 1 and 2 corresponds with the sum of the two angles adjoining side γ_3 in the triangle, and cyclically. As the triangle can be oriented randomly in space, these angles are fixed parameters for the particular 3-junction.

The 3-junction is completely characterized by the orientation of its axis $\vec{\ell}$ (two degrees of freedom) and the orientation of the interfaces around this axis (one degree of freedom). When considered as the (only) more-interface equilibrium case in the two-dimensional version of interface mechanics (fixed axis), we have only the angular degree of freedom. Note that in this case we have two mirrored ("isomeric") possibilities for the triangles (and for the interfaces): 123 or 132 in clockwise direction.

The 4-junction looks slightly more complicated but can be broken down to 3-juntions. In the 4-junction point, four 3-junction axes come together, each pair of them sharing a common interface.

Consequently, each pair of force triangles (corresponding with a pair of 3-junction lines) shares a common side. This implies that the four triangles can be fitted together to form a tetrahedron ("Gibbs' tetrahedron"). The 3-junction axes correspond with the normals of the tetrahedron sides, and their relative orientation is therefore uniquely fixed by the solution of the tetrahedron construction. Of course, also in the 4-junction case we have to distinguish between two stereo-isomeric cases.

Equations (4.1) are the basis of all the details to be considered in this chapter. In 4.1.2, we discuss the phenomenon of *wetting* of junctions (for simplicity, with OID γ). In 4.1.3, the phenomenology of 3-junctions with OD γ_i is catalogued, and in 4.1.4 - 4.1.6 this is extended to 4-junctions. D-junction equilibrium is discussed in 4.1.7.

4.1.2 Wetting of n-junctions and the topology of polygrain structures.

As we focus on principles rather than on details, we consider the simplest possible case, i.e. where in the 3- and 4-junctions we have only two possible values of (OID) γ. More variables, including orientation dependence, add only variations in the geometrical details, not in the number of topological cases to be discerned. (SMITH (1948))

Say that γ_1 corresponds with an interface between two grains of a crystal α and γ_2 with the interface between α and a second phase β. (fig. 4.1). For the 3-junction, the equilibrium condition for this case reads

$$\vec{n}_1\gamma_1 + (\vec{n}_2 + \vec{n}_3)\gamma_2 = 0 \tag{4.2}$$

This equation has a solution if and only if $\gamma_1 < 2\gamma_2$. The case $\gamma_1 \geqslant 2\gamma_2$ implies that a 3-junction cannot exist in equilibrium and that the interface between α_1 and α_2 is completely wetted by phase β. In other words, β will form a layer between α_1 and α_2. The layer thickness is unspecified, although for very small thickness (see page 78) a stabilisation is induced by the interaction of the parallel α/β and β/α interfaces.

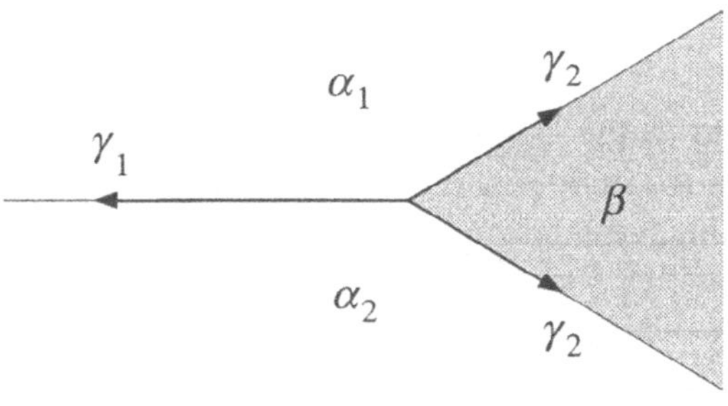

FIG. 4.1

Now consider *three* α grains and β in the middle (fig.4.2). As a thought experiment, consider the two-dimensional case (3-junction axis perpendicular to the drawing plane). The case of fig. 4.2b, i.e. $(p_\alpha - p_\beta) = 0$, corresponding with $\gamma_1 = \gamma_2\sqrt{3}$, marks the border between the cases of figs.4.2a and 4.2c where $(p_\alpha - p_\beta)$ is negative and positive, respectively.

The consequence of a negative $(p_\alpha - p_\beta)$, i.e. $\gamma_1 < \gamma_2\sqrt{3}$, is that the convex-triangular pipe containing the β phase is unstable and tends to break up into elongated-triangular droplets attached to the 3-junction. This is essentially an instance of the Rayleigh instability of a liquid cylinder: when the three 3-junction lines of fig. 4.2a approach the centre, the curvature is increased, the β overpressure is increased and the β phase is squeezed out into the neighbouring thicker parts of the pipe). In contrast, for positive $(p_\alpha - p_\beta)$, i.e. $\gamma_1 > \gamma_2\sqrt{3}$, the hollow-triangular β pipe is *stabilized* and tends to a constant cross-section. In analogy with the two-grain case we will call this *wetting of the 3-junction.* It means that, in the presence of phase β, an $\alpha_1/\alpha_2/\alpha_3$ 3-junction has no stable solution and is replaced by a hollow-triangular tube filled with β. Without any β phase, a real hollow core will result, quite analogous to a hollow dislocation (VAN DER HOEK 1982). (In agreement with this analogy, we may characterize a hollow-core dislocation as a *"wetted D-junction"*).

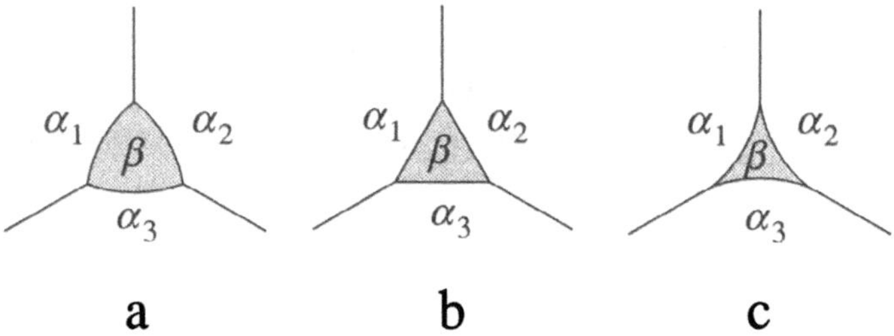

FIG. 4.2

Finally we extend to the 4-junction. With four α -grains and β in the tetrahedral position in the middle, we can repeat the reasoning given above. The critical case of flat tetrahedron sides, $(p_\alpha - p_\beta) = 0$, now corresponds with $\gamma_1 = \gamma_2\sqrt{8/3}$. For $\gamma_1 > \gamma_2\sqrt{8/3}$, a four-grain $\alpha_1\alpha_2\alpha_3\alpha_4$ 4-junction is unstable and is replaced by a convex-curved β tetrahedron (and four 4-junctions at the vertices). This, of course, is *wetting of the 4-junction.*

(*note:* the attachment of a concave-curved β tetrahedron to a *nonwetted* 4-junction is, of course, mechanically stable and in that sense such a 4-junction is still "wetted". The term "wetting" in this case refers to the relative overpressure of the β phase so that it tends to dissolve in the α grains, leaving the 4-junction "dry". If, instead, the 4-junction wetting condition applies, even a very slight amount of β, if present in the α grains at all, is sucked into the convex β tetrahedra).

Altogether, we have critical values for $(\gamma_1/\gamma_2)^2$ of 4, 3 and 8/3, respectively, above which interfaces (2-junctions), 3-junctions and 4-junctions are wetted. The condition for wetting implies at the same time a *stabilisation of size*. However, the stabilizing mechanism (and consequently the rate of approach to equilibrium) can rather differ between the three cases, as we will discuss now.

For the 3- and the 4-junction, the stabilizing force is the pressure difference $(p_\alpha - p_\beta)$, i.e. ultimately the Gibbs-Thomson shift. Being proportional to curvature, it is effective for fairly big β tubes and tetrahedra (as compared with the stabilizing mechanism for interface-layer thickness). How effective this Δp is in achieving homogeneity in tube-and tetrahedron size in a multigrain specimen, depends on the resistance to transport.

Consider liquid sintering at small β volume fraction, still in the two-γ model system. The grains to be sintered are α grains, whereas β is a liquid phase filling the space in between. The parameters characterizing the situation are the typical grain size (and size distribution) of the α phase, the volume fraction of β, and γ_1/γ_2. The liquid serves two functions: a) it promotes the mobility of wetted interfaces, and b) if the 3-junctions are wetted, it short-circuits the pressure in the liquid phase through the whole system; i.e. homogeneity of the pressure is ensured by viscous flow of the liquid through the triangular-tube network. If the 3-junctions are not wetted, pressure homogenisation can occur only by diffusion (including enhanced diffusion along interfaces).

The possibility of transport by viscous flow through a network of wetted 3-junctions may give rise to a "macroscopic" redistribution of the liquid phase in a sintering specimen of initially inhomogeneous composition and/or size distribution. Also, it may play a role in instability phenomena like anomalous grain growth during sintering. Of course, the actual situation is complicated by the occurrence of open pores and the effects of dissolution in polynary systems.

So for $\gamma_1 < \gamma_2 \sqrt{8/3}$, where the liquid phase is present in isolated droplets located at interfaces, 3- and 4-junctions (preference declining from 4-junction to 2-junction), the rôle of the liquid phase is not essentially different from that of any solid phase (except in the rate at which the equilibrium shape of the liquid-phase inclusions themselves are reached). For $8/3 < (\gamma_1/\gamma_2)^2 < 3$, this situation is not essentially changed. But for $3 < (\gamma_1/\gamma_2)^2 < 4$, the 3-junctions form a connected-tube network with the drastic change in pressure homogenisation kinetics mentioned above. How effective the triangular tubes are in homogenizing pressure depends on their size and therefore on the β volume fraction and on the α grain size (on the basis of power laws for the viscous flow in channels of given dimensions, and those for the Gibbs-Thomson effect and the volumes to be transported, power laws for the homogenisation rate can be derived by dimensional analysis).

Finally, for $\gamma_1/\gamma_2 < 2$, also the *interfaces* are wetted and a β layer creeps in between any two α grains. How thick is this layer? We are now confronted with the difference between the stabilisation mechanisms of 3- and 4-junctions (Gibbs-Thomson) and that of 2-junctions (interaction between parallel α/β and β/α interfaces). As a result of the curvature of the α/β interface in the 3-junction tubes, the β pressure is lower than the α pressure. As the β phase in the interface layer is in open contact with the tubes, the pressure of the α grains tends to squeeze out a *thick* β layer into the tubes. The layer thickness is finally stabilized when the interaction force is high enough to cancel this pressure. (Note: this implies that for a stabilized layer, the total free energy of the two interfaces α/β and β/α is no longer $2\gamma_2$, but somewhere between γ_1 and $2\gamma_2$. So in the overall equilibrium, the "elasticity" of the layer thickness is coupled with the underpressure in the open tubes).

Depending on the roughness of the α/β interface, the stabilized thickness may range between one and a few β monolayers. When the equilibration time is long enough for the viscous flow to reach equilibrium in the layer/tube/4-junction network system, then the layer thickness, as well as the tube size, is constant in the whole specimen. The situation is identical with that in a soap froth in the black-film region. This can have important consequences for certain physical properties of liquid-phase sintered materials, e.g. when the β/α ratio of electrical conductivities greatly differs from one. Unusually "constant" properties of the voltage-dependent resistance (VDR) characteristics in ZnO-Bi_2O_3 based materials, for instance, may be related with this phenomenon.

The distinction into four ranges of γ_1/γ_2 ratio has only this topological effect on the microstructure when the β phase has a small enough volume fraction. When this fraction is gradually increased, first the hollow-curved tetrahedra grow out to eliminate the straight triangular tubes which connect them, and then the flat interface regions. At a volume fraction of ± 26% (compacted spheres) the α grains are no longer effectively touching one another at a flat interface and therefore we are in the same situation as if they were lying individually in β liquid saturated with α. Then, only the α-grain size differences induce affinity (ΔA_c) differences and drive further α-grain coarsening by Ostwald ripening.

Solid phases.

What has been remarked about equilibrium configurations involving two or more γ values is not, of course, restricted to systems including a liquid phase. But for all-solid systems, liquid-pressure homogenisation is not available as a short-circuiting mechanism and accordingly the "range of approached equilibrium" (see ch.8) is many orders of magnitude slower (i.e. smaller, for a given reaction time). Therefore, for a *solid* β phase, the size of interlayers and triangular tubes will in general be much less homogeneous than for a liquid β phase.

We have discussed wetting phenomena at the simplest possible level of abstraction because they can easily be adapted to more general cases. So, when more than two phases are involved, we have to discern different types of interfaces, 3- an 4-junctions, all of which can be wetted or not. The same is true for orientation-dependent surface free energies: there the character depends on orientation. Therefore, we will leave to the reader to trace out the consequences of wetting in the following paragraphs.

In polycrystalline samples, the occurrence of wetted and nonwetted junctions will rather be a matter of statistics. The occurrence of wetted 3-junctions, and in particular the connectivity of 3-junction networks with its relevance for pressure homogenisation, may sensitively depend on the ratios of the interface free energies involved. Therefore, in the vicinity of these critical γ ratios, small chemical additions or changes in the sintering procedures can have relatively large effects on sintering kinetics and on the final structure and properties of the material.

4.1.3. 3-junctions between interfaces with OD surface energies.

a) 3-junctions:

For a 3-junction with OD surface energies (specified γ-plot orientations and 3-junction axis $\vec{\ell}$) the construction of the force triangle is complicated by the fact that the force vectors do not have a constant length as in the OD case but are variable according to their respective plots. When $\vec{\ell}$ is fixed, we are in the two-dimensional case, and we can obtain the relevant two-dimensional plots $(\vec{\mathbf{Q}}\gamma \wedge \vec{\ell}\,)$ by shadowing the three-dimensional $\vec{\mathbf{Q}}\gamma$ surfaces onto a plane perpendicular to $\vec{\ell}$. Indicating the shadows by $\vec{\mathbf{Q}}^{(\ell)}\gamma$, the force triangle can be constructed by arranging the three $\vec{\mathbf{Q}}^{(\ell)}\gamma$ plots in such a way that the origin of one plot coincides with the circumference of the next in a cyclic arrangement (fig. 4.3).

Geometrical solutions occur in pairs, only one of which is a physical solution. To show this, we adopt the notation used by HOFFMANN and CAHN

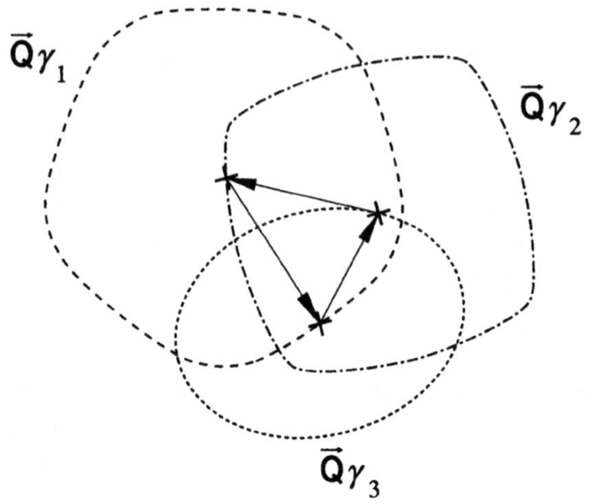

Fig. 4.3: Construction of Neumann force-equilibrium triangle for a 3-junction with OD surface free energy.

(1972) for the relative arrangement of the grains. When the grains are labeled 1,2,3 in counterclockwise order around the axis $\vec{\ell}$, the axis vector is given the suffix 123 (or cyclic equivalent). The γ 's are labeled as defined in the introduction to ch.2.

When the orientation on one interface (say φ_{12}) is given, we can draw the vector $\vec{Q}^{(\ell)}\gamma_{12}(\varphi_{12})$. The second plot $\vec{Q}^{(\ell)}\gamma_{23}$ is drawn fully, its origin coinciding with the point $\vec{Q}^{(\ell)}\gamma_{12}(\varphi_{12})$. Now when the third plot is made to follow $\vec{Q}^{(\ell)}\gamma_{23}$ in counterclockwise direction, (the origin of the $\vec{Q}^{(\ell)}\gamma_{31}$ plot following the $\vec{Q}^{(\ell)}\gamma_{23}$ curve), then after completion of a full cycle, its circumference will have crossed the origin of $\vec{Q}^{(\ell)}\gamma_{12}(\varphi_{12})$ in general $2n$ times, n on entering and n on leaving (when it has just touched, two of them coincide). It is easily verified that only the "entering" solutions correspond with physical solutions. (The others correspond with overlapping phases. If and only if all $\vec{Q}^{(\ell)}\gamma$ plots are π-invariant, they also correspond with the $\vec{\ell}_{213}$ arrangement of the grains).

Wetted 3-junctions.

For a wetted 3-junction, the equilibrium conditions for the surfaces beyond the central hollow-triangular core are identical with those of the nonwetted junction. So when a 3-junction is wetted by a fourth phase or grain, the only change is in the core region and in the *effective line energy* of the 3-junction.

Representation of the solutions.

The equilibrium configurations of the 3-junction (fixed axis) can be represented by a plot which specifies the orientations of the three interfaces for all possible positions in the full range 2π. Such a plot is sketched in fig.4.4a. One of the three interface orientations is considered as the imposed parameter, and plotted along the horizontal axis. The equilibrium positions of the other two are plotted vertically (for the OID case, we would obtain a set of two straight lines parallel with the diagonal).

Such a plot represents only the *orientation* of the interfaces obeying the equilibrium condition, not the force vectors. Unless we have facets as interfaces, there is a one-to-one relation (alternatively, we can draw the orientation of the *force vectors*).

As compared with the OID case, we can have the following complications:

a) Solution ranges.

As in the OID case, we have the possibility that no solution exists. This corresponds to complete wetting of one of the interfaces, i.e. its replacement by the two others. Now as existence of a solution can depend on orientation, the full 2π circle can be divided into regions without and with a solution. (fig. 4.4b).

In contrast with the OID case, we now have also the possibility of *more than one* solution. As a consequence of the fact that the shadow-plots are no longer circles, the number of solutions can change with orientation.

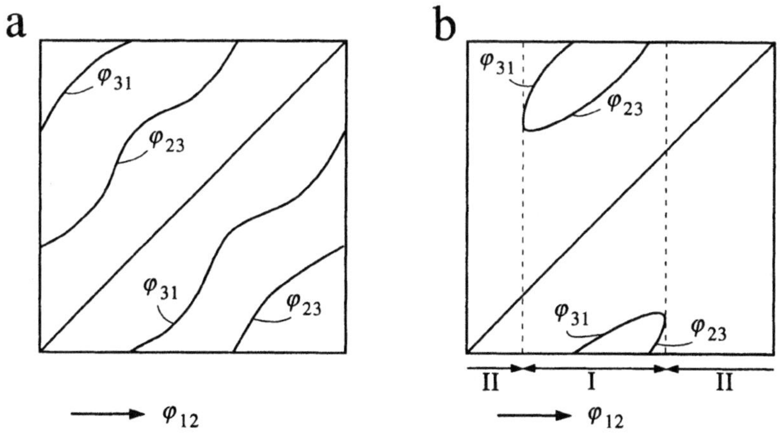

FIG. 4.4 a and b

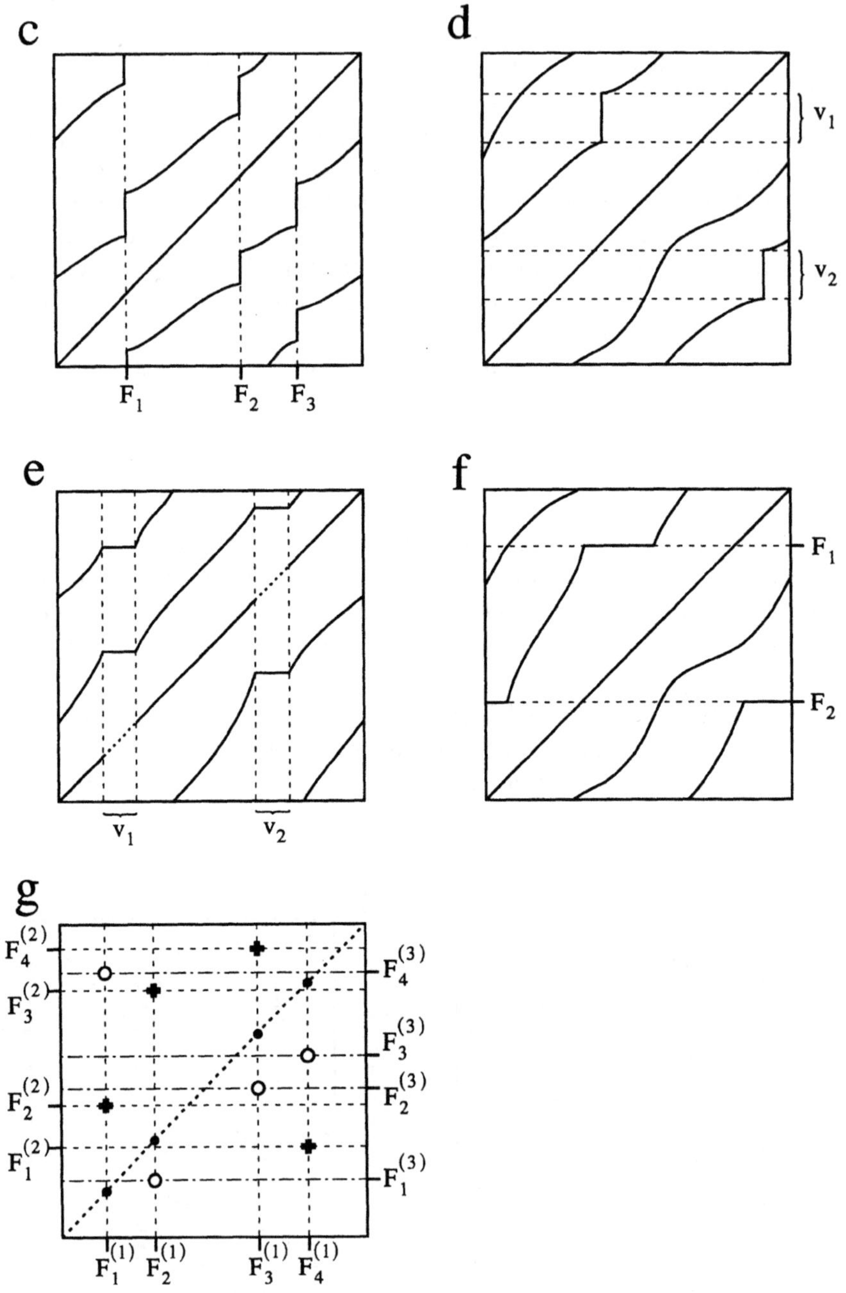
c
d
e
f
g
F1
F2
F3
v1
v2
v1
v2
F1
F2
F4(2)
F3(2)
F2(2)
F1(2)
F4(3)
F3(3)
F2(3)
F1(3)
F1(1)
F2(1)
F3(1)
F4(1)

FIG. 4.4 c-g

Note: The possibility of two different equilibrium positions for interfaces 2 and 3 when the position of interface 1 is fixed, i.e. a bistable situation, can induce relaxation oscillations in crystal growth experiments. For instance, zigzag patterns in the lamellar orientations of unidirectionally solidified eutectics may be related with this phenomenon (although CAROLI c.s. (1992) present an explanation which is unrelated with OD γ).

b) *Consequences of facets and edges in the* $\vec{Q}\gamma$ *shape*

When φ_{12}, the orientation of the *imposed* interface is in a facet direction, φ_{23} and φ_{31} are vertical in the plot of fig. 4.4c. When an edge in the $\vec{Q}^{(\ell)}\gamma_{23}$ or in the $\vec{Q}^{(\ell)}\gamma_{31}$ plot is passed, φ_{23} and φ_{31} also make a vertical jump, the orientations in-between corresponding with a zigzagged interface. (4.4). (v_1 and v_2 in fig. 4.4d are the corresponding "virtual ranges" of φ_{23} and φ_{31}).

When φ_{12} itself passes a zigzag ($\mathbf{S}^{(\ell)}\gamma_{23} < 0$) range between edge orientations, φ_{23} and φ_{31} are horizontal (v_1 - v_2 in fig. 4.4e). When we pass a facet in the $\vec{Q}^{(\ell)}\gamma_{23}$ or in the $\vec{Q}^{(\ell)}\gamma_{31}$ plots, the corresponding φ remains horizontal. (fig. 4.4f).

When all $\vec{Q}^{(\ell)}\gamma$ plots are completely faceted, the solutions, as far as orientations are concerned, can be represented as a discrete set of (φ_{12}, φ_{23}, φ_{31}) combinations. (fig. 4.4g).

4.1.4 *4-junctions between interfaces with OD surface energy*

In the 4-junction, four 3-junction axes $\vec{\ell}_1, \vec{\ell}_2, \vec{\ell}_3, \vec{\ell}_4$ meet in the junction point, together defining six interfaces. The axes are labeled in accordance with the phases or grains, $\vec{\ell}_i$ being the axis of the 3-junction between grains $\alpha_{j\neq i}$, taken in the direction away from the 4-junction point. So (fig. 4.5):

$$\vec{\ell}_i = \frac{\vec{n}_{jk} \wedge \vec{n}_{k\ell}}{|\vec{n}_{jk} \wedge \vec{n}_{k\ell}|} \varepsilon_{ijk\ell}\,; \qquad j,k,\ell \neq i \text{ (and cyclically)} \tag{4.3a}$$

$$\vec{n}_{ij} = \frac{\vec{\ell}_k \wedge \vec{\ell}_\ell}{|\vec{\ell}_k \wedge \vec{\ell}_\ell|} \varepsilon_{ijk\ell}\,; \qquad k,\ell \neq i,j \text{ (and cyclically)} \tag{4.3b}$$

where $\varepsilon_{ijk\ell}$ is the permutation operator, +1 for an even number and -1 for an odd number of permutations between i, j, k and ℓ. There are two "stereoisomeric" possibilities for the arrangement of the four phases. Each of the four 3-junction configurations should be in equilibrium on its own:

$$\left\{ \sum_{\substack{i,j \neq k \\ \text{cyclically}}} \vec{\mathbf{Q}}\gamma_{ij}(\vec{n}_{ij}) \right\} \wedge \vec{\ell}_k = 0$$

or

$$\vec{Z}_k \wedge \vec{\ell}_k = 0 \tag{4.4}$$

where $\vec{Z}_k$ is a shorthand notation for the sum of the three $\vec{\mathbf{Q}}\gamma$ vectors between the brackets.

Eq. (4.4) represents the set of four vector equations for the 3-junctions (k = 1 through 4). It can easily be verified that these four equations are equivalent to the mechanical-equilibrium condition for the 4-junction configuration (see, for example, HOFFMANN and CAHN (1972). Together they represent eight equations for the eight orientational variables implied in $\vec{\ell}_1 \ldots \vec{\ell}_4$. (Alternatively, the six *interface* orientations $\vec{n}_{ij}$ can be considered as the orientational variables, the common four axes bringing back the number of independent variables from twelve to eight).

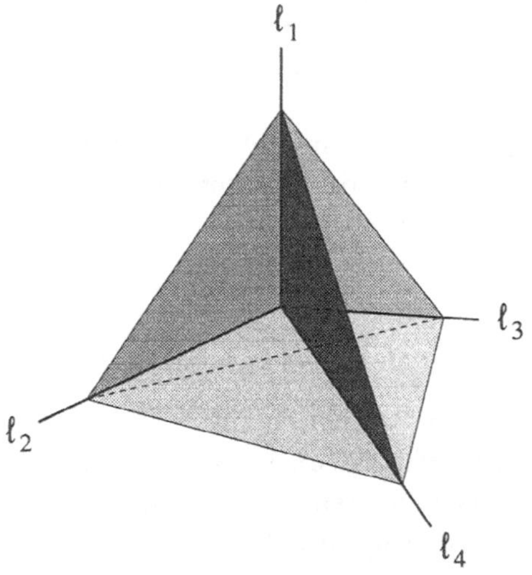

FIG. 4.5

However, an implicit relation between the $\vec{n}_{ij}$ variables implies that only five of the eight equations are independent. This relation results from the fact that the twelve $\vec{\mathbf{Q}}\gamma$ vectors in eq. (4.4) occur in opposite pairs, so that

$$\sum_{k=1}^{4} \vec{Z}_k \equiv 0 \tag{4.5}$$

This identity, considered as the vanishing of three projections of $\Sigma\, \vec{Z}_k$, is equivalent to three independent equations between the coefficients of the six $\vec{n}$ vectors (the equations are implicit because the $\vec{n}$ vectors involved are the arguments of the $\gamma(\vec{n})$ functions to which eq. (4.4) applies). Altogether, eq. (4.4) represents five independent equations between eight effective parameters, together leaving three degrees of freedom for the system. These can be used, for instance, to specify one 3-junction axis and one of the interfaces of this junction. All other variables of the 4-junction are then completely fixed by the equilibrium condition.

Note:

CAHN and HOFFMANN (1972), when considering the eight equations (4.4), overlooked the dependence of the $\vec{n}$ parameters and consequently considered the solution as unique. This would mean that for given grains $\alpha_1 \ldots \alpha_4$, all of fixed crystallographic orientation, all i/j interfaces in a 4-junction (and the $\vec{\ell}_i$ axes) *can have only one fixed orientation in space*. This result seems already hard to accept intuitively, as in a case where the grains would be attached "at the wrong (opposite) side" by some accidental boundary condition, the interfaces would have to be uneasily contorted in order to satisfy equilibrium.

Cahn and Hoffmann further hypothesized on the basis of the above conclusion, that 4-junctions in the OD case might either be in equilibrium (in which case they would act as inflexible pinning points in a polycrystalline microstructure) or not (in which case they might serve as sources of instability, e.g. as nucleation centres for recrystallisation or phase transformation).

This conclusion and hypothesis can be discarded in consideration of the arguments given above. Ergo, *there is no fundamental difference between the OD and the OID cases, insofar as the determinacy of the problem is concerned* (except complications due to multiple solutions).

4.1.5 The force-vector icosahedron of the 4-junction.

The geometrical meaning of equations (4.4) can easily be identified. Consider the vectors $\vec{\mathbf{Q}}\gamma_{ij}$ occurring in eq. (4.4), all with their origin in the 4-junction point (remember that they occur in pairs, $\vec{\mathbf{Q}}\gamma_{ij} = -\vec{\mathbf{Q}}\gamma_{ji}$). The twelve vectors can be considered as the vertices of an icosahedron (not in general a convex icosahedron, i.e. the triangles spanned by the $\vec{\mathbf{Q}}\gamma$ vector trios of eq. (4.4) can intersect mutually. These triangles can be numbered 1 through 4 in accordance with the k number). The sums $\vec{\mathbf{Z}}_i$ represent three times the vector connecting the origin with the centre of gravity of triangle i. In other words, equations (4.4) express the requirement that the axes of the icosahedron which pass through the centres of gravity of the four triangles mentioned, just coincide with the 3-junction axes defined by the corresponding equations (4.3a). So when the equilibrium is obeyed, the six physical interfaces of the 4-junction are defined by the four $\vec{Z}_i$ -axes of the icosahedron defined above.

(Note that the other sixteen triangles of the icosahedron have no physical meaning, as they do not correspond with any of the four equations (4.4)).

It is interesting to reconsider the special case of orientation-*independent* γ. Then the $\vec{\mathbf{Q}}\gamma$ vectors are oriented in the normal direction, and correspondingly the icosahedron-triangle planes are centered on the origin of the icosahedron. This implies that the medians of the triangles now represent the $\vec{\mathbf{Q}}\gamma$ vectors *themselves*, i.e. the same medians $\vec{\mathbf{Q}}\gamma_{ij}$ occur in the pairs of triangles k and ℓ (k,ℓ = i,j). (In the OD case, these were different *projections*). So the force triangles share common sides and consequently they can be fitted together to form a tetrahedron. This is *Gibbs' tetrahedron* as discussed by Cahn and Hoffmann, specifying the relation between the 3-junction axes and the *normals* to the sides of the tetrahedron (remember that in the icosahedron, the 3-junction axes correspond with the centre-of-gravity direction of the triangles spanned by the $\vec{\mathbf{Q}}\gamma$ vectors. These triangles are not the force triangles with the projections as sides!).

The *construction* of the icosahedron can be arranged as follows. Just as in the OID case, we have three degrees of freedom. For instance, we can choose one 3-junction axis (say $\vec{\ell}_{123}$) and the orientation of one of the corresponding interfaces. As discussed before, the choice of the latter suffices to determine the orientation of the other two interfaces in the same 3-junction uniquely (for the moment we exclude pathological cases such as multiple solutions, wetted 2-, 3- or 4-junction complications and degenerate solutions induced by zero or singular $\vec{\vec{\mathbf{S}}}\gamma$ eigenvalues).

Having thus satisfied one of the four equations (4.4), we have at the same time a) reduced the degrees of freedom for each of the remaining 3-junction axes to one (i.e., they have now to lie in a fixed plane) and b) fixed three of the six $\vec{\mathbf{Q}}\gamma$ vectors. The three remaining equations can thus be considered as three equations for three unknowns, together determining the solutions uniquely.

Choosing a second 3-junction axis, say $\vec{\ell}_{214}$, with provisionally random orientation in the α_1/α_2 interface, we can determine the α_2/α_4 and α_1/α_4 interfaces by the $\vec{\ell}_{214}$ 3-junction equilibrium. Thereby the last two axes, $\vec{\ell}_{432}$ and $\vec{\ell}_{341}$, are fixed as intersections of the $\alpha_2/\alpha_4 - \alpha_2/\alpha_3$ pair and the $\alpha_1/\alpha_4 - \alpha_3/\alpha_1$ pair, respectively. But the equilibrium of the 432 and 341 three-junctions has still to be verified. By varying the (only left) orientational variable of $\vec{\ell}_{214}$, we can find one direction for which the equilibrium condition of one of the other 3-junctions is satisfied. The equilibrium condition of the last 3-junction is then also automatically satisfied.

The construction procedure sketched above reflects only one way in which the three orientational degrees of freedom can be prescribed. However, it is not a likely way in which an actual 4-junction geometry will be arrived at in, say, the sintering kinetics in a polycrystalline sample. Therefore we sketch an alternative "construction" which can more readily be associated with the actual mechanism.

In sintering kinetics, the order of n-junction formation is $n = 2 \rightarrow 3 \rightarrow 4$. In other words, as the prenatal stage of a 4-junction we can consider a tetrahedron whose vertices are determined by the end-points of 3-junctions already formed (and relatively fixed in space). Now the alternative construction referred to above consists in considering the four vertices of such a tetrahedron as fixed points, and looking for that point within the tetrahedron which, together with the vertices, defines the six interface orientations in such a way that the total force on the tetrahedron equals zero.

Choose one of the vertices and imagine a 3-junction axis passing through it in a random but fixed orientation $\vec{\ell}_1$. This (together with the other vertices) determines the three interface orientations around this axis and the force exerted through them on the tetrahedron. Let the tentative 4-junction point walk along this axis. Thereby the three remaining interfaces tilt around the axes defined by the other tetrahedron vertices, and correspondingly the forces exerted by them on these axes, in the direction of the first axis $\vec{\ell}_1$, changes. In this way we can find a particular point $P(\vec{\ell}_1)$ where the two forces cancel. Performing this procedure for all $\vec{\ell}$ directions we find a *surface* Π_1, consisting of all points $P(\vec{\ell}_1)$.

Similar surfaces can be found for the other three vertices. Two of these surfaces intersect in a curve, three of them in a point. It might seem that the requirement that the fourth Π surface also contains this point would overspecify the problem. This is not true: the point defined as the intersection of Π_1, Π_2 and Π_3 determines six surfaces, together exerting a net force on the tetrahedron with zero components in three different directions. In other words, the force vector itself is zero, and thus has a zero component in any direction, including the fourth axis. So the four Π surfaces meet in a single point, the equilibrium location of the 4-junction.

Whereas the orientation relations of the 3-junction equilibrium (fixed axis, only one degree of freedom φ) can directly be plotted in two dimensions as discussed in 4.1.3, the 4-junction equilibrium, requiring three variables, cannot so easily be represented, and we will not try to develop a suitable method. Still, we want to mention that the pathological cases of 3-junctions, i.e. multiple solutions, wetted junctions and zero or singular stiffness of interfaces, all have their 4-junction analog.

4.1.6 The line energy of 3-junctions and its effect on 4-junction shape.

(see also the "oral discussion" in the paper by Hoffmann and Cahn (1972).

So far, junction equilibria have been considered as force equilibria between surface-related forces, not line-related forces. Still, a 3-junction almost certainly represents an excess free energy. (in a nonwetted junction related with the sharp edges in all three phases, in a wetted junction with the surplus energy in the interfaces and the underpressurized wetting phase). This *3-junction line free energy* can not be derived from the γ plots; it is inherently related with the delocalisation of the interface free energies and has to be considered as an extra quantity associated with the 3-junction. It is a function of the three orientation parameters of the 3-junction.

In a straight 3-junction, this line energy plays no visible role (except when it has a nonzero $d/d\varphi$ derivative causing a torque. We will neglect this for the moment). However, in a 4-junction, the sum of the four line tensions can result in a net force acting on the 4-junction point (we do not see any physical argument why the forces should cancel; the force is not related with expression (4.4) or (4.6)). Such a net force would be comparable, in its result, with the force exerted by a dislocation on an interface in which it ends. In the dislocation case, a hollow crater results, the interface obeying an equation like (4.10) or (4.11). The same will happen with the six interface sectors of the 4-junction. In general, all will react by a nonvanishing curvature to compensate torgether for the line-generated force, at the same time preserving such mutual orientations as to obey the surface-energy related equilibrium condition. The problem does not become overspecified by the appearance of the extra line forces, although it is certainly complicated by the coupling.

4.1.7 Dislocations and interfaces.

Although we are primarily concerned with surfaces in this monograph, the "junction catalogue" also includes line defects, especially where they interact with an interface and thereby modify the equilibrium shape and kinetic behaviour of this interface. This chapter is restricted to the discussion of *equilibrium* configurations of dislocation / surface structures. (In ch. 6 we discuss the configurations during growth and in particular the equilibrium/growth shape transition).

An important difference between interfaces and dislocations is the scale at which they can be considered as a mathematical surface and line, respectively. For grain boundaries and phase interfaces, the size is typically 10^{-7} - 10^{-6} cm; for single dislocations it is larger by roughly three orders of magnitude (~ 10^{-4} - 10^{-3} cm.). The size is defined here as the thickness of the layer and the diameter of the cylinder, respectively, where an appreciable fraction (say 50 %) of the excess free energy is located. (In the case of dislocations, this energy is mainly related with the stress field in the solid surrounding the dislocation). As a consequence, dislocations exhibit in general a much longer-range interaction, mutually and with an interface, than interfaces; also the critical radius below which curvature dependence of dislocation properties is appreciable is much larger than for interfaces.

The theory of dislocations is well established and the above remarks have been widely elaborated in the specialized literature, to which the reader is referred for specific information. (e.g. FRIEDEL (1964), HULL (1965))

As our discussions on interface properties are primarily based on "zero-thickness" parameters, we will use the similar viewpoint for dislocations. In view of what has been said above on sizes, this implies a more severe abstraction from reality for dislocations than for interfaces. At a large enough scale, however (typically $\geqslant 1\mu$) this abstraction is physically realistic. On a smaller scale, the model does not work and detailed models of the dislocation stress field have to be used instead.

4.1.7.1 Dislocation-induced force and the $\vec{\mathsf{Q}}$ *operator.*

In the zero-thickness model of a dislocation, its free energy is concentrated on a mathematical line (we use the unit line vector $\vec{\ell}$ as the orientation parameter). We use the symbol $\sigma(\vec{\ell})$ for the *line free energy;* (dimension: energy per length unit). In general it depends on the direction of $\vec{\ell}$ and a 3-dimensional "plot" of $\sigma(\vec{\ell})$ is generated by the $\vec{\mathsf{P}}$ operator ($\vec{\mathsf{P}}\sigma(\vec{\ell}) \equiv \vec{\ell}\,\sigma(\vec{\ell})$). By a reasoning which is essentially identical to that used in 4.1, we can see that the force exerted by the dislocation on any point to which it is attached is

$$\vec{F}_{\text{disl.}} = \vec{\mathsf{Q}}\sigma(\vec{\ell}) \tag{4.6}$$

(Here, $\vec{\mathsf{Q}}$ is used for $\vec{\ell} + d/d\vec{\ell}$; remember that for an interface, where the *normal* vector is used as the orientation parameter, the force is given by $\vec{\mathsf{Q}}\gamma(\vec{n}) \wedge \vec{\ell}$ where $\vec{\ell}$ is then the interface *boundary* on which the force works. Note also that for the dislocation, the parameter space for $\vec{\ell}$ is the unit sphere, i.e. 2-dimensional, whereas for the case of *ledges on a particular crystal facet* the parameter space is a circle, i.e. 1-dimensional).

In dislocation networks, we can construct the equilibrium configurations (angles) on the basis of the $\sigma(\vec{\ell})$ plots, when available, using a similar procedure as sketched in 4.1.3 for interface 3-junctions. The difference with the latter is that we have 2-parameter plots in the dislocation case, so that

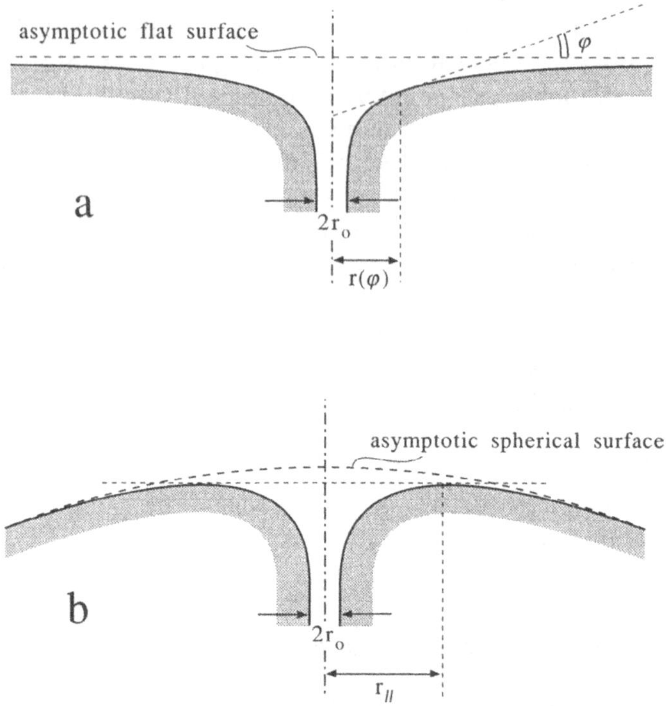

FIG. 4.6: D-junction shapes for the (unrealistic) case of concentrated dislocation line energy.

there is one degree of freedom left which has to be imposed by a suitable boundary condition.

A dislocation oriented in an "unfavourable" direction (i.e., having one or more negative eigenvalues of $\vec{\vec{\mathbf{S}}}\sigma(\vec{\ell})$) can break up into a *zigzag line* (actually, consisting of segments of two orientations when $\vec{\vec{\mathbf{S}}}\sigma$ has one negative eigenvalue, and of three orientations when it has two negative eigenvalues).

The equilibrium shape of a *closed dislocation loop* cannot simply be identified with a curve on the $\vec{\mathbf{Q}}\sigma(\vec{\ell})$ surface because the concept of a fixed pressure difference "across" the dislocation is meaningless (it might perhaps be applied to special cases such as loops surrounding large stacking faults, the stacking fault being considered as a "flat grain"). Accordingly, the $\vec{\vec{\mathbf{S}}}\sigma$ tensor cannot, as far as we can see, be correlated with anything like pressure, except again in the stacking-fault case.

Apart from dislocations, also *3-junctions* (including wetted 3-junctions with a uniform tube underpressure) have a line excess free energy. The "thickness" of a nonwetted 3-junction is much smaller than that of a dislocation and rather comparable with that of an interface. Of course, the line-force mechanics discussed above also applies to the 3-junction, although the latter is complicated by the inherent presence of the interfaces. (4.1.6).

4.1.7.2. Equilibrium shape of interfaces with a point force

A dislocation ending on an interface exerts a force on the point of intersection. The interface, when in mechanical equilibrium, is thereby distorted. We have two variables:

a) the point force $\vec{F}$ (assumed here to be aligned with the original interface normal).

b) the pressure difference $\Delta p = A/V$.

In order to simplify as much as possible, we restrict to the case of OID surface energy. In the cylindrically-symmetric case thus obtained, $\varphi(r)$ is the only geometrical variable left (see fig. 4.6a for the definition of φ and of r). The condition

$$1/r_1 + 1/r_2 = \Delta p/\gamma$$

can be transformed into (see, e.g. WOLF (1959))

$$\frac{1}{r}\,\frac{\mathrm{d}(r\sin\varphi)}{\mathrm{d}r} = \frac{\Delta\varphi}{\gamma}$$

with solution

$$r\sin\varphi = r\,\frac{\Delta p}{2\gamma} + C^* \tag{4.7}$$

where C^* is an integration constant to be matched with the constraints of the problem.

As the surface-induced force $2\pi r\gamma\sin\varphi$ has to balance the "point" force $\vec{F}$ it is obvious that instead of a real point, we have to define a finite radius of attachment

$$r_0 = F/2\pi\gamma \tag{4.8}$$

between dislocation and surface so as to prevent φ from becoming imaginary. (we have written F for $|\vec{F}|$) When r_0 is $\ll$ lattice constant there is no physical problem, when it is > lattice constant we may have a hollow dislocation core, although actually the zero-thickness concept of a dislocation is too crude to be applied here (see for a thorough discussion on hollow dislocations VAN DER HOEK, VAN DER EERDEN and BENNEMA, (1982). Remember that the hollow-core case can be considered as a "wetted D-junction"). Now C^* an be expressed in terms of r_0:

$$C^* = r_0\,(1 - r_0 \,.\, \Delta p/2\gamma) \tag{4.9}$$

and the complete solution is

$$\varphi = \arcsin\left\{ r\,\frac{\Delta p}{2\gamma} + \frac{F}{2\pi r\gamma}\left(1 - \frac{F\Delta p}{2\pi\gamma^2}\right)\right\} \qquad (4.10)$$

For $\Delta p = 0$, the solution is a catenoid ($\varphi = \arcsin F/2\pi r\gamma$), for the general case it is a "nodoid".

The equilibrium shape of the interface is given by (4.10), at least far enough away from the "point" of attachment. Note that for $\Delta p = 0$, the equilibrium shape is a *hollow* pit, whereas for a growing crystal with dislocations we are used to consider the dislocations as ending at the vertex of *protruding* pyramids (but they are growing on facets; i.e. we have to do with *OD* γ) In 6.2.6 we will discuss the transition from equilibrium shape to growth shape in more detail.

For $\Delta p > 0$, where the equilibrium shape *without* the dislocation would be spherical, we have a "cratered sphere" (fig. 4.6b). The radius r where the surface is perpendicular to the dislocation-induced force is given by

$$r_{\parallel} = \sqrt{\frac{F}{\pi\Delta p}\left(\frac{F\Delta p}{2\pi\gamma^2} - 1\right)} = r_0\sqrt{2\left(1 - \frac{\gamma}{r_0\Delta p}\right)} \qquad (4.11)$$

Note that, at least in the zero-thickness approximation, the dislocation ends in a hollow crater for any affinity A. This conclusion becomes suspect as soon as $r_{\parallel}$ (eq. 4.11) approaches $\sim 1\mu$.
However, under many technical conditions we have to expect the apices of growing dislocation pyramids to show a hollow (equilibrium-shaped) crater. This case is considerably complicated by the high orientation-dependent surface resistance occurring in ledge growth.

4.2 THE NEGATIVE– $\mathbf{S}\gamma$ INSTABILITY

Whereas a negative value of γ cannot occur as it is incompatible with a two-phase interface or a grain boundary, *eigenvalues* of $\overline{\overline{\mathbf{S}}}\gamma$ can be negative, although this corresponds with an unstable situation. (For $\gamma < 0$ the interface itself is unstable, for $\overline{\overline{\mathbf{S}}}\gamma < 0$ a *flat morphology* of the interface is unstable).

In 2.5 we have identified a negative $\overline{\overline{\mathbf{S}}}\gamma$ eigenvalue as corresponding with an orientation for which $\gamma(\vec{n})$ is situated in a virtual part of the γ-plot, i.e. where $\vec{\mathbf{Q}}\gamma$ is outside a self-intersection point or line, and we have discussed how the complete $\overline{\overline{\mathbf{S}}}\gamma < 0$ range could be short-circuited by a circle or sphere ("effective plot") for which $\overline{\overline{\mathbf{S}}}\gamma = 0$. Here we will rationalize this terminology as a minimum-free energy argument (for simplicity, in the 2-dimensional version).

Consider a macroscopically flat interface of orientation φ_0 . (fig.4.7). The direction φ_0 can be defined by two points, A and B. The straight line between A and B corresponds with a "smooth flat interface". The free energy of this interface equals the product of the distance AB and the energy $\gamma(\varphi_o)$, which is geometrically represented as the surface of the triangle ABO in fig. 4.7 (multiplied by 2).

However, A and B can also be linked by a kinked line A'C'B'. When the directions of A'C' and B'C' are chosen as the edge orientations, it can easily be shown that the total free energy of the surface ACB is now represented by the surface of the quadrangle A'C'B'O (again multiplied by 2). The latter surface is smaller than that of ABO, the ratio being that of the projection of C' in fig. 4.7 on the direction φ_o , to $\gamma(\varphi_o)$. Therefore, it is energetically favourable for the interface AB to assume the kinked shape A'C'B'.

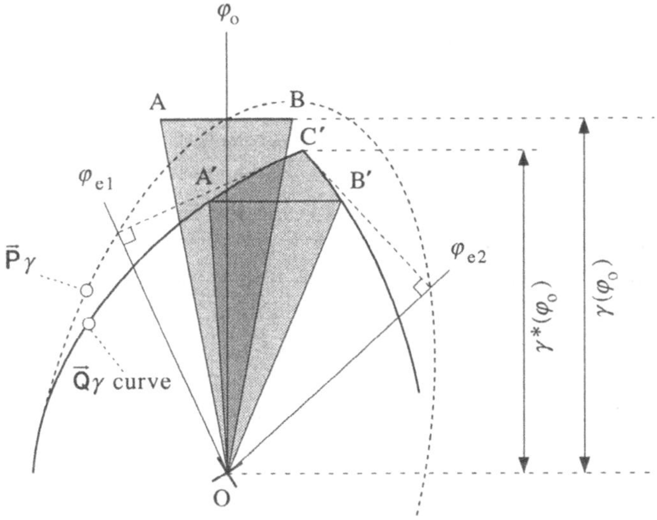

FIG. 4.7

(This is also the minimum free energy. Other directions than the edge orientations φ_{e1} and φ_{e2} result in a higher total free energy.)

At the point C', the force vectors of the interfaces C'A' and C'B' are equal and opposite (vector C' rotated by + and $-\pi/2$, respectively: namely $\vec{Q}\gamma(C) \wedge \vec{\ell}$ and $\vec{Q}\gamma(C) \wedge (-\vec{\ell})$. Therefore, although it is kinked, the surface A'C'B' is in true mechanical equilibrium.

The shape which the surface AB will assume is arbitrary in the following sense: a kinked line of smaller segments, but of the same fixed orientations φ_{e1} and φ_{e2} , has still the same free energy as A'C'B' (provided the energy contributed by the kinks can be neglected). An arbitrary curved line lying within the quadrangle A'C'B'O, and with tangents not exceeding φ_{e1} and φ_{e2} , can be approximated by such a kinked line. (The fact that the energy is independent of the shape, when the line is built up of segments of two fixed orientations, is sometimes referred to as the "Manhattan effect").

Extended to the three-dimensional case, we can have a corrugated surface (one negative $\vec{\vec{\mathbf{S}}}\gamma$-eigenvalue), or a surface broken up into pyramids (two negative eigenvalues). Note that in the first case, a transverse cross-section through the structure shows only two fixed orientations, whereas in the second case we find in general four or three different fixed orientations, corresponding with 4- or 3-sided pyramids.

4.2.1. The zigzag period in an affinity gradient.

The indeterminacy of the kinked line mentioned above only exists when $\vec{\nabla} A = 0$. We will now consider the structure of such an interface in a nonzero-gradient field of the thermodynamic potential. We will discuss the simple case $p = \mathrm{b}z$, where z is in the φ_s normal direction.

When the ($A = 0$) iso-affinity surface is flat and in an $\vec{\vec{\mathbf{S}}}\gamma > 0$ direction, the crystal/parent-phase interface will be smooth. But for negative $\vec{\vec{\mathbf{S}}}\gamma$, a kinked interface is preferred. What is the period of the structure?

When the kinks would really have zero free energy and no mutual interaction, the "Manhattan effect" would imply that the period could decrease to atomic size. In other words, the smooth-surface γ plot would coincide with the effective γ plot. If this is not so, there should be an extra energy contribution due to the kinks and their interaction, which for a periodic zigzag structure (period = λ) can be written:

$$\gamma(\lambda) = \gamma_0 + \frac{k}{\lambda} + \frac{\ell}{\lambda^2} + \ \ldots \tag{4.12}$$

Here γ_0 is the surface energy of the effective γ plot, the k term accounts for the extra energy *per period* (i.e., k = excess free energy of one reentrant and one protruding edge) and the following terms for the *interaction energy between edges*. All terms can in principle depend on the interface direction φ_s with respect to the edge orientations φ_{e1} and φ_{e2}.

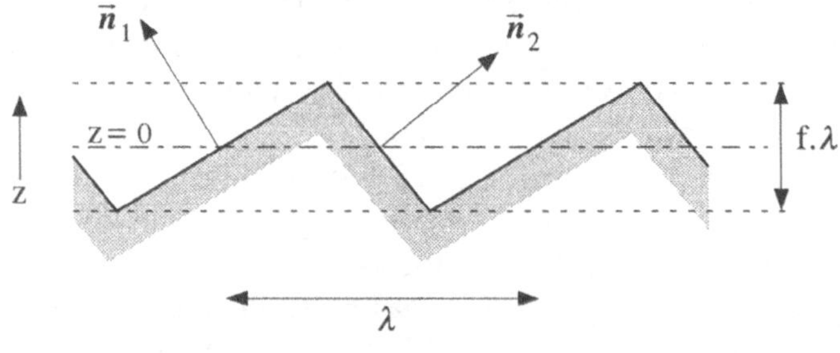

FIG. 4.8

In order to calculate the equilibrium spacing λ, we can proceed from the membrane-and liquid analog. The A-gradient corresponds with a pressure gradient, so the pressure across the interface is lower at the peaks than in the valleys. The energy due to the interface corrugation can thus be expressed as $\int p\mathrm{d}V$. (see 1.5.4 on the concept of a 3-dim. affinity field).

The situation is illustrated in fig. 4.8. It is practical to express the height of the zigzag as a function f of the spacing:

$$f = \frac{1}{\mathrm{cotan}\,(\varphi_{e1} - \varphi_s) + \mathrm{cotan}\,(\varphi_{e2} - \varphi_s)} \tag{4.13}$$

reducing to

$$f \simeq \tfrac{1}{2} \tan \tfrac{1}{2} (\varphi_{e1} - \varphi_{e2}) \tag{4.14a}$$

for orientations about mid-between φ_{e1} and φ_{e2}, and to

$$f = \tan (\varphi_s - \varphi_{e1}) \tag{4.14b}$$

for φ_s orientations near an edge orientation, $(\varphi_s - \varphi_{e1}) \ll (\varphi_{e2} - \varphi_{e1})$.

The mechanical-equilibrium condition at the interface at the edges requires that the average force perpendicular to the surface = 0:

$$\int_{\mathrm{bottom}}^{\mathrm{top}} p(z)\,\mathrm{d}z = 0$$

So, when we define $z = 0$ as the plane where $p = 0$, we have to locate the top (protruding) edges at $z = +\tfrac{1}{2} f\lambda$, the bottom (reentrant) ones at $z = -\tfrac{1}{2} f\lambda$. The energy $\int p\mathrm{d}V$ then becomes, per unit area:

$$\int p\mathrm{d}V = \frac{1}{\lambda} \int \mathrm{b}z \left(\tfrac{1}{2}\lambda - \frac{z}{\mathrm{f}} \right) \mathrm{d}z = \tfrac{1}{8} \mathrm{b}\lambda^2 f^2 \tag{4.15}$$

(Here the zig and zag sections have been considered as straight; however, see 4.2.2).

Summarizing, the A gradient tends to reduce λ, the extra energy due to the kinks tends to increase it. In equilibrium, the spacing adopts a value which minimizes the total free energy (i.e. the sum of (4.12) and (4.15)) with respect to λ:

$$\frac{\mathrm{d}}{\mathrm{d}\lambda} \left(\tfrac{1}{8} \mathrm{b}\lambda^2 f^2 + \frac{k}{\lambda} + \frac{\ell}{\lambda^2} + \ldots \right) = 0$$

or

$$\tfrac{1}{4} \mathrm{b}\lambda^3 f^2 = k + \frac{2\ell}{\lambda} + \frac{3m}{\lambda^2} + \ldots \tag{4.16}$$

When the kink-interaction terms are neglected ($\ell = m = \ldots = 0$), we obtain

$$\lambda_{eq.} = \sqrt[3]{\frac{4k}{bf^2}} \tag{4.17}$$

This result was mentioned by HERRING (1952), by WILCOX (1970) and by NOZIERES (1992).
For b we can read $\vec{\nabla} p$, $S_f \vec{\nabla} T$ (for a crystal-melt interface) or $\vec{\nabla} F$ (for the general case, where the size factor is F^{-1}).

In order to get an impression of the order of magnitude, let us calculate λ for $\vec{\nabla} T = 10°$ K/cm, $S_f = 0.02$ cal/cm³ °K, $f = 0.5$, $\gamma_0 = 100$ erg/cm². For the smooth-surface γ we might *guess* (!) 1.5 γ_0. For the smooth surface, we should take λ of the order of the lattice constant, say 0.4 nm. This corresponds with K = 2.10^{-6} dyne ($\ell = m = ... = 0$). Altogether, for these "technically reasonable" (?) values we arrive at

$$\lambda = \sqrt[3]{3.2 \,.\, 10^{-12}} \quad \text{cm.} = 1.5\,\mu$$

Thus it seems that under fairly common experimental conditions, the zigzag- or pyramid structure might just be observable optically. Indeed, LANDMAN c.s. (1986) report on an equilibrium zigzag structure, with $\lambda \sim 5\mu$, in the Si crystal/melt interface (macroscopic orientation ⟨100⟩, facet zigs and zags defined by (111) and ($\bar{1}\bar{1}1$) planes). The observation was made in an experiment on a 600 nm thick Si film in a quartz cell (quartz wafer below, oxidized Si on top), viewed optically in near infrared. Temperature and temperature gradient were controlled by two elliptically focused laser beams. Unfortunately, the authors presented neither data nor estimates of the temp. gradient involved, so that a comparison with eq. (4.17) cannot be made. On the other hand, they presented an additional theoretical argument for the equilibrium zigzag formation in the form of a Monte Carlo simulation of a comparable structure at "atomic" resolution. This model (although the effective $\vec{\nabla} T$ applied in this case is not evident either) also shows a zigzag structure, but with much smaller period (about 7 lattice units). Altogether, the Si crystal/melt interface may be a suitable model system to investigate zigzag structures both in equilibrium anmd during growth, but a better definition of experimental parameters is essential to check the theory.

I tried to verify zigzag structures myself in several salt and oxide crystal/melt interfaces in thin-cell microscopic experiments, but without success. (this was well before Landman's paper was published in 1986). There may be several reasons which prevent such structures to be observed:
a) Due to the technical difficulty of keeping an isotherm fixed to within an accuracy $\leqslant 1\,\mu$, over a time long enough for the structure to equilibrate and in a low temperature gradient, it may be practically impossible to provide an experimental environment in which an equilibrium zigzag structure

can develop and remain stable (VAN SUCHTELEN, 1978). Rather, the structure would correspond with a growth situation where the locally applied driving force A, statistically fluctuating around zero value, would be strong enough to overrule any equilibration. We would then observe a structure in fluctuating growth/meltback conditions rather than in real equilibrium.
b) Exactly the same argument as a), but with the intrinsic noise field instead of externally induced fluctuations.
c) The time required to reach equilibrium (starting from any random morphology) may be very long. In the first place, the driving force is a nonlinear function of $\Delta\lambda$, the departure of the equilibrium spacing. In the second place an interface for which $\mathbf{S}\gamma < 0$ is likely to be built up of faceted zig and zag sections, obeying a nonlinear $R(A)$ relation. In combination, the response of a zigzagged interface to a change in $\vec{\nabla}A$ is highly nonlinear. Therefore the evolution from an old (equilibrium) periodic structure to a new one with different period, is certainly not simply exponential.

4.2.2 *Stabilisation of* $\mathbf{S}\gamma < 0$ *interfaces by an affinity gradient*

Equation (4.15) is correct only for really faceted (mathematically planar) zig and zag interfaces. For less pronounced orientation dependence ($\mathbf{S}\gamma < 0$ but moderate) the interfaces between the sharp reentrant and protruding edges will be curved, their shape obeying eq. (3.7). This fact complicates both sides in eq. (4.16). Consequently, the calculation of the equilibrium interface morphology in this generalized case is rather complex and will not be further attempted here. (an intuitive picture is given in fig. 4.9). Provisionally, it seems that the period of the structure in this case can rapidly "explode" to infinity under the influence of a critical A gradient, so that an interface which is ($\mathbf{S}\gamma < 0$) - unstable in a $\vec{\nabla}A = 0$ situation, can be stabilized by a high enough positive A gradient. (inversely, a negative A gradient ($\vec{\nabla}A \cdot \vec{n} < 0$) might destabilize an interface with $\mathbf{S}\gamma < 0$, a phenomenon related with constitutional supercooling) We may expect that for even stronger A gradients, the stable/ unstable transition is pushed back to more negative $\mathbf{S}\gamma$ values so that ultimately the whole, originally unstable or "virtual", γ plot is stabilized. The "effective plot", in the case of an A gradient, would not be a circle or sphere through the plot origin, but rather one which is displaced in the direction of $\vec{\nabla}A$. Intuitively, we would expect that in this case a stability criterion might be given in the form

$$\mathbf{S}\gamma + g\,(\vec{\nabla}A \cdot \vec{n}) \begin{cases} > 0 : \text{ stable} \\ < 0 : \text{ unstable} \end{cases} \tag{4.18}$$

for the eigenvalues of the $\vec{\vec{\mathbf{S}}}\gamma$ tensor, where g is a factor relating the importance of $\vec{\vec{\mathbf{S}}}\gamma$ and $\vec{\nabla}A$.

For the moment this is an open question which requires further analysis. It is important to recognize that an essential ingredient in the calculation is the detailed information about the free energy of the pattern of edges, e.g. in the form of an expansion in $1/\lambda$. This information is not included in the $\gamma(\vec{\boldsymbol{n}})$ functions, and has to be supplied separately. The edges, in fact, are "interfaces" in the two-phase system of zigs and zags, as will be discussed below (and in ch. 7)

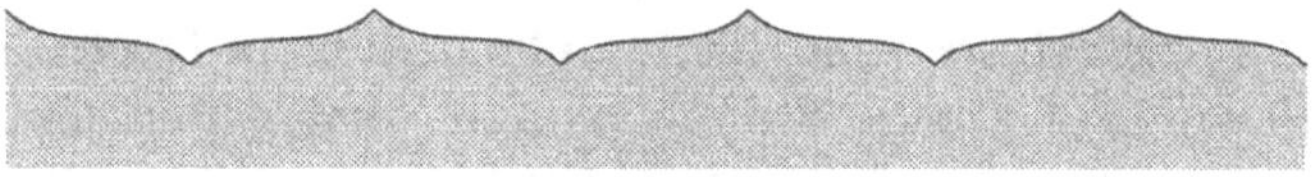

FIG. 4.9

4.2.3 The $\vec{\vec{\mathbf{S}}}\gamma < 0$ instability as a phase transition of the interface.

A zigzag structure can be considered as a two-phase structure, condensed from an originally homogeneous interface system (smooth surface in $\vec{\vec{\mathbf{S}}}\gamma \geq 0$ orientation) when a stability parameter has fallen below a critical level ($\vec{\vec{\mathbf{S}}}\gamma = 0$), bringing the interface system from a supercritical to a subcritical state. The "phases" are interfaces in the two edge orientations, the "interface" between these phases are the (protruding and reentrant) edges. These interfaces (edges) can be considered as new entities having a certain finite size (i.e., radius of curvature) and free energy, as well as interaction behaviour. According to the Landau phenomenology of critical phenomena, the curvature and free energy can be expected to depend on $\vec{\vec{\mathbf{S}}}\gamma$ by specific power laws. When an additional field (such as $\vec{\nabla} A$) works on the stability parameter, the transition can be shifted and a periodic pattern of zigzags can be stabilized, both depending on $(\vec{\nabla} A \cdot \vec{\boldsymbol{n}})$.

(*note:* A periodic zigzag or "spike" structure can actually be generated in a *fluid*/gas interface with artificially induced negative $\mathbf{S}\gamma$: i.e. a "magnetic fluid", a suspension of magnetic dipoles, in a perpendicular magnetic field. Also, the situation is quite comparable to magnetic-domain pattern formation in thin ferromagnetic layers, or to superconducting/normal domain pattern formation in type-I superconductor sheets, also both in the presence of a transverse magnetic field).

This concept can be further "materialized" when we do no longer consider the interface as a mathematical surface, but as a crystal surface below the roughening transition, where different orientations correspond with different ledge densities. Here, the ledges (themselves interfaces between low- and high-density surface layers) can be considered as the entities which can condense, so that the "ledge gas" undergoes a critical transition at $\overset{\Rightarrow}{\mathbf{S}}\gamma = 0$, passing from a supercritical to a subcritical state. In the subcritical state, two phases (i.e., regions of constant ledge density = orientation) can coexist in a certain range of overall ledge density, i.e. orientation. This is just the virtual range as defined in 2.5.

This comparison suggests the question whether the $\overset{\Rightarrow}{\mathbf{S}}\gamma < 0$ condition can and/or does only arise at temperatures below the roughening transition. The answer is affirmative if the surface free energy is related uniquely with the surface structure. However, other contributions may be equally important, such as the contribution by the free-electron or the phonon system, or by "chemical" layers (surface segregation, adsorption, double layers in the parent phase). As the latter contributions can be equally orientation-dependent, $\overset{\Rightarrow}{\mathbf{S}}\gamma < 0$ may well occur at $T > T_R$. More on this topic can be found in the chapter by BENNEMA and VAN DER EERDEN (1987) in the series "Morphology of Crystals".

The identification of the zigzag edges as interfaces (i.e., necessarily having a positive line free energy) has the immediate consequence that the condensation phenomenon, i.e. the transition from a flat to a zigzagged structure, involves a *nucleation barrier*. The height of this barrier, and the kinetics of its transition, have to be found from atomistic models rather than from a continuum model such as can be based on the $\gamma(\vec{n})$ function. This observation clearly illustrates the limitations of the $\vec{\mathbf{Q}}\gamma$-vector formalism. A quite similar limitation will be encountered in the application of the kinematic-wave theory, i.e. the $\vec{\mathbf{Q}}R$-vector formalism, to the kinetic-stability problem of zigzag structures (see 4.6.1). In ch. 7 we will discuss the $\mathbf{S}\gamma < 0$ instability in the wider context of "soliton physics".

Chapter 5:

GROWTH - KINEMATIC WAVE THEORY REVISITED

5.1 INTRODUCTION

In this chapter, on the "kinetics" of surfaces, the formal structure of the equations of motion (in various versions and representations) is developed. Applications and complications of these EOM will be discussed in subsequent chapters. In particular, ch. 6 will be devoted to transitions between equilibrium-determined and kinetically-determined geometrical details, mostly related with the restraints by boundary conditions at junctions and edges. Other applications are collected in a miscellaneous-topics chapter 8.

The structure of the present chapter is summarized in table 5.1. The rows in this table correspond with the three "representations" for EOM, mentioned in ch.1.6: i.e. as "normal" growth rate R, as vector velocity $\vec{\mathfrak{R}}$ and as spreading-rate tensor $\vec{\vec{\mathfrak{S}}}$. On the basis of a function $R(\vec{\boldsymbol{n}},m)$, supposed to be available as an analytic expression or in tabular form, but left unspecified, we will derive equations for $\vec{\mathfrak{R}}$ and for $\vec{\vec{\mathfrak{S}}}$ for three cases of increasing complexity. These cases (to be discussed in sections 5.2, 5.3 and 5.4) correspond with the columns in table 5.1:

section 5.2: R depends on the orientation parameter *only*. In this case, corresponding with the "classical" kinematic-wave theory, $\vec{\mathfrak{R}}$ can be identified with $\vec{\mathbf{Q}}R$ and $\vec{\vec{\mathfrak{S}}}$ with $\vec{\vec{\mathbf{S}}}R$. Kinematic wave trajectories (kwt's) are straight lines, and interface radii change at constant rate along the kwt.

section 5.3: R depends on (one or more) *other parameters* than orientation alone. In section 5.3 the other parameter(s) will be left unspecified and indicated by the "wild card" symbol m. (in practice, m may represent quantities such as temperature, concentration of chemical species or "affinity". Parameters such as temperature and concentration may affect the affinity A and the "mobility χ - in the formal relation $R = \chi . A$ - simultaneously!)

If the m field is homogeneous, we just have the case of 5.2; when it is inhomogeneous the expressions $\vec{\mathbf{Q}}R$ (for $\vec{\mathfrak{R}}$) and $\vec{\vec{\mathbf{S}}}R$ (for $\vec{\vec{\mathfrak{S}}}$) have to be supplemented with other terms involving the surface gradient of m.

	5.2	5.3	5.4
$R_\perp =$	$R(\vec{n})$	$R(\vec{n},A)$	$R(\vec{n},A+\Delta A_c)$
$\vec{\mathfrak{R}} =$	$\vec{\mathsf{Q}}R(\vec{n})$	$\vec{\mathsf{Q}}R(\vec{n},A) +$ expr.$<\vec{\nabla}_s A>$	$\mathsf{Q}R(\vec{n},A+\Delta A_c) +$ expr.$<\vec{\nabla}_s A, \vec{\nabla}_s(\Delta A_c)>$
$\vec{\vec{\mathfrak{S}}} =$	$\vec{\vec{\mathsf{S}}}R(\vec{n})$	$\vec{\vec{\mathsf{S}}}R(\vec{n},A) +$ expr.' $<\vec{\nabla}_s A>$	$\vec{\vec{\mathsf{S}}}R(\vec{n},A+\Delta A_c) +$ expr.' $<\vec{\nabla}_s A, \vec{\nabla}_s(\Delta A_c)>$
	case[A]	case [B1] ("imposed *A* map") case [B2] ("Stefan" case)	case [C1] ("imposed *A* map") case [C2] ("Stefan" case) case [D] ($A_e = 0$; ΔA_c only)

TABLE 5.1

It should be kept in mind that only in rather exceptional cases m can be given explicitly as a function of position. (e.g. a temperature field, undistorted by the interface, can be imposed when the thermal conductivities of both phases are equal). Mostly the m field results from temperature and/or concentration boundary conditions *imposed at some distance* from the interface. Then the m at the interface is the result of the combined effects of volume-diffusion processes and the interface kinetics - in other words, the interface EOM, with a priori unknown m, has then to be considered as a boundary condition for the volume-diffusion problem.

In the case of inhomogeneous m, kwt's can still be defined, but they are no longer straight. Also, they can be created or annihilated at inflection points of the interface, and finally they can trifurcate, whereby inflection points are created or annihilated. During the formation of sidebranches in dendritic growth, such a trifurcation is repeated periodically.

section 5.4: In 5.2 and 5.3, the effects of surface free energy were neglected. This omission can be remedied with the aid of ch.3, where explicit expressions for the curvature-induced affinity shift were given. In order to write down an "exact" EOM - i.e. including capillarity effects - we have to make the special choice $m = A$ (affinity) in 5.3. Then the exact EOM is obtained by applying the ΔA_c shift to the A_g parameter ($= A$ in $R(\vec{n},A)$) wherever appropriate. When another m parameter is used as a "second" parameter, the effect of affinity still has to be made explicit for the procedure of 5.4 to be practical.

"Casuistics":

For later reference it is practical to have a shorthand notation for reference to the various cases in table 5.1. This notation is given in the last row of the table. In principle, the cases represented by columns 5.2, 5.3 and 5.4 are indicated as A, B and C respectively; the special case of 5.4 without any external field (ageing and sintering) as D. Moreover, for cases B and C we discern between the "imposed-affinity" (B_1 and C_1) and the "Stefan" cases (B_2 and C_2). In the EOM (in $\vec{\Re}$ or in $\vec{\vec{\mathfrak{S}}}$ form) we can, of course, distinguish the extra terms which discern cases B from A and cases C or D from B.

Again we emphasize that the equations of motion in 5.2-5.4 are "formal" equations in the sense that $R(\vec{n})$, $R(\vec{n},m)$ or $\gamma(\vec{n})$ functions are left unspecified. For practical calculations we need explicit expressions for the $\vec{n}$ and for the m dependences of these OD parameters.

5.2 *THE VECTOR VELOCITY IN THE "CLASSICAL" KINEMATIC WAVE THEORY*

In this section the A and/or m parameters are considered to be constant, curvature effects included (i.e., the curvature shift ΔA_c is neglected), so that we have to do with one unique growth rate function $R(\vec{n})$, valid for the given A. The meaning of the vector $\vec{\mathbf{Q}}R(\vec{n})$ and the tensor $\vec{\vec{\mathbf{S}}}R(\vec{n})$ can now be demonstrated directly by constructing the shape of a crystal at time t, which has started at $t = 0$ from an infinitesimally small nucleus. As the interface with orientation $\vec{n}$ has advanced a distance $t.R(\vec{n})$ along the $\vec{n}$ direction, the crystal will touch the plane perpendicular to $\vec{n}$ which intersects the normal at a distance $t.R(\vec{n})$ from the origin. In other words, the growth shape is constructed from the $\vec{\mathbf{P}}R(\vec{n})$ plot by the very construction which yields the $\vec{\mathbf{Q}}R(\vec{n})$ shape, the size factor occurring between the $\vec{\mathbf{Q}}R(\vec{n})$ surface and the actual surface being simply the growth time t. Remembering the definition of the kinematic wave trajectory (a function of orientation) as the direction followed by a point of a certain fixed orientation in a curved interface, we can now, recognizing the similarity of the growth shapes at all times, identify this kwt with the direction of $\vec{\mathbf{Q}}R(\vec{n})$. The $\vec{\Re}$ vector growth rate is then identical with $\vec{\mathbf{Q}}R$. (The projection of $\vec{\Re}$ on the normal is the scalar growth velocity $R(\vec{n})$ again). The vector velocity was introduced by CHERNOV

(1963b). Further, $t.\vec{\vec{\mathbf{S}}}R(\vec{n})$ is the inverse of the second fundamental tensor of the growth shape; i.e. $\vec{\vec{\mathbf{S}}}R(\vec{n})$, its time derivative, is equal to the rate at which this tensor changes, (the growing surface being followed along a kwt). In other words, $\vec{\vec{\mathbf{S}}}R$ is the **S**_preading rate tensor_ $\vec{\vec{\mathfrak{S}}}$ introduced in 1.6.

The identities

$$\left.\begin{aligned} \vec{\mathfrak{R}} &= \vec{\mathbf{Q}}R \qquad (5.1)\\ \vec{\vec{\mathfrak{S}}} &= \vec{\vec{\mathbf{S}}}R \qquad (5.2) \end{aligned}\right\} \text{ if } R = R(\vec{n},m)\text{: } \textit{for homogeneous m only)}$$

apply to the growing crystal *independently of time*; i.e. also independently of the curvature of the interface. This implies that the identities (5.1) and (5.2) are also valid along a surface of *arbitrary* shape: the quantities $\vec{\mathfrak{R}}$ and $\vec{\vec{\mathfrak{S}}}$ can in this case be considered as surface properties dependent on orientation only. *This uniqueness applies exclusively to the case of uniform affinity and/or other growth-rate influencing parameters.*

Tilt rate:

As the growth rate of neighbouring points (having, in general, dissimilar orientations) is different, the interface, when followed along the original *normal*, will tilt. The rotation of the normal is directly obtained by taking the difference between R at the positions $\vec{x}$ and $\vec{x}+\mathrm{d}\vec{x}$, and dividing by $\mathrm{d}\vec{x}$ (see fig. 5.1) :

$$\left(\frac{\mathrm{d}\vec{n}}{\mathrm{d}t}\right)_{(\vec{n})} \left(= \frac{\mathrm{d}\alpha}{\mathrm{d}t} \text{ in fig. 5.1}\right) = -\frac{\mathrm{d}R}{\mathrm{d}\vec{x}} = \vec{\vec{K}}\cdot\frac{\mathrm{d}R}{\mathrm{d}\vec{n}} \qquad (5.3)$$

or, for the rotation rate:

$$\vec{\omega}_{(\vec{n})} = \vec{n} \wedge \vec{\vec{K}}\cdot\frac{\mathrm{d}R}{\mathrm{d}\vec{n}} \qquad (5.4)$$

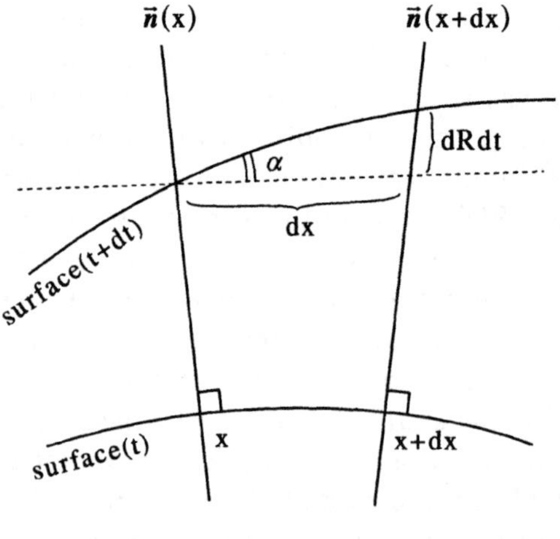

FIG. 5.1

(the suffix $(\vec{n})$ in a time derivative indicates that the evolution of the surface is followed along the normal). The vector $\mathrm{d}R/\mathrm{d}\vec{n}$ is the tangential component of $\vec{\mathsf{Q}}R(\vec{n})$, or $(\vec{\mathsf{Q}}-\vec{\mathsf{P}})R(\vec{n})$.

The kinematic wave trajectory can now be found by looking for the (tangential) velocity $\mathrm{d}\vec{x}/\mathrm{d}t$ at which we have to travel along the moving surface in order to keep the orientation constant, i.e. to cancel the tilt given by (5.3). Obviously this velocity is just

$$\frac{\mathrm{d}\vec{x}}{\mathrm{d}t} = \overrightarrow{\overrightarrow{K}}^{-1} \cdot \left(\frac{\mathrm{d}\vec{n}}{\mathrm{d}t}\right)_{(\vec{n})} = \frac{\mathrm{d}R}{\mathrm{d}\vec{n}} \tag{5.5}$$

This confirms again that $\vec{\mathsf{Q}}R(\vec{n})$ and the kwt have identical directions. (note that the kwt depends on $\vec{n}$ only, not on $\overrightarrow{\overrightarrow{K}}$. This independence is lost in the inhomogeneous-m case, as we will see in the next section).

(Note: In Frank's original version of the kinematic-wave theory, tan φ (= p in Frank's paper (FRANK 1958)) is taken as the orientation variable instead of φ itself. This choice was probably suggested by two circumstances: 1) the kwt theory was developed primarily for the study of crystal surfaces with near-facet orientations so that the range of interest in φ is small, and 2) the actual $R(\varphi)$ plots were correlated with models on the basis of the BCF paper (BURTON, CABRERA and FRANK 1951), where tan φ ($\varphi = 0$ for the facet direction) is a "natural" variable, being proportional to the ledge density. By this choice, the motion of the interface is represented in a fixed cartesian coordinate frame: $z(x,t)$. In this version, the problem of writing an EOM for the interface is equivalent to a Cauchy problem for a first-order partial differential equation, which can be solved by the method of characteristics (i.e., kwt's). Of course, the latter representation is fully equivalent to our vector representation, although we prefer the φ or $\vec{n}$ above the p variable because it does not suggest any distinctive orientation $p = 0$ and because the algebraic representation of eqs. (5.1)-(5.5) in cartesian coordinates is somewhat less transparent than a vector representation for the general case.)

5.3 THE INHOMOGENEOUS-AFFINITY CASE

5.3.1 Basic relations

When A and m fields are homogeneous, kwt's are straight lines and can be "annihilated" by collision only. In the present section we consider the general case of inhomogeneous A and/or m fields. We can then no longer derive expressions for $\mathfrak{R}$ and $\overrightarrow{\overrightarrow{\mathfrak{S}}}$ by just applying the geometrical similarity with the Gibbs-Wulff construction; instead we have to find the general expressions for the "gothic" symbols from first principles.

The extra terms in the expressions for $\vec{\mathfrak{R}}$ and $\vec{\vec{\mathfrak{S}}}$ are related with the typical features of curvature, generation/annihilation and trifurcation of kwt's in the inhomogeneous A and/or m case. In the following, we use m throughout as a symbol for the "second" parameter; it is irrelevant whether it stands for A, m or a combination. (m can also be a "vector" with components symbolizing more than one inhomogeneous parameters).

Consider again fig. 5.1. In order to calculate the tilt rate of the surface (just as in 5.2, but now for the general case) we start by writing out dR:

$$\mathrm{d}R = \mathrm{d}\vec{\boldsymbol{n}}\,\frac{\partial R}{\partial \vec{\boldsymbol{n}}} + \mathrm{d}m\,\frac{\partial R}{\partial m} \tag{5.6}$$

$$= \mathrm{d}\vec{\boldsymbol{n}}\,(\vec{\mathsf{Q}}-\vec{\mathsf{P}})\,R(\vec{\boldsymbol{n}},m) + \mathrm{d}\vec{s}\,.\vec{\nabla}_{\mathrm{s}}\,m\,\frac{\partial R}{\partial m} \tag{5.7}$$

resulting in a tilt rate

$$\left(\frac{\mathrm{d}\vec{\boldsymbol{n}}}{\mathrm{d}t}\right)_{(\vec{\boldsymbol{n}})} = \vec{\vec{\boldsymbol{K}}}\cdot\frac{\partial R}{\partial \vec{\boldsymbol{n}}} + \vec{\nabla}_{\mathrm{s}}\,m\,\frac{\partial R}{\partial m} \tag{5.8a}$$

Following the same procedure as in 5.1, we can find $\vec{\mathfrak{R}}$ by moving along the surface with a velocity d$\vec{s}$/dt which just compensates the d$\vec{\boldsymbol{n}}$/dt given by (5.8a):

$$\frac{\mathrm{d}\vec{s}}{\mathrm{d}t} = \frac{\partial R}{\partial \vec{\boldsymbol{n}}} + (\vec{\vec{\boldsymbol{K}}}^{-1})\cdot\vec{\nabla}_{\mathrm{s}}m\,\frac{\partial R}{\partial m} \tag{5.9}$$

so that

$$\vec{\mathfrak{R}}\ \ (= \vec{\boldsymbol{n}}R + \mathrm{d}\vec{s}/\mathrm{d}t) = \vec{\mathsf{Q}}R + (\vec{\vec{\boldsymbol{K}}}^{-1})\cdot\vec{\nabla}_{\mathrm{s}}m\,\frac{\partial R}{\partial m} \tag{5.10a}$$

In order to find the general expression for $\vec{\vec{\mathfrak{S}}}$ we have to consider the convergence or divergence of $\vec{\mathfrak{R}}$ vectors. In the basic relation (1.3), we thus have to substitute $\vec{\nabla}_{\mathrm{s}}\vec{\mathfrak{R}}$:

$$\vec{\nabla}_{\mathrm{s}}\vec{\mathfrak{R}} = \vec{\vec{\boldsymbol{K}}}\cdot\frac{\partial \vec{\mathfrak{R}}}{\partial \vec{\boldsymbol{n}}} + \vec{\nabla}_{\mathrm{s}}m\,\frac{\partial \vec{\mathfrak{R}}}{\partial m} \tag{5.11}$$

where we have to insert (5.10a) for $\vec{\mathfrak{R}}$. This results in

$$\vec{\vec{\mathfrak{S}}} = \vec{\vec{\boldsymbol{K}}}^{-1}\cdot\vec{\nabla}_{\mathrm{s}}\,\vec{\mathfrak{R}} = \ \ldots\ =$$

$$= \vec{\vec{\mathsf{S}}}R(\vec{\boldsymbol{n}},m) + \frac{\partial R(\vec{\boldsymbol{n}},\mathrm{m})}{\partial m}\,[\,\{\,(\,\vec{\vec{\boldsymbol{K}}}^{-1}\cdot\vec{\nabla}_{\mathrm{s}}\,\vec{\vec{\boldsymbol{K}}}^{-1})\cdot\vec{\nabla}_{\mathrm{s}}m\} + \{\vec{\vec{\boldsymbol{K}}}^{-1}\cdot(\vec{\vec{\boldsymbol{K}}}^{-1}\cdot\vec{\nabla}_{\mathrm{s}}^{2}m)\}] +$$

$$+\,2\,\frac{\partial^2 R(\vec{\boldsymbol{n}},m)}{\partial m\partial \vec{\boldsymbol{n}}}\,(\vec{\vec{\boldsymbol{K}}}^{-1}\cdot\vec{\nabla}_{\mathrm{s}}m) + \frac{\partial^2 R(\vec{\boldsymbol{n}},m)}{\partial m^2}\,(\vec{\vec{\boldsymbol{K}}}^{-1}\cdot\vec{\nabla}_{\mathrm{s}}m)^2 \tag{5.12a}$$

The expressions for $\vec{\mathfrak{R}}$ and $\vec{\vec{\mathfrak{S}}}$ and for the tilt rate for the "cylindrical" case (one orientation variable, for which we then can use the angular variable φ instead of $\vec{\boldsymbol{n}}$) are

$$\vec{\mathfrak{R}} = \vec{\mathbf{Q}}R(\varphi,m) + r\frac{\partial R(\varphi,m)}{\partial m}\frac{\mathrm{d}m}{\mathrm{d}x} \tag{5.10b}$$

$$\mathfrak{S} = \mathbf{S}\,R(\varphi,m) + \frac{\partial R(\varphi,m)}{\partial \varphi}\left(r\frac{\mathrm{d}r}{\mathrm{d}x}\frac{\mathrm{d}m}{\mathrm{d}x} + r^2\frac{\mathrm{d}^2 m}{\mathrm{d}x^2}\right) + $$

$$+\,2r\frac{\partial^2 R(\varphi,m)}{\partial m\partial\varphi}\frac{\mathrm{d}m}{\mathrm{d}x} + \frac{\partial^2 R(\varphi,m)}{\partial m^2}r^2\left(\frac{\mathrm{d}m}{\mathrm{d}x}\right)^2 \tag{5.12b}$$

$$\left(\frac{\mathrm{d}\varphi}{\mathrm{d}t}\right)_{(\vec{n})} = \frac{1}{r}\frac{\partial R(\varphi,m)}{\partial\varphi} + \frac{\mathrm{d}m}{\mathrm{d}x}\frac{\partial R(\varphi,m)}{\partial m} \tag{5.8b}$$

where x is the surface coordinate. Obviously, (5.10a/b) and (5.12a/b) reduce to (5.1) and (5.2), respectively, when the m field is homogeneous.

Note:

Of course we might also derive expressions for the evolution of $\vec{\vec{K}}^{-1}$ or $\vec{\vec{K}}$ when following a surface along the *normal* direction, $(\mathrm{d}/\mathrm{d}t)_{(\vec{n})}(\vec{\vec{K}}^{-1})$ etc. However, time derivatives along the normal are more or less to be considered as artefacts, mixing orientational derivatives of the growth rate function with those of the surface under consideration. The choice of following evolution along the kwt (definition of $\vec{\mathfrak{R}}$ and $\vec{\vec{\mathfrak{S}}}$) is a logical choice, from a "mathematically economic" point of view, when many *parameters occurring in the analytic expression of* $R(\vec{n},m)$ depend on $\vec{n}$ only.

5.3.2 Geometrical behaviour of kinematic wave trajectories

We assume that $\partial R/\partial m$ is nonzero and that m is a continuous function of the surface coordinate(s) with continuous first and second derivatives, which may have extrema in the region considered. The phenomenology of kwt behaviour can now directly be related with the $\partial R/\partial m$ terms in eqs. (5.10a/b) and (5.12a/b). For convenience, we take the "1-dimensional" equations as a point of departure:

1. *Curvature of kwt's* is related with
   a) the eventually inhomogeneous m variable, affecting the $\vec{\mathbf{Q}}R$ term in (5.10a/b);
   b) the rate of change of the second term in (5.10a/b), brought about by $r\,\mathrm{d}m/\mathrm{d}x$. (all terms to be taken at the kwt/interface intersection).

2. *Creation and annihilation*:
   Kwt's can be tangent to the surface, i.e. $(\vec{\mathbf{Q}}R)_{\mathrm{tang}} = \infty$ when $1/r = 0$. This can either be the case at an inflection point $(\mathrm{d}(r^{-1}/\mathrm{d}x \neq 0)$ or at a touching point $(\mathrm{d}(r^{-1}/\mathrm{d}x = 0)$.

At an inflection point, the interface orientation is an extremum within a finite range at both sides of this point; at a touching point it is not). As the sign of $1/r$ is opposite at both sides of the inflection point (whereas $\mathrm{d}m/\mathrm{d}x$, in general, is finite and does not change sign), we have two touching kwt's which meet at the inflection point. (fig. 5.2). This means that along the trajectory of the inflection point, kwt's are created or annihilated. (created when the tilt direction $(\mathrm{d}R/\mathrm{d}m).(\mathrm{d}m/\mathrm{d}x)$ extends the extremum orientation, annihilated when it regresses).

3. *Trifurcation or generation of inflection points*
   Consider a region of the surface where the curvature $1/r$ is everywhere either positive or negative. (i.e., which contains no inflection points). How are inflection points generated?
   A straightforward way to follow the evolution of curvature is to integrate $(\mathrm{d}/\mathrm{d}t)_{\mathrm{kwt}}\,(r^{-1}) = -r^{-2}\,\mathfrak{S}$:

$$\frac{1}{r}(t) = \frac{1}{r}(t=0) + \int_{t=0}^{t} \left(-\frac{1}{r^2}\,\mathfrak{S} \right) \mathrm{d}t \qquad (5.13)$$

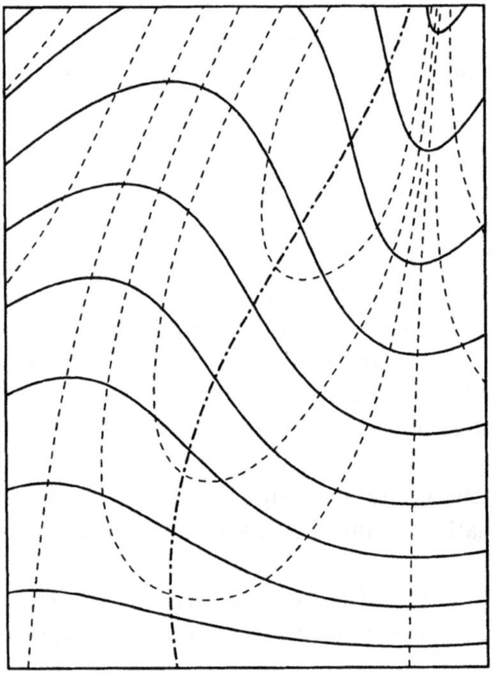

FIG. 5.2

——— surface
- - - - - kinematic wave trajectory
-·-·-·-·- inflection point trajectory

A straightforward way to follow the evolution of curvature is to integrate $(\mathrm{d}/\mathrm{d}t)_{\mathrm{kwt}}\,(r^{-1}) = -r^{-2}\,\mathfrak{S}$:

$$\frac{1}{r}(t) = \frac{1}{r}(t=0) + \int_{t=0}^{t}\left(-\frac{1}{r^2}\,\mathfrak{S}\right)\mathrm{d}t \qquad (5.13)$$

where the integrand is to be followed along the kwt. (for the single-variable case, we write $\mathfrak{S}$ as a scalar, i.e. a 1 x 1 tensor with just one element) This integral, apart from terms containing a factor $1/r$, has one term which is independent of $1/r$, i.e.

$$-\frac{1}{r^2}\,\mathfrak{S} = \mathrm{O}\left(\frac{1}{r}\right) - \left\{\frac{\partial R}{\partial m}\,\frac{\mathrm{d}^2 m}{\mathrm{d}x^2} + \frac{\partial^2 R}{\partial m^2}\left(\frac{\mathrm{d}m}{\mathrm{d}x}\right)^2\right\} \qquad (5.14)$$

Thus at first sight, it would seem that the value of $1/r$, following any kwt, would be able to pass through zero with finite speed, when this term keeps the appropriate sign long enough. However, as $1/r$ approaches zero, the tangential component of $\vec{\mathbf{Q}}R$ goes to infinity (for nonzero $\mathrm{d}m/\mathrm{d}x$). When we follow one particular kwt we can define its "lateral motion" $\mathrm{d}x^*/\mathrm{d}t$ resulting from the second term in eq.(2.11a) as measured from the $\vec{\mathbf{Q}}R$ vector. In doing so, we can calculate $1/r$ as a function of x^* by substituting

$$\mathrm{d}t = \frac{\mathrm{d}x}{(\vec{\mathbf{Q}}R)_{\mathrm{tang.}}} = \frac{1}{r}\,\frac{1}{(\partial R/\partial m).(\mathrm{d}m/\mathrm{d}x)}\,\mathrm{d}x \qquad (5.15)$$

This adds a factor $1/r$ in the integrand, and now *all* terms are $\mathrm{O}(1/r)$. The conclusion is:

In regions where both $1/r$ and $\mathrm{d}m/\mathrm{d}x$ are nonzero everywhere, $1/r$ can approach zero only asymptotically; i.e. inflection points cannot be generated.

However, the situation changes fundamentally when $m(x)$ has extrema in the region considered. (fig. 5.3). At the position of the extremum, $(\vec{\mathbf{Q}}R)$ is zero whatever the value of $1/r$. Among all the kwt's of the interface, there is one which keeps a nonzero angle with the interface all the time and where, consequently, the value of $1/r$ passes zero with finite speed.

At the moment when $1/r = 0$, the point where this is reached coincides with the extremum of $m(x)$. Hence, $(\vec{\mathbf{Q}}R)_{\mathrm{tang.}}$ is indeterminate, but to the left and to the right, $(\vec{\mathbf{Q}}R)_{\mathrm{tang.}}$ is infinite and of opposite directions. ($1/r \sim x^2$, $\mathrm{d}m/\mathrm{d}x \sim x$ as $\mathrm{d}^2m/\mathrm{d}x^2$ is in general finite).

This particular kwt proceeds in the same direction, now in a region of opposite curvature. To the left and to the right, there are inflection points (generating new kwt's). Beyond these inflection points, there are positions with the original orientation; i.e. the kwt's of these points are split off from the original kwt at the point where the value $1/r = 0$ was reached. In other words, we have a trifurcation of the kwt, and simultaneously the birth of two inflection points where new kwt's are generated as discussed in 2).

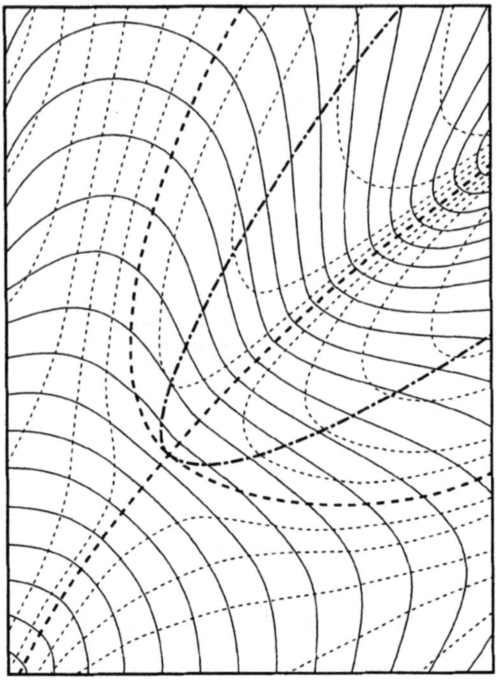

FIG. 5.3

————	surface
- - - - - - -	kinematic wave trajectory
—·—·—·—	inflection point trajectory
— — — —	trifurcating kwt

Of course, the "trifurcation procedure" can be followed in the opposite direction. A region of opposite curvature is thereby annihilated; the inflection points cancel at the trifurcation point together with two kwt's. The latter case corresponds with a maximum of $m(x)$, the "forward" trifurcation with a minimum ($\partial R/\partial m$ assumed to be positive).

Note that the trifurcation, i.e. the creation or annihilation of inflection points in a surface, or in other words the *growth or decrease of waves*, is closely related with growth or decay of instabilities.

5.3.3 Numerical simulation of growth or epitaxy: case A versus cases B/C/D

A striking difference between cases [A] on the one hand and cases [B],[C] or [D] (from table 5.1) on the other hand is made evident when we want to simulate epitaxial growth on a single-crystalline seed of given initial shape.

In case [A], R is only dependent on $\vec{n}$; i.e. independent of time and of curvature of the surface. We can therefore construct the surface of the growing structure *after an arbitrarily long growth time t in a single step,* just by displacing points on the surface by the vector $t.\vec{\mathrm{Q}}R(\vec{n})$. (for a perfectly flat area we need only one vector over which to displace this area). The only extra step we have to make after the displacement of the initial surface is the elimination of "virtual regions", i.e. the parts of the constructed surface shape outside the self-intersections (see ch. 2.7). The required "grid density" of points from which to start the $t.\vec{\mathrm{Q}}R(\vec{n})$ vectors is related only with the required accuracy; it can be expressed in terms of $\Delta\varphi$ of the surface and of the orientation dependence of the $R(\vec{n})$ function.

In contrast, in cases [B] and following, R also depends on an m parameter which can vary as a function of position (also over a flat area) and as a function of time. As a consequence, the $\vec{\mathfrak{R}}$ vector varies in length and direction as a function of time; kwt's are in general curved. This means that we have to construct the evolution by small time steps: Δt has to be chosen small enough for the curved kwt to follow the initial direction and velocity close enough. (as an alternative, we may *calculate* the first-order corrections in Δt for the components of $\vec{\mathfrak{R}}$ and so follow the kwt in a "first-order approximation"; but still in this case the Δt steps are limited). But not only the time intervals, also the grid density of surface points from which to start has to be high enough to ensure that m differences do not affect R too much, even on a flat area. This implies that, after a time step, we may have to interpolate new starting points on the surface when neighbouring kwt's diverge, or remove them when kwt's converge or collide.

After each time step, the m field on the (new) surface has to be determined as a function of position. In this way we can calculate $R(\vec{n},m)$ for any point on the new surface, and as we can also calculate r, $\vec{\nabla}_s r$ and $\vec{\nabla}_s m$, we can find the vector $\vec{\mathfrak{R}}$ for any point on the new surface from equation (5.10 a/b) or (5.17 a/b), and so work out the next time step.

The difference between the cases [B1]/[C1] and [B2]/[C2] lies in the procedure which is required to find the m value for each point on the interface. In the [B1] case, the m value can just be read from the "m map". In the [C1] case an additional calculation of ΔA_c on the basis of the (then known) orientation and curvature is required (here we anticipate on section 5.4).

Anyhow, in cases [B1], [C1] and [D] the expression for $\vec{\mathfrak{R}}$ as a function of position can *in principle* be considered as an explicit relation, although "tedious algebra is involved" (position and orientation coordinates $\rightarrow A$ value $\rightarrow R(\vec{n},A) \rightarrow (+r, \vec{\nabla}_s r, \vec{\nabla}_s A) \rightarrow \vec{\mathfrak{R}}$). The transparency of such an expression is likely to be poor and its applicability will practically be limited to numerical calculation rather than functional analysis for soliton solutions, except maybe in highly "stripped" versions. In contrast, in cases [B2] and [C2] we have essentially to do with coupled equations for the interface and for some diffusion field. One Δt step of interface advancement requires *two* alternating numerical simulations, one for the interface EOM and one for

FIG. 5.3

the diffusion "EOM" (the latter in agreement with the shifted interface and the appropriate continuity and preservation conditions). The time interval has to be short enough not only for the interface EOM to be precise enough but also for the diffusion-field simulation. Fig. 5.3 is an "artist's impression" (not a real numerical simulation) of such an iterative procedure for the simulation of dendritic growth, again illustrating the features of generation, annihilation and trifurcation of kinematic wave trajectories.

5.4 CORRECTION FOR (ORIENTATION-DEPENDENT) CAPILLARITY

The equations given in 5.3 can now be extended to include the effect of curvature-induced affinity shift by substituting eq. (3.7b) into the argument of the $R(\vec{n},A)$ function (where $A = A_g = A_l - \Delta A_c$) - at least, when the surface free energy is independent of the affinity or any other variable parameter except orientation. In this procedure, we can substitute for $\partial/\partial\vec{n}(\Delta A_c)$:

$$\frac{\partial}{\partial\vec{n}}(\Delta A_c) = V\frac{\mathrm{d}}{\mathrm{d}\vec{n}}(\overline{\overline{K}} : \overline{\overline{\mathbf{S}}}\gamma) = V(\overline{\overline{K}}\cdot\vec{\nabla}_s\overline{\overline{K}}^{-1})\cdot\overline{\overline{\mathbf{S}}}\gamma + V(\overline{\overline{K}} : \frac{\mathrm{d}}{\mathrm{d}\vec{n}}\overline{\overline{\mathbf{S}}}\gamma) \quad (5.16a)$$

or, for one orientation variable

$$\frac{\partial}{\partial\varphi}(\Delta A_c) = \frac{V}{r}\left\{\frac{\mathrm{d}\gamma}{\mathrm{d}\varphi} + \frac{\mathrm{d}^3\gamma}{\mathrm{d}\varphi^3} - \frac{\mathrm{d}r}{\mathrm{d}x}\left(\gamma + \frac{\mathrm{d}^2\gamma}{\mathrm{d}\varphi^2}\right)\right\} \quad (5.16b)$$

Note that this substitution is correct only when γ, although orientation-dependent, is independent of the affinity; $\partial\gamma/\partial A = 0$. If this condition is not fulfilled, i.e. if $\gamma = \gamma(\vec{n},A)$, then eq. (3.7b) is not an explicit equation for ΔA_c because $\mathbf{S}\gamma$ is affected by ΔA_c in a "feedback loop". Only for special $\gamma(\vec{n},A)$ functions can we then find an explicit expression for ΔA_c as a generalisation of (3.7b) which is then to be considered as a differential equation for ΔA_c . In view of the mathematical complexity of this case we will not further consider it in this chapter. However, we have to bear in mind that a $\gamma(A)$ dependence is not unrealistic physically: in particular when the surface conditions are close to the roughening transition. In that case, even the "exact" generalized EOM's 5.17-5.19 do not correctly account for the capillarity shift.

(Similar complications arise when γ depends on another parameter m which is nonuniform in the volume considered, but not necessarily connected with the growth affinity (i.e., the "Marangoni" case). We will not discuss this case either).

With these $\partial\gamma/\partial A = 0$, $\partial\gamma/\partial m_\gamma = 0$ restrictions, we have the following explicit equations:

a) *Vector growth rate:*

$$\vec{\Re} = \overline{\mathbf{Q}}R(\vec{n},A-\Delta A_c) + \frac{\partial R}{\partial A}(\overline{\overline{K}}^{-1}\cdot\vec{\nabla}_s A_e) +$$
$$- V\frac{\partial R}{\partial A}\left\{\left(\overline{\overline{K}}^{-1}\cdot\vec{\nabla}_s\overline{\overline{K}}\right) : \overline{\overline{\mathbf{S}}}\gamma + \overline{\overline{K}} : \frac{\mathrm{d}}{\mathrm{d}\vec{n}}\overline{\overline{\mathbf{S}}}\gamma\right\} \quad (5.17a)$$

For one orientation variable the expression reads: (x = surface coordinate)

$$\vec{\mathfrak{R}} = \vec{\mathbf{Q}}R(\varphi, A-\Delta A_c) + \frac{\partial R}{\partial A}\, r\frac{\mathrm{d}A_|}{\mathrm{d}x} + \\ - \frac{V}{r}\frac{\partial R}{\partial A}\left\{\frac{\mathrm{d}\gamma}{\mathrm{d}\varphi} + \frac{\mathrm{d}^3\gamma}{\mathrm{d}\varphi^3} - \frac{\mathrm{d}R}{\mathrm{d}x}\left(\gamma + \frac{\mathrm{d}^2\gamma}{\mathrm{d}\varphi^2}\right)\right\} \qquad (5.17b)$$

(In 5.17 a and b, all terms except the first are tangential. The third term accounts for the capillarity effect).

b) *Tilt rate* (in the normal direction):

$$\left(\frac{\mathrm{d}\varphi}{\mathrm{d}t}\right)_{(\vec{n})} = \vec{\vec{K}}\frac{\mathrm{d}R}{\mathrm{d}\vec{n}} = \vec{\vec{K}}\frac{\partial R}{\partial \vec{n}} + \vec{\nabla}_s A_|\frac{\partial R}{\partial A} + \\ + V\frac{\partial R}{\partial A}\left\{\vec{\nabla}_s\vec{\vec{K}} : \vec{\vec{\mathbf{S}}}\gamma + \vec{\vec{K}}\cdot(\vec{\vec{K}} : \frac{\mathrm{d}}{\mathrm{d}\vec{n}}\vec{\vec{\mathbf{S}}}\gamma)\right\} \qquad (5.18a)$$

or for one orientation variable:

$$\left(\frac{\mathrm{d}\varphi}{\mathrm{d}t}\right)_{(\vec{n})} = \frac{1}{r}\frac{\partial R}{\partial \varphi} + \frac{\mathrm{d}A_|}{\mathrm{d}x}\frac{\partial R}{\partial A} + \\ + \frac{V}{r^2}\frac{\partial R}{\partial A}\left\{-\frac{\mathrm{d}r}{\mathrm{d}x}\left(\gamma + \frac{\mathrm{d}^2\gamma}{\mathrm{d}\varphi^2}\right) + \frac{\mathrm{d}\gamma}{\mathrm{d}\varphi} + \frac{\mathrm{d}^3\gamma}{\mathrm{d}\varphi^3}\right\} \qquad (5.18b)$$

c) *Spreading rate:*

$$\vec{\vec{\mathfrak{S}}} = \vec{\vec{\mathbf{S}}}R(\vec{n}, A_|-\Delta A_c) + \frac{\partial R}{\partial A}\left[\{(\vec{\vec{K}}^{-1}\cdot\vec{\nabla}_s\vec{\vec{K}}^{-1})\cdot\vec{\nabla}_s A_| + \{\vec{\vec{K}}^{-1}\cdot(\vec{\vec{K}}^{-1}\cdot\vec{\nabla}_s^2 A_|)\}\right] + \\ + 2\frac{\partial^2 R}{\partial A\partial\vec{n}}(\vec{\vec{K}}^{-1}\cdot\vec{\nabla}_s A_|) + \frac{\partial^2 R}{\partial A^2}(\vec{\vec{K}}^{-1}\cdot\vec{\nabla}_s A_|)^2 + \\ + V\frac{\partial R}{\partial A}\left[(\vec{\vec{K}}^{-1}\cdot\vec{\nabla}_s\vec{\vec{K}}^{-1}\cdot\vec{\nabla}_s\vec{\vec{K}}) : \vec{\vec{\mathbf{S}}}\gamma + (\vec{\vec{K}}^{-1}\cdot\vec{\vec{K}}^{-1}\cdot\vec{\nabla}_s^2\vec{\vec{K}}) : \vec{\vec{\mathbf{S}}}\gamma + \right. \\ \left. + 2(\vec{\vec{K}}^{-1}\cdot\vec{\nabla}_s\vec{\vec{K}}) : \frac{\mathrm{d}}{\mathrm{d}\vec{n}}\vec{\vec{\mathbf{S}}}\gamma + \vec{\vec{K}} : \frac{\mathrm{d}^2}{\mathrm{d}\vec{n}^2}\vec{\vec{\mathbf{S}}}\gamma\right] \qquad (5.19a)$$

< continued on next page >

or for one orientation variable:

$$\mathfrak{S} = \mathbf{S}R(\varphi, A-\Delta A_c) + \frac{\partial R}{\partial A}\left(r\frac{\mathrm{d}r}{\mathrm{d}x}\frac{\mathrm{d}A_|}{\mathrm{d}x} + r^2\frac{\mathrm{d}^2A_|}{\mathrm{d}x^2}\right) +$$

$$+ 2\frac{\partial^2 R}{\partial A \partial\varphi}\left(r\frac{\mathrm{d}A_|}{\mathrm{d}x}\right) + \frac{\partial^2 R}{\partial A^2} r^2 \left(\frac{\mathrm{d}A_|}{\mathrm{d}x}\right)^2 +$$

$$+ \frac{V}{r}\frac{\partial R}{\partial A}\left[\left\{\left(\frac{\mathrm{d}r}{\mathrm{d}x}\right)^2 - r\frac{\mathrm{d}^2 r}{\mathrm{d}x^2}\right\}\left(\gamma + \frac{\mathrm{d}^2\gamma}{\mathrm{d}\varphi^2}\right) +\right.$$

$$\left. - 2\frac{\mathrm{d}r}{\mathrm{d}x}\left(\frac{\mathrm{d}\gamma}{\mathrm{d}\varphi} + \frac{\mathrm{d}^3\gamma}{\mathrm{d}\varphi^3}\right) + \frac{\mathrm{d}^2\gamma}{\mathrm{d}\varphi^2} + \frac{\mathrm{d}^4\gamma}{\mathrm{d}\varphi^4}\right] \qquad (5.19b)$$

Note: All equations (for $\vec{\mathfrak{R}}$, for $\vec{\vec{\mathfrak{S}}}$ and for tilt rate) mentioned in this chapter, in all different cases [A], [B] etc, are equations for a general point on a surface. The values of functions and derivatives are to be taken for the relevant point. Although this may seem a trivial remark, we want to emphasize that in the case of section 5.3 (equations (5.6) - (5.12)) this means that R in R, $\partial R/\partial\vec{n}$, $\partial R/\partial m$ etc. has to be read as $R(\vec{n},m)$, with $m = m(x)$, the local value of the field parameter m. In the case of section 5.4 (and equations (5.16) - (5.19)), however, R has to be read as $R(\vec{n},A_| - \Delta A_c)$, i.e. *with the local value of the growth affinity being shifted from the value corresponding with a planar interface* by the amount ΔA_c.

Chapter 6

KINETIC BEHAVIOUR OF JUNCTIONS, EDGES AND VERTICES

I have made this letter longer than usual because I lack the time to make it shorter.

Blaise Pascal (1623-1662)

6.1 INTRODUCTION

In 3.2.2 we characterized the subject of *equilibrium* in polygrain structures as that of "soap froth mechanics", with the complicating factor of OD soap-film free energy. Our present subject is the *interface kinetics* of the same type of structure, and in particular the rôle and the behaviour of the connective elements (CE's) between the interfaces: junctions, edges and vertices. (notice that we include edges and vertices, i.e. geometrical features in interfaces of *single (faceted) crystals*. We will explain later on in this chapter that their rôle in imposing boundary-conditions and the physical mechanisms of this rôle can be quite analogous to that of junctions).

Fig. 6.1 shows a crystal-growth situation illustrating all the BC-providing "elements" to be discussed in this chapter: in a growth cell (indicated by the shaded horizontal walls) we have a noncrystalline parent phase (A) and a number of crystalline grains (B_1, B_2, ...) growing from it. Different grains are discerned by different shades of grey. Grains may differ a) by their crystallographic orientation, b) by their composition (as in a eutectic), so that an "immobile" grain boundary or interphase boundary is growing between them (point J_3); c) because they are crystallographically different phases of the same composition. (B_1 may be a metastable and B_2 a stable phase of the same chemical compound, so that the A/B_1 affinity is different from the A/B_2 affinity and the B_1/B_2 affinity is not zero, even for a planar interface, so that also the B_1/B_2 interface can move (point J_2).

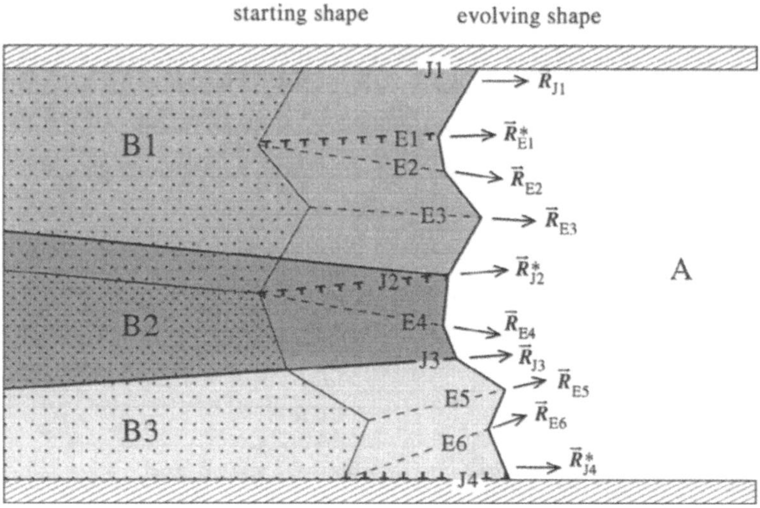

FIG. 6.1: Illustrating catalogue of "connective elements".

Here we will just give a short catalogue of these 1- and 0-dimensional elements, their physical and mathematical rôles to be specified in more detail below.

1-dimensional connective elements are 3-junctions and edges. Edges can be distinguished into reentrant and protruding (RE and PE, e.g. E_2 and E_3 in fig. 6.1). Junctions are just junctions, but they can differ in the number of participating (i.e, actually moving) interfaces. For instance, J_1 and J_4 in fig. 6.1 are junctions between a s/l interface and a "nonparticipating" wall and have only one mobile interface ("single- mobile 3-junctions"). J_3 represents a junction between two grains of the same phase, with different orientation: it can follow any direction but the B_2/B_3 interface does not move once it is formed. (when its interface energy is strongly orientation-dependent, e.g as for a twin plane, it will have a strong tendency to follow a particular direction). This type will be indicated as "double-mobile 3-junction". J_2 represents a junction between two grains of different phases- one stable and one metastable- and has three mobile junctions ("triple-mobile 3-junction").

A junction connecting more than three interfaces along a line is in general unstable, although linear 4-junctions may be stabilized by a sharply cusped γ-minimum in at least two of the participating interfaces.

0-dimensional connective elements cannot be represented in the cross-sectional drawing but are the obvious connections between the 1-dimensional connective elements. Vertices are to be characterized by the number of edges which they connect. We have three-edge re-entrant and protruding vertices, four-edge R, P and saddlepoint vertices, and hybrid combinations of edges and kinked 3-junctions. For 4-junctions (connecting four 3-junctions and six interfaces) we can again specify the number of actually moving interfaces. A single junction connecting more than four 3-junctions is generally unstable, but may be attached to a vertex.

A D-junction is another 0-dimensional connective element.

Restriction to surface kinetics

We have to emphasize that in most of chapter 6 (to be precise: the whole of ch. 6 except 6.4.4) *we restrict to the special case that growth rates are uniquely interface-controlled*; i.e. that bulk-diffusion fields can be neglected. (i.e., term (4) in eq. (1.8) = 0). So basically, in this chapter we extend the subject of the "classical" kinematic wave theory to include CE's. Although we do take the consequences of surface energy into account for junction equilibrium shapes and for the ΔA_c term, we assume a unique $R(\vec{n})$ function for each *planar* interface. The conclusions to be derived in this chapter, in particular those concerning the "velocity source" concept, are therefore applicable to this special case only.

In 6.4.4, however, we will (intuitively) consider how the velocity-source phenomenon is affected by bulk-diffusion fields. Generally speaking, we will find that for the other extreme (i.e., interface advance rates controlled by bulk diffusion only; term (3) in eq. (1.8) = 0) the velocity source concept is irrelevant. In the general case (term (3) $\neq$ 0, term (4) $\neq$ 0) it is relevant, but for relatively small interface "resistance" (Sherwood number) the geometrical effect of the velocity source (i.e., a different shape) is likely to remain trapped within a range ~ D/R from the CE.

Trajectories

During evolution of the structure of fig. 6.1 the connective elements follow trajectories (i.e. a line for a 0-dimensional, a surface for a 1-dimensional element). These trajectories can be drawn in the map of the evolving structure as has been done in fig. 6.1. Because trajectories have to be discerned from interfaces in a map, we have to draw them in a different style: in our drawings we will use dashed lines for trajectories and solid lines for interfaces. (moreover, we can use different dashing styles to distinguish between different types of trajectories, as we will do for "velocity source" trajectories - see 6.4).

In the special case when an immobile interface is involved, a trajectory can coincide with this interface (e.g., the B_2/B_3 interface coincides with the $B_2/B_3/A$ 3-junction trajectory in fig. 6.1).

Mathematical/physical characterisation of the rôle of "connective elements"

The most "basic" viewpoint may be the mathematical: considering the topology of a polygrain structure, we can recognize the CE's as the loci where *boundary conditions* can be imposed for the interfaces which meet in these junctions. Consequently, these junctions, edges etc. can be recognized as "alien" elements, where specific physical mechanisms are active which are absent or "dormant" in the smooth interfaces in-between. In particular, the connectivity of the interface structure can induce the γ-related terms in eq. (5.17) - (5.19) to become predominant in a small region near the junction, and/or the "fundamental restriction" of the continuum description can become manifest so that "statistical nucleation kinetics" has to be invoked.

So from a physical point of view, we may characterize the subject of this chapter as being related with the transition between geometrical details imposed by smooth-interface kinetics (the $R(\vec{n})$ functions) on the one side, and by energetics (the $\gamma(\vec{n})$ functions) plus "CE kinetics" ($\vec{\mathfrak{R}}^*_{junc}$, $\vec{\mathfrak{R}}_{edge}$) on the other side. The BC's related with the latter result in one out of two possibilities:

a) either the *macroscopic* ($\gg \rho_c$, critical-nucleus size) shape of the interface is identical with the shape it would have without the interaction of surface energy. We define this as "passive" behaviour of the CE with respect to the interface. However, in this case the *microscopic* shape reflects surface-energy effects in the sense that it approaches the *equilibrium* shape (derived from the $\gamma(\vec{n})$ function) within the range $\sim \rho_c$ from the CE. In other words, we have a *transition region* between the $\vec{Q}R$-shape (beyond $\sim \rho_c$) and the $\vec{Q}\gamma$-shape (within $\sim \rho_c$ from the CE).

b) Or the *macroscopic* shape is affected by the γ-equilibrium BC at the CE. This is defined as "active" behaviour of the CE with respect to the interface; we might say that the CE (or rather, the other interface(s) involved in the CE) *nucleates* the interface. In the latter case the microscopic shape of the interface remains planar; i.e. the equilibrium-shape influence is no longer "compressed" into a small region $\sim \rho_c$ around the junction but grows out to macroscopic size. *The consequence of this is that a curved or edged part of the $\vec{Q}R$ shape is moving from the initial starting point of the CE, the edges following different trajectories as the CE.* Because, in case b), *more than one* trajectory start from the same CE, it makes sense - for the sake of discussing principles - to define a *starting shape*, implying that we can follow shape evolution from "starting time" $t = 0$ when all affinities involved are "switched on". It also implies that this geometry can be chosen at will, as if, e.g., a 3-junction were assembled by fitting premachined wedges together. In the illustrations in this chapter, wherever appropriate, we have drawn these starting shapes with grey lines or planes, whereas the evolving shapes are drawn in black.

One of the questions to be considered is whether a given starting shape can also be a *stationary* shape, i.e. moving as a whole with uniform vector velocity $\vec{\mathfrak{R}}$. (in the discussion to follow, we will find that in the general case such a stationary shape will have a curved interface region within a range ~ ρ_c from a 3-junction in at least one of its three participating interfaces).

The velocity-source concept; active and passive edge trajectories

The recognition that possibility b) (above) can follow inevitably from the BC's can be adequately given shape in the "velocity source" concept. We define a connective element as a *velocity source for a given interface* when that interface is nucleated by it in the sense described above. In the example of a 3-mobile 3-junction, at least one and at most two interfaces are nucleated by the junction; i.e. it is essential to specify the interface(s) which are nucleated. For each of the nucleated interfaces, an extra "passive" edge trajectory is nucleated from the starting position of the junction.

We can eventually use an extra symbol in the dashed trajectory lines to indicate those interface boundaries where the interface is nucleated by a CE. As such, we will use perpendicular extensions from the dashes, at the side of the interface(s) concerned (see, e.g., the E_1, J_2, and J_4 trajectories in fig. 6.1). Those CE's which nucleate at least one interface (and thus are called "active") can be distinguished from passive CE's by adding an asterisk in the notation of their vector velocities: $\vec{\mathfrak{R}}^*_j$ versus $\vec{\mathfrak{R}}_j$.

STRUCTURE OF THIS CHAPTER

Junctions and edges

In this chapter we will separately discuss *junctions* (6.2) and *edges* (6.3). Both can occur in quite general structure types, without restriction to planar interfaces in-between. Still, in the discussions of 6.2 and 6.3, we will suppose the interfaces in the starting structures to be *planar and semi-infinite*. The reason for this restriction is to focus to essentials and avoid complications: for a semi-infinite planar interface, the orientation at later times always approaches the fixed starting orientation at sufficient distance from the junction. (For this reason we will label this "asymptotic" orientation with a suffix ∞).

This supposition *does not limit the validity of the conclusions* to planar interfaces: "soap-froth" structures with curved interfaces just have

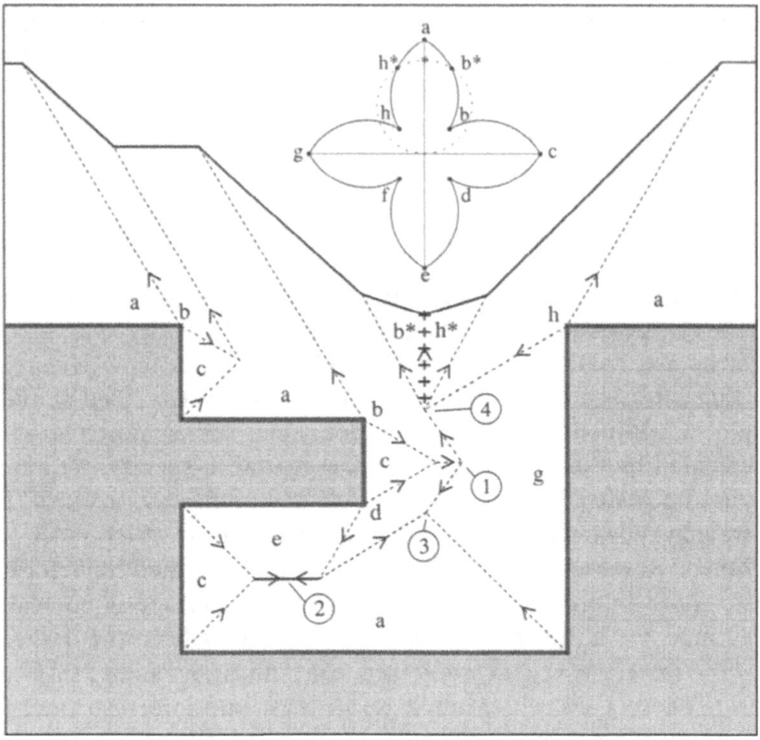

FIG. 6.2: "Multiplanar shape" with network of edge trajectories.

the extra complication that junction-orientation parameters are changing continuously, instead of adhering to asymptotic values. A similar complication results from inhomogeneity of the driving affinity. The catalogue of possible junction behaviour, however, is still properly evaluated by the discussions below.

The velocity source concept, although relevant throughout 6.2 and 6.3, will be illustrated mathematically by its most simple appearance: that of the "single-mobile 3-junction" in 6.2.2. The concept itself will be discussed in more geometrical detail in 6.4.

The emphasis on the topic of CE kinetics, involving a quite extended casuistics and resulting in a lengthy and maybe tedious chapter, may seem somewhat exaggerated in view of the sometimes hypothetical character of part of the theory. Still we thought it worth while to discuss it in so much detail, just because the subject seems to be somehow neglected (theoretically as well as experimentally). For instance, the $\mathbf{S}\gamma < 0$ "zigzag instability", in equilibrium as well as in motion, although predicted by many authors, has hardly been investigated systematically and only verified experimentally under rather poorly defined conditions. Also, the consequences of nucleation by edges and junctions may be important for many practical processes such as Bridgman growth, bulk crystallisation and sintering.

"Multiplanar shapes"

In this chapter we will also discuss (in 6.5) a technical application where planarity of the starting interfaces does have practical significance: the evolution of "multiplanar shapes". For such a structure we require not only the *starting shape* to consist of planar areas, but also the structure during evolution. A representative of such a structure is a 3-dimensionally structured semiconductor substrate (eventually with masks, heterolayers etc.) during epitaxial growth. In fig. 6.2 we have sketched a configuration which, although not likely to be characteristic for a technical application, is quite suitable to illustrate the principle and a number of details. We may follow the evolution of such a structure under two radically simplifying assumptions, which can nevertheless be fairly realistic for many practical cases (e.g., MOVPE of III-V compounds):

a) the starting geometry consists uniquely of *planar areas* connected by edges and vertices. Such a starting geometry results from a pre-growth processing step involving masking, etching, ion-milling etc. Also eventual mask layers should obey these conditions;

b) the $R(\vec{n})$ function is such that only planar interface sections grow from re-entrant as well as from protruding edges and vertices. This condition can be translated into the prescription that the $\vec{\mathbf{P}}R(\vec{n})$ plot is completely built up of parts of spheres (passing through the origin) fitted together in their (circle) intersections. This, again, is fairly realistic for crystals bounded by highly faceted interfaces. In the language of ch.12, it means $\varepsilon \ll 1$, $L \gg v_{\parallel}$, $M \gg v_{\parallel}$ and $a_{\text{comb.}} \gg 2$; in physical terms it means "high-α factor" interfaces and negligible bulk - diffusion resistance.

Under these conditions, the evolution of such a "multiplanar geometry" can be described by just accounting for the vector velocities of all edges, vertices and junctions involved. This means that we have only to calculate the 3-dimensional map of edge-, vertex- and junction-*trajectories* (plus velocities). Some of these trajectories may be "active" (nucleating new orientations), in which case they are accompanied by one or more passive edge trajectories between the new and the old orientations. Edge trajectories can collide, thereby annihilating the enclosed orientation. Colliding trajectories give birth to one or more new edge trajectories because a new combination of neighbouring orientations is generated (see fig. 6.2; point "4"). The same happens when a protruding edge hits an interface coming from an opposite direction. (point "1" in fig. 6.2).

The conditions guarantee that all trajectories are straight lines and/or planes. Therefore, the algorithm for evaluating the evolution of the structure can be relatively simple. Still, we have to be aware of possible velocity sources in the structure (such as reentrant edges between facets and junctions with mask layers), whose $\vec{\mathfrak{R}}^*$ vectors are independent of the $R(\vec{n},A)$ function.

Behaviour of junctions in an affinity gradient.

Complications arising from an inhomogeneous affinity field - either resulting from bulk diffusion or from an imposed gradient in T or A - are not considered explicitly in this chapter. A superficial discussion of such complications is given in 6.4.4; apart from that, the effect of interface kinetics on grain-boundary groove and related shapes was considered in 3.3, as an amendment on equilibrium structures and in connection with Marangoni and Gibbs-Thomson terms. Finally, in 8.4 we will consider the mutual *interaction between neighbouring junctions* in an (eventually moving) inhomogeneous affinity field, due to surface-energy and diffusion effects and resulting in an effective interaction potential between the junctions, which can have attractive as well as repulsive components. Such interactions are relevant in unidirectional-growth procedures of eutectic and cellular structures.

6.2 THE STATIONARY SHAPE OF MOVING JUNCTIONS

In order to build up complexity in a logical way, we will discuss junctions in the successive order 3-junctions – 4-junctions – D-junctions. As for the 3-junctions, the simplest case is the "single-mobile 3-junction" i.e. a mobile interface meeting an immobile planar wall (e.g., J_1 in fig. 6.1). Although this is conceptually the most simple case and as such a suitable prototype case to start with, it can demonstrate all features involved in the transition between "active" and "passive" behaviour of the junction with respect to the interface, and show the geometrical consequences of this behaviour. As this distinction between active and passive behaviour is the very basis for the "velocity-source" concept, it is appropriate to insert a section to introduce and define this concept directly after the discussion of this simplest-case 3-junction (in 6.2.2.3).

More complex cases of velocity sources will be met during the discussion of general 3-junctions, 4-junctions, D-junctions and reentrant and protruding edges in single-crystalline surfaces.

6.2.1 *Overspecification of stationary-growth conditions for an arbitrary 3-junction*

Prior to the discussion of the single-mobile 3-junction, we have to introduce some nomenclature on 3-junctions and edges in general (but under the restriction, explained in the introduction, that the starting geometry is built up of semi-infinite planar interfaces). For an n-mobile 3-junction, then, it makes sense to define the following three types of pair-related vector velocities and shapes: (all projected onto a plane perpendicular to the junction line, and written as $\vec{\mathfrak{R}}_{j\infty}$, $\vec{\mathfrak{R}}_{j\,pass.}$ and $\vec{\mathfrak{R}}_{jo}$).

a) *The vector velocity of the intersection of a pair of φ_∞-asymptotes*
For the intersection of these crossing lines we can write:

$$\vec{\mathfrak{R}}_{j\infty}^{(k\ell)} \cdot \vec{n}_{\infty}^{(i)} = R(i)(\vec{n}_{\infty}^{(i)}) \qquad i = k,\ell \tag{6.1}$$

Considering (6.1) as a set of vector equations in the plane perpendicular to the junction line, we see that the condition is overspecified when more than two interfaces participate ($i = k, \ell$ *and* m). This means that an arbitrary starting shape of a 3-junction with *three* mobile interfaces *cannot be a stationary shape unless by coincidence*. In other words, at least one of these interfaces is "nucleated", i.e. assumes an orientation different from the φ_∞ orientation in an expanding area between the junction and the asymptote. We express this by saying that a 3-mobile 3-junction acts as a *velocity source* for at least one of its interfaces. Only the junction with the new (nucleated) interface(s) can be stationary; the three original asymptotic planes (extended to full infinite planes) do not in general keep a common intersection line for $t > 0$.

Also 3-junctions with only one or two moving interfaces *can* be active, i.e. nucleate one or both interfaces. Similar considerations are applicable to 4- and D-junctions.

In connection with the overspecification of equation (6.1) for three interfaces, it is of interest to confront it with the condition of *force equilibrium* between three interfaces as given by eq. (4.1a) and as illustrated by fig. 4.3 (Neumann triangle). Whereas the latter construction is in general soluble for three $\vec{\mathbf{P}}\gamma$ plots, the first equation requirs *four* points to lie on a circle (the origin and three $\vec{\mathbf{P}}R$ vectors). This overspecified condition is illustrated by fig. 6.3.

b) *The vector velocity of the intersection of two "180-wedges".*
The vector $\vec{\mathfrak{R}}_{j\infty}^{(k\ell)}$ is a mathematical abstraction from the point of view of the 3-junction because it is concerned with intersecting *infinite planes.*

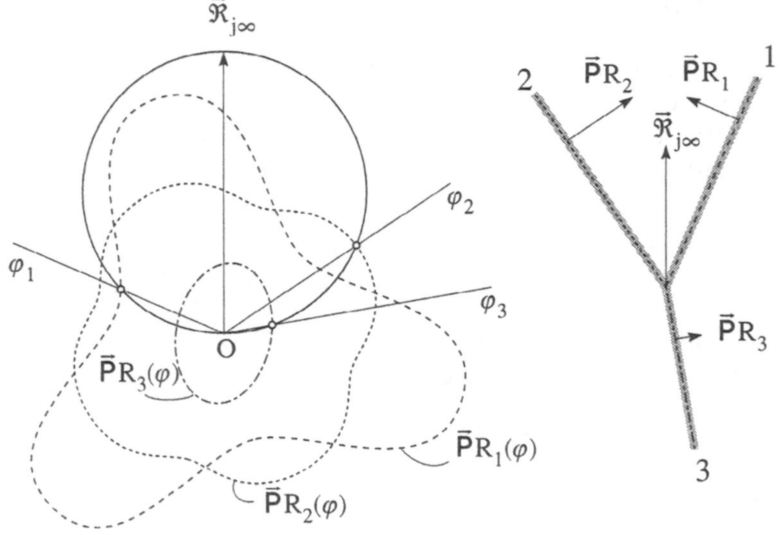

FIG. 6.3: Condition (6.1) for three interfaces; i = 1, 2 and 3.

The real interfaces meeting at the junction are *semi-infinite (half)* planes. How can we construct the shape evolution of these connected half-planes?

For a RE or a PE in a single-crystal interface, the construction is straightforward: from any point on the surface we can draw the $t.\vec{\mathsf{Q}}R$ vector, and the orientation jump at any edge is bridged by the corresponding section of the $t.\vec{\mathsf{Q}}R$ shape.

For the edge between two meeting interfaces in a 3-junction there is no such $\vec{\mathsf{Q}}R$ shape; rather we have to do with two *different* $\vec{\mathsf{Q}}R$ shapes, each related with one (semi-infinite) interface. (for the complete 3-junction we have three $\vec{\mathsf{Q}}R$ shapes, at least when all interfaces move: see fig. 6.3).

For each of the participating semi-infinite half-planes we can define the evolution shape of a hypothetical 180° wedge which is just the half-plane supplemented with a back side. Such an evolution shape is made up of interfaces oriented at φ_∞ and φ_∞+180° connected by that part of the $\vec{\mathsf{Q}}R$ shape which bridges the 180° gap. (if φ_∞ and/or φ_∞+180° are in $\vec{\vec{\mathsf{S}}}R < 0$ orientations, part of the connecting $\vec{\mathsf{Q}}R$ shape is "virtual"; see definition in ch.2).

We can define the vector velocity of the intersection line of any pair of developing 180°-wedges as $\vec{\Re}_{\mathrm{j\ pass.}}^{(k\ell)}$. Just as in case *a*), the requirement that

$$\vec{\Re}_{\mathrm{j\ pass.}}^{(k\ell)} = \vec{\Re}_{\mathrm{j\ pass.}}^{(\ell m)} = \vec{\Re}_{\mathrm{j\ pass.}}^{(mk)} \qquad (6.2)$$

leads to an overspecification, i.e. to the inevitable conclusion that at least one interface has to assume a new orientation at the junction side.

c) The vector velocity of the intersection of two φ_o -orientation.

The factors involved in the determination of the $\vec{\Re}_{j\infty}^{(k\ell)}$ and $\vec{\Re}_{j\,\text{pass.}}^{(k\ell)}$ vectors were unrelated with the (γ-equilibrium and/or kinetic) boundary condition imposed by the 3-junction itself. In the construction of the intersection points, data involving the $R(\varphi)$ functions were used for the half-planes separately without regard of eventual *interaction* of the half-planes in the 3-junction. The effect of such interaction can be expressed as a relation between the orientations φ_0 of the interfaces at the junction. The surface-energy related BC which is imposed at the 3-junction line is such a relation.

This crucial information can be embedded in a third pair-related vector velocity, i.e. that of the intersection of the φ_0 -orientations. These φ_0 - orientations are defined here as the orientations of the interface directly at the 3-junction, in agreement with the "mechanical" equilibrium as discussed in 4.1.3, but with the eventual amendment of a kinetic shift in the equilibrium, to be discussed in 6.2.2.1. Such φ_0 -orientation relations can be expressed by equations such as eq. (4.1a) or (6.6), in graphical form represented by fig. 4.4 a - g . Actually the degree of freedom left in the choice of φ_0 for a particular interface participating in a 3-junction is specified by eventual fixation of the other interfaces involved.

Altogether the pair-related φ_0 -orientation vector velocities are defined as

$$\vec{\Re}_{jo}^{(k\ell)} \cdot \vec{n}_o^{(i)} = R^{(i)}(\vec{n}_o^{(i)}) \qquad i = k,\ell \qquad (6.3)$$

The interesting point is that the argument on overspecification (for the three pairs of a 3-junction obeying the same vector velocity) which was used in a) and b) is also valid for the $\vec{\Re}_{jo}$ vectors, *but now the conclusion works in the opposite direction*, i.e. that a 3-junction with each of its three interfaces in equilibrium orientation cannot be stationary unless by coincidence; i.e. that in general a *moving* 3-junction (with 3 mobile interfaces) has also at least one *non*-nucleated interface. (according to b), it has also at least one *nucleated* interface). The geometric implications of nucleated and non-nucleated interfaces, i.e. of the "active" versus the "passive" rôle of the junction towards the interfaces, will be discussed below.

6.2.2 *Case 1: One mobile interface meeting a wall. (single-mobile 3-junction)*

As anticipated above, this case is relatively simple as compared with the general 3-junction. This simplicity results from the fact that two of the participating interfaces are immobile, so that only one growth rate function is involved and only one γ-equilibrium orientation φ_0 can be specified. Consequently, the three triples of pair-related vector velocities introduced in 6.2.1 just reduce to three *single* vector velocities, all aligned with the wall. It is therefore convenient to relate interface orientations (written as an angle φ) to that of the wall (φ_w). Choosing the zero-point for φ such that $\varphi = 0$ corresponds with an interface facing the wall, we have $\varphi_w = 180°$, φ in the range 0 - 180° implies a "crystal left/parent phase right" interface moving to the right, φ in the range 180° - 360° the opposite. Below we will, for convenience, only consider the range 0° - 180° ; the other case is similar by rotation (although the $R(\varphi)$ plots then can be different for the noncentrosymmetric case).

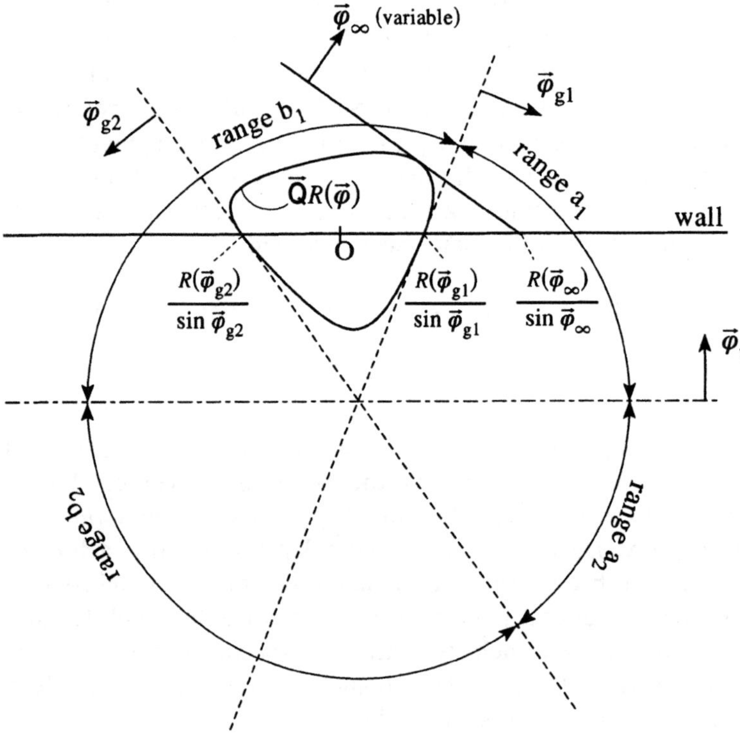

FIG. 6.4: Construction of the vector velocity for a single-mobile 3-junction.

The three relevant vector velocities are now: (vectors // wall)

a) $$\vec{\Re}_{j\infty} = \frac{R(\varphi_\infty)}{\sin \varphi_\infty} \tag{6.4a}$$

b) $$\vec{\Re}_{j\ \text{pass.}} \begin{cases} (\text{for } \varphi < \varphi_{g1}\text{, i.e. range } a_1 \text{ in fig. 6.4:}) = \dfrac{R(\varphi_\infty)}{\sin \varphi_\infty} \\ (\text{for } \varphi \geqslant \varphi_{g1}\text{, i.e. range } b_1 \text{ in fig. 6.4::}) = \dfrac{R(\varphi_{g1})}{\sin \varphi_{g1}} \end{cases} \tag{6.4b}$$

c) $$\vec{\Re}_{jo} = \frac{R(\varphi_o)}{\sin \varphi_o} \tag{6.4c}$$

where φ_∞ is the imposed asymptotic orientation of the starting shape, φ_o is the orientation imposed by the 3-j. BC and φ_{g1} is the orientation of the evolving 180°-wedge shape at the point of intersection with the wall.

In fig. 6.5, we have sketched an (arbitrary case of a) function $\vec{\Re}_j = R(\varphi)/\sin \varphi$ as a function of φ in the range 0°-180°. Such a curve can be constructed from the $\vec{Q}R(\varphi)$ plot by drawing this plot with its origin located on the wall plane and "rolling off" a plane (orientation φ) over its surface; the intersection point of this plane with the wall corresponds with $\vec{\Re}_j$ (again measured from the plot origin; see fig. 6.4).

The orientation φ_o is "given by nature" and determines the velocity $\vec{\Re}_{jo}$. (the position of φ_o can be anywhere between 0° and 180°). The orientation φ_∞ is imposed by the experimenter, and we can use the method indicated in fig. 6.4 to determine a plot for the actual 3-junction velocity as a function of φ_∞, such as in fig. 6.5.

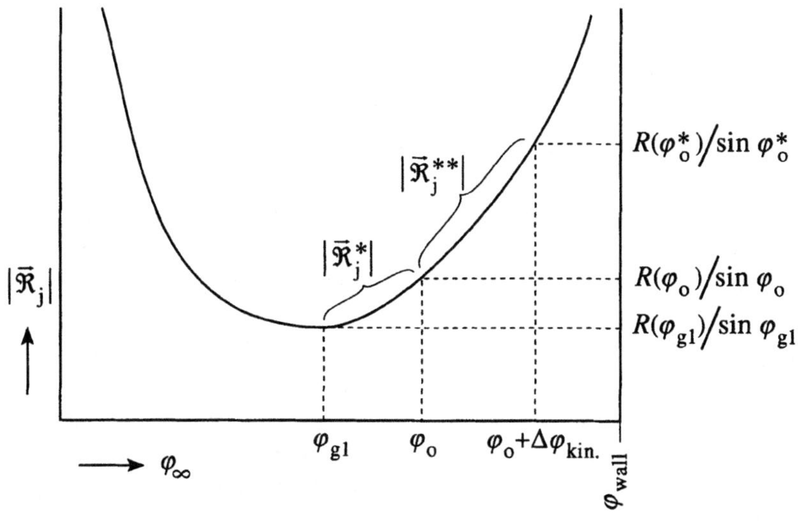

FIG. 6.5

The single-minimum curve of fig. 6.5 corresponds with a $R(\varphi)$ function having a positive $\mathbf{S}R(-\varphi_w)$ value. The minimum of this curve corresponds with φ_{gl}. (When $\mathbf{S}R(-\varphi_w)$ is negative, the curve exhibits two or more relative minima)

Now when the φ_o -imposing BC at the 3-junction would be inactive (so that indeed any interface orientation at the 3-junction would be physically realistic) we would only have to do with φ_{gl}. The interface orientation at the junction then would equal φ_∞ for $\varphi_\infty \leqslant \varphi_{gl}$ (implying that the interface would remain planar apart from a small ρ_c -sized area near the junction) and it would get stuck at φ_{gl} for $\varphi_\infty > \varphi_{gl}$ (implying that a curved part would develop, effectively a part of the $\mathbf{Q}R$ plot).

The φ_o orientation imposed by the BC actually removes this freedom. But we can discern two cases: φ_o can be either $<$ or $\geqslant$ φ_{gl}.

The effect of the combined motion of the φ_o -oriented interface at the wall and the φ_∞ -oriented asymptotic interface is illustrated by fig. 6.6. We have drawn four situations (together covering all possibilities):

a) $\varphi_o < \varphi_\infty$; $R_j(\varphi_o) > R_j(\varphi_\infty)$

b) $\varphi_o < \varphi_\infty$; $R_j(\varphi_o) < R_j(\varphi_\infty)$

c) $\varphi_o > \varphi_\infty$; $R_j(\varphi_o) > R_j(\varphi_\infty)$

d) $\varphi_o > \varphi_\infty$; $R_j(\varphi_o) < R_j(\varphi_\infty)$

Ergo, the intersection between a φ_o -oriented and a φ_∞ -oriented plane moves *towards the 3-junction* if $\varphi_o - \varphi_\infty$ and $\{\vec{\Re}_j(\varphi_o) - \vec{\Re}_j(\varphi_\infty)\}$ have *opposed sign* (cases 1 and 4) and *away from the junction* if they have *the same sign* (cases 2 and 3). Apparently, the decisive quantity for the behaviour of (this particular) 3-junction, (with non-participating planar wall) is

$$\frac{R_j(\varphi_o) - R_j(\varphi_\infty)}{\varphi_o - \varphi_\infty} = \frac{\{R(\varphi_o)/\sin\varphi_o\} - \{R(\varphi_\infty)/\sin\varphi_\infty\}}{\varphi_o - \varphi_\infty} \tag{6.5}$$

= slope of the connecting line between the points $(\varphi_o, \vec{\Re}_j(\varphi_o))$ and $(\varphi_\infty, \vec{\Re}_j(\varphi_\infty))$ in fig. 6.5, independent of the relative positions of φ_o and φ_∞.

If expression (6.5) is <0, the transition region between the two orientations φ_o and φ_∞ will tend to conform itself to part of the equilibrium shape. For instance in case a, when the (protruding) edge between φ_o and φ_∞ moves towards the junction, the curved transition zone between it is "compressed" into the angle between the φ_∞ -oriented surface and the wall so that the curvature is increased and the effective affinity acting at the junction intersection ($= A_| + \Delta A_c$) is decreased. The stabilized stationary situation is reached when the resulting $R(\varphi_o, A_| + \Delta A_c)$ has dropped to $R_j(\varphi_\infty)$. This situation can always be reached, because the "compression" can proceed to an arbitrarily small size of the equilibrium shape so that ΔA_c

is not limited, at least within the applicability range of eq. (3.7).

Similarly, in case d above, the *re-entrant* edge between φ_o and φ_∞ is compressed near the 3-junction so that the increasing ΔA_c is now positive. Now the junction's velocity $R(\varphi_o, A_l + \Delta A_c)$ is *speeding up* until it can keep pace with $R_j(\varphi_\infty)$ in a stabilized-stationary geometry.

In both cases (a and d), we end with a curved transition region connecting the φ_o orientation at the junction with the φ_∞ asymptote. The shape and size of this transition zone can be calculated from the EOM, as will be shown in 6.2.2.3. The typical size is of the order of the critical nucleus size corresponding with the affinity A_l.

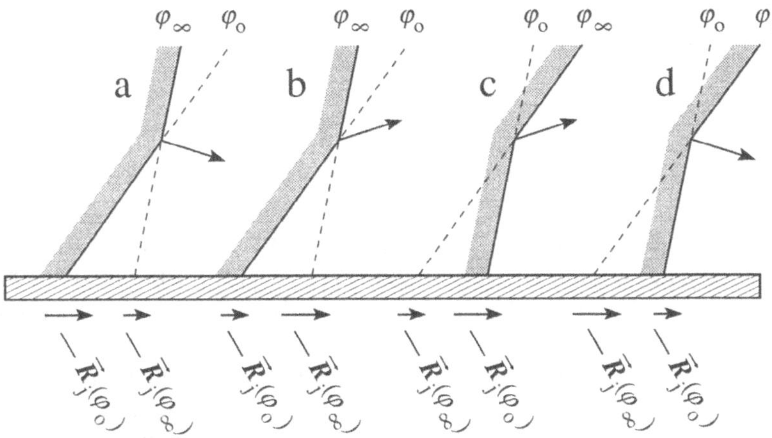

FIG. 6.6

In cases b and c, in contrast, the transition zone between φ_o and φ_∞ moves away from the junction, and the interface is *planar down to the junction on a microscopic scale*, even within a range ~ ρ_c (on a macroscopic scale the interface may be curved due to the growing transition zone between φ_o and φ_∞). This transition zone develops in agreement with the $R(\vec{n})$ function; i.e. as the evolution shape $(\overline{\mathrm{Q}}R(\vec{n}))$ of a wedge between half-infinite planes oriented at φ_o and φ_∞, according to the Gibbs-Wulff construction.

The criterion expressed by (6.5) can be translated graphically to illustrate the effect of the φ_o-imposing BC on the 3-junction velocity (fig. 6.7). We can associate a "mask" with the point $(\varphi_o, \vec{\Re}_j(\varphi_o))$ which moves along the $\vec{\Re}_j(\varphi)$ curve as we vary φ_o. This mask consists of two "grey" sectors as shown in fig. 6.7, corresponding with positive slopes in expression (6.5). In

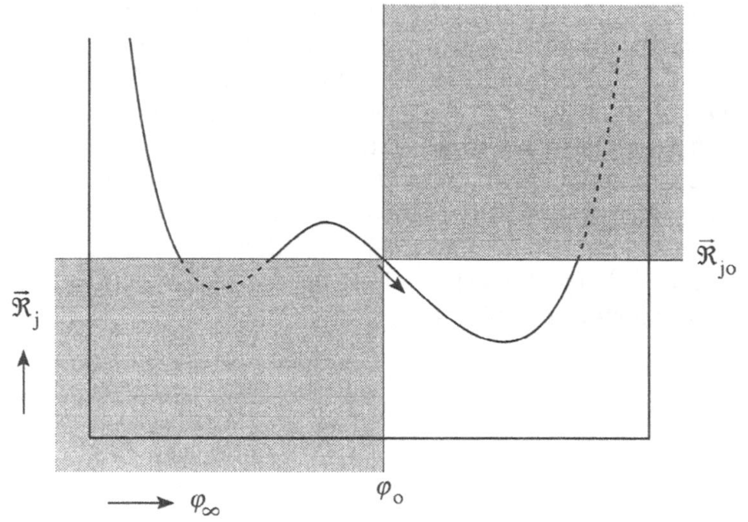

FIG. 6.7

other words, when a point $(\varphi_\infty, \vec{\Re}_j(\varphi_\infty))$ is covered by the mask, the junction velocity is "nucleated", i.e. dictated by $\vec{\Re}_j(\varphi_o)$: the covered part of the $\vec{\Re}_j(\varphi)$ curve is replaced by the horizontal boundary of the grey sectors. When the point is uncovered, the junction velocity is still $\vec{\Re}_j(\varphi_\infty)$.

If the boundary condition would be deactivated (so that any orientation at the interface would be realistic), then the orientation would "get stuck" on the 180°-wedge intersection velocity of eq. (6.4b) This is equivalent with an effective φ_o located on the *minimum* of the $\vec{\Re}_j(\varphi)$ curve, i.e. identical with φ_{g1} .

In fig. 6.8 we have sketched the evolution of a single-mobile 3-junction:
a) for a "passive" junction, corresponding with fig. 6.6 a or d;
b) for an active junction, b corresponding with fig. 6.6b and c with fig. 6.6c.

6.2.2.1 Complications:

We have to discuss two complications to this simple model:

a): Multiple minima

Above (6.2.2) we have discussed that the occurrence of multiple relative minima in a $\vec{\Re}_j(\varphi)$ curve corresponds with a negative $\mathbf{S}R(\varphi)$ range. Still, the criterion (6.5) and its graphical translation can be used without modification for the "construction" of the effective $\vec{\Re}_j(\varphi_\infty)$ curves. (we leave the verification to the reader). This implies that in certain cases, the

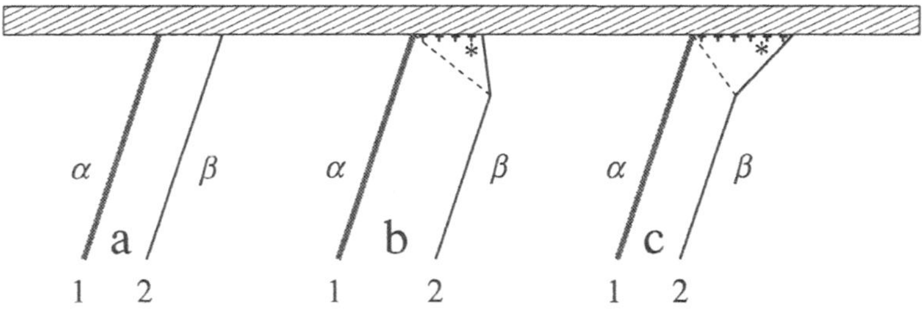

FIG. 6.8: Single-mobile 3-junctions

$\vec{\Re}_j(\varphi_\infty)$ curve can have horizontal sections, interrupted by local φ_∞-dependent regions where a local $\vec{\Re}_j(\varphi)$-maximum or minimum peeps up from behind the mask in fig. 6.7.

b): Kinetic shift in energetic boundary conditions

The 3-junction equilibrium condition, resulting in an imposed φ_0 BC at the junction, is not strictly applicable to a *moving* junction. The most obvious argument to illustrate this is to consider the case where φ_0 corresponds with a facet or near-facet orientation ($\gamma(\varphi)$-function cusp, as in fig. 4.4 c-g). The corresponding $R(\varphi_0)$ would be low as compared with $R(\varphi_\infty)$ for φ_∞ chosen in a non-facet direction, so that the 3-junction would actually "block" the interface.

Now the 3-junction is likely to "nucleate" the motion of the interface, not by affecting the equilibrium but rather by nucleating ledge trains from the RE between the wall and the interface. (see fig. 6.12 for this mechanism in a *double*-mobile junction.) If the wall were made of the same crystalline material (including orientation) as the advancing crystal we would have the case of a RE in a single-crystal interface and nucleation by "sharp-RE kinetics", even for a stationary-stable RE, would be obvious. (for an extensive discussion of this topic, see 6.3). But also in the case of a 3-junction we can have a similar ledge-nucleating activity by the other phase, in casu the wall material.

Nucleation of ledge trains by the RE has the effect that

a) the interface *orientation* is tilted away from the $A_| = 0$ equilibrium orientation φ_0. We will write the new orientation as φ_0*.

b) The junction velocity is increased above that of the $A_| = 0$ equilibrium orientation, $R(\varphi_0, A_|)$.

In other words, we still have to do with "nucleation by the 3-junction", but the mechanism tuns out to be slightly more complex than suggested in the (oversimplified) discussion of 6.2.2. Phenomenologically the situation

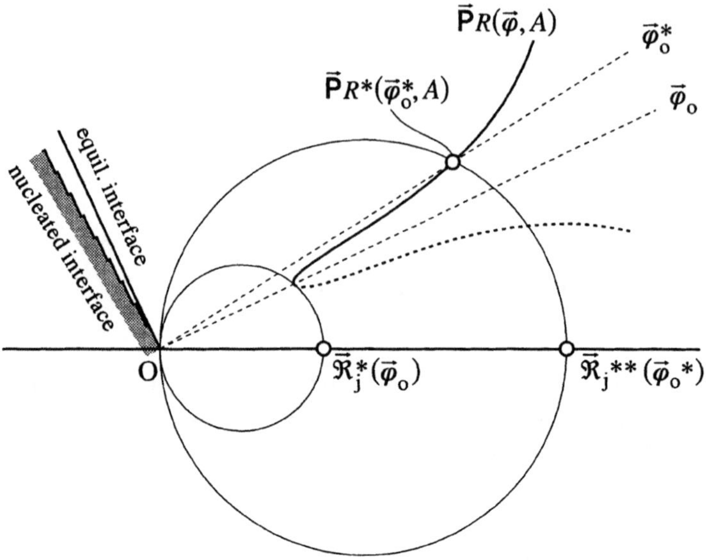

FIG. 6.9: Kinetic nucleation of the mobile interface in a single-mobile 3-junction.

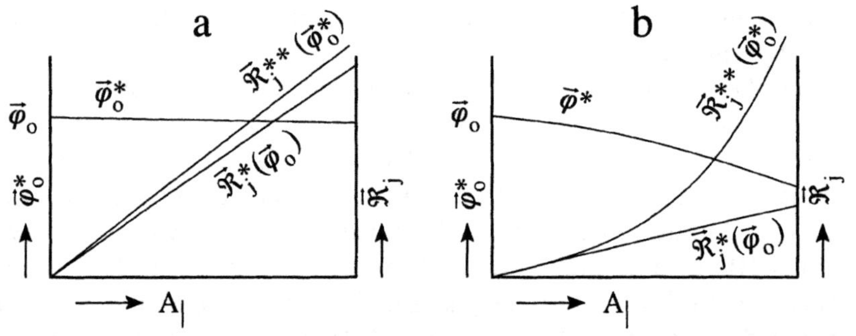

FIG. 6.10

is to be characterized as follows (again we start to consider a planar interface at the junction).

In the real *equilibrium*, i.e. $A_| = 0$ case, we have only to do with eq. (4.1 a/b). This means that φ_o is uniquely determined.

In the $A_| > 0$ case we have to do with two factors:

a) force equilibrium between the interfaces;

b) ledge nucleation by the RE of the junction.

Ad a): Even in the case that the γ_i involved are not affected by the applied affinity, we have to conclude that the orientation of the moving interfaces has changed from φ_o because the train of nucleated ledges, starting from the 3-junction, corresponds with an effective *tilt* from φ_o. *Ergo, eq. (4.1) seems to be violated: this implies that we have formally to extend eq.(4.1) with an extra term representing a drag force exerted by the 3-junction line under the influence of the affinity* $A_|$*:*

$$\sum \vec{\mathsf{Q}}\gamma_i \wedge \vec{\ell} + \vec{F}_\ell(A_|) = 0 \tag{6.6}$$

Ad b): Nucleation by the ledge is typically a case of "sharp RE kinetics" (see remark in 1.3) and can only be calculated by statistical procedures. However, it is evident that the new orientation φ_o^*, the velocity of the junction $\vec{\Re}_j^*$ and the affinity $A_|$ are coupled by a condition to be expressed as

$$\vec{\mathsf{P}}R(\varphi_o^*, A_|) = \vec{\Re}_j^{**} \tag{6.7}$$

whose graphical shape is qualitatively suggested in fig. 6.9 (remark: the effective-$\vec{\Re}_j^*$ circle (largest circle in fig. 6.9) will in general intersect the $\vec{\mathsf{P}}R(\varphi, A_|)$ plot of the moving interface *twice*; however, only one of these intersections is physically relevant, i.e. in agreement with a train of ledges moving *away from* the junction. Therefore, the "irrelevant" branch of the $\vec{\mathsf{P}}R$ plot is dashed).

In fig. 6.10 we have intuitively sketched the dependence of φ_o^* and $\vec{\Re}_j^*$ for the case of a "roughened", i.e. low-α factor interface (fig. 6.10a) and a "faceted", i.e. high-α factor interface (fig. 6.10b). In the first case, the kinetic component of RE nucleation will be hardly discernible; the equilibrium condition (4.1) may be expected to be only slightly affected by a kinetic drag component as in eq.(6.6) and the $\vec{\Re}_j^*(A_|)$ dependence will be almost linear.

Constructing (in fig. 6.9) the circle through $\vec{\mathsf{P}}R(\varphi_o)$ with diameter along the wall direction (i.e., constructing the $\vec{\Re}_j^*$ velocity as if it had to correspond with $\vec{\mathsf{P}}R(\varphi_o)$ and indicating this corresponding diameter as $\vec{\Re}_j^{**}(\varphi_o^*)$, the low-$\alpha$ factor case is characterized by

$$|\vec{\Re}_j^{**}(\varphi_o^*) - \vec{\Re}_j^*(\varphi_o)| \ll \vec{\Re}_j^*(\varphi_o) \tag{6.8a}$$

In other words, the velocity of the junction may be considered to be *imposed by the equilibrium condition (4.1).*

In contrast, in the case of the faceted interface:

$$|\vec{\Re}_j^{**}(\varphi_o^*) - \vec{\Re}_j^*(\varphi_o)| \gg \vec{\Re}_j^*(\varphi_o) \tag{6.8b}$$

This means that a mechanical junction-drag effect of the form of eq. (6.6) is an essential amendment to eq. (4.1) and that insofar as the junction velocity $\vec{\mathfrak{R}}_j^*$ is concerned, the equilibrium condition (4.1) is overruled (as a physical mechanism) by "sharp RE kinetics".
In accordance with this distinction we might discern "equilibrium nucleation" and "kinetic nucleation" as extreme cases. (actually, as suggested by fig. 6.10, both "components" are always mixed).

6.2.2.2 *The transition zone in junction-convergent interfaces*

In this subsection we consider the shape details of the stationary curved section (of a junction-convergent interface) near the junction. In order to be a stationary shape, all parts of the junction have to satisfy the condition $\vec{\vec{\mathfrak{S}}} \equiv \vec{\vec{O}}$; i.e. their vector velocities, drawn in plot representation, lie on the $\mathfrak{R}_j^*$ effective-plot circle. (fig. 6.9). This condition is equivalent to a relation between orientation and growth rate which can only be matched by the $R(\varphi, A_g)$ function of the interface for one particular value of A_g. In general this value is shifted from the applied affinity $A_|$; the shift is accomodated by ΔA_c (eq. (3.7b) so that ultimately we have a relation between orientation and curvature.

This procedure can be expressed by the following equations.
The junction velocity R_j satisfies

$$R_j = \frac{R(\varphi_\infty, A_|)}{\sin\varphi_\infty} \quad \text{or} \quad \frac{R(\varphi_{g1\,\text{oR}\,g2}, A_|)}{\sin\varphi_{g1\,\text{or}\,g2}} \tag{6.9}$$

depending on criterion (6.5). Here φ measures the angle between the interface normal and that of the wall plane, at least when the "immobile-wall" junction of 6.2 is considered. In the case of the general junction, φ is the angle between the interface normal and the $\vec{\mathfrak{R}}_j$ direction. As we are explicitly considering the case of junction-convergent interfaces here, we may symbolize either φ_∞ or $\varphi_{g1\text{ or }g2}$ (whichever applicable) by $\varphi_{\text{conv.}}$. The curved section has to obey

$$R[\varphi, A_g\{s(\varphi)\}] = R_j \sin\varphi = R(\varphi_{\text{conv.}}, A_|)\sin\varphi/\sin\varphi_{\text{conv.}} \tag{6.10}$$

This can be considered as an implicit equation for $A_g\{s(\varphi)\}$ or simply $A_g(\varphi)$:

$$A_g(\varphi) = (\arg R)_\varphi\,\{R(\varphi_\infty, A_|)\,.\,\sin\varphi/\sin\varphi_{\text{conv.}}\} \tag{6.11}$$

For linear $R(A)$ dependence, $R(\varphi, A) = A\chi(\varphi)$, this can be written explicitly as

$$A_g(\varphi) = A_| \frac{\chi(\varphi_\infty)}{\chi(\varphi)} \frac{\sin \varphi}{\sin \varphi_\infty} \tag{6.12}$$

so that

$$\Delta A_c = A_| - A(\varphi) = A_| \left\{ 1 - \frac{\sin\varphi}{\sin\varphi_\infty} \frac{\chi(\varphi_\infty)}{\chi(\varphi)} \right\} \tag{6.13}$$

On the other hand, $\Delta A_c = V\,(d\varphi/ds).\mathbf{S}\gamma(\varphi)$ (eq. (3.7) So finally we can write the differential equation for the curved section

$$\frac{d\varphi}{ds} = \frac{A_|}{V} \frac{1}{\mathbf{S}\gamma(\varphi)} \left\{ 1 - \frac{\sin\varphi}{\sin\varphi_\infty} \frac{\chi(\varphi_\infty)}{\chi(\varphi)} \right\} \tag{6.14}$$

with boundary conditions

$$\varphi \to \varphi_o \text{ for } s \to 0, \qquad \varphi \to \varphi_\infty \text{ for } s \to \infty$$

Note that the expression *without* the term between brackets represents the equilibrium-shaped critical nucleus for $A_|$. In other words, the term between brackets can be considered as the inverse of a φ -dependent size factor (relative to the critical-size equilibrium shape).

Starting with orientation φ_0 at $s = 0$, we can calculate the curved interface by integration following a similar procedure as with the grain-boundary groove shapes of section 3.3.3.

The term between brackets is equivalent with expression (6.5), and a stationary solution for the curved section can only be found when it is positive (in the opposite case, we have a junction-divergent interface, i.e. a nucleating junction).

6.2.2.3 The size of the transition region

In order to get a rough idea of the shape of the curved sections, we can consider the "size factor" of eq. (6.14) for OID surface mobility χ :

$$\left(1 - \frac{\sin \varphi}{\sin \varphi_\infty} \right) \tag{6.15}$$

For small φ_∞ , i.e. $\vec{\Re}_j \gg \vec{\Re}_g$, the factor can be $\ll 1$ and then the curvature near the junction is much higher (and the size smaller) than that of the critical nucleus for A_e. For "intermediate" φ_∞ , the factor is of order unity and the curvature is of the same order as that of the critical nucleus.

For $\varphi \to \varphi_\infty$, the size factor can be expanded to fist order:

$$\frac{d\varphi}{ds} = \frac{A_|}{V.\mathbf{S}\gamma(\varphi)} \cdot (\varphi_\infty - \varphi)\,\mathrm{K} \tag{6.16}$$

where $\mathrm{K} = \cot \varphi_\infty - \{(d/d\varphi) \log \chi(\varphi)\}_{\varphi=\varphi_\infty}$, or shortly cotan φ_∞ for

OID χ, so that (writing $\varphi_\infty - \varphi = \Delta\varphi$)

$$\frac{1}{\Delta\varphi}\frac{d(\Delta\varphi)}{d\varphi} = \frac{d}{ds}(\log \Delta\varphi) = \frac{K}{\rho_c} \text{ , or}$$

$$\Delta\varphi = \Delta\varphi_0 \exp(-K\, s/\rho_c) \tag{6.17}$$

i.e. an exponential asymptotic behaviour with "decay length" ρ_c/K. This decay length is a suitable indication of the size of the curved section.

Obviously, this estimate does not work for K = 0 ($\varphi_\infty = \pi/2$ for OID surface mobility; i.e. the moving interface is perpendicular to the wall). In this case we have to develop the size factor to second order. Writing $\psi = \pi/2 - \varphi$ and assuming OID γ and χ , we have

$$\frac{d\psi}{ds} = \frac{A_{|}}{2V\gamma}\,\psi^2(s) \qquad \text{(2nd order)} \tag{6.18}$$

resulting in the asymptotic behaviour

$$\Delta\psi \simeq 2\,\rho_c/s \qquad \text{for } s \to \infty \quad ; \tag{6.19}$$

i.e. a hyperbolic asymptotic behaviour which can again be characterized by the size ρ_c .

The single-mobile 3-junction as a velocity source prototype.

We have evaluated the single-mobile 3-junction in such ample detail because it can be considered as a velocity source prototype. In short, we have shown that the junction's velocity can either be imposed by that of the mobile asymptotic interface, $\vec{\Re}_j(\vec{n}_\infty)$ (= passive behaviour of the junction) or by that of the equilibrium-oriented interface, $\vec{\Re}_j^*(\vec{n}_0)$ (=active behaviour).

Although the transition criteria become more complex for the cases to follow (double- and triple-mobile 3-junctions, 4-junctions, D-junctions, edges and vertices) the characteristic differences between active and passive CE/interface pairs (which are universal) have all been demonstrated. These differences are summarized in table 6.1 in section 6.4, which is a survey of the velocity source concept from a geometrical point of view.

6.2.3 Case 2: The 3-junction with two mobile interfaces

The next step in increasing complexity above the "single-mobile" 3-junction discussed in 6.2.2, is the 3-junction with *two* mobile interfaces, the third (immobile) one just being defined by the trail of the intersection of the mobile interfaces as they move. Therefore, we have to do with only one "pair-intersection" $\vec{\Re}_{j\infty}$ vector, one $\vec{\Re}_{j\,pass.}$ vector and one $\vec{\Re}_{jo}$ vector, instead of the three pairs we would have for *three* mobile interfaces. (in spite of this relative simplicity, the increase in complexity above the "single-mobile" case is already considerable, as we will see below).

This case is very common in practice: it is represented, for instance, by grain-boundary grooves in solidifying polycrystalline, or by phase-boundary grooves in a solidifying polyphase material. (when two grains feel the same growth affinity with respect to the melt, there is no potential difference between them so that their interface is immobile- except, maybe, a transient, relaxing, ~ ρ_c-sized curved part near the junction).

Unfortunately, the increase in complexity (as compared with the single-mobile 3-junction) is so big that a thorough analysis of every possible complication is rather beyond our scope. The complexity results from the many degrees of freedom inherent in this case; as a consequence, "everything can be coupled with everything". In particular, in the general case the γ-equilibrium-prescribed $\vec{\Re}_{jo}$ vector velocity is only really autonomous when it nucleates both interfaces at the same time. When it nucleates only one of the interfaces, however, its magnitude and direction can be influenced by the other interface whose orientation φ_∞ approaches the 3-junction to a distance ~ ρ_c and provides a sort of elastic coupling between the non-nucleated interface and the velocity source $\vec{\Re}^*_{jo}$ vector of the nucleated interface. Only for the case of faceted mobile interfaces, this coupling can be so weak (or rather the 3-junction's $\vec{\Re}_{jo}$ vector so "stubborn") that the 3-junction can be considered as a velocity source independent of φ_∞.

We will treat the case on a rather intuitive basis, without pretending mathematical or physical rigour. In doing so, we will refer extensively to the discussion of case 1 (single-mobile 3-j.) of 6.2.2.

1) The definition of the two BC-independent vector velocities $\vec{\Re}_{j\infty}$ and $\vec{\Re}_{j\,pass.}$, as defined in 6.2.1, is straightforward. If no BC would be imposed, the 3-junction would follow $\vec{\Re}_{j\,pass.}$, and consequently the third (immobile) interface would be "drawn" in the direction of $\vec{\Re}_{j\,pass.}$.

2) In contrast with case 1, it is not directly obvious that the γ-equilibrium BC at the junction results in prescribed orientations φ_{o1} and φ_{o2} of the two mobile interfaces. (if this is the case, we may simply derive a fixed $\vec{\Re}_{j\,o}$ vector velocity from the $R_1(\varphi_{o1})$ and $R_2(\varphi_{o2})$ functions and follow the rest of the procedure of 6.2.2: this $\vec{\Re}_{j\,o}$ would or would not actually nucleate either interface according to a criterion similar to (6.5).

Rather, the γ-equilibrium BC prescribes a *relation between the three orientations* φ_{o1}, φ_{o2} and φ_{o3} (φ_{o3} is the orientation of the immobile interface, i.e. identical with the direction of $\vec{\Re}_j$ to be found as the result of the analysis). Such a relation can be represented as in figs. 4.4. (we will assume here that φ_{o3} is chosen as the imposed (diagonal) variable in such a representation). But in the present case the relation to be represented is not concerned with real equilibrium (zero affinity for all interfaces involved).

As discussed in 6.2.2.1, the *equilibrium* positions of the interfaces can be shifted by the affinity applied to the two mobile interfaces. This implies that the complete φ_{o1} and φ_{o2} curves are affected - actually the shift as measured from the real equilibrium, $A_| = 0$ positions would have to be determined for each value of φ_{o3} ; we may expect a range of behaviour types as suggested by fig. 6.10 a and b, but now dependent on orientation.

Now this orientation relation between φ_{o1}, φ_{o2} and φ_{o3} might seem to indicate that the orientations of the two *mobile* interfaces, φ_{o1} and φ_{o2}, together have a rotational degree of freedom instead of being fixed as in case 1. However, there is an extra condition: for a given $\varphi_{o2} - \varphi_{o3}$ combination, together with $A_{|2}$ and $A_{|3}$, we have interface velocities $R_2(\varphi_{o2}, A_{|2})$ and $R_3(\varphi_{o3}, A_{|3})$ which together determine $\vec{\Re}_{jo}$. But the direction of the third interface (φ_{o3}) has to equal that of $\vec{\Re}_{jo}$. For an arbitrary orientation φ_{o3}, this condition would be coincidental. There are two ways to match φ_{o3} and the direction of $\vec{\Re}_{jo}$:

1) Plotting the direction of $\vec{\Re}_{jo}$ (as determined by the above procedure) as a function of φ_{o3}, we will find a curve having (generally speaking, for the case of OD surface energies) one or more pairs of crossovers with φ_{o3}. For such crossover directions the condition is satisfied (intuitively we expect crossovers in one direction to represent stable situations, in the other direction unstable).

2) Allowing for curvature of the interfaces at the junction (and thereby introducing a ΔA_c shift in the effective affinity) we can release the imposed values of $A_{|2}$ and/or $A_{|3}$ while still observing the equilibrium. This implies a degree of freedom for R_2 and/or R_3 and consequently for the $\vec{\Re}_{jo}$ vector, which can so be made to coincide with φ_{o3} for *any* φ_{o3} direction.

Combining 1) and 2), we may conclude that the whole "double-mobile 3-junction" has complete orientational freedom, just as its immobile (real-equilibrium) counterpart, or just as a certain area of interface without a 3-junction at all. But this does not imply that the 3-junction has no preference for certain orientations (and corresponding $\vec{\Re}_{jo}$ directions). Such a preference results simply from the occurrence of relative minima in the total free energy of the 3-junction system, symbolically written as $\Sigma_{i=1}^{3} \gamma_i$. (of course the sum also involves areas; for the mobile interfaces, the "active" area is restricted to a width ~ ρ_c). When the sum is orientation-dependent, the junction's orientation can snap to the minimum positions, the "strength" and elasticity of the snapping being related with the depth and steepness of the $\Sigma\,\gamma_i$ minima. In general such minima correspond with minima in at least one of the γ_i and therefore with facet orientations in one or more of the participating interfaces. As the mobile interfaces of a 3-junction meet in a *re-entrant* edge, such facets are likely to be kinetically nucleated by the mating interface.

This rather intuitive reasoning results in the following picture of the "double-mobile 3-junction":

1) When no interaction with the outside world would exist, the 3-junction would choose the minimum-$\Sigma\,\gamma_i$ orientation closest to its starting shape and follow the corresponding 3-junction trajectory $\vec{\Re}_{jo}$. This $\vec{\Re}_{jo}$ vector is fully comparable with the $\vec{\Re}_{jo}$ vector of eq. (6.4c); the difference is that in the latter case the direction was imposed by the wall, whereas in the present case it is imposed by (one of the) $\Sigma\,\gamma_i$ minima.

For the moving interfaces "1" and "2", this choice of junction orientation implies an orientation φ_{o1} and φ_{o2} and a growth rate $R_1\,(\varphi_{o1}\,, A_{|1})$ and $R_2\,(\varphi_{o2}\,, A_{|2})$, respectively.

2) The discussion concerning the interaction between the φ_∞-oriented "starting orientation" and the φ_∞-oriented "BC orientation" is now fully analogous to the case of the single-mobile junction. The differences are a) that we now have to confront *two* interfaces with criterion (6.5); b) that there may be more than one possible $\vec{\Re}_{jo}$ vectors instead of the single one in 6.2.2 which was imposed by the wall direction.

The result can be one of three possible cases (see fig. 6.11):

a) For *both* mobile interfaces, the junction acts as a velocity source (i.e., expression (6.5) is > 0). This implies that a passive edge trajectory is generated for both interfaces at the starting position of the 3-junction separating the orientations φ_∞ and φ_o. The junction trajectory itself lies in between. (fig. 6.11d).

b) For both mobile interfaces, the junction trajectory and the φ_∞/φ_o edge trajectory *converge*. (i.e. expression (6.5) is <0). This means that the junction is "squeezed" between the two φ_∞ oriented interfaces; the junction's φ_o -geometry is reduced in size to within a range $\sim\rho_c$ and the ΔA_c resulting from the curvature makes the junction velocity match the velocity vector $\vec{\Re}_{j\infty}$. (fig. 6.11a)

c) Whereas cases a and b are straightforward and directly comparable with behaviour types of the single-mobile junction, case c) is a complication: expression (6.5) is <0 for one interface, >0 for the other. This means that

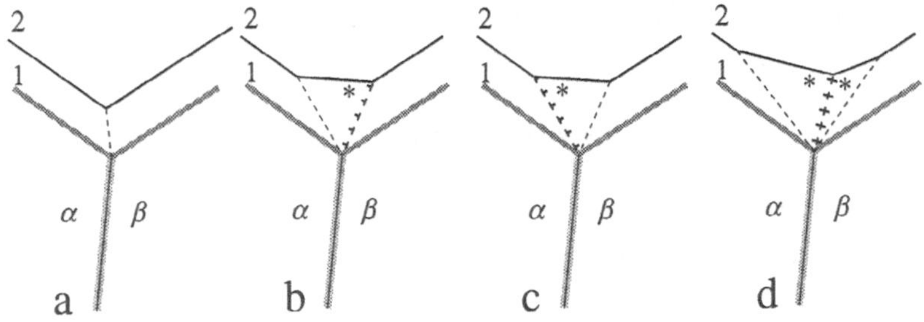

FIG. 6.11: 3-junction with two mobile interfaces; a) passive junction, b) nucleating α, c) nucleating β, d) nucleating both.

the first interface is nucleated whereas the other is "passive"; i.e. the φ_∞ orientation extends towards the junction. (fig. 6.11 c and d).

The complication implied in this case is that the "passive" side of the junction may affect the $\vec{\Re}_{j\,o}$ vector of the junction and thereby the nucleated interface (φ_o). This means that the vector velocities of the junction and of the passive edge at the nucleated side can depend on φ_∞ of the not-nucleated side.

In fig. 6.11, the "nucleated" interfaces are marked with an asterisk. Nucleating phase boundaries can sometimes be observed in unidirectionally solidifying eutectics (observation by the author; unpublished).

The case of fig. 6.11d (nucleation of both interfaces) has been sketched with "ledge resolution" in fig. 6.12. Both interfaces are assumed to be below the roughening transition.

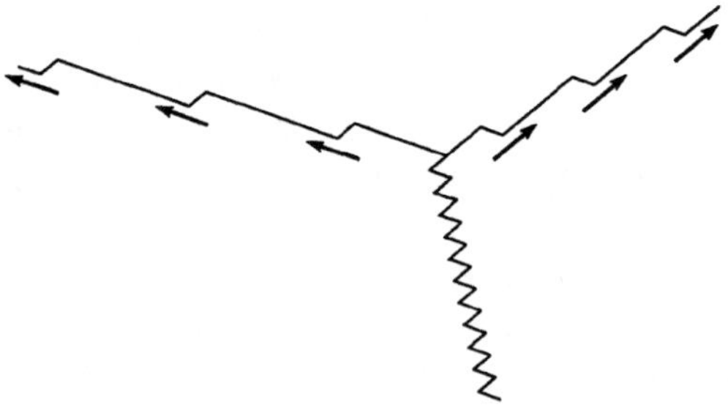

FIG. 6.12: 3-junction with two mobile faceted interfaces: kinetic nucleation by the junction.

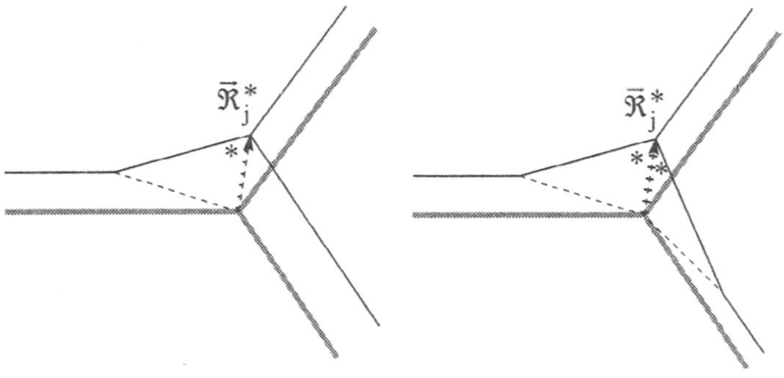

FIG. 6.13: "Triple-mobile" 3-junctions.

6.2.4 *Case 3: The 3-junction with three mobile interfaces*

That this case is not hypothetical has been argued in the introduction to this chapter: 3-junction J_2 in fig. 6.1 is an example (explained in the text).

For this "most general case", we have derived the conclusion that at least one interface is nucleated and at least one is not-nucleated. (for arbitrarily chosen φ_∞ ; see 6.2.1). Apart from this conclusion, the same considerations as in case b (6.2.1) are relevant concerning the triplet of φ_o orientations. Again, these will tend to snap into $\Sigma_{i=1}^{3} \gamma_i$ minimum positions. In this sense, the triple-mobile 3-junction behaves similarly as the double-mobile one. The difference is that the first has three moving interfaces, one of which is certainly *not* nucleated. (the two possibilities have been sketched in fig. 6.13, in perpendicular cross-section).

Note: In connection with the possibility of *three* mobile interfaces, we may remark that we have also to do with the relation $A_{\alpha/\beta} + A_{\beta/\gamma} + A_{\gamma/\alpha} = 0$ (for planar interfaces; for nonplanar interfaces, where we can have the complication of a pressure gradient in the bulk phases, this relation has to be replaced by the more general relation $\oint A = 0$ for a closed path around the junction). This relation does not prohibit the three affinities to be nonzero simultaneously. On the other hand, for the same reason a *single*-mobile 3-junction cannot be explained by simultaneous vanishing of two of the affinities. Ergo, such a case is related with $\chi = 0$ for one or two interfaces rather than with $A = 0$.

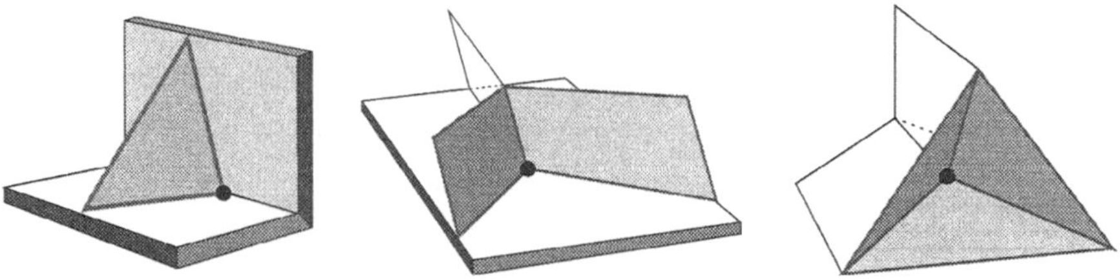

FIG. 6.14: "Starting shapes" of four-junctions with one, two and three mobile interfaces.

6.2.5 Four-junctions

The discussion of 6.2.1 - 6.2.4 on 3-junctions can readily be generalized to 4-junctions. Alternatively, the 4-junction can be considered as a connection between four 3-junctions or as a connection between six interfaces. However, as the 4-junction's possible rôle as a velocity source is concerned, the possibilities are not just exhausted by the combination of the nucleating activities of the combined 3-junctions. We have to consider the additional possibility (specific for a 4-junction, or eventually for a vertex between facet orientations) that the 4-junction itself nucleates one or more interfaces *independently of the 3-junctions*. (this is just a formal possibility; not an experimental observation).

A 4-junction nucleates an interface in a single point, whereas a 3-junction does so along a line. Consequently, a 3-junction velocity source generates an interface area in the shape of a ribbon attached to the nucleating 3-junction (the nucleated edge is marked with an asterisk as in figs. 6.15 - 6.17. A (passive) edge, parallel with the 3-junction, connects the ribbon area with the original ($\vec{n}_\infty$) orientation.

A 4-junction velocity source generates an interface area in the shape of a *triangle* led by the nucleating 4-junction (nucleating corner marked by ⊛ as in fig. 6.15 or 6.16). Two neighbouring interfaces may be simultaneously nucleated by a 4-junction, so that one of the resulting triangles is effectively truncated (fig. 6.16d).

All possible shapes of an evolving 4-junction are the result of the mutual interaction (i.e., eventual truncation) of

a) the original interfaces in their starting shape (orientation $\vec{n}_{\infty i,\, i=1 \text{ to } 6}$) moving in accordance with their $R_i(\vec{n}_i)$ (i = 1 to 6) functions;
b) the eventual ribbon-shaped areas generated by 3-junction velocity sources;
c) the eventual triangle-shaped areas generated by the 4-junction.

This mutual interaction (intersection) of 3-junctions and edges leads to a variety of hybrid 0-dimensional generalized vertices where edges and 3-

junction kinks meet. For instance, in all examples of figs. 6.15 - 6.17 we have one or more combinations of two edges and a kinked 3-junction.

Following the evolution of a starting structure amounts to calculating the *trajectories* of all these 0-dimensional plus interconnecting 1-dimensional features involved, in addition to the evolution of the curvature of interfaces in between, *if the latter are curved.* (In 6.5 we are going to discuss the special case of "multiplanar geometries" where all interfaces are flat from the start. This implies that the *trajectories only* contain all the information necessary to define the evolution of the whole structure).

Just as for 3-junctions, we can work out a catalogue of cases by systematically considering an increasing number of mobile or "participating" interfaces. Whereas we had single-, double- and triple-mobile 3-juctions, we have to discern from single- to sixtuple-mobile 4-junctions. (remember that the motion of an interface is not just a matter of affinity, but also of interface mobility. For the case of a sixtuple-mobile 4-junction we would require four grains all of different potential, e.g. a melt in contact with one (undercooled) stable and two *different* metastable phases. A rather exotic case indeed, but still *mathematically* possible. The catalogue to be discussed below is based on such a mathematical casuistics only. We have sketched starting shapes of the single-, double- and triple-mobile 4-junctions in fig. 6.14.

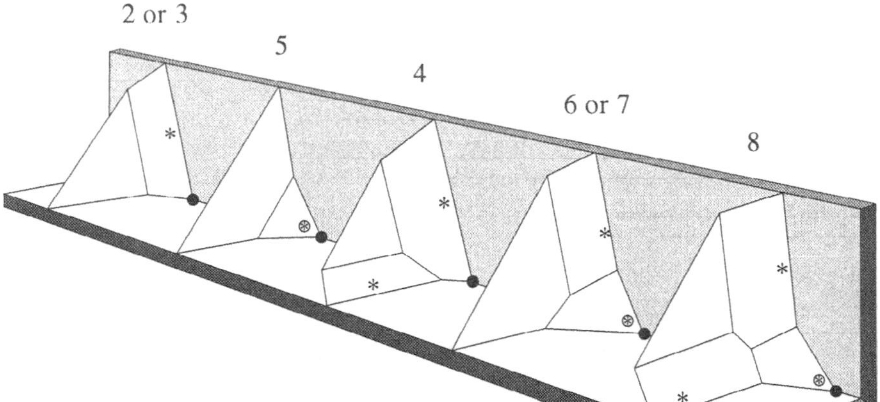

FIG. 6.15: Evolution of single-mobile 4-junctions.

In analogy to the 3-junction case, the simplest geometry to consider is the *single-mobile 4-junction*. Effectively this is an interface moving in a cell having two (nonparticipating) walls made of dissimilar materials which meet in a corner. Obviously, the combination of the interface with each of the walls on their own corresponds with two single-mobile 3-junctions. The extra element is the 4-junction itself. We will take it for granted here that a generalisation of section 6.2.2 (on single-mobile 3-junctions) to this case will lead to a distinction between "passive" and "active" or "nucleating" behaviour of the 4-junction - to be considered apart from the nucleating behaviour of the *3-junctions* participating in the 4-junction. Rather we will focus on the geometrical shapes which can result from the combination of 3-junction and 4-junction velocity sources.

The possibilities for the single-mobile 4-junction can be represented as the elements of a binary 2x2x2 matrix. The two relevant single-mobile 3-junctions are indicated as A and B. Both can be active or passive. Apart from that, the 4-junction can be active or passive. Together:

	A	B	4-junction
1	-	-	-
2	*	-	-
3	-	*	-
4	*	*	-
5	-	-	⊛
6	*	-	⊛
7	-	*	⊛
8	*	*	⊛

Cases 2+3 and 6+7 form symmetrical pairs. Case 1 is trivial (starting geometry is stationary). The cases 2/3, 4, 5, 6/7 and 8 have been drawn in fig. 6.15.

The *double-mobile 4-junction* is effectively a double-mobile 3-junction meeting a planar (nonparticipating) wall, i.e. now we have two single-mobile 3-junctions A and B, a double-mobile 3-junction C and a 4-junction. Indicating the mobile interfaces by <AC> and <BC>, respectively, we have for the nucleation possibilities:

	<AC> nucleated by		
	A	C	4-j.
1	-	-	-
2	*	-	-
3	-	*	-
4	*	*	-
5	-	-	⊛
6	*	-	⊛
7	-	*	⊛
8	*	*	⊛

	<BC> nucleated by		
	B	C	4-j.
1'	-	-	-
2'	*	-	-
3'	-	*	-
4'	*	*	-
5'	-	-	⊛
6'	*	-	⊛
7'	-	*	⊛
8'	*	*	⊛

Considering the nucleation of both interfaces as independent, we have 8x8=64 possibilities (many of which can be considered as symmetrical from a cataloguing point of view). Some of these cases have been drawn in fig. 6.16.

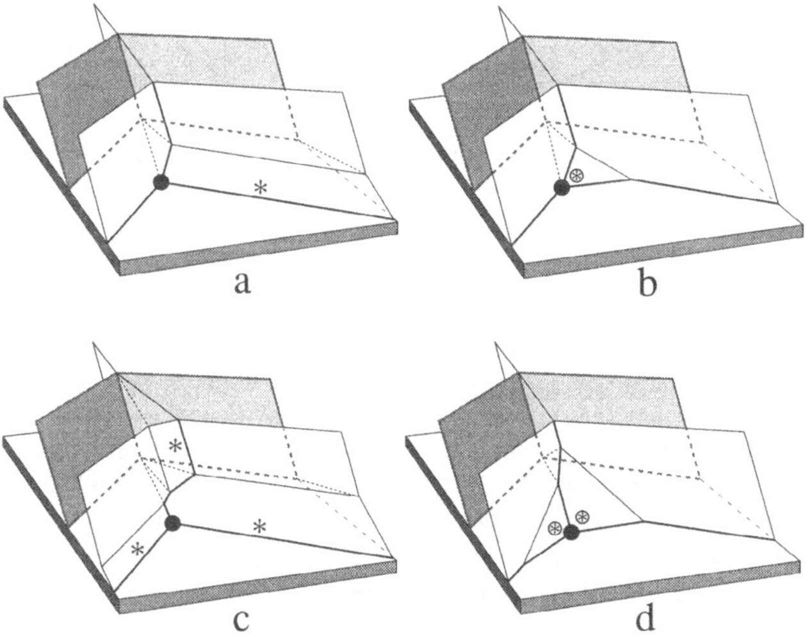

FIG. 6.16: Four cases of double-mobile 4-junctions.
a: case 1 / 2' ; b: case 1 / 5' ; c: case 4 / 2' ; d: case 5 / 5'.
(thick solid lines = 3-junctions; thin solid lines = edges)

The *triple-mobile 4-junction* consists of *three* double-mobile 3-junctions A, B and C. For each of the mobile interfaces <AB>,<BC> and <AC> a table as for the above cases applies. Some of the cases have been sketched in fig. 6.17 (note that figs 6.15 and 6.16 were attempts to perspective drawings; fig. 6.17 in contrast is a "projection" drawing, the mobile interfaces facing upwards).

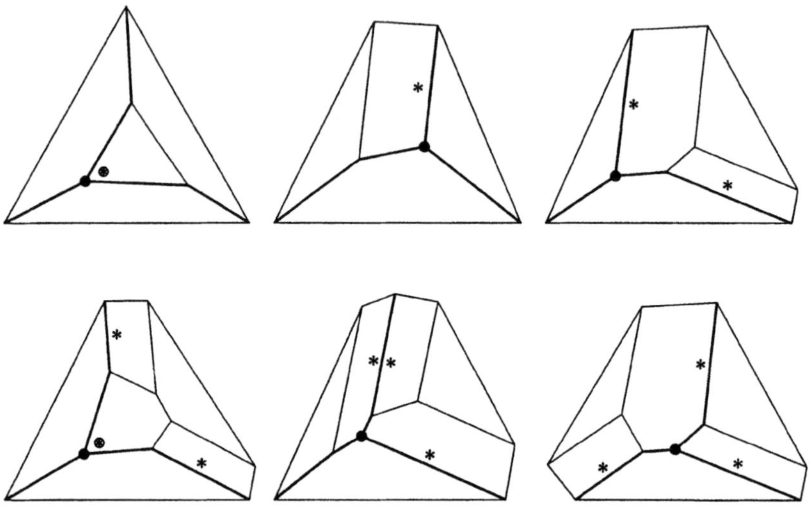

FIG. 6.17: Examples of triple-mobile 4-junctions in "head-on" view.

6.2.6 The D-junction

In this section on the kinetic rôle of junctions, the D-junction cannot be neglected. Now this D-junction is a "velocity source avant la lettre": its rôle in nucleating facet growth was recognized by FRANK c.s. (1949) and the mechanism of D-junction nucleation was worked out in considerable detail in the well-known paper by BURTON, CABRERA and FRANK (1951), with an explicit expression for $\vec{\Re}^*(A)$ as a result. Later, many authors have considered amendments to the model, leading to adapted $\vec{\Re}^*$ expressions more relevant to special cases (e.g., CABRERA and LEVINE 1956, CABRERA and COLEMAN 1963).

All above treatments are based on "atomistic" models of the D-junction (and as such they are the only viable way to find a reasonable explicit expression at all - but are beyond the scope of this monograph). Nevertheless, it is interesting- if only as a demonstration- to confront them with the continuum point of view represented here. In fact, the case provides a detailed example of the "fundamental restriction" to the continuum approach referred to in section 1.3, together with a way to escape from this restriction (in this special case) by applying a "ledge-continuum approach", i.e. something half-way between atomistic and surface-continuum.

In view of the fact that such an approach is a priori rather unrealistic from a physical point of view, we consider the oversimplified case that the dislocation's stress field is concentrated in a narrow region of a few lattice spacings diameter, and that the mechanical effect on the interface can therefore be represented as a force $\vec{F}$ acting on a small circle with radius $r = |\vec{F}|/2\pi\gamma$ (eq. (4.8)). In 4.1.7.2 we have seen that even for positive A but pinned interface, the dislocation pull generates a depression in the interface. This conclusion still applies when the interface is a facet, i.e. $\mathbf{S}\gamma \gg \gamma$, although the equations in ch. 4 then have of course to be replaced by equations accounting for orientation-dependent surface energy.

We may ask what the stationary shape of the pyramid apex (see fig. 4.6, pg. 90) would look like for the same unrealistic model but now with a free interface, i.e. a growth or dissolution situation. Superficially, we have a paradox for $A > 0$. Supposing that the $R(\vec{n},A)$ function has a zero-cusp for the facet orientation, the interface would (beyond the dislocation-induced depression) grow out to the tangent plane in the facet orientation and then stop. The depression would remain stable in accordance with the equilibrium shape compatible with the boundary conditions at the edge rim with the facet and the dislocation-attachment ring. We know, of course, that in reality the D-junction is a ledge source for $A > 0$.

The solution of the paradox is that the resolution of the $\vec{n}$ variable in the continuum $R(\vec{n},A)$ function is too crude to take properly account of the Riemann-surface character of the crystal interface around the dislocation. In reality, for a proper calculation we have to consider the detailed ledge shape. As seen "from above", this shape looks as indicated in fig. 6.18. In the central depression, which coincides with the equilibrium range, the ledge-spiral direction is opposite to that at the pyramid slope (growth-shape range). This implies that the ledge makes approximately a full U-turn on the vulcan-crater rim of fig. 6.18.

Stationary growth means that this whole ledge configuration rotates at constant angular speed, thereby advancing one ledge height per revolution.

Although this situation is stationary in a mathematical sense (in a co-rotating reference frame), it is not stationary at all in a reference frame which is fixed to the crystal, and in particular it is not stationary for the surface and bulk diffusion fields on and near the crater rim where the folded ledge passes periodically. The rotation rate of the whole structure, and therefore the vector velocity of the D-junction, is determined by the advance rate of the U-turn on the crater rim and by the radius of this rim. It

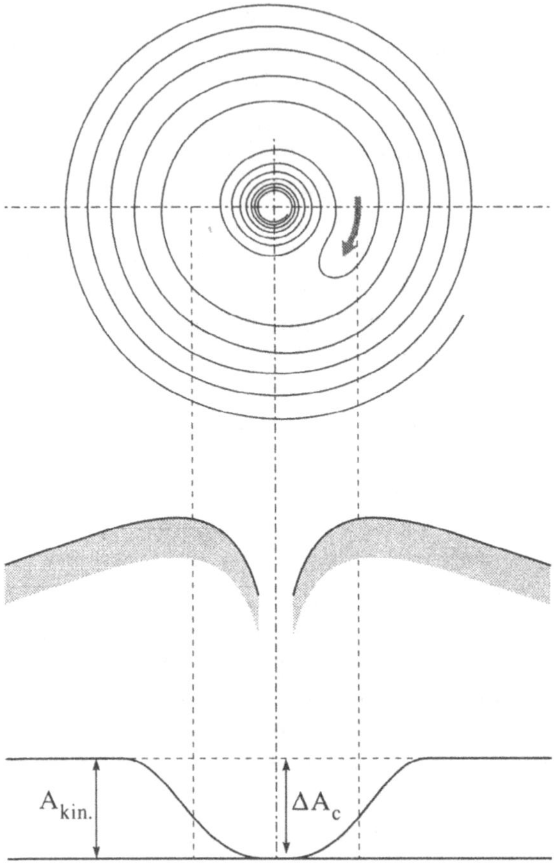

FIG. 6.18: Growth spiral on (nonrealistic!) dislocation pyramid with central dislocation force.

is obvious that for a proper calculation of this advance rate we have to start from an EOM of the *ledge*, i.e. an equation of the type (4.10a) where now A is the affinity for the ledge advancement, $R(\varphi,A)$ the single-parameter growth rate function for the ledge on that particular facet and $\gamma(\varphi)$ the *line* energy of the ledge. The coupled equation of the diffusion field is that of the *surface* diffusion field (which is in turn coupled with the bulk diffusion field). In the region of the U-turn, this surface diffusion field cannot be considered as stationary and it its coupled with neighbouring ledges which are "pushed away" by the advancing U-turn. Altogether, the total problem seems almost hopelessly complicated, but *in principle* it should be soluble from the two-dimensional ledge EOM plus diffusion equations for the surface and bulk fields.

The reason why we have so explicitly described this situation, in spite of its unrealistic assumptions, is to make clear that the screw-dislocation junction's vector velocity cannot, *in principle*, be derived from the 3-dimensional $R(\vec{n},A)$ and $\gamma(\vec{n})$ functions and corresponding EOM because these cannot properly describe the U-turn kinetics. In this sense the junction's vector velocity is really an extra and independent datum. However, when we go back one magnification level and start from the EOM's for the *ledge*, then the vector velocity may be calculated in principle.

Remarks:

1) The "thin" stress field model for the dislocation can be expected to be realistic only for extremely small driving affinities for which the critical-nucleus size $\rho_c \gg$ the actual radius of the dislocation's stress field. When this inequality is not obeyed, (i.e. in the majority of technical situations), the effect of the radially inhomogeneous stress field on the ledge kinetics has to be taken into account. Such an analysis has been worked out by VAN DER HOEK et al. (1982), although under several simplifying assumptions. The differential equation for the stationary spiral shape obtained by them yields solutions which also have a U-turn character, although the central D-junction depression is much shallower as a consequence of the distributed stress field. They provide a vector velocity as a function of A. In addition, this analysis considers the occurrence of hollow-core dislocation structures, and predicts "phase transitions" between hollow and solid dislocations as a function of A, accompanied by transitions in spiral shape and growth rate. One of the consequences is that for hollow-core dislocations, the effective line energy is a function of A too, which again complicates the $\vec{\Re}^*_{D\text{-}j}(A)$ dependence.

2) For negative affinity, the ledge spiral has the same direction everywhere and consequently no U-turn or radial part which acts as a bottleneck for the rotation rate. Consequently, the screw-dislocation D-junction mobility $\chi^*(A)$ has a "rectifying" character, being high for negative A and low (with complicated $\chi^*(A)$ dependence) for positive A. The negative-A case can be characterized as "equilibrium nucleation" by the equilibrium-shaped central part of the D-junction depression. This means that the distiction between edge- and screw-type dislocations, essential for the velocity-source behaviour for positive affinity, disappears for negative affinity (etching).

Altogether, the subject of the kinetic interaction between D-junctions and facets seems far from exhausted. For a proper discussion af all aspects, both the 2-dimensional (ledge curve) and the 3-dimensional shapes (equilibrium and growth shape of D-junction depression) have to be taken into account.

6.2.7 Kinetics of wetted junctions

In 4.1.2 we discussed the phenomenon of wetting of junctions. Roughly speaking, it implies that an extra ("wetting") phase penetrates the single, non-wetted n-junction so as to expand or inflate it to a cluster of n n-junctions which are kept in a stable arrangement by surface forces.

Motion of a wetted junction implicates motion of the whole cluster; i.e., *all of the n n-junctions* have to move with the same vector velocity, if the wetted junction is to remain stationary as such. Of course, the geometrical arrangement of the cluster may be different from the equilibrium (motionless) situation.

Altogether, wetting is a severe complication in the mathematical description of junction kinetics. Although explicit consideration of these complications is quite beyond our scope, we may give an intuitive survey:

Apart from the condition of the collective motion of the n junctions instead of one, we have to account for the equilibrium BC's at the n junctions and for the curved interconnecting interfaces. Wetting of our simplest prototype (the single-mobile 3-junction) blows it up to a conglomerate of two single-mobile and one triple-mobile 3-junctions. As far as the phenomenology of kinetic junction behaviour is concerned, we still have the possibility of active and passive behaviour of the wetted junction as a whole. But on a microscopic scale, the catalogue of possibilities is complicated. In particular, the wetting condition itself is now dependent on the junction velocity too. Junctions may change in character between wetted and nonwetted when changing velocity regimes. Such a transition may result in a simultaneous transition between active and passive behaviour of the junction.

6.3 STABILITY OF STATIONARY ZIGZAG/PYRAMID PATTERNS AND THE ROLE OF EDGES AND VERTICES.

6.3.1 Zigzag/pyramid structures versus junctions

In section 4.2, we discussed the (hypothetical) zigzag instability in an interface at rest, related with the $\mathbf{S}\gamma < 0$ condition. This subject, from the chapter on equilibrium structures, suggests the analog question whether or not the condition $\mathbf{S}R < 0$ can induce a similar (stabilized, stationary) zigzag structure in a *moving* interface - and, if not, what is the effect of $\mathbf{S}\gamma < 0$ related terms in the EOM, eventually in combination with $\mathbf{S}R < 0$ conditions. This question, in short, is the subject of the present section.

It may not be obvious at first sight why we choose to discuss this subject in the context of ch. 6 which is so specifically dedicated to the interaction between *equilibrium-specific* and kinetics-specific geometrical details. However, in the course of this section we will show that the $\mathbf{S}\gamma$-related terms in the EOM (at least for $\mathbf{S}\gamma < 0$) provide the most obvious mechanism for stabilisation of zigzag patterns; more specifically of the edges and /or vertices in such patterns. As a consequence, the $\mathbf{S}\gamma < 0$ -stabilized edges and vertices play a rôle as velocity sources, in a way which is directly comparable with the 3 - and 4 -junctions (with their γ - equilibrium imposed boundary conditions) of section 6.2.

There is, however, an important distinction between the cases of "junction kinetics" and of "zigzag kinetics". A junction, once present in a structure, cannot be extinguished; i.e. the locus where the BC is to be applied and the BC itself are obvious a priori. The reasoning resulting in the conclusions about active versus passive behaviour - including the velocity source concept - could therefore be built up by applying rather direct "construction principles" (see ch. 6.2).

In contrast, the present subject is typically related with *instability* which can lead from an initially smooth to a finally kinked or edged structure. At the start, the loci where edges might appear are uncertain - edges are triggered by noise. Consequently, the discussion has in this case rather to be set up as an investigation on "nontrivial" (i.e. nonplanar) stabilized solutions of the stationary-structure equation $\vec{\vec{\mathfrak{G}}} = \vec{\vec{O}}$. In this discussion the rôle of all physical effects (terms in the $\vec{\vec{\mathfrak{G}}} = \vec{\vec{O}}$ equation) can be considered, and in particular we can try to distinguish those which lead to stability of a ("nontrivial") solution.

In connection with this distinction, we might characterize a junction-related velocity source as a "heterogeneous", and an edge- or vertex-related one as a "homogeneous" velocity source.

6.3.2 Small-scale versus large-scale structure

In the discussion on stationary structures we have to discern between small-scale and large-scale structures. As a small-scale structure we define a "small" surface domain with nonzero curvature (zero- curvature solutions being considered as trivial stationary solutions of the EOM) and consequently defined by boundaries with inhomogeneous surface orientation. In this case, "stability" concerns the question whether this surface element (edge, vertex etc) can obey the definition of a stationary structure *and* be stabilized against perturbations.

As a large-scale structure we define a "large" surface area of constant macroscopic orientation. (e.g., a substrate in an epitaxial-growth experiment). Stationary nontrivial solutions in that case are in general periodic structures (cellular, terrace-like etc.) and the question of stability involves, apart from the stability of the *microscopic details* in the structures (the small-scale stability referred to above) also the stability of the *period*.

In ch.7 we will develop this distinction in more detail in terms of the soliton concept, as a comment on the Mullins-Sekerka-Coriell instability theory. Small-scale (nontrivial) stationary structures will be identified as single solitons in the solution of a nonlinear EOM, whereas the behaviour of large-scale structures is related with the *interactions* between the solitons (i.e., with the collective behaviour of the many-soliton system, eventually resulting in periodic structures). In the present section 6.3, however, we are uniquely concerned with small-scale stability.

An intermediate aspect between small- and large-scale structures is concerned with the occurrence of edges of opposite "sign": in a zigzag or pyramid structure built up of planar interfaces, any edge or vertex has its opposite edge or vertex (reentrant and protruding edges, RE and PE. For two orientation variables we have also the mixed type (saddlepoint) vertices). A prerequisite for the stability of zigzag structures is the stability of the edge-plane orientations, and we will consider this aspect as part of the small-scale stability.

(The situation is somewhat different in the inhomogeneous-affinity case. For interfaces with finite $\mathbf{S}\gamma$ and $\mathbf{S}R$, we may have cellular structures where only the RE's have singular curvature).

6.3.3 Intuitive "geometrical" discussion of the stability of edges.

At this point it is relevant to remind the reader that an edge in a crystal surface is covered by the "Fundamental Restriction" to the applicability of a continuum EOM. (the case to be discussed here was the very motive to call attention to this F.R. in ch.1). In short, the message of this F.R. was that even if an exact soliton solution of an EOM might be found, its physical relevance for the real crystal surface would be at most qualitative, not quantitative.

Having thus a priori degraded the status of the EOM for this case, we can feel free to approach it in an intuitive way, i.e look merely for the influence of separate terms in the EOM rather than attempt to find exact solutions. This includes the choice of the simplest possible representation. First, we choose to consider only the "cylindrical" case (only one orientation variable), assuming that extension to the 2-variable case does not involve essentially different physical principles. Second, we discuss only in a qualitative way the influence of $R(\vec{n})$ and $\gamma(\vec{n})$-related terms in the

EOM. We do this by considering plots of $\varphi(s)$ and $\varphi_s(s)$ (i.e., φ and $d\varphi/ds$ plotted vertically as a function of s) *) In such plots, the behaviour of parameters and terms in the EOM - as they depend primarily on φ and φ_s- is simply related with the vertical distance to the axis in these plots.

In the following, we will consider three situations (all with homogeneous external affinity $A_|$:

1) In the total φ range considered, **S**R is either > 0 or < 0 . *Surface energy effects are neglected.* The EOM shows directly that the "trivial solution" (planar interface) is the only stationary solution, to which a nonplanar starting surface automatically decays.

2) In the total φ range considered, a negative-**S**R range is enclosed by positive-**S**R ranges; surface energy effects are again neglected. In this case, the EOM provides no hints - predictions are ruled out by the F.R.

3) As case 2), but now we assume that, more or less in congruence with the **S**$R(\varphi)$ behaviour, a negative-**S**γ range is enclosed by positive-**S**γ ranges too (such a congruence between R and γ behaviour is to be expected on physical grounds; see 9.2). *The surface energy effects are now included*. The γ-related terms in the EOM hint to stabilisation of the re-entrant edge in a zigzag structure, and this edge will effectively behave as a velocity source.

(it may be added that any other curvature-related effect which acts in the same way can also help to stabilize the zigzag structure. Among such effects are diffusion in a boundary layer and the "Schwoebel effect").

Ad 1): Case of **S**R > *0 or* **S**R < *0; surface energy effects neglected.*

Fig. 6.19 shows a sketch of an arbitrary starting surface and the derived $\varphi_s(s)$ curve. A "nontrivial solution" implies that φ_s is not identically zero; a stationary solution that φ_{st} is zero. (φ_{st} can be represented as an arrow from the curve indicating the evolution direction of the curve, according to the EOM). Stability implies that a $\varphi_s(s)$ function which is perturbed from the stationary solution, approaches that solution again.

In the case considered here, the equation for φ_{st} is simply

$$\varphi_{st} = -\varphi_s^2 \ \mathbf{S}R(\varphi) \qquad (6.20)$$

*) For the "soliton mathematician", a NLPDE written in the Cartesian (i.e, the $y(x)$) form is probably the more familiar. Of course, the two representations are equivalent and can be translated into another, y_x being equivalent with φ and $\sqrt{1+(y_x)^2}$ with s. (the $\varphi(s)$ representation approaches the $y(x)$ representation when the φ range of a surface is narrowed to $\ll 1$). However, the $\varphi(s)$ representation seems the more "natural" for our EOM because φ_s occurs frequently as a factor. Single terms related with a particular physical phenomenon would become scattered over several terms in the $y(x)$ form, so obscuring their meaning.

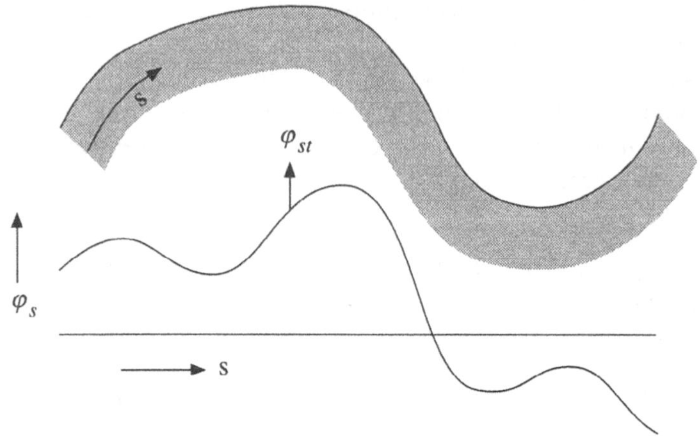

FIG. 6.19

which shows immediately that the φ_{st} arrow has the same direction everywhere (**S**R has the same sign everywhere by assumption). Of course, the flat interface ($\varphi_s = 0$) is the asymptotic situation. It is easy to verify that **S**$R > 0$ leads to sharp protruding edges, **S**$R < 0$ to sharp reentrant edges. However, the trivial solution is the only macroscopically stable one: although φ_s becomes singular for one sign (sharp RE or PE), the edge *angle* is not stabilized: it returns asymptotically to zero together with the opposite areas in $\int \varphi_s \, ds$ (the positive areas in $\int \varphi_s \, ds$ are to be compensated by the negative ones).

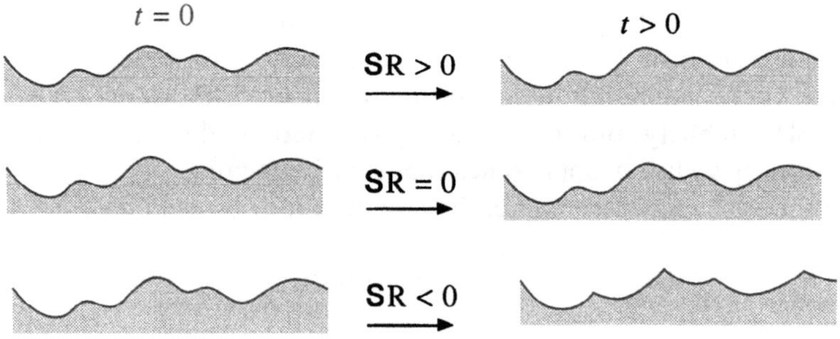

FIG. 6.20: Short-term evolution of a growing surface with either positive or negative **S**R. The long-term asymptotic situation is a planar interface, for **S**$R < 0$ as well as > 0.

The evolution of an undulated starting surface, obeying the assumptions of 6.3.3, case 1) and with the sign of $\mathbf{S}R$ as a parameter, is sketched in fig. 6.20. If $\mathbf{S}R \neq 0$, the $t \rightarrow \infty$ asymptotic situation is a *planar* interface.

Ad 2): Case of $\mathbf{S}R$ *variable over* >0 *and* <0 *ranges; surface energy effects neglected.*

In fig. 6.21 we have sketched the situation "2" as characterized above. The $\mathbf{S}R>0$ regions (for $\mathbf{S}R \gg R$, "facets" in the growth shape) now separate $\mathbf{S}R<0$ regions of opposite curvature. For $\varphi_s > 0$ (region A in fig. 6.21.) the situation is simple: the curved interface evolves into a passive PE. For the region where $\varphi_s < 0$ (B in fig. 6.21) the situation is less trivial. When the initial configuration has a finite curvature, then we have within B three distinguishable parts of finite size, where the vector velocities converge, diverge and converge again. As a result, two passive RE edges develop whose vector velocities diverge. But it is less obvious what happens when we let the initial configuration approach a sharp edge between flat regions of positive $\mathbf{S}R$ orientation. The central part of the curved region where the vector velocities of the passive RE diverge becomes smaller, whereas it remains surrounded by regions of converging vector velocity. When the central part could be squeezed out altogether ($r \rightarrow 0$), the edge might be stabilized by the neighbouring positive-$\mathbf{S}R$ regions.

What becomes apparent from this intuitive and rather vague argument is that this approach is blocked by the "fundamental restriction" of ch.1: conclusions cannot be drawn.

Ad 3): Stabilisation of RE by $\mathbf{S}\gamma<0$ *range.*

Roughly, the argument on the $\mathbf{S}\gamma < 0$ - induced stability proceeds as follows. Assuming linear $R(A)$ dependence ($\mathbf{S}R(\varphi,A) = A\mathbf{S}\chi(\varphi)$), we have simply

$$\varphi_{s\,t} = -\varphi_s^2 \,\{A_| - V\varphi_s \;\mathbf{S}\gamma(\varphi)\}\,\mathbf{S}\chi(\varphi) \tag{6.21}$$

+ terms accounting for the gradient in the capillarity shift;. i.e. the capillarity shift introduces a $(\varphi_s)^3$ term. This means that for negative $\mathbf{S}\gamma$, $|\varphi_s|$ can increase for both signs of φ_s (A_e assumed to be positive). For an originally unstable edge, φ_s has first to exceed a critical or nucleation barrier. In principle, the edge is now stabilized by the free-energy minimalisation of the surface in the vicinity of the edge: the edge orientations are approximately those of the negative-$\mathbf{S}\gamma$ zigzag structure discussed in 3.7.

According to the above EOM, φ_s grows to infinity, i.e. a mathematically sharp edge and obviously not a physical reality. The "fundamental res-

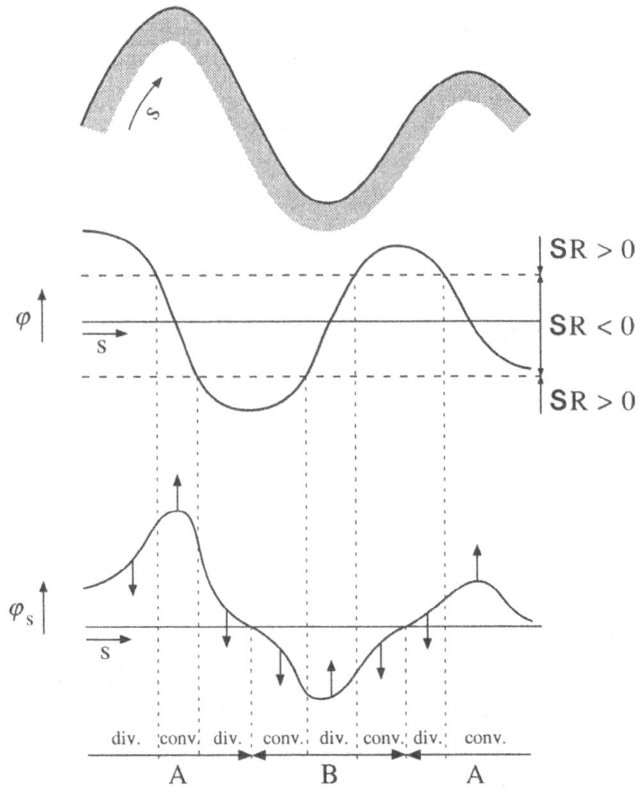

FIG. 6.21

triction" implies that we have to consider this result as a hint that the edge is stabilized and that we have to start a statistical (Monte Carlo) analysis to determine its vector velocity as a function of affinity: $\vec{\mathfrak{R}}^*_{\mathrm{edge}}(A)$. The same analysis will also provide the orientations of the nucleated interfaces.

In the same way as the capillarity term, any other effect which introduces a term proportional with curvature can destabilize the trivial solution. Still, it does not necessarily stabilize a nontrivial solution (with stabilized zigzag orientations) in itself. (If not, the limit of a 180° reentrant stabilized edge can automatically take over the rôle of the stabilized edge, but then the rest of the surface has to be curved. This is what happens in cellular growth for low-α factor materials).

In this connection, the delocalisation effects mentioned in 2.8 are to be reminded. In the first place, a diffusion boundary layer gives rise to an effective curvature dependence of the growth rate function (eq. (8.8)). In this case the limited validity of the φ_s expression is even more evident.

The basic mechanism is identical with that of constitutional supercooling.

In the second place, "asymmetric delocalisation" of the growth rate as a result of the Schwoebel effect gives rise to a φ_s term in eq. 2.19. Unlike the boundary-layer case, this effect is orientation-dependent, so that it might in principle stabilize a finite zigzag angle. If so, this could mean that a nontrivial zigzag solution could be stabilized by kinetic effects alone, without the help of negative-**S**γ effects. (This mechanism is not implicit in the EOM as it cannot be expressed as a feature of the $R(\varphi)$ function. It should automatically be considered in any atomistic model, however).

6.3.4 The **S**γ *< 0 instability hypothesis and the evolution of zigzag structures*

Experimental evidence on $R(\vec{n})$ functions, on many crystal/parent phase interfaces, shows abundantly that "fast-growing orientations" (with **S**$R < 0$) do occur in practice. This proves that such interfaces have been stable at least for the duration of the experiment: if they had broken down *immediately* into a zigzag shape as hypothesized in the previous section, the measured velocity would have been that of the stabilized RE, i.e. the observed growth rate plot (with a zigzag interface) would have been a "soap-froth plot" with **S**R identically zero. Large and "sharp" maxima in R plots would simply not be found experimentally.

Conclusion: if the arguments on **S**$\gamma < 0$ -stabilisation (in 6.3.3, case 3) are valid so that zigzag instability is a realistic effect, then the "stability" of the experimentally found **S**$R < 0$ orientations is in fact a *meta*stability. Stabilized RE's, if not present in the starting shape of the experiment, are to be *nucleated* (and, once they are, can eventually be annihilated again). If the nucleation barrier is high enough, then the metastable flat interface can persist as such during the whole growth experiment.

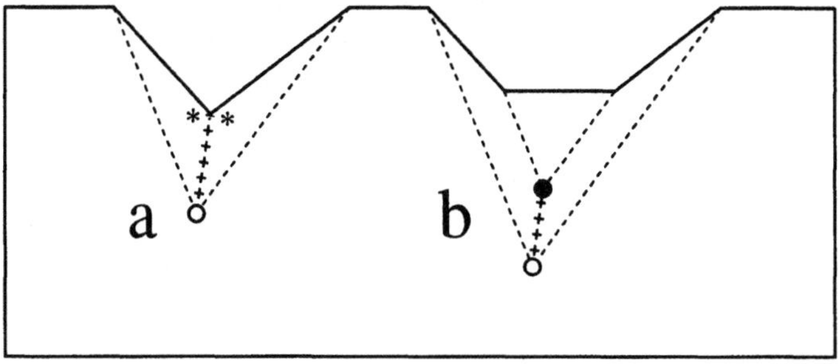

FIG. 6.22: Generation (a) and subsequent generation + annihilation (b) of a (hypothetical) low- growth rate RE velocity source in a high-growth rate oriented interface.

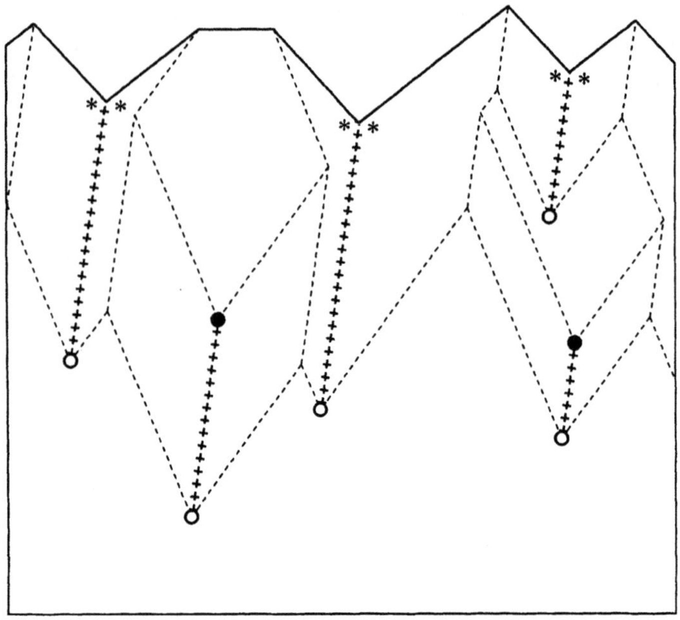

FIG. 6.23: Zigzag corrugations in a $\mathbf{S}R < 0$ oriented interface associated with statistical "field" of RE generation and annihilation events.

Generation and annihilation of stabilized RE's is typically an atomistic phenomenon; estimates of generation/annihilation occurrence have to be based on statistics, e.g. Monte Carlo simulation. *) This in contrast with the $\mathbf{S}\gamma > 0$ case, where the EOM-based Mullins-Sekerka approach is valid.

In fig. 6.22a and b, we have sketched the evolution of interface shape of an originally flat $\mathbf{S}R < 0$ interface upon nucleation and subsequent annihilation of a stabilized RE. As nucleation of a new orientation automatically generates a passive edge between new and old orientations, the dually nucleating RE trajectory is enclosed by two "passive" PE trajectories. After annihilation (marked by a black dot in the figures) the stabilized RE breaks up into two passive RE's (i.e., zig/original and original/zag), with trajectories parallel to those of the passive PE's.

Neighbouring nucleation and/or annihilation events give rise to interactions of inclined passive trajectories. Such interactions result in the formation of a kinked surface. In fig. 6.23, we have sketched an interface as resulting from a (hypothetical) "field" of generation and annihilation events

*) As to the annihilation of RE's, a possible complication may be that interaction between stabilized RE's and D-junctions trapped by them may result in an additional annihilation mechanism.

(i.e., "active" RE trajectory sections in the drawing). Such generation-annihilation maps (observable as RE trajectory maps) have been found experimentally in zone-melted thin Si monocrystalline films (PFEIFFER c.s. (1985); see for details also 6.3.5).

When spontaneous annihilation is absent, the zig and zag orientation regions grow out to fill the whole area (on collision they form the corresponding PE so that ultimately no interface of the original orientation is left). As generation can only occur in the original $\mathbf{S}R<0$ orientation (horizontal in fig, 6.23), no new RE's can be formed in a completely zigzagged interface. Therefore, such a structure is "stable" (once it is formed) when annihilation is practically nonexistent, metastable when annihilation probability is finite. (note that generation probability of RE's is proportional to the total area of the *surface* of $\mathbf{S}R < 0$ orientation, annihilation probability is proportional to the *length* of RE present in the interface).

Case of inhomogeneous affinity field

Although the bulk of sections 6.1 through 6.4 is concerned with the homogeneous-affinity case (case "A" as defined in ch. 5, table 5.1), we make a short excursion to the inhomogeneous-affinity case here, as a counterpart to the corresponding section 4.2 on equilibrium zigzags. For simplicity, we consider the imposed-affinity field (case "B" as defined in ch. 5) first.

For positive $\vec{\nabla} A$, the originally flat groove slopes tend to become concave (fig. 6.24). So, also the passive PE trajectories become curved and ultimately the whole RE groove is stabilized to a certain width (related with $\vec{\nabla} A$). When the RE angle, (flattened by $\vec{\nabla} A$) becomes very blunt, the annihilation probability increases: the RE trajectory may be broken off and the whole groove disappears. Summarizing: a positive $\vec{\nabla} A$ tends to limit the nucleated RE grooves both in size and duration. When nucleation events are rare enough, the resulting interface morphology may look as a surface with emerging "bubbles" which disappear sooner or later after formation. A negative $\vec{\nabla} A$ has the opposite effect: each nucleated RE grows out infinitely.

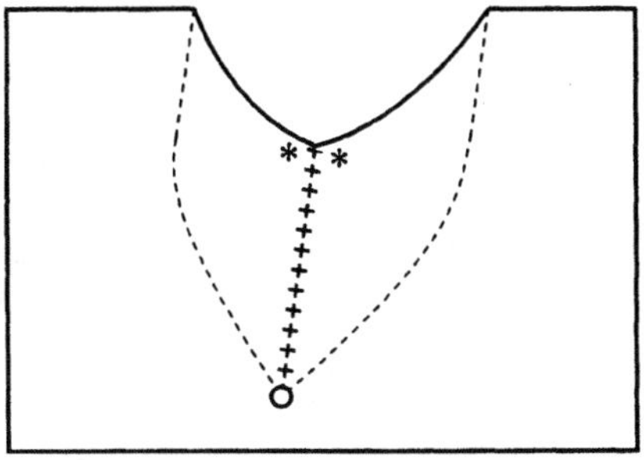

FIG. 6.24

The "constitutional supercooling" (CS) condition (i.e., imposed growth rate in a Bridgman or Czochralsky situation, resulting in a negative-$\vec{\nabla}A$ layer of a size related with D/R) typically results in a zigzag period also stabilzed to a size related with D/R, as discussed in the next subsection.

6.3.5 "Integration" of CS and $\mathbf{S}\gamma < 0$ *instability*

The validity of existing theories on CS instability (Mullins-Sekerka, Wollkind-Segel) in a unidirectional-growth, case C_1 or C_2 (ch. 5, table 5.1) situation, is restricted to the condition $\mathbf{S}\gamma > 0$. Extension to the range $\mathbf{S}\gamma < 0$ implies that generation/annihilation of sharp edges has to be considered within the same theory. The basic mechanisms have been sketched above, but the "case C" situation implies a further complication as compared with the "case B" point of departure we used there.

From the mathematical point of view, the basic difference between the ($\mathbf{S}\gamma \geqslant 0$ only) and the general ($\mathbf{S}\gamma < 0$ or > 0) cases can be characterized in that the first one can be fully described within the continuum approach, whereas the second case essentially requires a statistical description of RE nucleation kinetics (as a consequence of the "Fundamental Restriction"). We may try to make an intuitive prediction of interface behaviour as a function of $\mathbf{S}\gamma$. Experimental parameters are externally applied growth rate R and $\vec{\nabla}A$; $S\gamma$ can be considered as a system parameter. (of course, it is a very crude parameter; in fact the detailed shape of both $\gamma(\vec{n})$ and $R(\vec{n})$ functions are important). Experimentally, the evidence is that for stable enough conditions (from the Mullins-Sekerka point of view) the interface is flat, for unstable enough conditions it is cellular (smoothly curved for $\mathbf{S}\gamma > 0$, zigzag for $\mathbf{S}\gamma < 0$ conditions) or even dendritic. As a rough criterion to distinguish between stable and unstable conditions, we can determine whether or not a supersaturated/undercooled layer exists ahead of a (supposedly flat) solidification front. This is the CS criterion as first described by RUTTER and CHALMERS (1953):

$$
\begin{aligned}
&R/\vec{\nabla}_n T < K_{\text{crit.}} \quad \text{or} \quad \vec{\nabla}_n A_| > 0 \quad \rightarrow \quad \text{interface stable} \\
&R/\vec{\nabla}_n T > K_{\text{crit.}} \quad \text{or} \quad \vec{\nabla}_n A_| < 0 \quad \rightarrow \quad \text{interface unstable}
\end{aligned}
\tag{6.22}
$$

where $K_{\text{crit.}}$ is a critical value depending on system parameters such as phase diagram data and diffusion constant, and $\vec{\nabla}_n A_|$ is the externally imposed A gradient (see 1.5.4 (page 20) on the "3-D $A_|$ field" concept).

Plotting the thickness d_{CS} of this layer as a function of the "CS parameter" $R/\vec{\nabla}_n T$ provides - very roughly - an order of magnitude for the possible stabilized wavelength of cellular structures formed after break-up of the planar solidification front, although without the refinements due to capillarity effects as discussed in the MS theory (fig. 6.25a). In agreement, we have sketched in fig. 6.25b the period of a cellular structure as it might be expected to develop, as a function of **S**γ. (we emphasize that the sketch is hypothetical and intuitive, and just serves to illustrate principles):

* For **S**$\gamma = \gamma = 0$ and $\chi^{-1} = 0$, we are in the original Rutter-Chalmers case and the critical value of $R/\vec{\nabla}_n T$ represents the transition between planar and cellular situations *exactly*.
* For **S**$\gamma > 0$, we are in the Mullins-Sekerka case. The capillarity effect tends to stabilize the shorter wavelengths, so that the onset of instability is delayed to a $R/\vec{\nabla}_n T$ value larger than the Rutter-Chalmers limit.
* For **S**$\gamma < 0$, in contrast, the transition is smeared out. Even for $R = 0$ we have the (hypothetical) zigzag instability as described in 4.2; eq.(4.16) or (4.17). For increasing R, complications arise because interface kinetics (the $R(\vec{n})$ function) acquire an active rôle *and* because $\vec{\nabla}_n A_|$ is affected. Roughly, λ will tend to increase slightly, until the Rutter-Chalmers limit is reached - after which it will approach the size typical for **S**$\gamma > 0$ cells.
* The *nucleation frequency* of RE's plays a decisive role in the "incubation time" of instability. As such, it may be compared with the $(1/\delta)(d\delta/dt)$ versus λ plots in the MS theory.

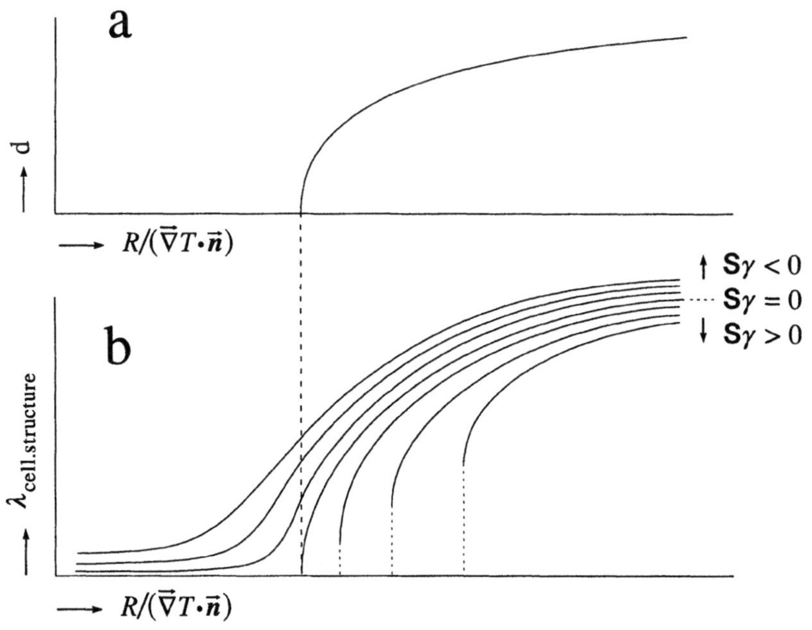

FIG. 6.25: Hypothetical behaviour of cellular size in a unidirectional-solidification situation.

For the case of an $R(\vec{n})$ function with sharp cusps for minima and maxima, the interface is "polyplanar" anyhow and the nucleation approach has to replace the continuum approach completely. This case, although basically different from the MS case, is still relatively simple.

But for smoother $R(\vec{n})$ and $\gamma(\vec{n})$ functions, nucleation of sharp RE's has to be considered *alongside* with the "continuum" instability. This implies, for instance, that nucleation in the valleys ($\vec{\nabla} A < 0$ case) is more probable than on the hill tops, so that the nucleation frequency should be known *as a function of A and of orientation.* This promises to complicate the description considerably.

Experimentally, faceted-cellular structures have often been observed in thin Si films, e.g. by L. PFEIFFER c.s. (1985). Microscopic observation of the RE trajectory maps resulting from this phenomenon is facilitated because impurities (present in the melt) heap up in the RE, so that after some time the lattice is distorted and the RE trajectories develop into low-angle grain boundaries. The authors present micrographs of low-angle grain-boundaries = RE trajectories in Si specimens unidirectionally solidified in a CS situation, whose behaviour agrees with the right side of fig. 6.25. However, the transition to the CS-stable situation (left side) has not been followed. (equilibrium and near-equilibrium zigzag structure in thin Si layers was reported by U. LANDMAN c.s. (1986) - see page 96). Unfortunately, experimental data concerning important parameters such as $\vec{\nabla} T$ and the nature and concentration of impurities are not mentioned in the paper.

The authors also present a mathematical model for the RE trajectories which, as worked out into a computer simulation of a trajectory map, shows a striking similarity with experiment. I would like to comment, however, that such experimental trajectory maps are by no means *necessarily* linked with this particular mathematical model. (which was based on a rather questionable relation between facet size and facet growth rate, and did not account for $\vec{\nabla} T$, nor for impurity diffusion effects). In contrast, I expect that a very wide class of mathematical models (relating $\vec{\Re}_{RE}$ with $A_|$) will result in quite similar results in terms of RE trajectory maps. Only the details of such maps - e.g., the precise relation between imposed growth rate $R_{mech.}$, $\vec{\nabla} T$ and impurity content - will depend on the model.

6.4 THE VELOCITY SOURCE CONCEPT: A SUMMARY

Now we have illustrated the velocity source concept in the particular context of junctions (6.2) and stabilized edges (6.3), it is appropriate to recapitulate and define this concept formally in geometrical terms. For this purpose, we use again the term "connective element" (CE) for junctions, edges and vertices together.

In section 6.4.1 we discuss the kinetics of CE's "phenomenologically": i.e. we consider their possible behaviour, either observed by experiment or deduced from theory such as in previous sections, as given facts - without caring for the specific type of EOM or for the physical mechanism responsible for their behaviour. We just distinguish, on a geometrical basis, the cases in which a CE acts as a velocity source from the "passive CE" case.

Then in 6.4.2, we relate this distinction with (types of) equations of motion. We emphasize the fact that the velocity source is incompatible with the "classical" kinematic-wave theory, i.e. with an EOM classified as "case A" or "case B" in ch. 5, table 5.1, where curvature- induced affinity shift is neglected. Those boundary conditions (imposed by CE's) leading to velocity source behaviour are simply absent in case A or case B equations.

Finally, in 6.4.3 we illustrate the concept by typical examples of velocity sources.

6.4.1 Formal definition of a velocity source

In a structure of interfaces such as illustrated in fig. 6.1, CE's were defined as those joints between smooth interfaces and/or line defects where orientation changes "abruptly". (this abruptness will be specified more precisely in 6.4.2). The catalogue of CE's consists of 1-dimensional (3-junctions and edges) and 0-dimensional ones (4-junctions, D-junctions and vertices). A *single* CE is associated with *at least two different* interface orientations.

It should be emphasized that the velocity source concept is associated not with a CE as a whole, but with a *(CE-interface) pair.* (an edge has two such pairs, a 3-junction three, a vertex three or four, a 4-junction six. The D-junction is associated with a possible continuum of orientations). So the CE can be a velocity source *for one or more of its interfaces;* not necessarily for all or none of them. Alternatively we can say that a CE can "nucleate" one or more of its interfaces.

The interface partner in such a pair is not infinitely extended: it is an interface with a boundary; i.e. the CE. Therefore, the relative position of interface and CE is important. For 1-dim. CE's we can distinguish "left-bounded" and "right-bounded" interfaces as illustrated in fig. 6.26. (left and right are defined by referring to the interface normal as vertical; an alternative nomenclature is counterclockwise versus clockwise).

For convenience, we first consider the case of 1-dim. CE's. (one orientation variable). If the CE is a 3-junction, we have to do with three different CE-interface pairs, i.e. three different $\vec{\mathsf{P}}R$ plots; if it is an edge, with only one $\vec{\mathsf{P}}R$ plot. In a plot figure, the vector velocity of a CE is just a single vector $\vec{\mathfrak{R}}_{3\text{-j}}$, $\vec{\mathfrak{R}}_{RE}$ etc. The stationarity condition for the (CE, interface) pair

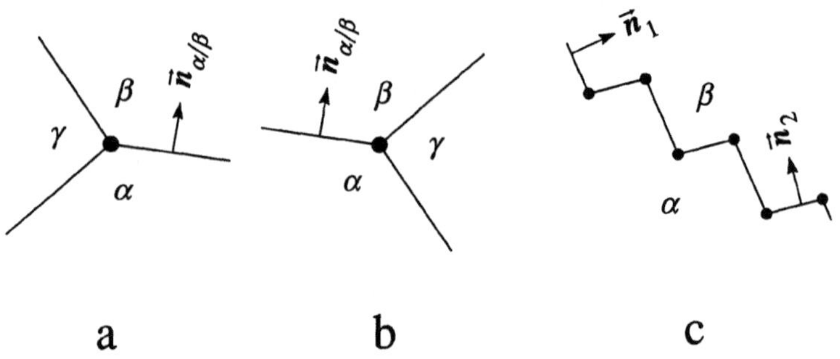

FIG. 6.26:
a: left-bounded or ccw-bounded (CE- α/β interface) pair in a 3-junction
b: right-bounded or cw-bounded (CE- α/β interface) pair in a 3-junction
c: for a zigzag structure in a single α/β interface, PE's and RE's are associated with "opposite" pair types: the four (CE-interface) pairs are (PE, $\vec{n}_{1\,\mathrm{left}}$), (PE, $\vec{n}_{2\,\mathrm{right}}$), (RE, $\vec{n}_{1\,\mathrm{right}}$) and (RE, $\vec{n}_{2\,\mathrm{left}}$)

implies that $\vec{\mathfrak{R}}_{CE} - \vec{\mathfrak{R}}_{\mathrm{interface}}$ is a vector in the plane of the interface, which is equivalent with an equation of the type (6.1). In a plot drawing, we can draw the "effective growth rate plot" circle (see ch.2) associated with $\vec{\mathfrak{R}}_{CE}$, on which all $(\vec{n}, \vec{\mathrm{P}}R_{\mathrm{interface}})$ positions obeying the above condition should be located. The intersection point of this " $\vec{\mathfrak{R}}_{CE}$ circle" and a $\vec{\mathrm{P}}R_{\mathrm{interface}}$ plot defines orientation ($\vec{n}_{\mathrm{nucl.}}$) and velocity ($R(\vec{n}_{\mathrm{nucl.}})$) of the interface, but it is irrelevant whether we refer to to a left- or to a right-bounded interface. When we want to express the latter in a plot drawing we have to do this by direction or "attachment" symbols (←, →) as in figs. 6.28, 6.30 and 6.31.

The essential criterion in the velocity source concept lies in the location of the vector $\vec{\mathfrak{R}}_{CE} - \vec{\mathfrak{R}}_{\mathrm{interface}}$. *When this vector lies in the relevant part of the interface, we say that* $\mathfrak{R}_{CE}$ and $\vec{\mathfrak{R}}_{\mathrm{interface}}$ *converge, (fig. 6.27a) otherwise that they diverge (fig. 6.27b).*

This "in/out" distinction is *not* related with equation 6.1 which just expresses that $\vec{\mathfrak{R}}_{CE} - \vec{\mathfrak{R}}_{\mathrm{interface}}$ is located in the *plane* of the interface; rather it is related with the presence of the boundary (CE) itself. Note that the convergence/divergence distinction is applicable to 1- and 0-dimensional CE's.

For a 1-dimensional CE, where we have just *complementary* positions of left- and right-bounded interfaces of a given orientation, convergence of an $(\vec{\mathfrak{R}}_{CE}, \vec{\mathfrak{R}}_{\vec{n}\,\mathrm{left}})$ pair automatically implies divergence of the complementary $(\mathfrak{R}_{CE}, \mathfrak{R}_{\vec{n}\,\mathrm{right}})$ pair, and vice versa. For (0-dimensional) 4-junctions and vertices we have to do with interface *sectors*, each defined by two 1-dimensional CE lines meeting in the 0-dim. CE.

	$\vec{\Re}_{CE}$ and $\vec{\Re}_{interface}$ convergent	$\vec{\Re}_{CE}$ and $\vec{\Re}_{interface}$ divergent
mathematical meaning	$\vec{\Re}_{CE} - \vec{\Re}_{interface}$ *in* interface (as drawn from CE) 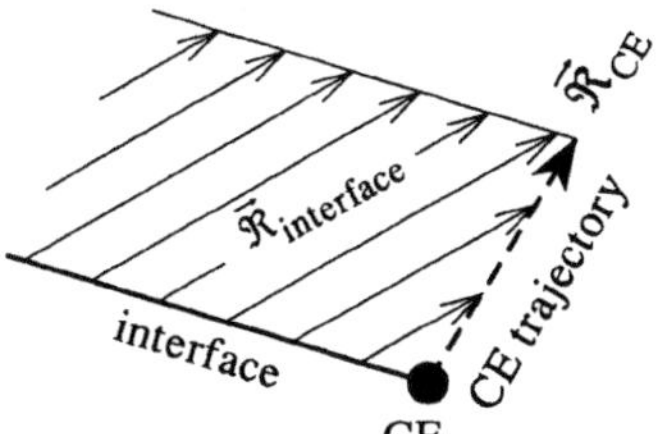 FIG. 6.27a CE trajectory is a *sink* for interface velocity vectors	$\vec{\Re}_{CE} - \vec{\Re}_{interface}$ *outside of* interface (as drawn from CE) 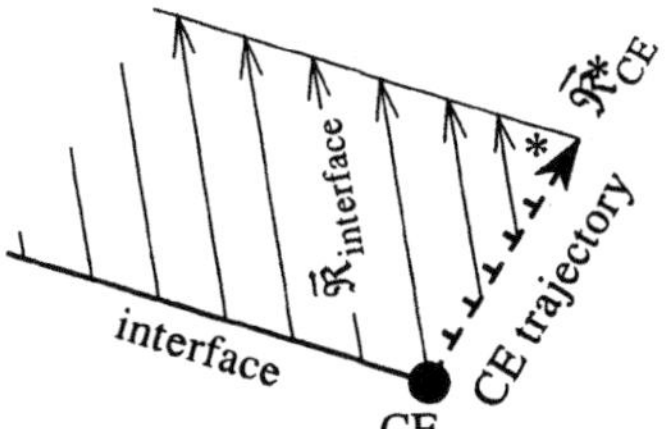 FIG. 6.27b CE trajectory is a *source* for interface velocity vectors ("velocity source")
terminology	*passive* behaviour of CE with respect to interface	*active* behaviour of CE with respect to the interface; CE *nucleates* the interface = *is a velocity source* for the interface
phenomenology	a) ρ_c- sized *microscopic* shape detail, in accordance with capillarity terms in EOM, in interface near CE = region of orientation change + ΔA_c b) a change in asymptotic orientation $\vec{n}_\infty$ is possible and is absorbed in a change of $\vec{\Re}_{CE}$	a) no microscopic shape details interface is flat up to CE on a scale < ρ_c. b) a change in asymptotic orientation $\vec{n}_\infty$ is not permitted. Such a change is answered by a split-up of the CE in two; one of these keeps the original interface orientation and $\vec{\Re}_{CE}$; the other is passive.

TABLE 6.1

The different aspects of the divergent/convergent distinction are summarized in table 6.1 (including figs. 6.27a/b).

Consider fig. 6.28, where we have drawn one CE velocity vector $\vec{\Re}_{CE}$ and its associated effective-plot circle, in connection with two different $R(\vec{n})$ plots A and B. Together with the two possible alternatives of a left-

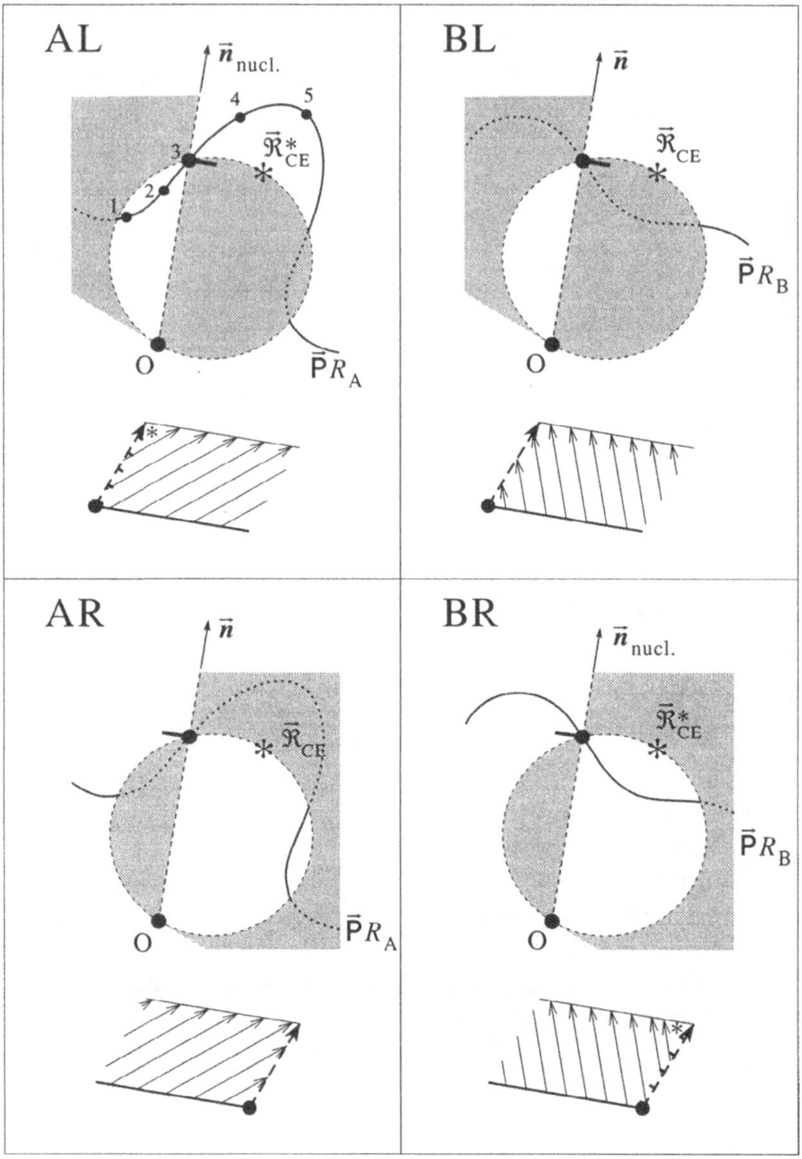

FIG. 6.28: Illustrating velocity-source criterion for CE- interface pair.

and a right-bounded interface (L and R) this gives four different situations AL, BL, AR, BR. (in the drawings, the left- and right-bounded interfaces are symbolized by "handles" representing the position of the interface).

1) As eq. (6.1) has to be satisfied anyhow, the possible (CE-interface)

pairs are represented by intersections of the $\vec{P}R(\vec{n})$ plot and the $\vec{\mathfrak{R}}_{CE}$-effective plot circle. (an effective plot circle which does not intersect the $\vec{P}R$ plot at all is an incompatible situation). In fig. 6.28, the cases AL, BL, AR and BR each represent just one CE-interface pair, as drawn in the inserts. (the other intersections between plot and circle are not related with that particular CE-interface pair!).

2) The divergent/convergent criterion is related with the relative position of the $\vec{P}R(\vec{n})$ plot and the $\vec{\mathfrak{R}}_{CE}$-effective plot circle: AR and BL represent "convergent" situations, AL and BR "divergent" ones. The difference can be visualized by distinguishing "white" and "grey" areas in the plot drawings in agreement with fig. 6.7 (section 6.2.1). White and grey change rôles between right- and left-bounded interfaces of the same orientation.

Starting interfaces with $\vec{P}R(\vec{n})$ in the white zone (full-drawn plot) are nucleated by the CE, while those with $\vec{P}R(\vec{n})$ in the grey zone (dashed plot) are not. In fig. 6.29 we have sketched the evolution of five CE-interface (left-bound) starting pairs with orientations/ $\vec{P}R$ positions marked with the numbers 1 through 5 in fig. 6.28, AL. Only in case 3 the interface motion can be considered as parallel translation of the starting surface along $\vec{\mathfrak{R}}_{CE}$

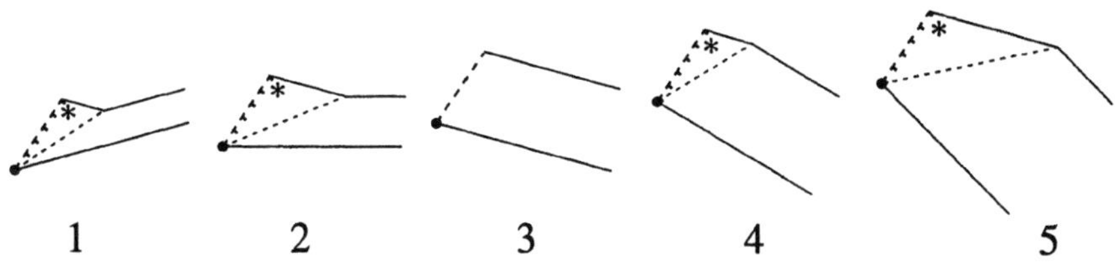

FIG. 6.29: Evolution of CE-interface pairs corresponding with the positions 1 through 5 on the $\vec{P}R$ plot "AL" in fig. 6.28.

(in agreement with eq. (6.1). In the other cases (not satisfying eq.(6.1)) a passive edge (with convergent $\vec{\mathfrak{R}}_{interface}$ pairs) is generated between the original and the nucleated orientation. Note that case 3 obeys the $\vec{\vec{\mathfrak{S}}} \equiv \vec{\vec{O}}$ criterion for stationary motion, *with translation direction identical with* $\vec{\mathfrak{R}}_{CE}$, although the interface vector velocity is at an angle with $\vec{\mathfrak{R}}_{CE}$.

In contrast, when the plot is in the grey zone at both sides of the $\vec{P}R(\vec{n})$ plot / $\vec{\mathfrak{R}}_{CE}$ circle intersection, we can not connect the CE (moving at velocity $\vec{\mathfrak{R}}_{CE}$) to the interface by a connecting interface obeying the $\vec{P}R$ plot. Any change in starting orientation then affects $\vec{\mathfrak{R}}_{CE}$ itself. This situation corresponds with a *converging* CE-interface pair.

6.4.2 *Correspondence with EOM types*

In the "classical kinematic-wave theory" case, i.e. "case A" as defined in ch. 5, the evolution of any starting shape is constructed by translating every point of it over $t.\vec{Q}R$. This procedure can result in self-intersections of the translated starting surface, i.e. to abrupt orientation changes which therefore are to be regarded as CE's in the general sense.

Trajectories of these CE's are implicitly determined by this construction. However, the principle implies that $\vec{Q}R$ vectors of the interfaces connected in the CE either converge or are parallel; *they cannot diverge.* Ergo, the velocity source concept is incompatible with the "type A" EOM. This conclusion is not basically affected when we extend to *junction* CE's, at least when we allow for complete "orientational flexibility" of the interfaces connected in the CE's (in agreement with the absence of surface energy effects, whose equilibrium would impose *fixed* mutual orientations). Also, the extension to "case B" of ch. 4 - i.e., inhomogeneous affinity - can be ruled out to affect the conclusion: on the small scale of the dimension of the CE, affinity can be considered to be constant.

In conclusion, we can say that CE's can behave as a velocity source for an interface *only when they impose a boundary condition* on this interface which is not already implicit in a case A or case B EOM and which, consequently, can induce the interface to depart from the "classical kinematic-wave theory" behaviour. In sections 6.2 and 6.3 we have extensively discussed how *surface energy effects* and *ledge nucleation kinetics* can (both or separately) be responsible for such BC's. Therefore, the "case C" EOM of ch. 5 - including capillarity correction - can in principle induce velocity source behaviour. But also "nucleation kinetics" - not implicit in any continuum EOM because it escapes the validity of continuum descriptions a priori - can affect the CE's boundary conditions, as can other curvature-related effects.

Summarizing we can sketch the following picture:

1) On the basis of "case A or B" EOM's alone, all CE's behave as "passive elements": only the *left* column of table 6.1 applies (without the details about curvature in a $\sim \rho_c$-sized region near the CE).

(Note: in connection with *junction* CE's, the meaning of "case A or B" EOM's implicates that there is no equilibrium orientation relation, imposed by surface energy effects, of the type discussed in ch. 4).

2) Extension of the EOM by γ-related terms or direct calculation of CE vector velocity on the basis of nucleation statistics can result in divergence of $\vec{\Re}_{CE}$ and $\vec{\Re}_{interface}$. (this implies that shape details such as edges are stabilized which would follow a different evolution according to the $R(\vec{n})$ function alone). The possible behaviour of CE's is therefore extended to *both* columns of table 6.1.

The effect of surface energy terms in the right-hand column is *macroscopic:* by nucleating a fixed orientation, an extra passive edge trajectory is induced in an arbitrary starting geometry (fig. 6.29). But also in the left-hand column the surface energy term has an effect, but only on a scale $\sim \rho_c$ (critical nucleus size). Consequently, the meaning of the expression "abrupt change" (in the definition of a CE, 6.4.1) can now be specified to include those gradual changes within a range $\sim \rho_c$ from the CE which are induced by capillarity and other curvature-related effects.

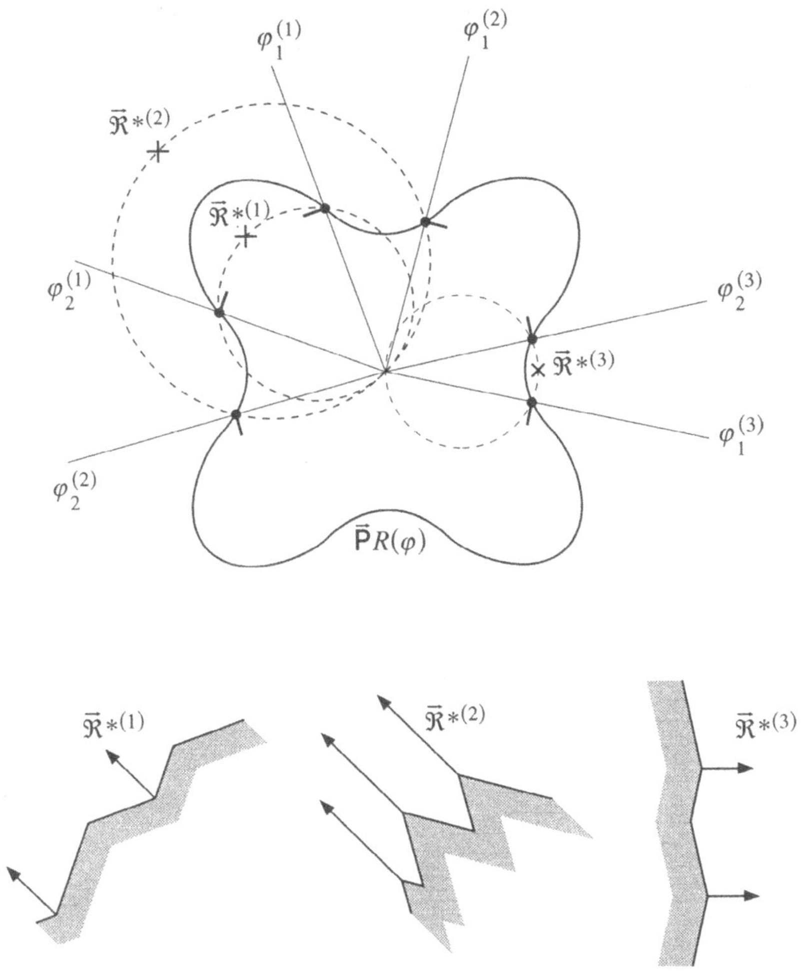

FIG. 6.30: Examples of velocity sources.

6.4.3 Examples.

In fig. 6.30 we have brought together three examples of divergent (and symmetric), i.e. "active" CE's, associated with one $R(\vec{n})$ plot:
case 1: nucleating re-entrant edge
case 2: nucleating protruding *sharp* edge
case 3: nucleating protruding *blunt* edge

The physical mechanisms leading to the nucleating property can be quite different. Case 1), for instance, typically represents the (hypothetical) nucleating RE in a $\mathbf{S}\gamma < 0$ orientation. (6.3). Case 2) can be seen in Néel-wall motion in piezoelectric crystals and is associated with the mechanical stress field involved in the polarisation reversal. Case 3), finally, is the tyical case of dislocation pyramids; the velocity source in that case is a D-junction whose vector velocity is represented as $\vec{\mathfrak{R}}*^{(3)}$ (i.e., an "heterogeneous" velocity source). For low affinity, where the $R(\vec{n})$ function of the dislocationless crystal has a very deep cusp indeed, the crystal would not grow at all without the dislocations (fig. 6.31).

In all cases of nucleating CE's, the enclosed orientation range in the $\vec{\mathbf{P}}R(\vec{n})$ plot is "obscured from view" in a growth experiment.

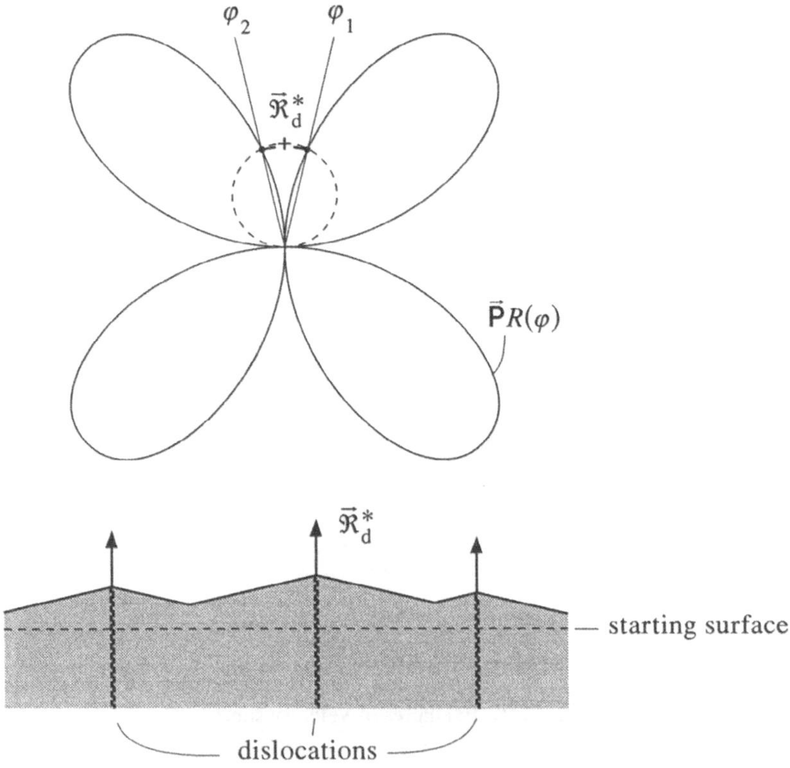

FIG. 6.31: D-junction velocity source (dislocation pyramids)

6.4.4 The velocity source interacting with a bulk diffusion field.

In the bulk of ch. 6 we have deliberately restricted to the case that interface velocities are surface-controlled: i.e. term (4) in eq. (1.8) = 0, and all interfaces involved obey a unique $R(\vec{n})$ function. This restriction enabled a straightforward demonstration of the principles involved in the velocity-source concept.

We now consider the complications which arise if the restrictions are relieved. To start with, we discuss the situation in the opposite extreme case: surface resistance is absent, interface velocities are *bulk-diffusion* controlled: i.e. now term (3) in eq. (1.8) = 0, term (4) $\neq$ 0. ("case 1").

Case 1:

This condition implies that interfaces (and their CE's) are *in equilibrium*. (in fact this is the assumption which is taken for granted in most descriptions of eutectic, cellular and dendritic growth of low-α factor crystals). It also implies that equations (6.1) are irrelevant: because $\chi = \infty$ for all interfaces, there are no driving affinities across the interfaces. The interface shapes simply correspond with zero-affinity equipotential surfaces of the bulk diffusion fields, which in turn are kept in motion by their "far" BC's, and the interface continuity and equilibrium conditions, including the CE equilibrium. For the shape of the interfaces connected in a CE we can now sketch the following characteristics:

1) For a stationary planar interface, we have the usual exponential diffusion profile of size ~ D/R ahead of the interface (D for thermal or for chemical diffusion, whichever applicable). At sufficient distance($\gg D/R$) from the CE, there is no $A_{|}$ gradient tangential to the interface.

2) Within a distance ~ D/R from a "macroscopic" deviation of the planar front (such as a 3-junction or an edge) the bulk diffusion field is in general affected by this deviation. This implicates that the interface cannot remain planar: in that case a tangential T or C gradient would be induced. It also implicates that the interface cannot remain stationary unless a moving T and/or C gradient is *imposed* by external BC's. Altogether, if we restrict to stationary situations, we are forced (in case 1) to choose "unidirectional solidification" conditions: i.e. to impose a positive A gradient in the direction of the interface normal and move the whole field at constant rate.

To be more specific, let us consider again our "simplest possible example" of the single-mobile 3-junction (fig. 6.32). The unidirectional solidification conditions referred to above imply that the interface, far enough away from the junction with the wall, orients itself in the zero-$A_{|}$ surface (supposed to be planar; and made to move with velocity R as measured in the interface-normal direction). In fig. 6.32 we have sketched three possible interface orientations with respect to the wall.

In the stationary situation, the diffusion profiles (and therefore also the interface shapes) deviate from planar over a distance ~ ($D/R \sin \alpha$) as measured from the junction. (ρ_0-sized details due to the γ-equilibrium conditioned φ_0 orientation are neglected here). Only for $\alpha = 90°$ the diffusion field is not affected and the interface remains planar. (Note that the interface shapes are not to be compared with groove and tip shapes of a cellular interface: the latter would correspond with $\vec{\nabla} A$ *perpendicular to the interface* instead of parallel with the wall).

Case 2:

As a second case we choose the same situation as sketched in fig. 6.32, but now with the slightly changed condition that the interface requires a small A_g to sustain the growth rate. (small: i.e. $A_g \ll A_{\text{bulk diff.}}$; or "Nusselt number" $\ll 1$).

For this situation we can roughly define a range $\ll D/R \sin \alpha$ from the CE where conditions are roughly equivalent with those in the rest of ch. 6:

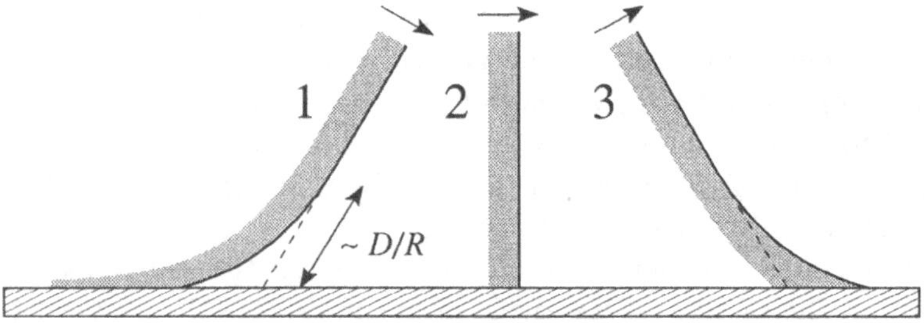

FIG. 6.32

i.e. the whole discussion about nucleation by the junction, the velocity-source concept etc. are valid, *but only within this range*. (in a way analogous to the size limitation of RE grooves by the A gradient in fig. 6.24).

Case 3:

The case more or less opposite to case 2 is $A_g \gg A_{\text{bulk diff.}}$, but the latter is not zero. This implies a small "diffusion-field amendment" to the discussions in the main body of ch. 6.

Again on an intuitive basis, we may expect that the interfaces (whether nucleated or not) may be "slightly" affected by such a diffusion-field amendment. *This may shift the active/passive transition of the junction.*

Conclusion: A full description of the active/passive transition of junctions, edges and vertices is quite complicated as compared with the constant-affinity case discussed in the rest of this chapter. To start with, stationary situations without any A_g require unidirectional-solidification conditions, i.e. imposed $\vec{\nabla} A_l$ and R. For interfaces with a certain interface resistance, imposed $\vec{\nabla} A_l$ and R are not *necessarily* required for stationary situations.

We will not attempt an analysis of the combined result of diffusion fields, interface kinetics and nonplanar junction shapes. Rather we hypothesize on an intuitive basis that the effect of a diffusion field on "velocity source" activity is twofold:

a) the transition between active and passive behaviour can be shifted;

b) in the "active" case, the shape effect can either be "trapped" within a range related with D/R, ot it can "escape" and so affect the shape of the interface on a larger scale, just as in the case without bulk diffusion. (the latter only in the absence of imposed $\vec{\nabla} A$: this would again prevent outgrowth). This would mean that in the general case we have to do with two quite distinct transitions and related criteria: the first related with the velocity source transition, the second with the escape possibility.

6.5 EVOLUTION OF MULTIPLANAR SHAPES

The numerical simulation of crystal growth under homogeneous-affinity conditions requires the $\vec{Q}R$ vectors to be determined and positioned on a grid of the starting surface for "all" orientations occurring on it, including those within the ranges jumped over in edges and vertices. This procedure is quite general; there are no restrictions to the initial shape of the surface or to the $R(\vec{n})$ function.

However, a much simpler and faster procedure can be followed in the case of "multiplanar" structures (consisting of planar areas connected by edges, vertices and eventually 3- or 4-junctions; in the 2-dim. case, we might speak of "polylines"). Under the restriction that a) the starting surface is multiplanar and b) the $R(\vec{n})$ function is such that only planar areas develop from reentrant as well as protruding edges and vertices, the multiplanar characteristic is maintained during growth. *The evolution of such a structure is fully defined by the velocity vectors of the CE trajectories only.* On their way, the trajectories can combine and/or split up and so form a network (this trajectory network itself is also multiplanar). When growth is continued until the layer thickness is much larger than the original substrate-detail dimensions, edge-trajectory meetings occur frequently, in general resulting in a decrease of the number of orientations. (and, for polycrystalline samples, a decrease of the number of grains too- e.g. in "columnar growth").

However, by deliberately choosing suitable substrate profiles in combination with equally suitable R-plot features (cusped minima and maxima) it is possible to provoke the generation of new (facet) orientations;

in the latter case more than one edge trajectory leaves the meeting point. Altogether, regions of different surface orientation are created and/or annihilated when edge trajectory planes (2D) or vertex trajectory lines (3D) meet. Anyhow, the calculation of the CE trajectory network, with in practice a relatively small number of straight trajectory elements, is a much simpler task than the general procedure.

The multiplanar case is not as exotic as it may seem at first sight: the growth of fully faceted crystals obeys the conditions, as well as low-pressure MOVPE epitaxy of, e.g., III-V semiconductors on 3D-structured "multiplanar" substrates (grooves, mesas, masks etc.) In the latter case, faceting can be exploited for precision shaping of extremely small details such as quantum wires and dots. (e.g., NAGAMUNE, TSUKAMOTO and ARAKAWA (1993); NISHIDA, SUGIURA, NOTOMI and TAMAMURA (1993)). Also, polycrystalline or polyphase crystal/parent phase interfaces are "multiplanar", provided the interfaces of the (monocrystalline) constituing grains are so.

For simplicity, we will restrict to the 2-dimensional ("polyline") case in most of the following discussion. Extension to the 3-dim. case does not provide any essential complications, apart from the occurrence of both 1- and 0-dimensional CE's and their trajectories.

6.5.1 Conditions for multiplanar growth

1) Starting shape:

The multiplanarity condition for the starting geometry is obvious. One remark has to be made on RE velocity sources: we have hypothesized (in 6.4) that both a planar $\mathbf{S}\gamma < 0$ interface and a RE with the same macroscopic orientation can be metastable. The metastable RE is certainly "sharp" down to atomic scale, as it is maintained by sharp-RE nucleation kinetics. When such RE's are present in the starting geometry we have an uncertainty between two possibilities:

a) either the starting RE is not sharp enough to be stabilized. In that case, the RE "velocity source" is annihilated right at the start and a *fast* planar interface section is generated according to the classical $\vec{\mathbf{Q}}R$-vector construction principle (identical with the annihilation event sketched in fig. 6.22,b). Two passive edge trajectories then start from the RE.

b) or the starting RE is sharp enough. It is then maintained as a (slow) velocity source, following a single (active) edge trajectory. If the original RE orientations do not correspond with the stabilized-RE zigzag orientations, two passive edge trajectories proceed along with it to adapt the orientation (this situation is identical with the sketch in fig. 6.22,a)

In case b, the vector velocity of the stabilized RE should be known. (it cannot be read from the $R(\vec{n})$ function). Note that (re)stabilisation of a RE may be strongly enhanced by constitutional-supercooling conditions, i.e. by bulk-diffusion resistance in the parent phase. In practice, the choice will have to be made by agreement with experiments. (which, at least for III-V MOVPE, seem to suggest that stabilisation of a RE is a rare phenomenon indeed).

2) $R(\vec{n})$ function.

The fact that any shape developing after the collision of two edge trajectories is approximated as a multiplanar one, corresponds with the assumption that the growth rate plot has sharp maxima and minima only (connected by $v_{//}|\sin(\varphi-\varphi_1)|$ curves). Moreover, the only $R(\vec{n})$ data which are relevant for the construction are the $\vec{\mathbf{P}}R(\vec{n})$ vectors corresponding with these minima and maxima. (as these extrema are sharp, each has two $\tilde{\mathbf{Q}}R(\vec{n})$ vectors: one at each side). In a $\vec{\mathbf{P}}R$ plot, we can eventually represent the maxima as + and the minima as - symbols on their relevant $\vec{\mathbf{P}}R(\vec{n})$ positions. The connecting $v_{//}|\sin(\varphi-\varphi_1)|$ curves can be omitted because these circles are fully defined by the + and – points. *)

Minima and maxima are *relative* minima and maxima. For the 2-dim. case this means simply that, following φ from 0 to 2π, we have *alternating* max. and min. directions. For the 3-dim. case it means that for any maximum-orientation on the unit sphere we can find a spherical triangle of neighbouring minimum-orientations which contains no other maximum (and inversely for minima).

(as we do not explicitly require that the (multiplanar) *starting* orientations $\vec{n}_1$, $\vec{n}_2$ etc. happen to correspond with either maxima or minima, their growth rate and orientation may be indicated by open circles on the plot).

6.5.2 Calculation of the trajectory network

The complete algorithm for evaluation of the trajectory network requires the following four ingredients: (the 2D case is discussed here and illustrated by fig. 6.2; the extension to the 3D case is obvious although intricate. The configuration of fig. 6.2 has been chosen to show the events to be discussed explicitly, although from a technical point of view it is an unlikely situation. The $R(\vec{n})$ function which is supposed to describe the evolution of the shape of fig. 6.2 and which satisfies the "multi-circle" condition 2) mentioned in 6.5.1, is sketched as an insert in fig. 6.2.

*) The "sharp-cusp restriction" can be made more specific in terms of $|\mathbf{S}R/R|$ maxima for the cusps. After growth of a layer thickness d on a substrate with details (grooves, mesas, masks etc) of typical size λ, the radii occurring in the epilayer are of the order Max $|\mathbf{S}R/R|$.d. This implies that d $\gg$ $|R/\mathbf{S}R|$ is a sufficient condition for the multiplanar structure to be a good approximation of the real one. As in practice $|\mathbf{S}R/R|$ values can be fairly high ($\gg 10^2$ or so), this condition can often be met already when d is only a fraction of λ.

1) Determination of all *edge-trajectory vector velocities* starting from an edge in the starting structure. (orientations $\vec{n}_1$ and $\vec{n}_2$). There may be one or more trajectories for one edge, depending on plot data. (the complication of eventual metastable velocity-source RE's was discussed in 6.5.1, sub 1). This part of the problem is equivalent with the Gibbs-Wulff construction in the sector between $\vec{n}_1$ and $\vec{n}_2$.

2) Determination of the *meeting points (and times) of trajectory pairs* which have started from neighbouring edges. When trajectories meet, the enclosed orientation is eliminated from the "multiplane" and a new edge is formed, which (starting from the meeting point) can be calculated by the procedure of 1). Some bookkeeping is required to keep track of starting times of these higher-generation trajectories.

In coincidental cases, two planar interface areas can collide as a whole (encircled (2) in fig. 6.2). Also, in enclosed areas, the remaining edge trajectories will ultimately meet in a single point (encircled (3) in fig. 6.2). The *mathematical* situations corresponding with such events (2) and (3) may be *physically* unrealistic, because the assumed constant-$A_{|}$ situation is unlikely for such areas where the enclosed parent phase is gradually depleted.

3) According to 2), *new edges can form by collision* of neighbouring trajectories. But we have also to account for another possibility of formation of new edges: i.e. by *impingement of a protruding edge onto an interface approaching from the opposite direction* (fig. 6.2, encircled point (1). By such an event, the *topology* of the multiplanar shape is affected.

In order to search for such events, collision between PE trajectories and all interfaces involved has to be scanned for systematically. Also the freshly formed RE's resulting from such collisions have to be accounted for by the procedure according to 1).

4) The conditions for *nucleation* of extra orientations (and the orientations themselves, i.e. the $\vec{\mathfrak{R}}^*$ vectors) should be known. In the $R(\vec{n})$ plot in the insert of fig. 6.2, the points b* and h* are assumed to be nucleated by an RE enclosed by orientations b and h. These orientations meet at encircled (4) in fig. 6.2.

Once we have calculated all trajectory elements of a network up to a certain growth time t_{end} (including the start and finish times of each section), we can simply draw the interface shape at any time $0 < t < t_{\text{end}}$ by calculating the point on each trajectory element where the corresponding edge has arrived at that time.

Chapter 7

STRUCTURAL PHASE TRANSITIONS OF A CRYSTAL SURFACE AS A BRANCH OF SOLITON PHYSICS

In our universe, matter is arranged in a hierarchy of structures by successive integrations.

François Jacob (1920-1977)

7.1 INTRODUCTION: INTERFACE INSTABILITY AND PERIODIC STRUCTURES

As pointed out in ch.1, the work presented in this monograph was at least partly motivated by the inability of the available (Mullins-Sekerka) theory of interface instability to account for experimentally observed features of unstable interfaces, in particular facet-oriented corrugations. Now we have discussed detailed aspects of orientation dependence in chs. 3 to 6, we are in a position to consider interface instability theory from a more general point of view. A generalized theory will have to account for constitutional-supercooling-induced *and* $\mathbf{S}\gamma < 0$ or $\mathbf{S}R < 0$ induced instabilities. (see also 6.3.5 on the anticipated behaviour of interface corrugations),

In this introduction we first argue why the *linear* perturbation approach by Mullins, Sekerka and Coriell (MSC) (MULLINS and SEKERKA (1964), CORIELL and SEKERKA (1976)) falls short to provide an instability criterion in the $\mathbf{S}\gamma < 0$ case. An appropriate theory including OD γ and R effects has necessarily to be, in mathematical terms, a theory on the stability of solutions of *nonlinear* partial differential equations (NLPDE's). These PDE's (equations of motion for the surface) have been reviewed in ch. 5. In 6.3 we discussed the stability of edges and vertices in connection with certain terms in these PDE's.

Interface instability and the resulting (mostly) periodic structures will be discussed here on the basis of these same PDE's. However, as compared with the discussion on zigzag stability in 6.3 we proceed one step further. Whereas a single stable edge can be considered as a soliton solution of the relevant nonlinear PDE, a periodic structure results from the *mutual interaction between these solitons* and can therefore be considered as a stationary state of the many-soliton system (i.e. a "soliton gas" or even a soliton crystal). What is more, the interaction between solitons can be expressed in new PDE's for the soliton continuum which in turn can have *their own* soliton solutions (which, as compared with the original solitons, are then "supersolitons"). In the same sense, terraces and zigzag patterns are supersolitons resulting from ledge condensation: the advancing ledges themselves are also solitons (of the field equation for the density of atoms adsorbed on the crystal-parent phase interface).

Somehow in contrast with this principle of *superstructure formation* is the reciprocal concept of *substructure formation* which is possible in $\geqslant 1$-dimensional physical or mathematical "objects":

In the statistical mechanics of many-particle systems, interactions between the particles can be expressed as deviations from the "ideal" equation of state, leading to condensation below some critical value of an ordering parameter depending on temperature, growth rate etc. Condensation implies the formation of an *interface* between the two bulk phases; this interface is, mathematically, a soliton as mentioned above. Now the dimension of the interface is decreased by a unit step as compared with the original bulk system. This same concept can in turn be applied to a two-dimensional interface, considered as a subsystem. The "condensation" is then the roughening transition and the "interface" is a ledge - which reveals a ledge as another type of soliton.

In this case, because the ledge is an interface in a lower-dimensional subsystem, we can speak of a series of *substructure* formations. So, the zigzag or pyramid interface can be considered as a superstructure of substructures of the original interface. Again, the edge is a soliton (of the equation for the ledge-continuum field).

Altogether, this way of looking at the interface with successive magnifications results in the concept of a *tiered thermodynamics* of interface structures. The units to which statistics is applied at one level of this tiered structure are the two-phase interfaces at a lower level. Generalizing, we can identify all types of periodic or repeated structures as occurring in crystal growth (and, in fact, in the whole physical and biological world) as systems of units for which a (quasi) thermodynamics can be defined. In crystal growth this is relevant for cellular, terraced, zigzag, dendritic and eutectic structures.

Of course, the mathematical (soliton) and he thermodynamic (interface) pictures are closely related or rather identical. A *critical point* in the thermodynamic picture corresponds with *marginal stability of a soliton solution* and vice versa.

Summarizing, in this chapter the subject of surface morphology and its successive phase transitions is recognized as a branch of soliton physics. We support this view by considering the tiered structure of parentally related phase transitions from two complementary points of view:

a) (statistical) thermodynamics; i.e. the *repeated procedure* of making a step from a many-"particle" or many-object system to an equation of state for a continuum which may be subcritical (7.2.1 and 7.2.2).

b) mathematics; i.e. the *repeated procedure* of describing the evolution of a continuous medium by an EOM which is effectively a NLPDE or a system of coupled NLPDE's which may have soliton solutions (7.2.3 and 7.2.4).

In the discussion of 7.3, the concepts developed in 7.2 are applied to the several morphological features occurring on crystal surfaces. In view of the mathematical complexity of the subject, quantitative results cannot be given; therefore this chapter is to be considered as a conceptual framework rather than as a developed theory.

Limitation of the linear-perturbation (MSC) approach

The "classical" theory of interface stability originally published by RUTTER AND CHALMERS (1953) and later developed in more detail by MULLINS AND SEKERKA (1964) is focused on the "unidirectional solidification" situation. The approach consists essentially of a linear perturbation analysis. A small-amplitude wave $\delta(t)\ \sin\omega x$ is superposed as a perturbation to the planar interface, which is a trivial solution but may be unstable. By proper calculation of the diffusion fields for heat and matter, using boundary conditions which define thermodynamic equilibrium at the interface, the rate of change of δ is determined as a function of ω. Stability of the interface requires $\mathrm{d}\delta/\mathrm{d}t$ to be $\leqslant 0$ for all ω.*)

(Although the proper calculation of $\mathrm{d}\delta/\mathrm{d}t$ requires the analysis of time-dependent diffusion equations, the $\mathrm{d}\rho/\mathrm{d}t$ terms can be omitted for the sake of calculating the *zero-transitions* of $\mathrm{d}\delta/\mathrm{d}t$, i.e. the only essential results to be used in the stability criterion. This stationary-field approximation greatly simplifies the procedure).

In the Rutter-Chalmers analysis, γ and χ^{-1} are both taken zero. Mullins and Sekerka, in their original paper, still put $\chi^{-1}=0$ but include the effect

*) Strictly taken, the criterion $\mathrm{d}\delta/\mathrm{d}t = 0$, not being identical with $\vec{\vec{\mathfrak{S}}} = \vec{\vec{O}}$, does not correspond with marginal stability, but the difference is second order in δ/λ.

of OID surface energy. In the paper by Coriell and Sekerka these restrictions are relieved somewhat and finite OD interface mobility and surface energy are substituted. (For both parameters, however, *linear dependence on the orientation parameter around the flat-interface value* is assumed; this excludes the behaviour of interfaces with near-facet orientation). But the linear-perturbation approach is maintained and consequently they find, for the zero-transitions of $d\delta/dt$, travelling waves, shifting along the interface in accordance with what in this book is defined as the vector velocity. As a consequence of the diffusion field, the $\vec{\mathbf{Q}}R(\omega)$ behaviour is dispersive.

The point to be emphasized here is that a linear perturbation approach falls inherently short to describe short-wavelength behaviour. This becomes evident when we question the meaning of a $d\delta/dt$ versus ω curve for ω approaching $2\pi/a$, a being the lattice constant of the crystal. A "small-amplitude perturbation", i.e. $\delta \ll \lambda$, would imply that the crystal units are shifted by a small fraction of a from their equilibrium position in the lattice, i.e. represent something like a surface *fonon* rather than an interface in equilibrium where atoms can choose between the solid and the parent phase. In other words, the "quantisation" of lattice population implies that for $\omega \rightarrow 2\pi/a$, necessarily the orientation range to be considered becomes large, and as a consequence the nonlinearity of the differential equations representing the EOM can no longer be neglected.

Now in the case of OID surface energy, although the above remark remains correct, it has little practical meaning because the surface-energy related term in $d\delta/dt$ is negative. Being proportional to ω^2, it dominates for $\omega \rightarrow 2\pi/a$ and ensures stability of the planar interface, so that large-amplitude fluctuations are damped anyhow. But for orientation-dependent surface energy the situation is drastically changed. Remembering that, in the OD case, γ has to be replaced by $\mathbf{S}\gamma$ in the expression for the capillarity shift, we can see that, in the linear approximation, the expression for $d\delta/dt$ can be maintained with only this replacement.

For a negative-$\mathbf{S}\gamma$ interface orientation, the capillarity term (dominating for $\omega \rightarrow 2\pi/a$) is now *positive*, enhancing instability. This means that the above remark on the essence of nonlinearity is quite relevant for $\mathbf{S}\gamma < 0$ interface orientations, and that the MSC criterion is not applicable to this case (an intuitive discussion on the evolution of the EOM was given in 6.3.3). Summarizing, we conclude that proper consideration of OD γ and R effects necessarily implies a nonlinear stability problem, and attempts to draw conclusions from any linear approach are a priori suspect.

7.2 INTERMEZZO: ON TIERED PROCEDURES IN THE PHYSICS OF STRUCTURAL DIFFERENTIATION

At this point it seems appropriate to discuss, very roughly, the phenomenology of structural and morphological differentiation from a general point of view. Key words are: critical phenomena in many- particle systems, the analogy between the regularity of many so-called "dissipative structures" of irreversible thermodynamics (NICOLIS and PRIGOGINE, 1977) and those of crystals or liquids in "equilibrium" thermodynamics; sub- and superstructure formation; transition from many-particle to continuum representations by statistical procedures; soliton physics, and the identity, in subcritical systems, of interfaces, solitons and generalized particles.

On the basis of this survey, we will correlate phenomena such as surface roughening, ledge formation, terracing instabilities of several origins, cellular and dendritic growth, zigzag formation etc. as a number of successive morphological phase transitions. As a central theme, we emphasize the new concept of a repeatable procedure to be applied to the description of evolving structures in nature in general, leading to a "tiered" structural differentiation.

7.2.1 Critical phenomena and structural differentiation by the formation of sub- and superstructures

Any disproportionation phenomenon can be considered either as the formation of a sub- or as a superstructure. Consider the condensation of a nonideal gas. The liquid droplets forming in the vapour are a substructure of the original homogeneous phase as a continuum, and at the same time a superstructure of the particles constituting the vapour. However, in the following we will use the prefixes "sub" and "super" in a more restricted sense.

In the BURTON, CABRERA and FRANK (1951) or the JACKSON, UHLMANN and HUNT (1977) description of the series of critical phenomena in the Kossel model leading to the series interface-, ledge- and kink-formation, we will typically speak of a series of *substructures* because in this series the condensing particles are identical, but the dimensions of the condensing systems decrease by units (bulk phase/interface-adsorbed phase/ledge-adsorbed phase). By the ledge formation below T_R , the interface formed in the preceding condensation develops a substructure, and by the kink formation the ledge develops a substructure (actually, only in the limit $T \rightarrow 0$). The maximum number of steps in such a series equals the dimension of the first, homogeneous system. (By the way, the lower-dimensional system does behave as a more or less isolated system by the very fact that an interface constitutes itself an energy barrier between itself and the rest of the system). A quite similar series of substructures can be found in elementary-particle physics, based on "internal" dimensions or degrees of freedom such as spin, colour and charm.

In contrast, the term *superstructures* will be reserved for those condensation phenomena which are made possible because the interfaces themselves (resulting from a preceding condensation) can act as new entities, showing collective behaviour. Thus, attractive interaction between ledges will result in "ledge condensation", where the ledges disproportionate into dense and dilute phases, corresponding with interfaces of different orientation. So the "two-phase" ledge system corresponds with a zigzag structure macroscopically, and the "interfaces" formed in this condensation are *edges*. Depending on the mechanism of interaction, condensation of ledges can either result in macroscopic zigzag patterns, or in multi-unit terrace structures of stabilized periodicity. In the latter case, the terraces themselves (being macro-ledges) can again form super-terraces in a next step as a result of a different mechanism causing attraction between the macro-ledges. Such a many-generation superstructure has experimentally been observed by TOLKSDORF and WELZ (1978) in an LPE situation (fig. 7.3).

Note that we have avoided to specify the nature of the "attractive interaction" in the above discussion. Actually, such an interaction can originate as well from a static interaction as from a kinetic process.

An example of the first category is the effective potential well between kinks or ledges at a distance which corresponds with a periodic bond chain (PBC) of sufficient energy, matching the surface. Such an interaction corresponds with a cusp in the γ plot, and the resulting zigzag structure, in equilibrium, is described by the negative-$\mathbf{S}\gamma$ instability (4.2).

An example of the second category is the interaction between the (2-dim. or 3-dim.) diffusion fields of neighbouring advancing ledges (e.g. CHERNOV (1961); BENNEMA and GILMER (1973)). This interaction only occurs at $R \neq 0$, and its description involves the growth rate function (plus eventual curvature-related complications), and the diffusion field equations. In actual crystal surfaces during growth, the interaction will generally have both potential- and kinetics-induced contributions.

The concepts of sub- and superstructures are further illustrated in table 7.1 by examples both from crystal growth and from other systems. Substructures are represented by horizontal arrows, superstructures by vertical ones. Of course, a prerequisite for the advancement of a (horizontal or vertical) step in table 7.1 is that the parental system is subcritical, i.e.

that the "interfaces" to be sub- or superstructured exist at all. This implies that the degree of branching of the "family tree of structural differentiation" (7.2.4) depends on temperature and other order-influencing parameters.

Sub- and superstructure formation lead to a "tiered" hierarchy of structural differentiation. In the following, we will refer to the discrete systems considered in this hierarchy (i.e., those separated by arrows in table 7-1) as to the different levels in this hierarchy.

7.2.2 *Stationary dissipative structures: equilibrium and nonequilibrium aspects*

In table 7.1, we deliberately combined typical "equilibrium" systems (such as the crystal-surface and the ferromagnetic-layer system) and typical

---

TABLE 7.1: ***Sub- and superstructures***

Crystal growth:

a) interface ($T < T_R$) → ledges → kinks
ledges ↓ zigzags; ledges ↓ terraces (= macro-ledges) ↓ super-terraces
kinks ↓ zigzag ledges

b) interface ($T > T_R$) → cellular structure, lamellar → cellular structure, acicular
cellular structure, lamellar → dendritic structure (sidebranch instability)

Domain structures in ferromagnetic layers:

Bloch wall → Bloch lines
Bloch wall ↓ magnetic domain pattern, ribbon-type → magnetic domain pattern, bubble-type

Convection cells:
homogeneous system → convection rolls → hexagonal convection cells

Gas discharges:
travelling striations in low-pressure gas discharge

(Further examples in the list on page 194)

nonequilibrium systems such as the Bénard and the constitutional-supercooling system. (the former can be extended to nonequilibrium by including magnetic-domain interfaces *moving* with respect to the particles or spins).

Thermodynamically, the first category is familiar enough; it can be identified with closed systems, to be dealt with by the appropriate statistical methods leading to the definition of entropy, free energy etc. The second category are open systems, through which fluxes are forced by affinities; a conventional equilibrium description is therefore not applicable. Still, the phenomenologically interesting aspects of the "dissipative structures" (NICOLIS and PRIGOGINE, 1977) occurring in the second category is their (mostly) regular stationary character when observed in a suitable reference frame moving along with the structure. This suggests that the latter aspect might rather be considered at another level of description as the original nonequilibrium system; at this other level we should regard only those characteristics of the dissipative structures which are distinguishable as separate units (cells, interfaces) and describe *their* collective behaviour as that of a new system. At this more macroscopic level, the sight on many details of the original nonequilibrium system is lost; *in particular the details of the entropy production density in the original have to be translated into an effective interaction potential between the units of the macroscopic level.*

In terms of what has been said before, this step amounts to the consideration of a super- or a substructure. We thus have to recognize that, when the concept is extended to nonequilibrium systems, the character of the system can (but does not necessarily) change from nonequilibrium to equilibrium on advancing a level.

As a relevant example from crystal growth, consider ledge motion. Microscopically, a growing crystal is obviously a nonequilibrium system. The interface can develop a ledge substructure: i.e. the interface subsystem (at $T<T_R$) is subcritical and develops a two-phase structure. This is, approximately, an equilibrium phenomenon: the velocity of the ledges is too small to affect their structure or the instablity itself, unless at very high affinity when kinetic roughening occurs.

The next level of description is the system of moving ledges. Now the character of this system may be either equilibrium or nonequilibrium. Consider fig. 7.1. At the ledge-system level of description, the ledges (seen in transverse cross-section as zero-dimensional objects) are "particles" moving from boundary 1 to boundary 2. In fig. 7.1a the ledge-gas density is homogeneous, i.e. the ledge-gas pressure is constant in the region considered and the boundary conditions are such as to allow this. In other words, in the ledge transport mechanism there is no diffusion resistance with res-

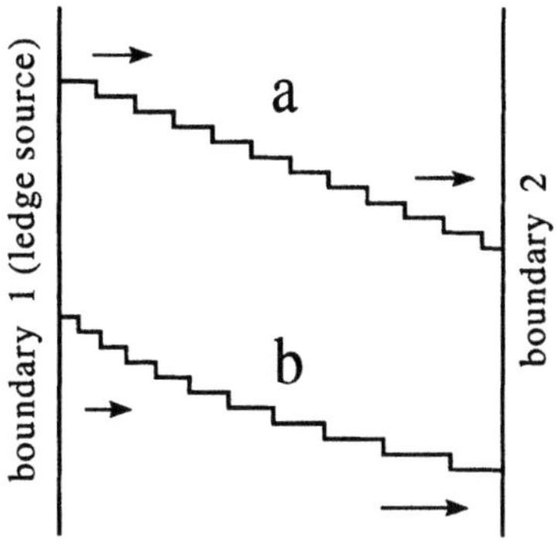

FIG. 7.1

pect to the crystal, i.e. the potential of the ledge gas (not to be confused with the thermodynamic potential of the interface atoms) is constant between boundaries 1 and 2. The motion of the ledges is driven by the interface-atom potential difference across each ledge, but in the ledge system this motion is to be considered as a "wind velocity". Although the boundaries 1 and 2 are open, the ledge potentials cancel and the ledge system between them can be considered as a system in equilibrium.

In contrast, fig. 7.1b suggests a (hypothetical) situation where ledges do experience a diffusion resistance on moving from 1 to 2. The ledge-gas pressure (and potential) is higher at boundary 1 (source) than at 2 (sink). Now the ledges move under the combined influence of the ledge-gas pressure gradient and of the interface-atom potential drop across the ledge. Clearly, fig. 7.1b represents a *nonequilibrium* system.

In both cases, the ledge system can be either subcritical or supercritical (note that this distinction is of a different nature and unrelated with that between equilibrium and nonequilibrium). When it is supercritical, there is no superstructure level and the ledges form a homogeneous phase, whose character may range between "low-density gas" (for near-facet interface orientations) to "crystalline" (for neighbouring facet orientations). Note that this homogeneous "single phase", although interfaces with another coexisting phase cannot exist, may well support travelling waves, i.e. "fonon fields" in this phase (MULLINS and HIRTH (1963)). When it is subcritical, superstructures occur (terrace- or zigzag patterns). A complication can be that in a case where the ledge gas is subcritical, but the ledge supply rate at boundary 1 is very constant, the incubation time needed to develop the instability may be too short for the superstructure to develop. This can happen at a dislocation-induced velocity source and give rise to a smooth top region in an otherwise terraced dislocation pyramid. (see fig. 7.3 which is taken from TOLKSDORFF and WELZ, 1978)

We have used the ledge system as an example because it is familiar, although the concepts of ledge pressure and -potential may require some imagination. However, nothing new is told here, as the effects mentioned are all implied in the macroscopic EOM presented in ch. 5, where the interface is regarded on a macroscopic scale (the equilibrium and kinetic interactions between ledges are reflected in the behaviour of the γ and R functions, respectively, *plus* the extra data of the stabilized RE vector velocities $\vec{\Re}^*$, which we have shown to be essential in 6.3). We merely presented the example because it can be transferred to further levels. In particular we emphasize the possibility that a superstructure developing in a nonequilibrium system, given homogeneous boundary conditions for the superstructure-level of description, may be represented as an equilibrium system.

7.2.3 A short introduction to soliton mathematics

For those readers who are not familiar with the soliton concept we insert a rather superficial introduction. In accordance with the rest of this intermezzo, it is aimed at an understanding of concepts without claiming mathematical rigour. For the latter the reader is referred to monographs on this subject (e.g., EILENBERGER (1981), SCOTT, CHU and MAC LAUGHLIN (1973), BISHOP and SCHNEIDER (1981), INFELD and ROWLANDS (1990)).

The objects of nonlinear physics are systems or "fields" obeying evolution equations (equations of motion, EOM) which can be written as nonlinear partial differential equations (PDE's) or sets of coupled PDE's of the form:

$$\Phi_t \text{ or } \Phi_{tt} = F(\Phi, \Phi_x, \Phi_{xx}, \ldots)$$

where Φ is a one- or more-component field $\Phi(x,t)$. The field can also be defined in a more-dimensional space: $\Phi(t, x, y, \ldots)$. Examples can be found in classical continuum physics (hydrodynamics, plasma physics) as well as in areas such as solid-state physics, nonlinear optics, surface physics and elementary-particle physics. Once the system of EOM's is established, the subject is actually transferred into the theory of systems of nonlinear PDE's. The successive abstractions of interface-system EOM's discussed in ch. 5 all belong to this type. To be specific, Φ represents the distance of an interface from some fixed plane with coordinates x and y (only x for 2-D models) and accordingly, when this plane is chosen in a facet orientation, the slope Φ_x represents ledge density. Curvature involves Φ_{xx}, etc. Alternatively, we can use a $\varphi(s)$ or $\vec{n}(\vec{s})$ representation with their inherent advantages for our OD functions, as explained in 5.2 and the footnote in 6.3.3 (pg 155).

In a situation which is labeled "Stefan-type" in 1.2 and 5.4 ("case C2"), we have to do with two coupled PDE's. The second is the EOM of the affinity field, which in simple cases can be a thermal or a chemical diffusion equation or in the general case a set of $n+1$ coupled diffusion equations for n chemical constituents plus "entropium". (the coupling between the equations is effectuated by the continuity condition(s) at the interface, for heat and/or matter flux).

Travelling waves, solitary waves and solitons.

In general, nonlinear PDE's can have several types of solutions. For the purpose of this section, we are specifically interested in those types which have a "permanent" or stationary character, i.e. which are immobile in a coordinate frame which moves at a suitable uniform speed u. Such solutions are called "*travelling waves*", $\Phi_T(\xi)$, depending on x and t only through $\xi = x - ut$, where u is a fixed constant. (the relation $\vec{\vec{\mathfrak{S}}} = 0$ for stationary structures, introduced in 1.6, is a generalisation of this definition).

A *solitary wave* is usually defined as a *localized* travelling wave, i.e. a travelling wave approaching asymptotic values $\Phi_T(\xi \to +\infty)$ and $\Phi_T(\xi \to -\infty)$ beyond a certain range in ξ. Although from this definition it seems that the distinction between solitary waves and travelling waves in general depends on the possibility of defining an asymptotic value, the concept can be extended by including those solutions where an n-th derivative of Φ exhibits such an asymptotic behaviour: see fig. 7.2.

(Although this distinction is usually not expressed explicitly in the mathematical litterature, it is of practical relevance for the crystal-surface morphological features discussed here. The Φ, $\int\Phi$ and $\int\int\Phi$ "solitons" of fig. 7.2 a, b and c can be identified, respectively, on two levels:

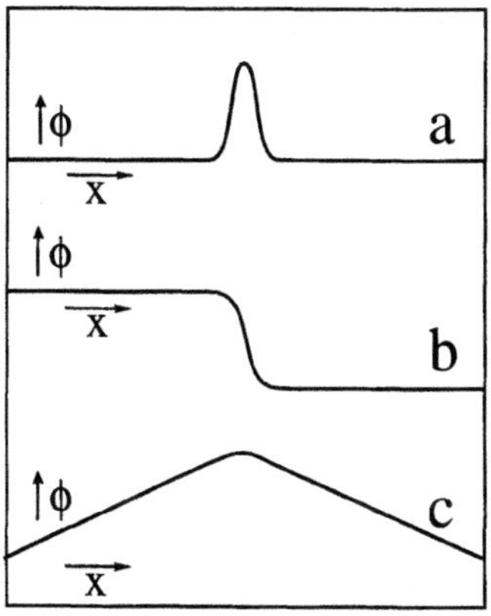

FIG.7.2

level 1: 7.2a = interface between 2 monolayer phases above.
(Φ represents ledge)
7.2b = density of adatom-monolayer system; 2 coexisting phases with monatomic step. (Φ represents shape)
level 2: 7.2a = interface between 2 coexisting phases of ledge system.
(Φ represents edge)
7.2b = density of ledge system (Φ represents slope)
7.2c = height of interface (Φ represents shape)

So at the next level of description, where interfaces are considered as particles, we have to differentiate with respect to x and the soliton types of the corresponding field equations regress one step in fig. 7.2).

Now a *soliton* is defined as a special type of solitary wave which asymptotically preserves its shape and velocity upon collision with other waves, including other solitons. This implies that shape and velocity are stabilized for perturbations, i.e. solitons have particle-like characteristics, including the possibility of collision and interaction with other solitons. The asymptotic preservation of shape and velocity means that shape and velocity return to a standard form within some fixed decay time once other solitons are far enough away, i.e. beyond some interaction range δ.

An initial-value problem of the nonlinear PDE can be solved in terms of solitons and continuous "radiation". The latter corresponds with low-amplitude sinusoidal waves, generally with dispersive characteristics, satisfying the linearized PDE. This combination can be considered as a generalisation of Fourier analysis (sinusoidal waves only), applicable to linear equations. More than one soliton can be involved in a solution (N-soliton solutions). When two (or more) solitons, having different velocities, come within close ($\leqslant \delta$) distance, their shape and velocity may change from that of the single- soliton solution under each other's influence. This change, and the accumulated effect on each soliton when they have separated again, can be computed from the PDE. In terms of particles, this amounts to the analysis of two- or more-particle collision and interaction kinetics.

Certainly, not all nonlinear PDE's have soliton solutons. As a soliton solution does not spread, it cannot be considered as a simple superposition of dispersive low-amplitude waves. *The essential feature of solitons is that the effect of dispersion is cancelled by the effect of nonlinearity*: both effects are required.

Mathematically, two types of soliton-exhibiting nonlinear PDE's have to be discerned:

1) those which exhibit "true" or "exact" soliton solutions, i.e. solitons which exhibit no interaction at all. (from the particle point of view, systems

of such solitons behave as an ideal gas thermodynamically). Nonlinear PDE's of this type are rare; mathematical conditions are given, e.g., in EILENBERGER (1981).

Only a few of such PDE's have been analyzed systematically, such as the Korteweg-de Vries equation, the sine-Gordon equation and the nonlinear Schrödinger equation.

2) A less restricted class of nonlinear PDE's, which are less "ideal" mathematically but more common physically, exhibiting a wide variety of stable, finite-energy particle-like field patterns which retain some central physical nonlinear properties but not all of the "exact"-soliton properties referred to above. As compared to category 1), the solitons of category 2) do exhibit interaction, which implies that they behave as non-ideal systems thermodynamically (so that, e.g., they may "condense" at higher density).

Nonlinear PDE's belonging to the second category can often be related with one in the first, "ideal" category in that difference is brought about by the addition of perturbation terms in the category-1 PDE or in the spatial modulation of a parameter which is constant in the category-1 PDE. In such a case, the "physical" PDE can be attacked by a perturbation analysis in which the ideal PDE is used as a zeroth approximation. (Eilenberger suggests the illustrative comparison with celestial mechanics, where the unperturbed elliptic Kepler orbits, rather than force-free motion of planets in a straight line, are to be preferred as a starting point for the calculation of interplanetary interactions).

7.2.4 A tiered hierarchy of soliton-exhibiting PDE's.

Although the author could not find reference to it in the literature, the considerations of 7.2.1 (on a tiered structure of phase transitions) suggest the concept of a similar tiered hierarchy of soliton systems. This hierarchy can be considered on a strictly mathematical basis, for the moment without reference to physical systems.

Suppose we have a nonlinear PDE, defined in a certain domain or "volume" with boundaries where BC's for the equation are prescribed, with a known many-soliton solution. The next level in the hierarchy suggested above is obtained by looking only at the solitons, consider *them* as a continuum (by proper statistical means), thereby losing resolution but also economizing on the number of variables, and writing down the EOM for this continuum (i.e. the soliton density) under the given boundary conditions.

The next-level EOM thus obtained, although related with the original EOM, is a new and simpler equation. It may or may not have soliton solutions ("supersolitons"). If the original EOM is of the first category mentioned in 7.2.3, i.e. yielding only exact solitons, the next-level EOM (describing a system without any internal interaction) is linear and certainly exhibits no solitons. If the original EOM is of the second category, the next-level EOM may or may not exhibit soliton solutions. (Marginal stability of a soliton solution can be considered as the "critical point of Fourier-component condensation"). If it does and the volume is large enough to support many solitons of the next-level EOM, then the procedure can be repeated and a third- level EOM written down. Etcetera, until the last-level EOM does not yield soliton solutions any more.

Note that subsequent levels concern densities of solitons at previous levels, so that the resolution character, for climbing levels, moves on as in fig. 7.2 a,b,c. How the complete solution is written depends on the desired accuracy or "magnification" of the description. When we are satisfied with an accuracy at which the n-th level solitons can be considered as a continuum, the complete solution of the n-th level EOM suffices (i.e. not only the solitons of this solution).

In chapter 5, we discussed EOM's based on γ and R functions defined with an accuracy where the ledges (3-dim.) or the kinks (2-dim.) were considered as a continuum (or, alternatively, where they really were a continuum, above the corresponding T_R). So the solutions of these EOM have the corresponding resolution. However, eventual *solitons* in these solutions show up as "macroscopic" features such as stabilized RE, PE or RE/PE pairs (= terraces). (macroscopic as compared with atomic size; the features may be in the micron-size range). For a completely terraced or zigzagged interface, we may construct a next-level EOM where these terraces, in turn, are considered as a continuum.

In the procedure sketched above, the first-level PDE and its BC's contain all information required for the establishment of all next-level solutions. This undoubtedly makes all but the last EOM's uneasily (and also unnecessarily) complicated. (it may be compared, for instance, with the explicit accounting for the gravity-induced potential field in the equation of state of a dilute gas in a small volume). Therefore, the sketched procedure is to be considered as a principle rather than as a practical approach. For the latter, all terms, or spatial modulations of parameters in the n-th level EOM which are not directly related with the soliton character of the corresponding solutions, have to be omitted - to be reintroduced only in the EOM at the $(n+m)$th level where they are essential for the soliton character *of that particular level*. In the example given above, gravity plays an essential rôle only in the equation for the two-phase interface, once the temperature is so low that the original gas has become a subcritical vapour-liquid system. Without gravity, the condensed vapour forms a mist in the volume; gravity makes the droplets fall down and ultimately results in a horizontal liquid surface.

Remarks:

1) A soliton solution of an EOM can be considered to be *parentally related* to two already existing instabilities, i.e. solitons in lower levels. The first and most trivial one ("mother") is the instability implicit in the equation of state, implying that we have to do with an interface at all. The second ("father") parent is hidden in the inhomogeneous boundary conditions or in the perturbation terms which induce the nonlinearity of the EOM and bring about the soliton-type solutions. In general, these can occur once some parameter exceeds a critical value. Now the very existence of this parameter, i.e. the existence of fluxes or potential gradients, is connected with some already existing structure (inhomogeneity) in the outside world, and ergo with a previous condensation in some other, more macroscopic system. For instance, the existence of a graviational field is a prerequisite for the occurrence of convective instability in a Bénard experiment. But the gravitation field itself is a consequence of the local inhomogeneity of matter distribution in the universe, in particular with the "condensation" of matter into a planetary system where the experiment takes place. Other structures lead to fluxes, resulting in the possibility of the dissipative structures of irreversible thermodynamics.

In this sense, combinations of instabilities (with their own structural differentiation) can generate new instabilities (i.e. further differentiation), so that a real family tree of structural differentiation can be discerned, all starting from the "big bang". All structure in the universe can be considered as to be members of this family tree: the family of elementary particles → atomic nuclei → condensed structures in plasmas → atoms → chemical compounds → phase separation in polynary systems → crystal growth phenomena; other branchings concern condensation of galactic and star systems and morphogenesis in biological systems. Our subject of structural transitions in crystal interfaces is just a very small part of this universal evolutionary tree. It can be embedded in the somewhat wider field of transport phenomena in general (wave formation on interfaces, Bénard cells, "dissipative structures"). Various aspects of structural differentiation are discussed, e.g., in Nicolis and Prigogine (1977), Haken (1977), Thom (1972), Pelce (1988), and Cross and Hohenberg (1993).

conditions specifying (im)permeability for the species involved and exchange with reservoirs. In accordance with the exchange possibilities, we can define the type of ensemble required to calculate a partition function. Then we can write down this partition function and hope to find a method to work out the integration or summation involved. If successfull, we end with an expression for the entropy, free energy etc.; in general with an "equation of state". This equation provides macroscopic information on the basis of parameters such as temperature and chemical composition, which can be used to establish the EOM describing the system in the next level.

For nonequilibrium many-particle systems, concepts such as temperature, partition function and free energy are not applicable. Instead, the boundary conditions provide other reservoirs or imposed fluxes, and the essential start of analysis is to realize "what is in equilibrium with what" in a nonequilibrium system (most BC's can be characterized as potential or flux sources, e.g. a velocity source (see ch. 6.4) as a ledge-flux source).

Although a partition function cannot be defined for a nonequilibrium system, we may in many cases write equations describing the evolution of the density in a suitably defined phase space (Fokker-Planck equations). Even a nonequilibrium generalisation of a partition function is (in particular cases) possible in the form of the Onsager-Machlup path integral. Using such techniques, reduced equations of motion can be derived for the continuum version of the many-particle system which are mathematically equivalent to "equations of state", or (together with the boundary conditions) to the "field" EOM referred to in our survey of soliton physics. As an elaborate account of such techniques exceeds the scope of our monograph, the reader is referred to the specialized literature.

The units considered in statistical mechanics are usually atoms (with all sorts of interactions). In higher levels of our tiered hierarchies, units are ledges, terraces and so on (i.e. >0-dimensional entities). Although definition of a phase space for such higher-level units is possible, the determination of a distribution function is not straightforward because the >0 dimensional system is extended and its exchange with the thermal noise field does not lead to a simply predictable distribution function.

As an alternative, an EOM for higher-level continua can be derived from first principles, i.e. by starting with describing the two-unit interaction and subsequently translating this interaction into a continuum EOM by summation or integration.

7.3 *DISCUSSION: APPLICATION TO INTERFACE MORPHOLOGY*

A crystal/parent phase interface, in equilibrium or in motion, can show a series of instabilities or morphological phase transitions which together provide a typical example of the "tiered hierarchy of structural differentiation" referred to in 7.2. We will complete our comments on the general interface-instability problem by a review.

We will discuss three successive levels in this hierarchy: a) the roughening transition (7.3.1); b) terrace-, zigzag- or cellular growth, i.e. generalized Mulins-Sekerka instability (7.2.3); and finally c) eventual superterraces. Simultaneously we emphasize the possibility of different interpretation of γ or R function data, depending on the "resolution" of the measurements.

7.3.1 Ledges as solitons.

For the present discussion, we have to return to atomistic models (for which the reader is referred to BENNEMA and VAN DER EERDEN, 1977), just in order to adstruct our tiered-thermodynamics concept. The reason is that the soliton interpretation of ledges can be considered as the first step in the tiered structure.

The approach followed in the cited chapter is to describe the interface, on an atomic scale, by a solid-on-solid model and calculate its thermodynamics and kinetics by relating this with a two-dimensional Ising system. The roughening transition is then identified with the order-disorder transition of this system.

An alternative approach was proposed by GUYER (1979). Instead of the "binary" (Ising-type) nearest-neighbour interaction, he proposed the sine-Gordon chain as a more detailed interaction model. In the SOS model, the crystal is formed of neighbouring columns with an integer- multiple unit height (Kossel model). In the sine-Gordon model, (where Guyer only considers the one-dimensional version) the height of neighbouring columns is a continuous variable, but their energy is sinusoidally modulated. The neighbour interaction is represented by a force proportional to the height difference of neighbouring columns. Together:

$$E(h_1,h_2,\ldots,h_{n+1}) = \sum_{i=1}^{n} \{\tfrac{1}{2}E_2(h_{i+1}-h_i)^2 - E_1 \cos\frac{2\pi h_i}{a} + E_3 h_i\} \qquad (7.1)$$

This model is equivalent with an array of pendula which are coupled by torsional springs: the sine-Gordon chain. E_3 represents a constant extra (gravitational) torque working on all the pendula and is equivalent with an affinity for crystal growth.

(by the way, the idea to consider the free energy of a surface as a continuous, periodically modulated fuction of the height as measured from an F plane was already suggested by CAHN c.s. (1960)).

The thermodynamics and kinetics of the sine-Gordon chain have extensively been analysed because of its validity as a model for several physical systems. The discrete-element ("chain") model can be replaced by a continuous model. Obviously, for equilibrium the parameter E_3 has to be taken zero.

Soliton solutions are stable below a critical temperature T_C (for equations see GUYER, 1979) and for $T < T_C$ we have an equilibrium density of solitons. Guyer c.s. exclusively consider an interface parallel to an F face. (interface orientation $\varphi = 0$ and consequently $n_+ = n_-$ (density of solitons and antisolitons, or _┌ and ┐_ ledges, respectively). However, GUPTA and SUTHERLAND (1976) and CURRIE c.s. (1977) consider the general case. The last authors argue that the thermodynamics of the (1-dimensional) sine-Gordon field is equivalent to the thermodynamics of a chemical equilibrium reaction:

$$(n_+) + (n_-) \rightleftarrows C \tag{7.2}$$

where n_+ represents the soliton species, n_- the antisoliton species and C is continuous "radiation" or fonons of the sG chain.

(note: the concept of a nonzero equilibrium density of ledges was also the basis for the description of roughening in parametrized $R(\vec{n})$ and $\gamma(\vec{n})$ functions (chs. 9 -12). In particular, eq. (7.2) corresponds directly with the (2)✕ F, ε' ✕○ assembling step expressed in eq. (11.5). The parameter K in that expression represents the $\sqrt{n_+ . n_-}$ equilibrium density for $\varphi = 0$).

An essential shortcoming of all cited literature is that discussions are based on the analogy between a ledge and the one-dimensional sine-Gordon chain (either discrete or continuous). On an actual, i.e. 2-dimensional surface, a one-dimensional sG soliton corresponds with a *straight* ledge. In the one-dimensional sG chain, a creation / annihilation event corresponds with the "condensation" of an originally extended fonon mode into a soliton. Extension of that concept to the 2-dimensional surface would imply that birth or collapse of a ledge on an otherwise planar interface would take place *simultaneously along a straight (infinitely long) line.* This, obviously, is physically unrealistic. Actually, birth of ledges takes place by the nucleation of 2-dimensional "crystals" (islands of monatomic height), and an essential ingredient in the description of nucleation kinetics is the capillarity-shift barrier brought about by the curvature of the original ledge.

So, whereas in the cited discussions the energy barrier for the creation / annihilation process is the maximum energy of the intermediate configurations for a straight ledge, the *actual* barrier is the combined result of this energy for a critical island plus the capillarity term. Accordingly, the one-dimensional estimates of this barrier are not applicable.

This is not necessarily so for the equilibrium density ($n_+ + n_-$). For zero applied affinity, the average ledge curvature can be so low that the capillarity shift can be neglected: the equilibrium density as derived from the 1-dim. model would then be a fair estimate (this is the Arrhenius factor mentioned above). However, at high affinities, corresponding with the "birth-and-spread" mechanism of crystal growth for high-α factor interfaces (NIELSEN 1964) the curvature of the expanding islands can still be appreciable at the moment of their coalescence. In that case the 1-dim. result is also in error.

The irreversible thermodynamics of the 1-dim. sine-Gordon field, corresponding with the kinetics of moving ledges, has been considered by GUYER and MILLER (1978). Roughly spoken, their analysis covers the sine-Gordon analog of the Burton-Cabrera-Frank discussion of crystal growth by the ledge-motion mechanism, including those cases where the surface-diffusion fields of the ledges overlap. (however, the sG terminology is in terms of damping parameters for the coupled pendulae rather than in terms of a surface-diffusion process).

A full discussion of the ledge- sineGordon-soliton analogy falls beyond the scope of this monograph. Still the soliton character of ledges has been clearly established. Besides, the physical equivalent of a soliton (an interface) is also an obvious result of the Ising-model description.

7.3.2 The second tier: periodic terrace-, zigzag and cellular structures

We have already mentioned the equivalence of the EOM's presented in ch. 5 and the equations for the collective behaviour of ledges (7.2.1). In 6.3, where we discussed the stability of sharp edges on a small scale, we were thus implicitly discussing the condensation of the ledge gas or the occurrence of interface solitons in the ledge-density field. These are the solitons of the second tier (those of the first being defined as the ledges themselves) An attractive interaction between the ledges is a prerequisite for such an interaction. We discussed the possibility of an attractive *energetic* interaction ($\mathbf{S}\gamma < 0$ case), and of an attractive *kinetic* interaction (Schwoebel effect, boudary layer diffusion or $\mathbf{S}R < 0$ plus curvature effect). Altogether, the EOM's of 5.3 and 5.4 could be identified as the nonlinear PDE's describing the evolution of the ledge-gas continuum. We repeat here that, of the several different mechanisms leading to attractive interaction, not all are fully described by an EOM as given in ch.5. As to the $\mathbf{S}\gamma < 0$ case and the case of interaction via the bulk-diffusion field, the equation (resp. the coupled interface- plus diffusion-field equations) can be considered as complete: they should result in soliton solutions without further additions. For the Schwoebel effect and the boundary-layer diffusion mechanisms, however, which cannot in detail be represented by a $R(\vec{n},A)$ function, the EOM has to be extended by the appropriate terms accounting for the curvature effects (if not available in exact expressions, by series-expansion terms).

So far for the single-soliton solutions of the interface EOM's, corresponding with single stabilized edges. What is of interest next is the macroscopic distribution and density of the solitons (if any), i.e. a) the homogeneity and periodicity of the zigzag (terrace, cellular) structure and b) the eventual instability of a homogeneous distribution, leading to superstructures.

In order to investigate this, we have to go to the next (third) tier, i.e. consider the interaction between the *edges* (not the ledges) and to derive equations of state for their behaviour as a continuum. Now at the magnification level of the edge structure, the inhomogeneity of the affinity field becomes appreciable (unless we are in the rather exceptional case of homogeneous affinity). The quantity $\vec{\nabla} A \cdot \vec{n}$ is an important parameter in the edge-interaction potential. Although an exact mathematical description of the interaction is not available at the moment, we can qualitatively discuss the principles leading to repulsion or attraction by comparing the zigzag/terrace/cellular structure with the poly-grainboundary interface structure considered in 8.4. The stabilized edges or solitons play the same role as the 3-junctions in fig. 8.9 (the zigzag and terrace cases are complicated by the fact that the edges occur in RE/PE pairs; the cell-groove cusps (180°) are directly comparable to the grain-boundary grooves). As argued in 8.4, the interaction between grain-boundary 3-junctions is due to the overlap of their groove regions (which depend, in turn, on $\vec{\nabla} A$ as well as on the growth rate) and to the interaction of the groove-induced depression and the bulk-diffusion field. As all effects are represented by the EOM (with suitable boundary conditions for the 3-junction equilibria), the interaction can *in principle* be calculated from the EOM. By analogy, the same can be said of the edge-soliton interaction (where the soliton structure, once calculated, takes the place of the 3-junction configuraton of the former case). We may expect, therefore, that a pair-interaction potential may *in principle* be calculated for edge-soliton pairs, with attractive as well as repulsive components (sensitively depending on $\vec{\nabla} A \cdot \vec{n}$ and on eventually imposed velocity of the A field).

For a purely repulsive interaction, the edge distribution will always be homogeneous; i.e. superstructures cannot occur and the structure will be (statistically) periodic (a soliton "supercritical gas"). The period, i.e. the inverse equilibrium density of solitons, is determined by the soliton creation/annihilation equilibrium also to be derived from the EOM. Hysteresis of the period (adaptation transient after a change of the period-determining parameters) is to be calculated from the soliton nucleation (creation) kinetics.

7.3.3 Superstructures

An attractive interaction in the edge-soliton interaction can *in principle* destabilize the homogeneity of the periodic soliton structure so that a third tier of structural differentiation occurs ("terrace condensation"). When the force is attractive for any distance, this condensation just results in coalescence and thus in a similar structure of larger period. But when the force changes sign to become repulsive at a short distance whereas the second-tier edge solitons remain stabilized, a superstructure can result. Roughly speaking, stabilisation of the size (and correspondingly the interaction range) of terraces can be identified with some kind of boundary-layer mechanism. The occurrence of a superstructure, then, would suggest that two types of boundary layer (with at least an order-of-magnitude difference in thickness) are operative simultaneously.

For crystal growth from a stirred flux or melt, we can expect the chemical-diffusion boundary layer and the (hydrodynamic) laminar boundary layer to play such rôles. This might be an explanation for the three-tiered structure observed by TOLKSDORF and WELZ (1978) on a flux-grown garnet crystal (fig. 7.3). (The fact that the superstructure only occurs on one of the three dislocation-pyramid sides suggests that the mechanism is somehow connected with the *relative* motion of terrace vector velocity and hydrodynamic flow direction, i.e. with coupling in a hydrodynamic boundary layer).

7.3.4 Further examples

Apart from the ledge/terrace/superstructure series, crystal growth phenomenology provides many other examples of sub- and superstructure forming phase transitions. We mention some without further discussion:

1) *Surface reconstruction*, i.e. re-arrangement of the surface layer of a crystal facet into a multiple-periodic "surface lattice", is a widely investigated phenomenon. As a phase transition of the interface system, it can interact with the ledge system: this is beautifully shown in a demonstration

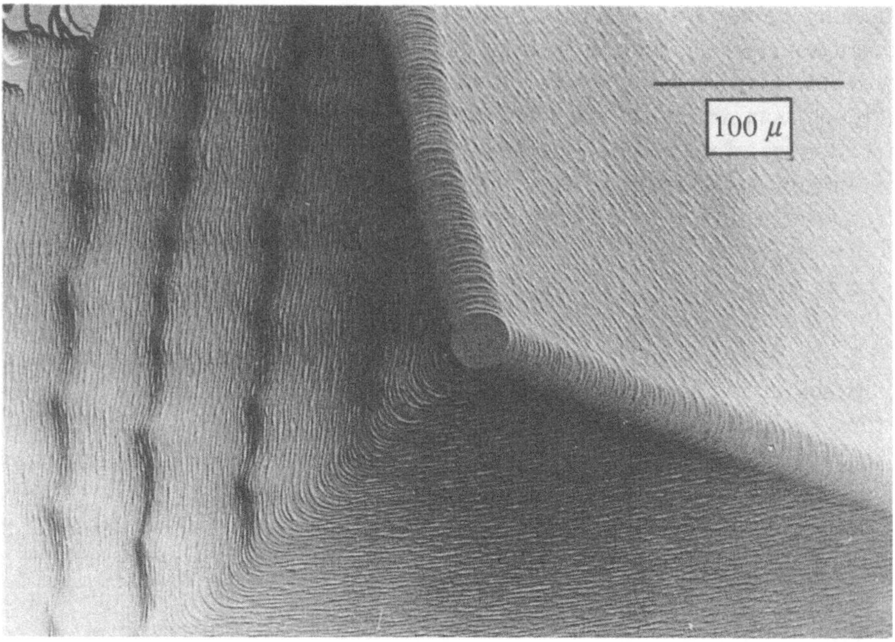

FIG. 7.3: Dislocation pyramid on flux-grown garnet crystal showing "family tree" of structures: terraces (all three sides), superterraces (left) and a substructure of superterraces.
The cental plateau (about 30 μ across) seems flat and featureless, indicating that the monatomic ledges have not yet bunched into terraces there. These ledges therefore form the "first generation" of the family tree.
From TOLKSDORF and WENZ (1978); courtesy Philips Forschungslabor Hamburg.

film by Honjo (see also HONJO, TAKAYANAGI, KOHAYASHI and YAGI (1979)) where the surface- reconstruction areas can be shown to be "heterogeneously nucleated" by the (monatomic?) ledges.

2) Unidirectional disproportionation of eutectics (JACKSON and HUNT (1966)), eutectoids (CARPAY and VAN DEN BOOMGAARD (1971)) and supersaturated solid solutions (CARPAY and CENSE (1974)) results in *periodic two-phase structures* which have extensively been investigated as a consequence of their technological importance (HALE (1976), ALBERS (1973)). Their structure can be considered as the trail of of the advancing disproportionation interface and as such reflect the morphological phase transitions of this interface.

In the first place, the regularity of the two-phase structure itself (lamellar or rod-like) is an instance of periodic arrangement of the 3-phase junctions (8.4). The details of the 2-phase pattern, in particular the relations between crystallographic orientations of the phases and the 2-phase interface orientations, result from the orientation dependence of the interface energies and mobilities, in combination with a process of growth competition between grains of different orientation.

Between the lamellar and the rod-type structure, a substructure transition is sometimes possible (HUNT and CHILTON 1962). Superstructures can result from instability of the two-phase interface with a wavelength ≫ the two-phase period. When both phases are nonfaceted, a macroscopic cellular pattern develops. When one phase is faceted, the other nonfaceted, the macroscopic cell forms can be imposed by the faceted phase, so that complex structures result (e.g. Al-Si eutectic; CROKER, BARAGA and SMITH 1975)

3) The nucleation, growth and interaction of *>1-dimensional defects in crystals* can be considered as condensation and growth phenomena. Generation of dislocations can be considered as condensation of a "supersaturated" mechanical stress field (mostly heterogeneously nucleated at an interface); subsequent formation of low-angle grain boundaries as a superstructure formation of the dislocation "continuum" and finally the precipitation of second-phase particles or pores as an ordinary 3-dimensional nucleation phenomenon, only complicated by the fact that the parent phase can support nonisostatic stress.

4) In binary and polynary systems, the extra *compositional* degree(s) of freedom give rise to further differentiations and/or interactions with the purely morphological structures. Examples are (apart from ordinary bulk phase disproportionation): surface segregation, phase separation in monoatomic adsorbed layers, and "decoration" of dislocations, grain boundaries and ledges on a surface.

5) *Side-branching instability in dendritic growth.*
In a moving inhomogeneous affinity field, a cellular interface can be stable. The transition to a dendritic pattern in which the dendrite tips, all of the same crystallographic orientation and vector velocity, are kept stabilized on an iso-affinity plane by a macroscopically positive $\vec{\nabla} A . \vec{n}$ can then be considered as a morphological phase transition (sidebranching instability) and therefore as a substructure formation.

Concluding remark:

The "soliton approach" is the obvious complement of linear instability theory of crystalline interfaces. Whereas the linear theory provides a satisfactory instability criterion for a planar interface, it cannot predict the ultimate, re-arranged and restabilized, but "very nonplanar" interface resulting after this instability has reshaped it. (i.e. cellular, terraced, zigzag etc.) It seems an interesting challenge to work out the interface EOM's (or coupled interface/diffusion field EOM's) by the techniques of soliton mathematics and to confront the results with experimental structures.

As we have pointed out in this chapter, the explanation of sub- and superstructures (which are experimentally observed) may be attempted by defining sub-level and super-level "continuum" EOM's which do not describe the interface as such, but rather the solitons in the surface.

Chapter 8

MISCELLANEOUS TOPICS

8.1 NONEQUILIBRIUM SHAPES

8.1.1 Notes on dendritic growth

The theory of dendritic growth is a highly specialized subject, and it is quite beyond our scope - as well as our ability - to give a serious contribution to this field (although we do expect that a rigorous iterative simulation of dendritic growth on the basis of prototype R and γ functions such as developed in part II of this monograph, as suggested in 5.3.3, would produce realistic dendrites). Still, we may present a few elementary remarks concerning the relevance of orientation dependence in γ and/or R functions for the structure of dendrites and for sidebranch instability.

Plane instability versus sphere instability

In the classical Mullins-Sekerka theory of instability of planar and spherical interfaces (MULLINS and SEKERKA (1963, 1964)), γ was considered to be OID (and growth resistance was considered to be absent at all, so that orientation dependence in χ was neglected a fortiori). As a consequence, there was no basic difference between either the mathematical approach or the resulting phenomenology of the planar in comparison with the spherical case. It may therefore be relevant to point out here that a) introduction of orientation dependence (in γ, in χ or in both) introduces one "basic difference" between the planar and the spherical case; b) introduction of OD γ and χ *together* introduce a second "basic difference".

Ad a)

For a planar interface with a sinusoidal perturbation, the orientations at the maxima and the minima are identical. Ergo, their $\mathbf{S}\gamma$ and $\mathbf{S}R$ are identical too (in the case of OD γ and R). In contrast, for a spherical interface with any spherical-harmonic perturbation, the orientation at the position of the maxima is *different* from that at the minima, so that for OD γ and/or χ, $\mathbf{S}\gamma$

and $\mathbf{S}\chi$ will differ too. More in particular, the rate of increase of such a perturbation will be sensitive for a "favourable" coincidence of perturbation-minima with γ or χ-minimum orientations, and perturbation-maxima with γ or χ-maximum orientations. In the rest of this subsection, we will use the informal epitheta "stiff" and "soft" for γ and/or χ- minimum- and maximum-orientations, respectively, even when the difference between minima and maxima is very slight. (without distinguishing between γ and χ: in practice their minima and maxima coincide).

Consider a very small nucleus in an unstable parent phase ($A_{|}>0$). When the noise field has just brought the nucleus to the critical size ρ_0, its shape coincides with the $\vec{\mathbf{Q}}\gamma$ shape. When it continues to grow, $A_{|}$ is initially constant over the interface, and for some time the nucleus' shape gradually shifts from the $\vec{\mathbf{Q}}\gamma$ to the $\vec{\mathbf{Q}}R$ shape corresponding with this $A_{|}$. The subsequent evolution depends on the value of the Peclet number, Rr/D. (R and r should be read as averaged growth rate and radius, respectively, in the case of orientation dependence). The condition Pe > 1 would implicate that the growth rate is slowing down, i.e. it is not a quasistationary condition such as we necessarily have in the vicinity of, e.g. a dendrite tip. Still in that case (Pe > 1), different parts of the growing crystallite are not coupled by the diffusion field and can only draw their material supply from a region of size $\ll r$ in the adjacent parent phase. This implies that the A at the interface is lowest in the regions of maximum interface mobility: an inhomogeneous A distribution is induced at the interface with maximum A in the direction of minimum R, and inversely. As compared with the original constant-A situation, the spreading rate is increased at the protruding parts of the $\vec{\mathbf{Q}}R$ shape (smaller radius) and decreased at the flatter parts, i.e. the A field pushes the crystal shape back to spherical. (the spherical shape is asymptotic, because then the A inhomogeneity disappears). So the actual shape will be stabilized somewhere between the $\vec{\mathbf{Q}}R$ shape and a sphere, with positive curvature everywhere. A technical application of this stabilizing mechanism is the "gel-growth" method.

In the opposite case, Pe < 1 (the realistic case for dendritic growth) the A-field inhomogeneity is inverted. The range of coupling by the diffusion field is larger than the crystal size, and A inhomogeneities are smoothed out as much as possible: i.e. the A gradient is primarily radial. So now the protrusions in the $\vec{\mathbf{Q}}R$ shape coincide with the *maxima* in A and the minimum-R directions with the minima. (the surface of the $\vec{\mathbf{Q}}R$ shape at the minimum-R orientations, although it has positive curvature radius, is *below* the iso-A surface which adapts to the average radius). In other words, the difference between the $\vec{\mathbf{Q}}R$-shape and a sphere is the very kind of perturbation which will grow out by the constitutional-supercooling instability: *the orientation dependence of R (and/or γ) provides a finite initial perturba-*

tion, i.e. a triggering mechanism. Additional statistical fluctuations would have to be very strong and extended indeed to overrule the perturbation already preset by the $\vec{Q}R$ shape, even for slight orientation dependence.

Ad b)

In the argument given above, orientation dependence in γ and in R play a rather similar rôle because both provide a similar "ab initio" favoured perturbation mode. However, when evolution is followed for a longer time, we discover a second basic difference between the plane and the sphere cases:

As long as the nucleus size is only slightly larger than ρ_o, the ΔA_c terms are effective to maintain the $\vec{Q}\gamma$ shape: i.e. even although the γ-minimum orientations are favoured-perturbation minima from the CS- point of view, the actual *outgrowth* is prevented by the ΔA_c terms. But for $r \gg \rho_o$, where the crystal would grow according to the $\vec{Q}R$ shape in a homogeneous-A environment - and therefore the favoured perturbation mode is still the same (because R-minima coincide with γ-minima) - the ΔA_c terms are negligible. In other words: although the orientation dependence in γ predetermines the orientation of the maxima which will ultimately break through, the outgrowth of this instability is effectively "launched" only from a nucleus size $\sim 2\rho_o$.

Altogether, the breakthrough of the favoured perturbation mode, i.e. of the protrusions which later develop into the dendrite axes, can be expected to start from a certain size of the crystal, to be expressed as a multiple of ρ_c. (of course, also related with $\vec{\nabla}A \cdot \vec{n}$, i.e. ultimately with A_∞) In the MS case with OID γ, small enough spheres are also stabilized by capillarity, but a nonzero launching speed of a favoured mode is not predicted.

Core and sidebranch regions of the dendrite

The spherical case discussed above explains only the *birth* of the dendrite axes (in the preferred "soft" crystallographic directions). This is a singular event; subsequent sidebranch split-offs are different in character because the symmetry of the diffusion field around the dendrite tip is paraboloid (Ivantsov) rather than spherical. Still, the cases have in common that the "degeneracy" of the instability-perturbation spectrum is relieved.

In the literature on dendritic growth theory, the selection of growth *rate* seems to be emphasized as the major problem. (for a survey, see e.g. LANGER (1980), LANGER and MÜLLER-KRUMBHAAR (1978), MÜLLER-KRUMBHAAR (1987); POMEAU and BEN-AMAR (1992)) Still, the selection of growth *direction* may be just as intriguing. The latter suggests that orientation dependence plays an essential rôle - an obvious experimental fact - and when it does so in selecting a direction, it may also select the *length* of the growth rate vector.

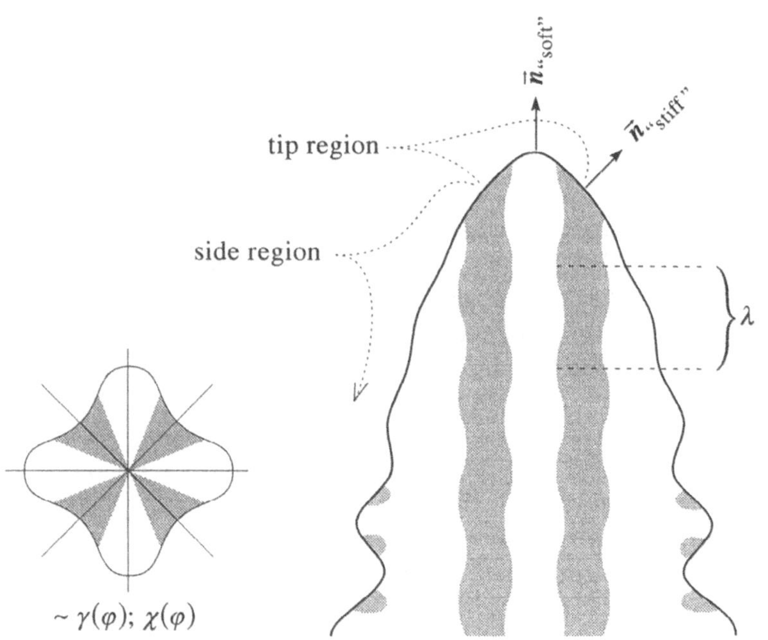

FIG. 8.1: Traces of "soft" (white) and "stiff" (shaded) interface orientations in a dendrite.

In fig. 8.1, we have sketched a cross-section through a dendrite tip for the case of a crystal with slight orientation dependence in γ and/or χ. In the crystallized material, we have grey-shaded those regions which crystallized with an orientation in the "stiff" orientation sectors (see the insert which suggests the shape of γ as well as χ plots qualitatively but exaggerated). The dendrite axis tends to align with a soft direction. So in the cross-section, we can distinguish a central core grown in the soft orientation range, enclosed by a more or less tubular structure grown in the stiff orientation range. (in a cross-section *perpendicular* to the dendrite axis, the grey zone will have further orientation-dependent structure). Beyond this grey tube, soft directions are predominant again (although in the transverse cross-section pronounced maxima and minima may occur).

Comparing the dendrite tip shape with the Ivantsov iso-A equiaffinity surface, we can see (by similarity with the spherical-symmetry case discussed above) that the "white" tip front sticks out above the iso-A surface, the front of the grey zone stays below it. The dendrite tip is not a paraboloid; it has a slightly sharper tip.

For a discussion of the mechanisms involved in dendritic growth, it may be relevant to distinguish between the "tip region" of the interface (roughly defined as the interface of the white core plus the inner half of the grey tube) and the "downwards" regions. This distinction represents, apart from a mere spatial separation, also a separation between predominant physical mechanisms:

a) Tip region:

1) Curvature (for both curvature directions together, for a 3D dendrite) is at least an order of magnitude higher than "downwards". This suggests that surface energy effects, as far as direction and velocity selection mechanisms are concerned, are probably essential in this zone only.
2) Orientation range includes *soft (core) and stiff (grey tube) orientations within close* ($Pe \ll 1$) *distance.* (in analogy with sphere case, and in contrast with planar case). This provides a "rudder mechanism" which tends to steer the growing dendrite in the soft direction (see below).
3) Parametric oscillator mechanism: Experimental evidence as well as intuitive reasoning suggest that the same combination of stiff and soft regions, coupled by the diffusion field *and* by the surface itself, may result in a parametric oscillation (or rotation) of the dendrite tip. Such an oscillation would result in a fluctuating (or spiraling) cross-section of the grey zone (as suggested in fig. 8.1) and in an initial wave on the downward surface. (wavelength $\lambda = R_{\text{tip}}/$ frequency of par. oscillator).

b) Downward region:

As compared with the tip region:
1) curvature and ΔA_c terms are at least an order of magnitude lower (until side instabilities break through still further down);
2) absolute growth rate is at least an order of magnitude lower;
3) $\vec{\nabla} A \cdot \vec{n}$ is at least an order of magnitude lower.

The rudder mechanism in the tip region

The combination of a soft tip surrounded by a stiff cone in the tip region of the dendrite provides a rudder mechanism for the dendrite tip (steering it towards the soft direction). This can be understood as follows:

As a point of departure, consider the symmetric case: i.e. the macroscopic vector velocity of the dendrite is aligned with the symmetry axis of the soft direction (surrounded by the stiff directions, supposed that the spatial distributions of the $\gamma(\vec{n})$ and $R(\vec{n})$ functions indeed provide such a symmetry axis).

Suppose that the dendrite deviates to the right. In response, the centre of the (negative-)affinity "bow wave" (where, however, affinity is highest) shifts to the "slower" cone. Just because it is slower, affinity can *increase*- and, as Pe $\ll$ 1, the increased affinity wave spreads out over the dendrite surface. Thus it accelerates the faster tip and side regions: the area of stiff orientation grows out further, and the tip's kinematic-wave trajectory is deflected to the left, restoring the original direction.

Additional asymmetries in the external affinity field (which may be induced, for instance, by enforced flow of the fluid parent phase or by an electric current) will interact with this mechanism. As such asymmetries can be quantitatively estimated, such effects may provide a useful experimental approach to investigate orientation dependence in γ and R and their influence on dendritic growth.

In connection with the rudder mechanism, we may call attention to the *possibility* that the grey and white regions are actually *oscillating* around some stabilized average rather than being stabilized on this average itself. (i.e., that the dendrite tip region is effectively a parametric oscillator). Such a mechanism might contain the following elements:

1) In the shape of the interface in the central zone (white core), γ-related terms in the EOM are predominant. The curvature of the (near-) paraboloid tip, and therefore the velocity of the whole structure, is ultimately imposed by surface energy.

2) The period of oscillation is (downward) limited by the inertia of the kinematic wave trajectories, i.e. their resistance to a rapid change of direction. This "stiffness" is again related with surface energy. The persistence of the interface to keep a certain area exposed within an orientation range results in oscillating minima and maxima in the diffusion field in front of the dendrite, corresponding with the oscillations in the dendrite-tip shape. Whereas the inertia of the kinematic wave trajectories introduces a "selfinductance" analog, the relaxation in the diffusion field can be considered as an "RC" analog. The parameter which is mainly affected in the oscillation is the selfinductance. As Pe $\ll$ 1, the effect of the oscillations in the side "grey cone" region is felt at the tip of the paraboloid (feedback loop). In terms of the rudder action mentioned above, the oscillator mechanism might be said to correspond with a "flapping" of all rudders on the grey cone together.

For very slight orientation dependence, the curvature in the tip region remains positive in spite of eventual oscillations. For somewhat stronger orientation dependence, oscillation in the tip region corresponds with repeated trifurcation of the kwt's in the stiff directions. (see fig. 5.4). The sidebranch outgrowth region then shifts further towards the dendrite tip.

Sidebranching

From the CS point of view, the downward region is unstable (and will effectively produce sidebranches, unless neighbouring dendrites suppress them). But in contrast with the tip region, we are in the case of instability of a planar (2D dendrite) or cylindrical (3D dendrite) surface again. And in contrast with the classical MS case, $\vec{\nabla} A$ and R are functions of position and time in a coordinate frame which is fixed to the crystal (and the sidebranch positions). Although these dependences complicate the analysis considerably, the ultimate result may in general be expressed in terms of a fastest growing perturbation wavelength Λ. As the conditions determining Λ are the $\vec{\nabla} A \cdot \vec{n}$, $\vec{v}_{\text{dendrite}} \cdot \vec{n}$ and A conditions prevailing in the downward region, it would be quite coincidental indeed if this Λ would agree with the λ resulting from an eventual oscillation in the tip region (see above).

In combination, our intuitive picture of sidebranch instability is the following: Without any ab initio perturbation, the downward region would develop a (cellular or dendritic) instability by the CS mechanism, of some fastest growing wavelength Λ. Now when waves (wavelength λ) originating from a self-oscillatory tip come down, they provide a finite initial perturbation. When $\mathrm{d}\delta(\lambda)/\mathrm{d}t$ is positive, the λ wavelength may grow out in the beginning just because they are ahead from the start. After some time, interference with the Λ period selects only a few of the λ-wave maxima to grow out. (still, tip oscillation is not essential for sidebranching).

Still further down, and after the branches (including those from neighbouring dendrites) have consumed all available material, A gradients die out and a dendritic structure in a homogeneous solution or melt remains. Such a structure is unstable due to the difference in ΔA_c between positively and negatively curved interface. From then on, we are in the stage of the "Ostwald ripening" process, eventually leading to dendrite branches melting off from the main stem (section 8.3)

Concluding, in the analysis of the dendrite growth mechanism it may be essential to distinguish between two or even three stages of evolution. These stages are separated both spatially and temporally:

1) tip region: orientation range including **S**γ, **S**R-minimum core and **S**γ, **S**R maximum-surrounding cone. Both surface energy effects and orientation dependence of γ and χ are essential. Growth rate and -direction are selected in this zone; i.e., the tip region plus interacting diffusion field around it are an (extended) velocity source for the dendrite, in the sense of section 6.4. Possibility of oscillatory growth mode.
2) Sidebranching region: time-dependent version of unstable cylinder.
3) Ageing stage: after completion of solidification, dendritic structure is unstable. Ostwald ripening.

8.1.2 A note on whiskers and platelets.

The most dramatic deviations between actual shapes of growing crystals and growth forms derived from perfect-crystal $R(\vec{n},A)$ functions, are provided by whiskers and platelets (unless the $\vec{\mathbf{Q}}R$ shapes themselves are already needles and platelets, which can occur for highly anisotropic crystals). They result from the combined action of a perfect-crystal R function and one or more velocity sources.

For whiskers this velocity source can be either a point source (a and b) or a 3-junction source (with a closed-loop 3-junction); (c):

a) screw-dislocation D-junction;
b) protruding vertex stabilized, e.g., by $A_|$ inhomogeneity resulting from a bulk diffusion field (Pe << 1) or boundary-layer diffusion in combination with nonlinear $R(A)$ dependence near the tip ("faceted dendrite" or "hopper crystal"). This is just the tip region of a dendrite (which was recognized as a velocity source on page 211) condensed into a point, as a consequence of the $\mathbf{S}\gamma < 0$ instability which now acts to stabilize the *protruding* vertex.
c) VLS (vapour-liquid-solid) whisker growth mechanism; the whisker grows as a column from a liquid droplet and is actually directed (and its diameter stabilized) by the crystal/liquid/ambient 3-junction whose diameter is controlled by the available volume of the liquid.

For platelets the mechanism can either be the cooperation of two point sources (d) or one line source (e and f):

d) "web crystals" spanned by two (type a or type b) whiskers or needles,
e) TPRE (twin-plane reentrant edge) growth or growth nucleated by any other "stiff" interface (grain boundary). Bi-phase growth can also occur in eutectics, nucleated by the 2-phase interface. This corresponds with a nucleating "double-mobile 3-junction" as discussed in 6.2.2.
f) stabilized protruding edge, i.e. the 2-dim. analog of case b). Apart from a flat platelet, hollow needles can grow from a polygon-shaped edge.

Mechanisms b and f can be identified with dendritic growth where sidebranch instability is suppressed by a strong orientation dependence of γ and/or R: high "stiffness" of the stabilized planes combined with relatively low stiffness of the advancing tip or edge. Such needle crystals and platelets grow in a direction of *maximum* growth rate (and γ); their shape is in the small direction(s) defined by facets. In contrast, a *dislocation*-induced whisker grows in the direction of the D-junction velocity vector, i.e. in general a *minimum*-growth rate (and γ) direction of the perfect crystal.

8.1.2 A note on the theory of nucleation of crystals

Whereas in whiskers and platelets the deviation from a less anisotropic $\vec{Q}R$ shape is obvious, this is hardly so for crystals with velocity sources in *three* directions (e.g., screw dislocations). Such a crystal may at first sight be recognized as geometrically similar with the growth shape of the perfect crystal, but its growth rate is much faster (corresponding with the D-junction vector velocities).

This difference between a perfect and a dislocation-bearing crystal is most obvious when the driving affinity is so low that the perfect crystal does not grow at all, e.g. for a slightly supercritical nucleus. Nuclei with dislocations then grow out much faster than the perfect ones; the latter may even not grow out at all. This raises the question whether, in comparing nucleation theory with experiment, the theoretical prediction of nucleation rate would not have to be made for dislocated or twinned nuclei rather than for perfect ones. Determination of the perfection of counted nuclei seems essential in order to verify what is in agreement with what (or not). For the same reason, experimental determinations of perfect-crystal growth rate functions should be considered very critically. The actual "perfect" functions may be completely hidden by the combined effective-growth rate plot generated by the D-junction velocity sources (as the latter velocities depend on the dislocation Burgers vector and the dislocation density, a considerable statistical spread can be expected).

8.2 BOUNDARY-LAYER EFFECTS

In this section we discuss, in a qualitative way, some consequences of a (resistive) *boundary layer (BL)* in the parent phase. (we do not specify the nature of the BL which can be of many kinds: diffusion/viscous/thermal/chemical etc.) Implicit in the BL concept is that the potential drop is supposed to be evenly distributed over a layer of thickness δ and that this δ is supposed to be constant and independent of orientation. Also, the parent phase is supposed to be isotropic.

The following three effects of the BL can be discerned: growth rate limitation, dispersion of the vector velocity and curvature dependence of the growth rate.

8.2.1 Growth rate limitation

Let the crystal interface obey the growth rate function $R(\vec{n},A) = A\chi(\vec{n},A)$. The "resistance" of the BL may be characterized by a similar relation $R(A) = A\xi$, where ξ is an (OID) "conductance" and the flux through the BL is written directly as a growth rate; i.e. translated into growth rate

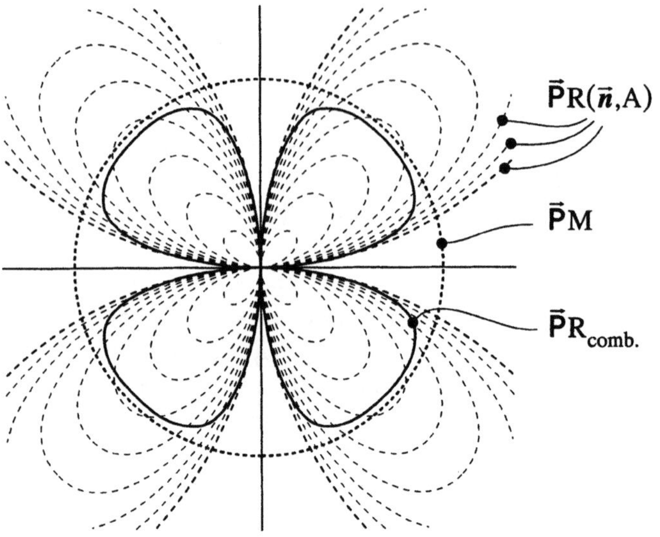

FIG. 8.2: Limitation by an orientation-independent maximum M.

units instead of conventional flux units by an appropriate multiplication factor incorporated in ξ.

With the (planar) interface and the BL in series, an externally applied affinity $A_{ext.}$ is split up into A_{BL} and $A_{|}$:

$$A_{ext.} = A_{BL} + A_{|}$$

$$R_{comb.} = A_{BL}\xi = A_{i}\,\chi(\vec{n},A)$$

When ξ and $\chi(\vec{n},A)$ are known, the growth rate function $R_{comb.}$ of interface + BL can be solved as a function of $\vec{n}$ and $A_{ext.}$. In the special case that the interface growth rate is linear in A, i.e $\chi(\vec{n})$ independent of A, we have simply

$$R_{comb.}(\vec{n},A_{ext.}) = A_{ext.}\,\chi(\vec{n})\,\xi\,/(\chi(\vec{n}) + \xi)$$

$$= A_{ext.} \textcircled{-1} \langle \chi(\vec{n}), \xi \rangle \qquad (8.1)$$

in the notation adopted in ch. 11.

In any case, the $R_{comb.}$ function is limited to a maximum $M = A_{ext.}\,\xi$ allowed by the boundary layer. The situation is illustrated by fig. 8.2. The interface growth rate plot is drawn as the set of dotted clover-leaf plots for increasing A. The circle is the $M(A_{ext.})$ limit, and the drawn limited clover-leaf plot is $R_{comb.}(\vec{n},A_{ext.})$.

When $R_i(\vec{n},A_{ext.})$ varies between deep cusps ($\ll M$) and high maxima ($\gg M$), then in $R_{comb.}$ the cusps are relatively unaffected as compared with R_i whereas the maxima are limited (see fig. 8.2). This implies that $A_|$ varies between values $\simeq A_{ext.}$ (where the interface orientation corresponds with a cusp in $R_i(\vec{n},A_{ext.})$ and $\simeq$ 0 for orientations where $R_i(\vec{n},A_{ext.})$ is $\gg M$. Consequences of the resulting surface gradient in $A_|$ are discussed in 13.5.2c.

8.2.2 *Dispersion*

The M limitation mentioned in 8.2.1 has nothing to do with the thickness of the BL; it would also occur when the OID "resistance" were concentrated in an infinitesimally thin sheet. In contrast, the fact that the resistive layer has a *thickness* implicates that structural details on the surface which are different in size but similar in shape follow a different evolution. As an example, fig. 8.3 shows two similar grooves or RE's in a surface: the size of case A is $\ll \delta$, that of case B is $\gg \delta$.

Consider an interface with a corrugation Δz = a sin $2\pi x/\lambda$, a $\gg \lambda$. (small orientation range around average orientation $\vec{n}_0$). This corrugation follows a certain kinematic-wave trajectory, but the trajectory is *dispersive* in the range $\lambda < \delta \rightarrow \lambda > \delta$. The limiting cases $\lambda \ll \delta$ and $\lambda \gg \delta$ are the only ones which can be considered as "classical" kwt cases (i.e., "case A" in the sense of ch. 5.1). In between, we have essentially inhomogeneous surface affinity and consequently curved kinematic-wave trajectories.

1) For $\lambda \ll \delta$, the affinity at the interface $A_|$ is uniform:

$$A_| = A_{ext.} - \xi R_{comb.} \tag{8.2}$$

(when the corrugation amplitude increases, i.e. when a $\ll \lambda$ no longer applies, then the averaged surface χ may be affected, but even then $A_|$ remains uniform).

The corrugation follows the kinematic-wave trajectory belonging to that particular A:

$$\vec{\Re}_{\lambda\ll\delta} = \vec{\mathbf{Q}}R(\vec{n},A_|) = (\vec{n} + \partial/\partial\vec{n})R(\vec{n},A_|). \tag{8.3}$$

2) For $\lambda \gg \delta$, ($\gg$a), the interface affinity is not homogeneous, but depends on orientation. For the vector velocity of the long-wavelength corrugation we now have directly

$$\vec{\Re}_{\lambda\gg\delta} = \vec{\mathbf{Q}}R_{comb.}(\vec{n},A) = (\vec{n} + \partial/\partial\vec{n})R(\vec{n},A_{ext.}). \tag{8.4}$$

Fig. 8.2 illustrates the difference between expressions (8.3) and (8.4). For a

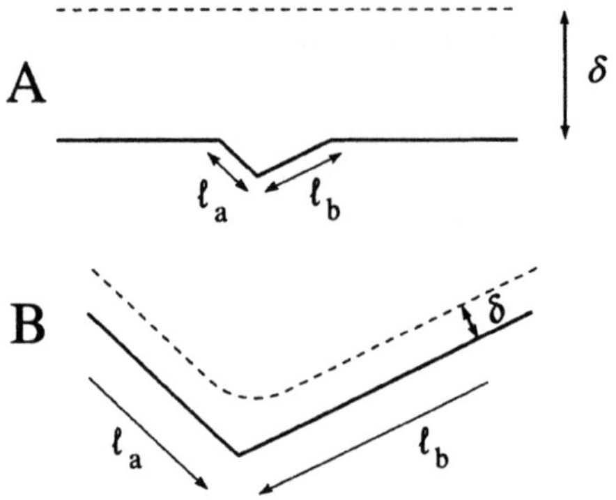

FIG. 8.3

certain macroscopic orientation, we have a fixed operating point on the drawn $R_{comb.}(\vec{n},A_{ext.})$ plot (the "flattened" clover-leaf). Expression (8.4) just expresses the Gibbs-Wulff $\vec{\mathbf{Q}}R_{comb.}$ vector corresponding with this plot and operating point. However, the "local" plot which is relevant at the surface is the dotted (unlimited) $R(\vec{n},A_|)$ clover-leaf plot passing through the operating point. The $\vec{\mathbf{Q}}R_{local}$ vector corresponding with it (represented by eq. (8.3) is quite different, because the slope of the dotted $R(\vec{n},A_|)$ plot differs from the $R_{comb.}(\vec{n},A_{ext.})$ slope.

This can be made explicit for a particular case. Suppose that we have a linear m-symmetric "flower-leaf" R plot

$$R(\varphi,A) = A\chi_o|\sin m\varphi| \tag{8.5}$$

so that

$$R_{comb.}(\varphi,A_{ext.}) = A_{ext.}\frac{\chi_o|\sin m\varphi|\,\xi}{\chi_o|\sin m\varphi| + \xi} \tag{8.6}$$

and

$$A_| = A_{ext.}\frac{\xi}{\chi_o|\sin m\varphi| + \xi} \tag{8.7}$$

Then eqs. (8.3) and (8.4) have the explicit form

$$\vec{\Re}_{\lambda\ll\delta} = A_{ext.}\,\xi\chi_o\,\frac{1}{\chi_o|\sin m\varphi| + \xi}\,\vec{\mathbf{Q}}\,|\sin m\varphi| \tag{8.3a}$$

$$\vec{\Re}_{\lambda\gg\delta} = A_{ext.}\,\xi\chi_o\,\vec{\mathbf{Q}}\left\{\frac{|\sin m\varphi|}{\chi_o|\sin m\varphi| + \xi}\right\} \tag{8.4a}$$

(the $\vec{\mathbf{Q}}$ operator now has to be read as $\vec{\varphi} + \partial/\partial\vec{\varphi}$).

The difference between the two limiting cases $\lambda \ll \delta$ and $\lambda \gg \delta$ shows that the vector velocity is now a function of wavelength: it shows dispersive behaviour around $\lambda \sim \delta$. This is purely a consequence of the thickness of the boundary layer. This dispersion, in combination with the nonlinear effects which can either result from $\mathbf{S}\gamma < 0$ or $\mathbf{S}R < 0$ conditions or from the curvature dependence of M which also results from BL thickness, can lead to "soliton" structures (terraces, faceted cells).

8.2.3 *Curvature dependence of the growth rate function*

In 2.7 we have shown that the thickness of a BL results in an effective curvature dependence of its resistance. For $\delta \ll r$, this can be expressed in terms of an expansion in δ/r. From eq. (2.30) we obtain, taking $f = 1$ for full asymmetry:

$$\xi = \xi_0 \{1 + \frac{1}{2}\frac{\delta}{r} - \frac{1}{3}(\frac{\delta}{r})^2 + \ldots\} = \xi_0 \frac{r}{\delta} \ln (1 + \frac{\delta}{r}) \qquad (8.8)$$

i.e., the hill tops grow faster than the valleys. This is just the onset of the constitutional-supercooling mechanism, a typical consequence of bulk diffusion which, as such, starts to show up in the boundary layer.

The condition $\delta \ll r$ is relevant when we want to investigate the transition $\lambda \ll \delta \rightarrow \lambda \gg \delta$, i.e. the dispersion region. Near the $\lambda \gg \delta$ limit, the δ/r correction is relevant. Near the $\lambda \ll \delta$ limit we are rather in the "ordinary" bulk-diffusion (MS) case. Anyhow, the *nonlinearity* induced by the δ/r term gives rise to the evolution of a "cellular" pattern. In combination with the *dispersion* discussed in 8.3.2, such patterns can be stabilized in the form of soliton-like features as the edges in terraces or macrosteps.

8.2.4 *Interaction with junctions.*

In 6.2 we discussed the kinetic behaviour of junctions, and in 6.3 that of edges and vertices, on the basis of a uniform driving force at the interface. The presence of a boundary layer can, of course, complicate matters because affinity gradients can occur over a typical length scale δ. One example of such a complication is the facet neck formation mentioned in 13.5.2, sub c.

When δ is larger than the critical nucleus size ρ_0, an extra complication arises because, due to the inhomogeneity of the affinity $A_|$ felt at the interface, ρ changes with position too.

8.3 TRANSIENTS TO EQUILIBRIUM SHAPE: THE RAE CONCEPT

8.3.1 Definition of the range of approached equilibrium (RAE)

A useful concept in the discussion of nonequilibrium-equilibrium transients is that of the "Range of Approached Equilibrium", RAE or $\Phi(t)$.

Consider an interface in a homogeneous external zero-affinity field; (i.e. for a planar interface the driving force is zero everywhere). When the interface has a nonequilibrium shape, the driving force for shape changes exclusively consists in the differences in the capillarity shift. The *mechanism* for shape changes can be diffusion (bulk or surface), motion controlled by finite interface mobility, viscous flow, or a mixture of these.

Now we define $\Phi(t)$ *as the range within which deviations from thermodynamic equilibrium have been smoothed out to less than 1/e of their original value.* It has therefore the dimension of length and it increases monotonically with time. Especially in the case of OID γ and infinite χ, the basically simple laws for $\Phi(t)$ can be derived by dimensional analysis.

The RAE concept can be used to characterize ageing and sintering phenomena, meniscus kinetics etc. and to estimate the size of those details in growing interfaces ($A_l \neq 0$) where surface energy is predominant in determining the shape. The essential procedure is to compare the size of typical details in the initial structure with $\Phi(t)$. Examples of such details are the initial grain size in sintered materials, the two-phase period in eutectics and eutectoids, the interparticle distance in Ostwald ripening and the evolution of dendrite branches.

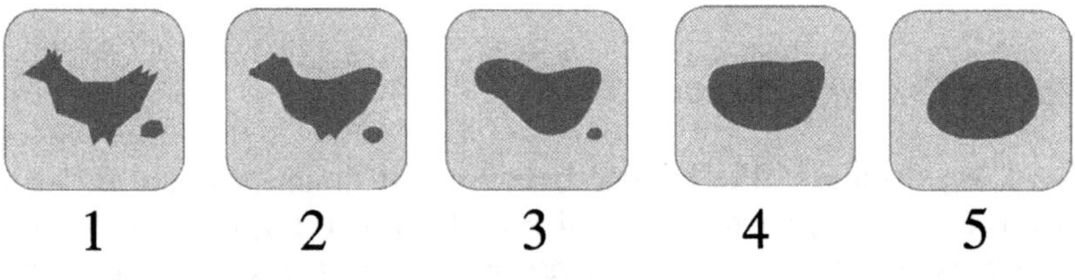

FIG. 8.4

We will first explain the underlying principle of the concept and give the explicit functions $\Phi(t)$ for several cases of interest. Let fig. 8.4(1) be the initial configuration of an arbitrary nonequilibrium interface. (the configuration may even imply separate particles). The bulk "phases" at both sides are in chemical equilibrium; they may be actually different phases in the thermodynamic sense or the same (in the case of a grain boundary). Successive stages (film frames) in the ageing process are indicated in fig. 8.4(2-5).

The important point is that the "cinematographic pictures" of fig. 8.4 (1-5) represent the evolution *independently of the scale factor used in the drawings*; only the *speed* of the film depends on this scale factor. This observation (originally made by HERRING in 1950) is the basis of the RAE concept. Two initial structures with exactly similar shape but a different scale factor α, so that they are represented by the same fig. 8.4(1), will also follow the same path of evolution 8.4(2-5), but with a speed proportional to some power $(-n)$ of α, dependent on the transport mechanism.

We use the term "scale factor" throughout for the (dimensionless) factor relating the size of objects of similar shape. With reference to drawings it is the ratio of the length *represented* by a size bar to the *actual length* of this size bar in the drawing, and with reference to similar objects it is just the ratio of the sizes. Of course, instead of the dimensionless factor α we can also use the size L of some characteristic structural detail, with the dimension of length, as a parameter: then the evolution rate of change will contain the factor L^{-n}.

(Strictly speaking, the above is only correct when in the bulk of the two phases the $\partial C/\partial t$ terms in the diffusion equation can be neglected against the $\partial C/\partial t$ or $D\vec{\nabla}^2 C$ terms at the phase interfaces; i.e. when the diffusion fields can at any time be considered as near-steady-state solutions for the particular boundary conditions at that moment. In the appendix of my paper on eutectic ageing (VAN SUCHTELEN 1978) I have shown that this assumption is always justified in the case of a simply connected interface dividing the two phases. When one phase is dispersed in topologically separate particles, however, the "steady-state assumption" can be translated into conditions to be met by the geometrical characteristics of the structure beyond which the steady-state approximation is no longer valid and the simple power law no longer applies. An example is the stage in Ostwald ripening processes of small crystals dispersed in a solution, immediately before a particle is completely dissolved).

In order to have a quantitative tool in the description of ageing processes, we can apply a Fourier analysis to any nonequilibrium shape and so obtain a spectrum of sinusoidal perturbations all on their way to equilibrium. (Formally this limits the analysis to a topologically restricted class of surfaces, but the general concept can be used in a wider sense). In this case, the wavelength λ is the natural size parameter to play the rôle of L mentioned above.

In the linear ($f \leqslant \lambda$, small-amplitude) approximation of the evolution process for a sine perturbation of an originally flat interface $z = 0$,

$$z(x) = f \sin 2\pi x/\lambda \tag{8.9}$$

the shape remains a sine during evolution, with $f = f(t,\lambda)$:

$$f(t,\lambda) = f(0,\lambda)\exp\left(\frac{-c_n}{\lambda^n}\right)t \tag{8.10}$$

The exponent n and the factor c_n depend on the transport mechanism:

for viscous flow:	$n = 1$	$c_1 = 2\pi\gamma\eta$	(8.11a)
for interface resistance:	$n = 2$	$c_2 = (2\pi)^2\gamma\,\chi\,V$	(8.11b)
for bulk diffusion:	$n = 3$	$c_3 = (2\pi)^3\gamma\,MCD/(R_gT\rho)$	(8.11c)
for surface diffusion:	$n = 4$	$c_4 = (2\pi)^4\gamma\,MCD^*/(\delta R_gT\rho)$	(8.11d)

For the derivation of eq. 8.10/11, the reader is referred to MULLINS (1959). In the expressions for c_n, M is the molecular weight of the diffusing phase, C the fractional solubility of the diffusing species, D the diffusion coefficient, R_g the gas constant, ρ the solid density of the diffusing phase, D^* the virtual boundary diffusion coefficient in an interface layer of thickness δ, and χ is the interface mobility as defined by eq. (1.7).

For convenience it is assumed that the transport process is brought about by the diffusion of the species of one of the phases through one medium only, whereas the other phase adapts its boundaries without resistance. Strictly this picture is a good approximation only for crystals in a liquid solution. In the general case, e.g. two solid phases in equilibrium, we have two diffusing species at least, each diffusing through two different phases. The exact description of this process is formally complicated, but conceptually there is no difference with the simple case. In fact, all difficulties can be taken into account, mainitaining equations 8.11 c and d for $n = 3$ and $n = 4$, by properly defining effective averages of M, C, D and ρ.

As to the applicability of the bulk- and surface- diffusion mechanisms to the case of grain-boundary motion, we have to realize that the same atoms which constitute a crystal lattice can diffuse *through* the lattice (or along an interface) as interstitials or by a vacancy diffusion mechanism. This implies that the same element has to be considered as to occur in two species: a) the immobile lattice species, b) the "dissolved" (interstitial or vacancy) species which can diffuse. Ergo, the symbols C and D refer to the latter "species".

The range of approached equilibrium is now defined by equating the exponent to −1 for $\lambda = \Phi$:

$$\Phi(t) \equiv (c_n\,t)^{1/n} \tag{8.12}$$

The definition is illustrated in fig. 8.5.

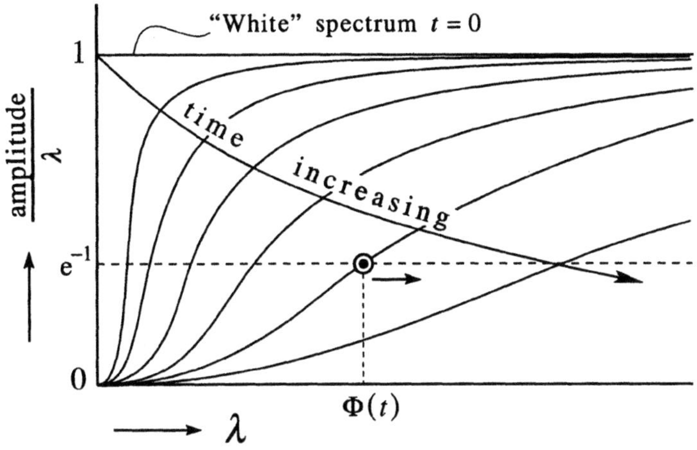

FIG. 8.5: Illustrating the definition of the range of approached equilibrium.

Simultaneous occurrence of several transport mechanisms.

In the case of more than one simultaneously operative transport-rate-limiting mechanism, $\Phi(t)$ is to be defined as the maximum of the corresponding Φ's (drawn curve in fig. 8.6). So, along the time axis, regions can be discerned where mechanisms with decreasing n are predominant.

8.3.2 The description of nonisothermal processes using the RAE concept.

The constants c_i are temperature-dependent. The *shape similarity* principle remains valid, however, during any temperature programme: only the "film speed" is accelerated at higher temperature.

It is easy to generalize the RAE concept to nonisothermal processes: the acceleration can be used to define a sort of "biological time" for the structure under consideration. Starting from a certain "normal" or reference temperature T_1 (room temperature or operating temperature) we can collect the temperature-dependent part of the constants c_n in a function $g(T)$ which is normalized to 1 at $T = T_1$:

$$\Phi = [\, c_n(T)\, t\,]^{1/n} = [\, c_n(T_1).\, g(T).\, t\,]^{1/n} \tag{8.13}$$

The effect of a temperature history $T(t)$ on the process of ageing can now be described by introducing the *effective age* t^*:

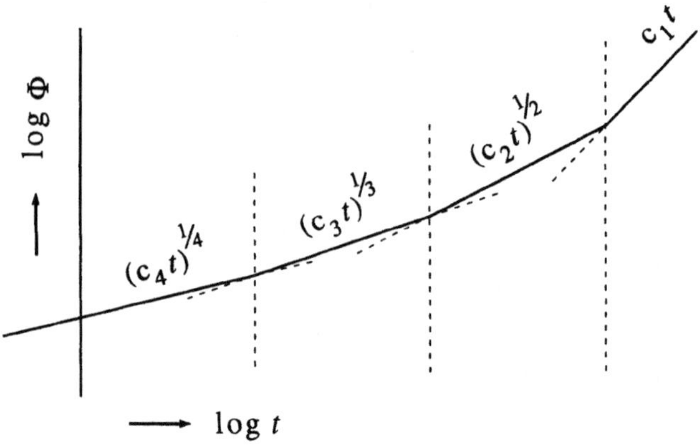

FIG. 8.6 RAE(t) behaviour when several transport mechanisms are active simultaneously.

$$t^* = \int_{t=0}^{t} g[T(t)]\, dt \tag{8.14}$$

A particular temperature treatment corresponds with a certain extra age $t^* - t$ of a structure. A special case is the extra age of a crystal or a polycrystalline aggregate grown by solidification of a melt or by a high-temperature sintering process. This "manufacturing age" can in some cases amount to much more than the useful or technical life of the structure, which in that case will not appreciably change further after manufacture.

8.3.3 Orientation dependence of γ and/or χ .

Although the RAE concept is typically a "linear" concept, we can extend the definition to OD γ and χ, provided the cases to which the concept is applied are considered carefully.

The evolution laws (8.9 - 8.11) are first-order approximations and exact only for $f \ll \lambda$. Nevertheless, the evolutionary stage of an ageing or sintering process can be characterized by comparing the size of certain typical geometrical features of the structure (e.g. grain size, eutectic period) with $\Phi(t)$. Even when the $f \ll \lambda$ limit is exceeded by the initial structure, such a comparison provides a correct order-of-magnitude argument.

Now when during a whole evolutionary period the range of orientations in a structure does not exceed a range where $\mathbf{S}\gamma$ or $\mathbf{S}\chi$ can be considered as constants except for a fractional deviation only, we can apply definition (8.12), but now with $\mathbf{S}\gamma$ and $\mathbf{S}\chi$ substituted in the c_n expressions 8.11 a/d instead of γ and χ, respectively. However, when during this period the orientation range is large, in particular so large that an appreciable fraction of the equilibrium shape, or even a complete equilibrium-shaped particle will ultimately result (e.g. a broken crystal fragment or a dendritc side-branch), then the average γ and χ should be substituted because the latter are indicative of the *overall* size of the detail under consideration. The choice depends both on the size and spatial boundary conditions, and (sometimes) to the considered time span, of the structure. It is obvious that only qualitative order-of-magnitude estimates can be derived from such a comparison between size and $\Phi(t)$.

8.3.4 *Extension of the RAE concept to nonzero affinity.*

Consider nucleation and growth in an otherwise arbitrary crystal growth situation. It is obvious that a critical nucleus has exactly the equilibrium shape, apart from statistical fluctuations due to the noise field, because its growth rate is zero by definition. (By the way, for small critical nuclei , i.e. high affinity, the fluctuations induced by noise may become comparable with the overall size and then a more detailed kinetic description of the fluctuating shape on the basis of an EOM which includes the noise field, becomes necessary). For twice the critical size, i.e. $\Delta A_c = \frac{1}{2} A_|$, the growth rate is half its ultimate value, and the overall tendency of ΔA_c to keep the growing crystallite in equilibrium shape just equals the effective driving force to bring it in the growth shape (presuming that a difference can be distinguished at all). For crystal size $\gg \rho_c$, only shape details of size $\leqslant \rho_c$ are in agreement with the equilibrium shape. Because the only locations where eventual deviations between growth and equilibrium shape are observable is in regions of high curvature, this applies specifically to the detailed shape of edges, vertices and junctions, which have been discussed in some detail in 6.2 and 6.3.

The critical size $\rho_c = V\gamma/A_|$ can be considered as an asymptotic limit. Consequently, we may extend the RAE definition to nonzero-affinity situations by

$$\Phi(A \neq 0, t) = \left(\frac{1}{\rho_c} + \frac{1}{\Phi(A=0, t)} \right)^{-1} = \left(\frac{A_|}{V\gamma} + (c_i\, t)^{-1/i} \right)^{-1}$$

$$= \ominus\!\!1 \times \Phi(t)\,,\, V\gamma/A_| \times\!\!\circ \tag{8.15}$$

where $A_|$ is the external affinity for a planar interface The second expression is written in the "assembled function" notation to be introduced in ch.10.

8.3.5 Applications of the RAE concept.

Ageing.

The RAE concept is in particular useful for the consideration of ageing effects. As an example, the author has investigated the coarsening of eutectic and eutectoid structures during and after unidirectional growth. In principle, this process results in periodic lamellar or acicular two-phase structures with periods in the micron range, with imperfections caused by grain boundaries and unavoidable thermal fluctuations. (VAN SUCHTELEN 1978)

Once solidified, these structures can slowly modify by ΔA_c-driven interface motion. Three main types of modifications can be discerned, leading to different changes in the microstructure:

a) decay to local equilibrium near interruptions in lamellae or needles or near grain boundaries, whereby sharp discontinuities are rounded off.
b) Touching and coagulation of neighbouring grains of the same phase, which come in contact as a result of an event of type a) or c).
c) spheroidisation of needles in an acicular structure. This is an effect of the Rayleigh instability and can effectively be described by an equation of type (8.12) but with *negative* c_n (CLINE 1971)

For each of these three types, typical transient or "incubation" times can be calculated as a function of the two-phase period λ and the volume fractions ζ and $(1-\zeta)$ of the phases. One of the conclusions is that coarsening during growth imposes a minimum period which can technically be grown, whereas thermal noise caused by imperfections of the growth apparatus imposes a maximum period. The range in-between widens for increasing temperature gradient in the solidification zone.

In a quite similar way, the RAE concept can be applied to the ageing of dendritic structures, multilayer structures and even to small details in integrated circuits.

Sintering.

Sintering processes are a typical instance of capillarity-driven morphological changes - in fact, these are just another case of (intentional) "ageing". Hence, the RAE concept can be a useful tool, but we have to be aware of the fact that, during the complete process, the predominant mechanism (and the constants involved) can change several times.

In the initial stage of neck formation and compacting, the curvature of the relevant interfaces in motion is high as compared with the size of the grains to be sintered, and we have in general to do with solid/solid/gas

3-junctions. In this stage, a transition from surface-diffusion to bulk-diffusion transport predominance can be expected. When the temperature is high enough for a liquid phase to form, the situation can change drastically as the diffusion constants in the liquid can be many orders of magnitude higher. Apart from this, the "connectivity" of an eventual liquid phase can be important, as indicated in 4.1.2.

We have to do with different diffusion constants and/or mechanisms for pore diffusion kinetics, grain boundary motion and two-phase interface motion. Hysteretic motion and pinning of interfaces can result from interaction with pores and segregations. In the final stage of grain growth, a transition in evolution speed can be expected when the RAE exceeds the grain size. From then on, grain growth can be considered as Ostwald ripening for grains in a densely packed structure, because all interfaces, as spanned by their 3-junction frames, are in equilibrium. Only the 3-junction "skeleton" structure is still in evolution then.

Meniscus kinetics and hysteresis.

When a crystal, grown from a liquid parent phase (melt or flux) is pulled from the liquid, it has to pass the liquid surface. This means that a parent phase/crystal/ambient gas 3-junction sweeps across the interface. The motion of this "meniscus" can be intermittent and then it can leave traces on the surface ("meniscus lines") which are of technical relevance when a smooth surface is required. A notorious example are the meniscus lines left on LPE grown semiconductor device layers made by sliding-boat techniques (CLINE (1971), SMALL, BACHEM and POTEMSKI (1977)). We will try a rough sketch of the physical situation. For ease of discussion, we assume identical and OID free energy of the three interfaces.

In fig. 8.7 and 8.8, the original surface is assumed to be planar; it is drawn horizontally. In fig. 8.8, the parent-phase/gas interface (p/g) deviates by an angle φ from the vertical position. Macroscopically, the force on the 3-junction as a whole is $\gamma \sin \varphi$ (under the assumption of identical γ's).

First we consider the growth of a meniscus ridge for an immobile 3-junction (i.e. $\varphi = 0$; fig. 8.7). Suppose that at $t = 0$ the p/g interface is brought to some position $x = 0$ on the crystal surface with $\varphi = 0$. Because the crystal surface is not yet curved, a very high underpressure (of the order of γ/a ; i.e. 10^3 - 10^4 kgf/cm^2) is induced in the crystal below the 3-junction which is identical with a positive affinity for growth. Immediately a meniscus ridge starts to form, thereby increasing the radius and decreasing the affinity. The size of the ridge is essentially $\Phi(t)$. (note that $\Phi(t)$ for the p/c interface and that for the g/c interface are entirely different; when we assume the latter to be identically zero, the 3-junction moves along the 60° slope in fig. 8.7).

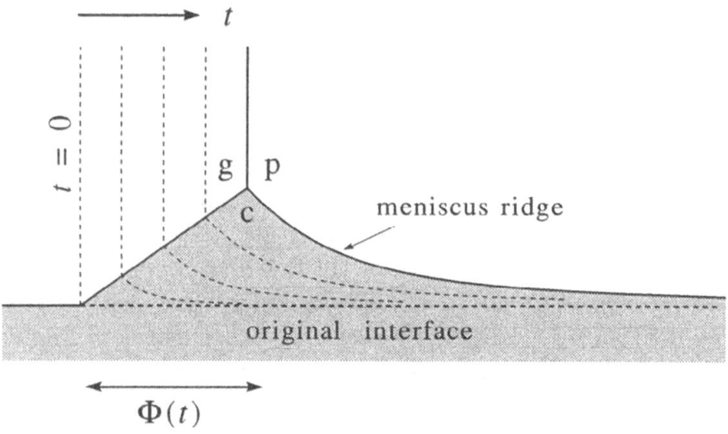

FIG. 8.7: Growth of an "anchoring" meniscus ridge.

The important point is that the meniscus ridge, once formed, is effectively a pinning feature which prevents the 3-junction to move when φ is changed. This is the origin of the hysteretic behaviour of meniscus lines. In order to obtain a full understanding of meniscus kinetics, we would have to solve the problem of the stationary shape of a meniscus 3-junction moving at constant velocity (driven by the force $\gamma \sin \varphi$), including its stability.

This stationary shape is sketched, intuitively, in fig. 8.8, together with the associated pressure and affinity distribution. Necessarily, the c/p and c/g interfaces coincide asymptotically with the original interface behind and ahead of the junction. In order to satisfy equilibrium with the tilted p/g interface, the c/p interface forms a trough. Because this trough is moving, its positive and negative slopes are a sink and a source, respectively, for the diffusion field of solvent and heat. At the same time, the curvature of the trough bottom corresponds with a pressure and a capillarity shift in the affinity, driving the diffusion field. The stationary situation implies that the trough shape and the corresponding affinity boundary condition agree with the stationary diffusion field; in that case $R\gamma \sin \varphi$ corresponds with the total dissipation by the diffusion field. This promises to be a rather difficult variational problem even for interface-equilibrium conditions.

Even more difficult is the stability problem. Noise and small perturbations on the c/p interface (e.g., terraces or even single ledges) induce fluctuations in the junction velocity. When left alone, the trough configuration would grow out with the velocity corresponding with the $\Phi(t)$ function, for the argument t^* corresponding with $\Phi(t^*)$ = trough size. When the fluctuations exceed this "intrinsic" velocity, the junction can be kept momentarily immobile and got trapped.

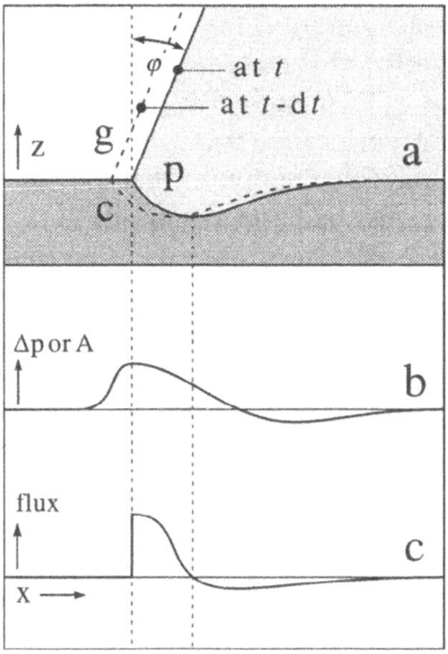

FIG. 8.8: *Stationary shape* of a moving p/g/c 3-junction or meniscus (unstable!)

8.4 INTERACTION BETWEEN NEIGHBOURING JUNCTIONS IN AN INHOMOGENEOUS AFFINITY FIELD.

In a crystal/parent phase interface which is stabilized by an affinity gradient, grain boundaries form grooves. In 3.3.3 we discussed equilibrium groove shapes for "isolated" junctions such as menisci and grain-boundary junctions widely separated from other junctions. Boundary conditions can then be characterized as $s = 0$ and $s \to \infty$ conditions. We are now going to consider, in a rather qualitative way, the case of junctions at a mutual distance which is comparable with their own "size" as following from (3.19), so that their curvature fields overlap. *This overlap results in an effective mutual force.* This force, in turn, can influence the evolution of the pattern of grain-boundary (and inter-phase boundary) grooves in a nonequilibrium situation. In unidirectional solidification, it can either lead to a stabilized and more or less *regular cellular-grain pattern* (when the effective mutual force is repulsive), or to the gradual expulsion of grains, i.e. *increase of grain size* (when the force is attractive). In view of the relevance of this mechanism for Bridgman and Czochralsky growth techniques we discuss this interaction here.

1) Attractive interaction

The interaction is attractive in the simple case of an *imposed affinity field* in which the iso-affinity surfaces are *planar* and without interaction with the interface (case "C1" in table 5.1, pg 102). Consider, for simplicity, an interface with OID surface energy and two junctions with a grain boundary having the same free energy, so that the equilibrium angles are all 120°. In fig. 8.9, both a one- and a two-junction solution of an equation of the type (3.19) have been sketched, other junctions being far away.

The single-junction solution (drawn line) is characterized by "horizontal" symmetry, i.e. by the $\varphi\,(s=0)=60°$ boundary condition. The other boundary condition is $(s \to -\infty\ ,\ \varphi \to 0)$. The "$s \to -\infty$" BC is satisfied by the curve ABB' (which can be shifted arbitrarily in the x direction). The (single)-junction point B is found as the point where the curve ABB' has a 60° orientation, i.e.the "s = 0" BC. (the other interface is the mirrored curve BD, and the grain boundary, BF, is in the z direction).

For the two-junction solution (dotted line) the $s \to -\infty$ BC is still $\varphi=0$ and therefore the interface coincides with the AB curve; only the solution is to be continued to a larger tangent orientation, say $60° + \varepsilon$. Consequently, both the grain boundary B'F' and the tangent of the connecting curve B'C' are tilted by an angle ε . The curve B'C' is fully specified by eq. (3.19) and one BC; i.e. the angle at B' and the "depth" z(B'). The mirror symmetry plane OO' between the two junctions is reached as the B'C' curve has reached orientation $\varphi = 0$.

Altogether, the tilting angle ε , the depth $z(\mathrm{B'}) = z(\mathrm{C'})$ and the horizontal distance between B' and C' are uniquely related, and the curves AB', B'C' and C'E are all fully specified by only one of these data (and of course $R_{\mathrm{mech.}}$ and $\vec{\nabla} A$). Inversely, the tilting angle ε is a function of the distance B'C' between the junctions.

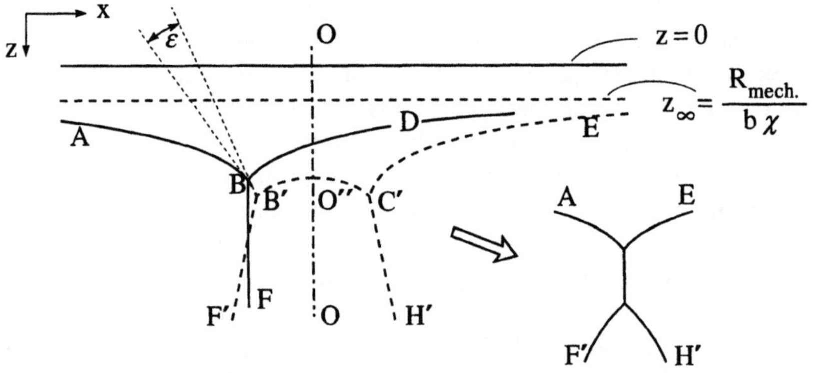

FIG. 8.9: Neighbouring grain-boundary grooves in a $\vec{\nabla} A$-stabilized interface and the mechanism of attractive interaction.

Because the grain boundaries B'F' and C'H are tilted in opposite directions $+\varepsilon$ and $-\varepsilon$, they feel a mutually *attractive* force $\gamma \sin \varepsilon$. As we have not yet specified boundary conditions for the grain boundaries other than at the junction side, we cannot say without further data whether the situation is stable or unstable. Generally speaking, when they are "fixed" at points at a distance $\gg$ B'C' from B' and C' or only have vertical asymptotes, it is unstable: the junctions B' and C' are pulled together, whereas the force increases to a maximum γ so that they finally coalesc and the grain boundaries form a new 3-junction (insert of fig. 8.9). This "grain expulsion phenomenon" can easily be observed microscopically in a Jackson-Hunt type set-up where a thin transparent cell with solidifying material can be pulled through a temperature gradient. By this mechanism, the number of grain-boundary 3-junctions in the solid-liquid interface gradually decreases.

2) Repulsive interaction

Now the case of an imposed field with planar iso-affinity surfaces is a limit case: in the general "Stefan" case ("C2" in table 5.2), only the boundary conditions of the diffusion field at a large distance from the interface are imposed. Repulsive interaction between the junctions agrees with the situation that the junctions are *effective sinks* for an impurity, so that they correspond with maximum-A_{l} regions in the diffusion field. Consequently, the interface regions in-between the junctions can be bent to *concave* shape - which corresponds with repulsion between the junctions. (Note: The *mechanism* by which junctions can become sinks is left aside here. We just may mention the fact that the grain-boundary grooves are regions of effective generation of "fresh interface" (which is subsequently buried in the form the lengthening grain boundary). When an impurity is preferentially adsorbed onto the c/p interface, this interface-generation region is an effective sink in the 3-D parent-phase diffusion field).

The transition between a regular (stabilized) junction distribution and "condensation" of junctions

The dynamical behaviour of a certain distribution of (3- or D-) junctions in a US situation - in terms of the alternative posibilities of a stabilized equi-spaced distribution of repelling junctions versus the collapse or condensation of attracting junctions - is a complex problem. In principle, the interaction between the diffusion field(s) and the interface has to be taken into account, just as in the Mullins-Sekerka instability analysis. The result would have to be translated into a two-junction interaction potential, and from this potential the collective behaviour of a multi-junction "continuum" would have to be predicted by the quasi-thermodynamic procedure to be sketched in ch. 7. Under certain conditions, this repulsive component might dominate for interjunction distances exceeding a critical length, and thus *stabilize a homogeneous distribution of grain-boundaries or of D-junctions*, thus leading to periodic lamellar or cellular groove-or dislocation patterns. (if such a homogeneous distribution can be stable after all, its maximum density (eventually exceeding the initial density of junctions) can be expected to depend on $\vec{\nabla}_n A$. We emphasize that this (hypothetical!) stable situation is characterized by a geometry in which the junctions are stabilized *on top of protrusions of the interface*, in contrast with the unstable situation where they combine and disappear in grooves and craters.

Interaction with the cellular-instability transition

The stable-unstable transition of a regular junction distribution, although intimately associated with $\vec{\nabla}_n A$, is to be considered apart from the cellular ("Mullins-Sekerka") instability (the latter is usually described without considering complications by junctions). Both the stable and the unstable junction distribution can be compatible with stability of a planar interface in the MS sense. (planar: apart from junction grooves or craters).

However, the presence of 3-junctions will certainly interact with the interface stability as such. As a consequence, we can expect that the instability criterion of a general interface will depend on the presence of grain boundaries and/or D-junctions. Another way of expressing this interaction is to say that a) the junctions and b) the solitons occurring as the cell-boundary grooves of the unstable interfaces (see 7.3.2) - which are both stable, but different, "nonlinear" features of the interface, form (ab) complexes (the solitons being trapped by the 3-junctions or vice versa).

This type of interaction wil lead to a "fine structure" of the cellular-instability transition as sketched in fig. 6.25 (where interaction with junctions was neglected). In the original fig. 6.25, the $R/\vec{\nabla} T \cdot \vec{n}$ axis was split up into two regions for the $\mathbf{S}\gamma = 0$ case, by the critical "Rutter-Chalmers" condition for CS. The presence of junctions might split up this critical value into two: a lower critical transition between stable and unstable *junction distribution* (both still to be considered as "MS-stable") and a higher critical transition corresponding with the actual *cellular instability*. This picture can be complicated still further because we can have to do with *two* junction "species": 3-j and D-j. Note that the solidified structure changes from stabilized parallel 3- and D-junctions, to increasing-grain size, to parallel 3-junctions again, in the latter case corresponding with cell grooves.

Interactions between grain-boundary grooves, dislocations and cellular structures during the onset of instability have been investigated, for instance, by BILONI, BOLLING and COLE (1966), BILONI (1967), MORRIS and WINEGARD (1969) and RUKWIED, RUFF and WILLARD (1971). An instance of a low-angle grain-boundary groove interaction on a *faceted*-cellular structure

(Si) was given by PFEIFFER (1985). Although a relation between cellular depressions and dislocations and/or grooves was reported by some of these authors, a detailed analysis of the situation has not yet been attempted.

Three-junctions in eutectic and eutectoid decomposition.

A very special case of interacting 3-junctions is presented by *eutectics and eutectoids*. Geometrically this case can be represented by the junction-pair geometry of fig. 8.9, or rather by a periodic sequence of such pairs, but now the boundaries B'F' and C'H' are *mirrored phase boundaries*. (In the OD γ case, their surface energies need not to be equal even if they are parallel). The difference with the single-phase, grain-boundary case is that the affinities of the interfaces B'C' and of AB',C'E are different (more or less mirrored, in accordance with the liquidi in the phase diagram) functions of the composition. This radically changes the coupling between these interfaces: now it is dominated by the chemical-composition field; the predominant function of the latter is the "horizontal" exchange of the two phase species. Because of the technical interest of unidirectionally decomposed eutectics and eutectoids, this case has extensively been investigated (e.g. JACKSON and HUNT (1966)., CARPAY and VAN DEN BOOMGAARD (1971), NASH (1976).

However, the consequences of finite χ have hardly been considered. In general, finite χ diminishes the interaction range of junctions. Therefore, it seems quite likely that this can affect both the $\lambda^2 R$ = const. relation (between growth rate and structure period) and the morphology, especially of eutectoids and high-α factor eutectics, because in those cases χ is likely to be smaller.

8.5 LAYOUT FOR A GEOMETRICAL CHEMISTRY OF CRYSTAL GROWTH

Although we have explicitly restricted the subject of this monograph to the *geometrical* aspects of crystal growth, i.e. to the unary case, it is instructive to consider how this might in principle be embedded in a *multi-component* description. This extends the subject to a "chemical geometry" of crystal growth: anyhow, it describes how the chemical composition of a crystal, especially when growing with faceted interfaces, can become orientation-dependent. This, of course, would be incompatible with growth with an interface at equilibrium.

The orientation dependence of chemical composition is of considerable technological relevance. In bulk crystals, growth sectors with different composition can often been discerned, giving rise to mechanical stresses. Also in "3D" structured epitaxial layers, especially in the case of ternary or quaternary III-V semiconductors, the composition and the doping level can depend sensitively on orientation. In principle, it is possible to include this compositional aspect in the formal continuum description of crystal growth as discussed in this monograph, by an extension as suggested above.

A consideration of this wider subject is instructive for several reasons. In the first place, the concepts of growth affinity and growth pressure can be given a more specific physical definition. In the second place, we can show - although superficially- what is needed to expand our subject to include compositional aspects. On our way, we can illustrate the different rôle of the interface in the temperature- and in the chemical- diffusion fields. Finally, the exercise can show more clearly which practical situations can really be considered as (quasi) "unary cases", even although we have (for $T > 0$) always to do with a combination of fluxes of heat *and* matter.

8.5.1 Species vectors and their interrelations

In the polynary case we have to do with n chemical species and with heat flow. The description of a transport situation therefore implicates n or n+1 fluxes $\vec{j}_1, \vec{j}_2, \; \ldots \;, \vec{j}_{\text{n or n+1}}$, and similarly n or n+1 potentials μ_k, fractional compositions C_k, affinities A_k etc. For convenience we adopt the habit to use suffix $(n+1)$ for the heat-related component, if relevant.

We can symbolize such lists of species-related quantities as "species vectors", with the notation $\tilde{J}, \tilde{\mu}, \tilde{C}, \tilde{A},$ etc. Notice that quantities such as $\tilde{J}$ and $\vec{\nabla}\tilde{A}$ have a conventional ("direction") vector character too, which would actually require a combination of the $\sim$ and the $\rightarrow$ symbol in our notation. For typographic reasons we had to omit the arrow, but the reader should be aware of this.

There are thermodynamic relations between chemical potentials and compositions, and irreversible-thermodynamic relations between fluxes and "thermodynamic forces" $X = A/T$ or affinities A. Such relations can be considered as relations between species vectors. As an example, the relations between fluxes and the "forces" of irreversible thermodynamics can be written as

$$\tilde{J} = \tilde{\tilde{\mathcal{L}}}(\tilde{X}) \tag{8.17}$$

where $\tilde{\tilde{\mathcal{L}}}$ represents, quite generally, a set of $n+1$ functions, each of $n+1$ variables (a "vector map"). Such a relation is usually referred to as a "constitutive relation". When the relation is linear this takes the form of a tensor multiplication

$$\tilde{J} = \tilde{\tilde{L}} \cdot \tilde{X} \;\text{ or }\; \tilde{J}_0(\tilde{X}_0) + \tilde{\tilde{L}} \cdot \Delta\tilde{X} \tag{8.18}$$

where $\widetilde{\widetilde{L}}$ is a square matrix or "species tensor" - which illustrates why we have also labeled the *function* $\widetilde{\widetilde{\mathcal{L}}}$ with a double-wave symbol. In our case, the relation expressed by $\widetilde{\widetilde{\mathcal{L}}}$ or $\widetilde{\widetilde{L}}$ will in general be associated with some physical activity of the interface and will in principle be dependent on $\vec{n}$.

8.5.2 *Multiport capacitance: $\widetilde{\mu} - \widetilde{C}$ and $\widetilde{A} - \widetilde{C}$ relations*

The concepts of growth affinity and growth pressure as defined in 1.5.1 can naturally be extended to the polynary case. For a mixed crystal with n chemical constiutents (molar fractions $C_1, C_2, \ldots, C_n$; $\Sigma_k C_k = 1$; together an n-dimensional vector $\widetilde{C}_c$) growing from a parent phase with corresponding vector $\widetilde{C}_p$, we can define n-dimensional species-fraction chemical-potential vectors $\widetilde{\mu}_c$ and $\widetilde{\mu}_p$. Taking both at the interface at identical positions "cis" and "trans", we can define the interface affinities $\widetilde{A}_|$ (for planar interface) and $\widetilde{A}_g$ (for curved interface) as

$$\widetilde{A}_| = \widetilde{\mu}_p(T_i, p_p, \widetilde{C}_p) - \widetilde{\mu}_c(T_i, p_p, \widetilde{C}_c) \tag{8.19}$$

$$\widetilde{A}_g = \widetilde{\mu}_p(T_i, p_p, \widetilde{C}_p) - \widetilde{\mu}_c(T_i, p_c, \widetilde{C}_c) \tag{8.20}$$

(again, we neglect the term due to momentum transfer at the interface (see footnote on page 16) Notice that we neglect to mention the driving force $1/T$ for the entropy as the $(n+1)$th species. Here we anticipate on eq. (8.31), stating that $T_{\text{interface,c}} = T_{\text{interface,p}}$ anyhow, as a consequence of the practical absence of a thermal interface resistance. This simplification implicates that only interaction between the *chemical* species have to be considered in the interface kinetics and that interface "Soret effects" can be neglected.
The relations

$$\widetilde{\mu}_p = \widetilde{\widetilde{\mathcal{M}}}_p(T_i, p_p, \widetilde{C}_p) \quad \text{and} \quad \widetilde{\mu}_c = \widetilde{\widetilde{\mathcal{M}}}_c(T_i, p_p, \widetilde{C}_c) \tag{8.21}$$

i.e. the "constitutive relations" between the n-dim. $\widetilde{\mu}$ and $\widetilde{C}$ vectors (still dependent on p and T), are relations between potentials and "charges"; i.e. they are *multiport (n-port) capacitances* in the terminology of Network Thermodynamics as defined by OSTER, PERELSON and KATCHALSKY (1973). When these vector maps can be linearized for a certain range $\Delta\,\widetilde{C}$ in the $\widetilde{C}$ space, $\widetilde{C}_p = \widetilde{C}_{p,0} + \Delta\,\widetilde{C}_p$:

$$\widetilde{\mu}_p = \widetilde{\mu}_{p,o}(T_i, p_p, \widetilde{C}_{p,o}) + \widetilde{\widetilde{M}}_p \cdot \Delta\,\widetilde{C}_p \tag{8.22}$$

(and similarly for $\widetilde{\mu}_c$), where $\widetilde{\widetilde{M}}_p(T_i, p_p)$ and $\widetilde{\widetilde{M}}_c(T_i, p_c)$ are n-dim species tensors, then the $\Delta\,\widetilde{C}$ -variable part of the $\widetilde{\mu}$ vectors can be represented by an *linear* n-port capacitance, which may eventually be simulated by an electrical analog built of capacitors.

The interface affinity $\widetilde{A}_|$ is thus a function of T_i, p_p, $\widetilde{C}_p$ and $\widetilde{C}_c$. Now in the special case that diffusion fluxes in the bulk of the growing crystal can be neglected against the crystallisation rate, so that eq. (8.34) applies (see below), $\widetilde{C}_c$ is just a function of $\widetilde{C}_p$ and $\widetilde{A}_|$, so that $\widetilde{A}_|$ can be considered as a function of T_i, p_p and $\widetilde{C}_p$ *only*:

$$\widetilde{A}_| = \widetilde{\widetilde{\mathcal{M}}}^* (T_i, p_p, \widetilde{C}_p) \tag{8.23}$$

i.e. another n-port capacitance (as compared with two of such capacitances for the general case).

In 1.5.1, eq. (1.2a), we have distinguished between the actual growth affinity A_g and the "abstract" planar interface affinity $A_|$. The reason for the split-up was mainly procedural : as $A_|$ can be correlated with the T,C field, the deviation ΔA_c can be correlated purely with the curvature, which therefore can be calculated from it.

When we want to extend equation (1.2a) to the polynary case, we have to consider ΔA_c and ΔP_c as species vectors too:

$$\Delta \widetilde{A}_c = \widetilde{A}_g - \widetilde{A}_| \tag{8.24}$$

This agrees quite naturally with the definition of a partial molar volume, i.e. the volume which correlates with the fraction $(\widetilde{C})_k$ of species k, which is obviously just $= (\widetilde{C})_k \,.\, V(\widetilde{C})$:

$$\Delta \widetilde{A}_c = \widetilde{C}_c \,.\, V(\widetilde{C}_c) \,.\, \vec{\vec{K}} : \vec{\vec{\mathbf{S}}}\gamma \tag{8.25}$$

$$\Delta \widetilde{P}_c = \widetilde{C}_c \,.\, \vec{\vec{K}} : \vec{\vec{\mathbf{S}}}\gamma \tag{8.26}$$

In a way, equations (8.25 / 8.26) may be considered equivalent with the concept of partial surface energies γ_k, so that $(\Delta \widetilde{P}_c)_k = \vec{\vec{K}} : \vec{\vec{\mathbf{S}}}\gamma_k$. To complicate matters, γ is in general a function of $\widetilde{C}_c$ too.

8.5.3 Geometrical growth affinity versus partial affinities

Obviously we can define total affinities, growth pressures and curvature-induced shifts:

$$\left.\begin{array}{ll} A_| = \Sigma_k(\widetilde{A}_|)_k & A_g = \Sigma_k(\widetilde{A}_g)_k \\ P_| = \Sigma_k(\widetilde{P}_|)_k & P_g = \Sigma_k(\widetilde{P}_g)_k \end{array}\right\} \tag{8.27}$$

$$\left.\begin{array}{l} \Delta A_c = \Sigma_k(\Delta \widetilde{A}_c)_k \\ \Delta P_c = \Sigma_k(\Delta \widetilde{P}_c)_k \end{array}\right\} \tag{8.28}$$

These sums over species determine the *geometrical* behaviour of the interface, just as in the unary case. The interesting complication of the polynary case is the possibility of having *mechanical equilibrium* ($P_g = A_g = 0$) without *chemical equilibrium* (individual $(P_g)_j \neq 0$, but compensating each other). This, again, is directly comparable with a porous membrane separating fluids of different composition. A (mixed) crystal can have zero growth rate geometrically while still exchanging matter with a solution (a "diffusion couple" with barrier). More in general, the geometrical growth rate (and direction) of a crystal is related with the *total* growth affinity rather than with the partial ones. The total growth affinity is the effective driving force; in the phase transition the different species are *coupled by the interface* (i.e., by the phenomenological relation (8.33)). It is quite possible that some of the species have zero or even an opposite partial affinity, but are nevertheless "trapped" into the growing crystal as a consequence of (8.32) (for the case of linear flux/force relations as expressed by (8.31), trapping of a species for zero partial affinity is induced by the off-diagonal tensor elements of $\tilde{\tilde{\ell}}$).

8.5.4 *Multiport transformers: stoichiometric coupling*

As more or less special multi-component cases we may mention

a) crystallisation of an ionic crystal from an electrolyte solution in which the ions are "independent" species (LUNDAGER MADSEN (1991));
b) chemical vapour deposition in which atoms are deposited onto an interface by a *heterogeneous* chemical reaction on the interface.

In both cases, fluxes of different species (ions in case a), reactants and reaction products in case b) are coupled or "geared" by their stoichiometric coefficients. In chemical reaction kinetics, the affinity of a (coupled) chemical reaction is defined as the sum of the partial affinities of the species involved (see, e.g., DE GROOT and MAZUR (1969)). As a consequence of the stoichiometric gearing, the effective dimension of the species vectors $\tilde{J}$ and $\tilde{X}$ involved in eq. (8.29) is brought back to the number of *independent* fluxes in the reaction.

The affinity of the electrolyte-solution growth case is "simply" associated with incorporation kinetics of the ionic version of a Kossel crystal. That of the CVD case has an additional contribution due to the chemical (surface) reaction (which can also be orientation-dependent). Correspondingly, the "resistance" of the CVD case is composed of the resistance of the chemical reaction as such plus that of subsequent incorporation kinetics (surface diffusion, ledge diffusion etc.). To complicate matters, the "resistance" of the chemical reaction may also depend on the number of surface sites available for the heterogeneous reaction to take place, i.e. it may depend on adsorbate density of *other* species.

In network thermodynamics, the stoichiometric "gearing" between a number k out of n different species is represented by *n-port transformers* (to be considered as ideal transformers or "gearboxes", which are also effective in coupling "DC currents" or fluxes).

8.5.5 *Multiport conductance: multicomponent growth kinetics*

Up to here, only the *driving force* for interface motion was considered, not this motion itself. It is relevant to consider this (dissipative) motion of the interface and the entropy produced by it from the viewpoint of irreversible thermodynamics.

In the discussion of total and "partial" growth affinities for the polynary case, we have compared the crystal/parent phase interface with a porous membrane: this "analog" of a crystal interface is appropriate to illustrate the coupling between the geometrical growth of a crystal and the individual species fluxes. However, there are important *phenomenological* differences between the crystal interface behaviour and a technical membrane in their rheological properties. For an ordinary membrane (to which we can also attribute a thermal resistance) we may expect linear relations between fluxes $(\tilde{J})_i$ and "thermodynamic forces" $(\tilde{X})_i = (\mu_{cis,i}/T_{cis} - \mu_{trans,i}/T_{trans})$:

$$\tilde{J} = \tilde{\tilde{L}} \cdot \tilde{X} \tag{8.29}$$

where $\tilde{J}$ and $\tilde{X}$ are the species vectors and $\tilde{\tilde{L}}$ is the matrix of linear phenomenological coefficients. The $\tilde{J}$ vector includes a component J_{th} for the heat flux; the corresponding X_{th} component is $(1/T_{cis} - 1/T_{trans})$. Fluxes of different species, including matter- and heat fluxes, can be coupled (implicating that off-diagonal elements of the $\tilde{\tilde{L}}$ "species tensor" are non-zero). The off-diagonal $L_{i\neq j}$ elements are coupled by the Onsager relations, $L_{ij} = L_{ji}$. The entropy production by the irreversible transport through the membrane :

$$dS_{irr}/dt = \sum J_i X_i \tag{8.30}$$

is quadratic or bilinear in the driving forces.

Now a crystal interface is *not* an "ordinary membrane". The two important differences are

1) The thermal resistance (Kapitza resistance) is zero for almost all practical cases, *including* crystals with faceted interfaces;
2) The $\tilde{J}(\tilde{X})$ relation for the *chemical* species is *nonlinear* (and orientation-dependent) for a crystal with faceted interfaces, even under most technical growth conditions. Consequently, the interface kinetics of (faceted) crystal growth is similar with heterogeneous chemical reaction kinetics rather than with diffusion kinetics.

1) is a *simplification* as compared with the general case: it implies that $T_{cis} = T_{trans}$ so that a relation between the $(n+1)$ -dimensional flux and force vectors such as (8.29) reduces to a relation between (n) - dimensional *chemical* flux and force vectors, which can then also be expressed as a relation between flux- and *affinity* vectors $(A_k = X_k . T)$ If the relation is linear, it can be written as

$$\tilde{J} = \tilde{\tilde{l}} \cdot \tilde{A}_g \quad \text{with } (\tilde{\tilde{l}})_{ik} = (\tilde{\tilde{L}})_{ik} / T \; ; \quad \text{i,k} = 1 \, .. \, n. \tag{8.31}$$

The heat flux across the interface is not dissipative.(in the bulk phases, it is dissipative, of course)

2) is a severe *complication*: although the phenomenon of coupling between different species does exist, it cannot simply be expressed in a linearized form as (8.29) or (8.31) and the j → i and i → j effects cannot be related as simply as by the Onsager relations. At most we can state that there is some functional relationship, which is nonlinear in the species components of $\tilde{A}$:

$$\tilde{J} = \tilde{\tilde{\mathcal{L}}}(\vec{n}, \tilde{A}_g), \quad \text{i.e. a generalisation of } R(\vec{n}, A). \tag{8.32}$$

The entropy production can still be written as $\mathrm{d}S_{irr}/\mathrm{d}t = \sum J_k \,.\, A_k/T$, but it is no longer bilinear in the affinities. (see remark on the consequence for the minimum- REP principle in 1.5.2).

When the roughening transition is passed (for all orientations), the linearized relation (8.31) will be approximated again, the validity range in A_g growing as $(T - T_R)$ increases.

Relations such as (8.31) and (8.32) represent, in terms of network thermodynamics, an *n-port resistance*. When the relation can be linearized such as in (8.31), with $n \times n$ matrix $\tilde{\tilde{l}}$ (obeying the Onsager relations), then the corresponding linear n-port resistance can be simulated by an electrical circuit composed of "ordinary" resistances.

8.5.6 Flux-conservation BC's and geometrical growth rate

Finally, in order to make the link between the interface-crossing flux $\tilde{J}$ of eq. (8.31) or (8.32) and the *geometrical* growth rate, we have to write down the (species-specific) mass conservation BC's at the interface. Subsequently, we can find the compositions $\tilde{C}_c$ and $\tilde{C}_p$ at the growing interface, but the calculation implies the relevant diffusion fields also. When these compositions are plotted in a phase diagram, we will find that they are shifted from the (equilibrium) solidus and liquidus lines as a consequence of the "resistive" properties of the interface. This shift is orientation-dependent, because $\tilde{J}$ is orientation-dependent.

We discuss two situations:

a) Diffusion in the bulk of the crystal can be neglected (not necessarily a realistic assumption for the multicomponent case)

In this case, the relation between $\tilde{\boldsymbol{J}}$, $\tilde{\boldsymbol{C}}_c$ and $\tilde{\boldsymbol{C}}_p$ is straightforward. Both the composition $\tilde{\boldsymbol{C}}_c$ and the growth rate R of the growing crystal can directly be related with $\tilde{\boldsymbol{J}}$:

$$R(\vec{\boldsymbol{n}}) = V(\tilde{\boldsymbol{C}}_c) \sum_k \{\tilde{\boldsymbol{J}}(\vec{\boldsymbol{n}})\}_k \tag{8.33}$$

$$\tilde{\boldsymbol{C}}_c(\vec{\boldsymbol{n}}) = \frac{\tilde{\boldsymbol{J}}(\vec{\boldsymbol{n}})}{\sum_k \{\tilde{\boldsymbol{J}}(\vec{\boldsymbol{n}})\}_k} = \tilde{\boldsymbol{J}}(\vec{\boldsymbol{n}}) \frac{V(\tilde{\boldsymbol{C}}_c)}{R(\vec{\boldsymbol{n}})} \tag{8.34}$$

At the parent-phase side of the interface (where diffusion is supposed to be possible; otherwise $\tilde{\boldsymbol{C}}_c = \tilde{\boldsymbol{C}}_p$ would be imposed) the mass-conservation relation can be written as

$$\tilde{\boldsymbol{C}}_p(\vec{\boldsymbol{n}}) = \frac{\tilde{\boldsymbol{J}}(\vec{\boldsymbol{n}}) + \tilde{\boldsymbol{J}}^*_{\mathrm{diff,p}}}{\sum_k \{\tilde{\boldsymbol{J}}(\vec{\boldsymbol{n}}) + \tilde{\boldsymbol{J}}^*_{\mathrm{diff,p}}\}_k} \tag{8.35}$$

where $\tilde{\boldsymbol{J}}(\vec{\boldsymbol{n}})$ is the interface-crossing flux and $\tilde{\boldsymbol{J}}^*_{\mathrm{diff,p}}$ the extra flux reaching (or leaving) the interface by bulk diffusion through the parent phase. On the other hand, $\tilde{\boldsymbol{J}}^*_{\mathrm{diff,p}}$ is obviously identical with $R(\vec{\boldsymbol{n}})(\tilde{\boldsymbol{C}}_p - \tilde{\boldsymbol{C}}_c)/V(\tilde{\boldsymbol{C}}_c)$.

In a US situation, R is imposed and $\tilde{\boldsymbol{C}}_c$ and $\tilde{\boldsymbol{C}}_p$ can be calculated. In a homogeneous-T situation, $(\tilde{\boldsymbol{C}}_p)_\infty$ (at distance $\gg D/R$ from the interface) is given and the stationary $R(\vec{\boldsymbol{n}})$, $\tilde{\boldsymbol{C}}_c(\vec{\boldsymbol{n}})$ and $\tilde{\boldsymbol{C}}_p(\vec{\boldsymbol{n}})$ can be calculated.

b) Diffusion in the bulk of the crystal can not be neglected

This situation is rather more subtle because the straightforward relation (8.34) at the crystal side of the interface now has to be replaced by a relation similar to (8.35) (suffix p instead of c, of course). In fact, equation (8.34) is suspect already from the fact that R occurs in the denominator in the second equality: even if the geometrical growth rate is zero we can still have a net diffusion transport across the interface in a multicomponent case.

This conclusion makes us recognize that the total flux $\tilde{\boldsymbol{J}}$ across the interface, *in the case that bulk diffusion in the crystal is allowed*, is actually composed of two contributions due to different physical mechanisms: a) a flux $\tilde{\boldsymbol{J}}_{\mathrm{gr.kin.}}$ related with "ordinary" growth kinetics, as due to ledge motion and birth-and-spread growth, and b) an interface-crossing diffusion flux $\tilde{\boldsymbol{J}}^*_{\mathrm{diff.\,i}}$ due to direct exchange of atoms between both phases, *without* ledge motion. The latter mechanism can require a finite affinity for a faceted interface, but the $\tilde{\boldsymbol{J}}^*_{\mathrm{diff.\,i}}(\tilde{\boldsymbol{A}})$ relation is certainly different from the $\tilde{\boldsymbol{J}}_{\mathrm{gr.kin.}}$ $(\tilde{\boldsymbol{A}})$ relation. $\tilde{\boldsymbol{J}}^*_{\mathrm{diff.\,i}}$ and $\tilde{\boldsymbol{J}}_{\mathrm{gr.kin.}}$ are parallel and independent fluxes driven by the same affinity $\tilde{\boldsymbol{A}}$.

Altogether, in case b) the relation between $R(\vec{\boldsymbol{n}})$, $\tilde{\boldsymbol{C}}_c$ and $\tilde{\boldsymbol{C}}_p$ as expressed by the species-specific continuity conditions at the interface are essentially more complicated than in case a). Formally we have to split up this BC into a crystal side and a parent-phase side. Cases a) and b) are compared in fig. 8.10, where the "concentration profiles" symbolize the behaviour of (multicomponent) concentrations. The corresponding continuity conditions are sketched below the C profiles ($\tilde{\boldsymbol{C}}$ and $\tilde{\boldsymbol{J}}$ represent values *at the interface position*). In case a) we have the situation as described by equations (8.33), (8.34) and (8.35) (from which $\tilde{\boldsymbol{C}}_c$ and R can be solved). In case b), $\tilde{\boldsymbol{C}}_c$, $\tilde{\boldsymbol{C}}_p$, and R are coupled by *two* continuity conditions which we have visually split up in fig. 8.10b and which each couple four rather than three fluxes. (*note:* we can also consider fig. 8.10 as a representation for the 2-dim. conservation condition in the *ledge-advance* EOM. In that case, even more than four fluxes are involved: also the adsorption/ desorption exchange with the bulk parent phase and eventual exchange with the crystal bulk phase have to be taken into account). In principle, also in case b) $\tilde{\boldsymbol{C}}_c$, $\tilde{\boldsymbol{C}}_p$, and R can all be calculated, but a simple expression like (8.34) can no longer be given because R can no longer be simply related with $\tilde{\boldsymbol{J}}_{gr.kin.}$: the bulk diffusion fields in the c and p phases now determine the rate too.

All this is implicit in the multiport network representation of case b), which has to contain a distinct multiport conductance element for $\tilde{\boldsymbol{J}}^*_{diff.\ i}$ parallel to $\tilde{\boldsymbol{J}}_{gr.kin.}$. We will not work this out further. Still we thought it relevant to point out how the mass-conservation boundary conditions, which are quite straightforward in the unary case, lead to intricate couplings in the multi-component case.

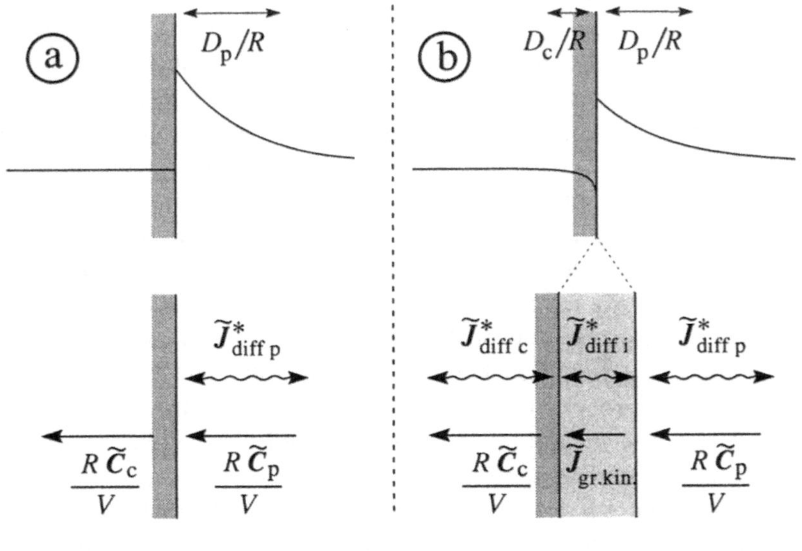

FIG. 8.10

8.5.7 Surface diffusion and Tangential Adsorbate Drag

In the description of unary growth which forms the main body of this monograph, we have neglected *surface diffusion* as a relevant mechanism in the transport process. In other words, “interface growth kinetics” is considered as a transport process from the parent phase to the crystal *perpendicular* to the interface, not parallel to it. Inclusion of a parallel component would complicate our equations of motion considerably, but in the unary case this neglect seems to be plausible.

However, this might change drastically for the polynary case, in particular with faceted interfaces. This is because in the 2-dimensional surface diffusion field of more than one adsorbed species, a quite specific interaction mechanism can be effective, different from the mechanism involved in the “ordinary” collision statistics of multicomponent diffusion. We may call it “tangential adsorbate drag” (TAD). As it seems that this effect has not yet been discussed in the literature, we may grant it a short explanation.

Consider, as an extreme case, a crystal consisting mainly of A atoms with a low level of “impurity” B atoms. B has a high segregation coefficient; i.e. it is not readily incorporated into the A lattice, but it can be adsorbed onto the crystal surface, even at a higher relative concentration than in the parent phase.

Suppose that the adsorbed B atoms can readily diffuse over the (faceted) interface of the crystal. (Maybe the A atoms can do the same; in that case the surface (diffusion) fluxes of A and B interact rather like dissolved atoms in a 3-dim. liquid, and their local surface fluxes and potential gradients are related by the Onsager relation). However, apart from the eventual individual A atoms, the adsorbed B atoms meet another “species” on the facet: i.e. the *advancing ledges* of the crystal. When a ledge presents an effective potential barrier for the B atoms, the latter are pushed ahead of the advancing ledges (rather like a rejected impurity is pushed by an advancing solidification front) and an exponential surface diffusion field of characteristic width $D_{\text{surf.}} / v_{\text{ledge}}$ is generated. In the flux balance equation at the ledge front, we have a balance of four terms: incorporation into the ledge, climb *onto* the ledge (= effective “leak” to the next plateau), adsorption/desorption exchange with the bulk of the parent phase (another leak), and pushed surface diffusion. The overall effect is a net *tangential* transport component of B atoms in the direction of the ledge motion, i.e. in the direction of $\vec{\mathfrak{R}} - \vec{n}\,(\vec{\mathfrak{R}} \cdot \vec{n})$: a tangential drag effect working on the adsorbed species.

It seems that for a favourable combination of diffusion and ledge-interaction parameters, the TAD mechanism can be highly effective in generating surface fluxes. Such fluxes might have interesting consequences, such as the effective "cleaning" of a certain region around the top of a dislocation pyramid (D-junction velocity source) or other protrusions of the interface (characterized by net *divergence* of ledges). Conversely, impurities would be concentrated in depressions in a faceted interface, i.e. in regions of net ledge *convergence*. Altogether, this would provide a mechanism for cellular instability for $\mathbf{S}R \gg R$ interfaces, producing a *shallow* cellular morphology. (terrace formation, microstep instability). Further theoretical and experimental work is required to verify the mechanism.

8.5.8 Concluding remarks

1) Network thermodynamics as a tool for crystal growth theory

In this subsection we have sketched how, in principle, the coupled transport fields involving bulk diffusion, crystallisation and eventual heterogeneous chemical reactions on the interface for the polynary case can be described in terms of "network thermodynamics". The basic formalism of network thermodynamics has been developed mainly for the purpose of describing the complex reaction/diffusion phenomena involved in biochemistry. However, it seems also very suitable for the description of transport phenomena related with crystal growth. The essential extra relations for this are a) the constitutive relation for the multiport resistance describing the interface incorporation kinetics, (8.31) or (8.32) and b) equation (8.33) , relating the *geometrical* evolution with individual species fluxes. The latter evolution governs the shape evolution of the interface, and thereby the evolution of *orientations* which can have such a dramatic effect on the $\stackrel{\approx}{\mathcal{L}}$ function. As far as I am aware, the available techniques of network thermodynamics and generalized circuit theory have never been applied to crystal growth, and an effort in this direction might be highly rewarding.

The "construction" of $\stackrel{\approx}{\mathcal{L}}$ functions for the *unary* case is discussed in ample detail in part II of this monograph; this might be extended, along similar lines, for multi-component interface transport. It is to be stressed that the relevant mechanism which results in orientation dependence of the *chemical composition* is essentially related with the incorporation kinetics in the advancing ledge; i.e. the 2-D part of the "non-linear networks" (fig. 12.1a) is the one which has to be extended to a multiport version, rather than the 3-D part (fig. 12.1b). This agrees with the fact that the OD composition effect largely disappears above the roughening transition. For the geometrical and the chemical aspects of ledge kinetics we can roughly repeat the discussions of 8.5.2 through 8.5.6, now translated into a 2-D version. This translation implies that the conservation relations, as split up into "cis" and "trans" in agreement with fig. 8.10b, have to include extra terms expressing the (multi-component) exchange flux with the *bulk* phases - inceasing the number of coupled fluxes to five on each side. The other essential ingredient are the $\tilde{\tilde{\mathcal{M}}}$ (multiport capacitance) functions required to describe the $\tilde{A}$ - $\tilde{C}$ relation. This is a subject of *equilibrium* thermodynamics).

2) Formal unary or single-variable cases

What situations are effectively excluded by our limitation to the "unary case" ? To answer this, we have to consider the question under what conditions the affinity parameter can uniquely be related with either a T or a C parameter associated with a unary diffusion field, so that the $A_|$ parameter can simply be related with T *or* C by a functional relation.
Consider eq. (8.19):

$$A_| = \mu_p (T, C_p, p) - \mu_c (T, C_c, p)$$

where C_p and C_c represent the fractional composition of a (binary) mixed crystal and its parent phase (participation of more than two species cannot result in a "unary" situation anyhow).
1) for a *pure melt /crystal* system, C_c and C_p are irrelevant and the *temperature* is the only relevant field parameter. The unique relation between $A_|$ and T in this case satisfies the above condition. (this is relevant, for instance, for the case of "thermal dendrites".
2) In the case of a *solution at uniform temperature*, T is not a variable (effects of heat of crystallisation should be negligible). Then C_p and C_c are the relevant field parameters. For a unary crystal growing from solution (C_c = 1), therefore, we have again a unique $A_| (C_p)$ relation. (this is the relevant case for "solute dendrites").

In all other (>1-dimensional) cases, both T *and* C fields affect the affinity. In such cases we either have to write R directly as a function of the T, C parameters, or we have to translate T and C into $A_|$. However, it is not guaranteed that the $R(T, C)$ dependence can indeed be reduced to an $R(A)$ dependence on such a basis - if not, we need two parameters, and we have effectively to do with a polynary description.

Chapter 9

ANALYTIC REPRESENTATIONS OF $R(\vec{n},A)$ AND $\gamma(\vec{n})$ FUNCTIONS

The important thing in science is not so much to obtain new facts as to discover new ways of thinking about them.

Sir William Lawrence Bragg (1890-1971)

9.1 "PROTOTYPE" VERSUS "EXACT" FUNCTIONS

In part I of this monograph, the "geometry of crystals" is discussed in terms of equations involving $\gamma(\vec{n})$ and $R(\vec{n},A)$ functions which are not further specified. Although this may suffice to elucidate principles, it is not enough when we actually want to "work" with the equations: then we need an explicit, preferably analytic expression for the surface-energy and growth-rate functions, either based on measurements or on predictions from a theoretical model.

Whereas the experimental approach will be worked out in ch. 13, chapters 10, 11 and 12 are focused on mathematical construction methods based on a generalized theoretical model. *The basic "invention" put forward in these chapters is that such functions can be represented as the conductance of a multiply connected nonlinear network, independently of the details of the "atomistic" model.* Such details are represented partly by the individual elements of the network, partly by the configuration of the network as a whole.

As to the question to what degree of precision such functions would have to copy the behaviour of a real crystal, we might discern two extremes. At the high end of the scale we may place the "exact" description of an actual crystal, i.e. γ and R functions enabling to calculate all exact positions and sizes of facets, the shape of growth spirals or hillocks on these facets as generated by velocity sources, eventual consequences of roughening transitions etc. As we will show in ch. 12, it is quite possible to construct such functions in a step-by-step process, starting from PBC vectors and the data concerning bond energies, kink velocities, roughening etc. Even the analytical expression of such a detailed function can be rendered on a single line, when we use the compact assembling notation to be introduced in ch.10. Still, the actual calculation (e.g., for the purpose of numerical simulation of growth phenomena) will become quite complex unless in highly symmetric cases.

At the other end of the scale we would have to put relatively simple functions, (e.g., representing highly symmetric crystals which can have identical facets) which contain a minimum number of parameters, to represent characteristic behaviour of crystals in equilibrium or during growth. (e.g. the effect of thermal or kinetic roughening). For such functions the exact position and number of facets would be unimportant, but rather the typical roughening behaviour of one facet, the symmetry or the "degree of orientation dependence" would be the relevant item to be expressed as simply as possible. This latter type of function might properly be called a "prototype function" as it would serve to characterize crystal growth phenomena in general rather than the growth of one particular crystal. The "precision" or detail given by a prototype function, and consequently the complexity of the analytic expression, would have to be as low as possible, just enough to characterize the phenomenon of interest.

Analytic prototype functions for γ and R can serve the following purposes:

a) As a classification tool in a "systematic catalogue of crystal growth situations". Two major parameters classes are to be discerned in such a catalogue: 1) those expressing the "external" growth conditions and 2) those representing the behaviour of the crystal/parent phase interface. For the latter class, prototype functions, and in particular properly chosen parameters in these functions, can be used as a basis for classification. More explicitly, when the prototype functions are inserted in the equation of motion, dimensionless numbers can be defined, and critical values of these numbers can be used to discern between different situations, as is customary in the field of transport phenomena in general.

b) As a basis for the development of the theory of certain crystal growth phenomena, such as dendritic growth, stability of planar interfaces during epitaxial growth, terrace- and multiple-step formation etc. Simplified prototype functions which contain just enough detail are preferable for such theories above "exact" but more complex functions.

c) Narrowly related with b): as a basis for computer simulation of crystal growth phenomena. This might be attempted either for the simulation of technical growth situations of real crystals (rather "exact" functions would then be required) or for the testing of theories as suggested in b) and the computer-experimental investigation of the influence of parameters. For the latter type of application, prototype functions are preferable; the parameters of the prototype functions can be matched to the characteristic behaviour of real crystals, and critical values of these numbers can be used to predict instability behaviour.

Note: In connection with application areas a, b an c we want to stress the point that many, indeed maybe most crystal growth phenomena can (at least qualitatively) be as well understood from a two-dimensional as from a three-dimensional model. This greatly simplifies the description (compare the 2-D functions in 12.2 with their 3-D analogs). The same applies to "physical" growth experiments, as can be learned from the famous cuvette-growth demonstrations by K.A. JACKSON and J.D. HUNT (1967).

9.2 ON THE DEFICIENCY OF "BROKEN-PBC" TYPE PLOTS

Rather to the author's surprise, no systematic attempts seem to have been made to construct physically relevant analytic growth-rate functions $R(\vec{n})$, in contrast to $\gamma(\vec{n})$ functions. In "computational" attempts to simulate dendritic growth as published so far, orientation dependence in surface energy or -kinetics was limited to small fractional correction terms of the type $\varepsilon \cos n\varphi$ ($\varepsilon \ll 1$). Eventual nonlinearity in the dependence on driving force was neglected too. (In contrast, the network approach to be introduced here provides a natural way to introduce such nonlinearity in a $R(\vec{n}, A)$ function).

For surface-energy "plots", most descriptions found in the literature are variations of the broken-bond model. Essentially, estimating the surface energy by just counting the broken bonds and multiplying with a constant energy leads to a "raspberry" or "soapfroth" type of plot which results from the addition of spheres:

$$\gamma(\vec{n}) = \sum_i \gamma_i \, | \vec{n} \cdot \vec{p}_i | \tag{9.1}$$

where the $\vec{p}_i$ are unit vectors in the direction of the crystal's PBC's and γ_i are the corresponding partial surface energies of the i-type PBC bonds, ending in a plane oriented at $\vec{n}$. An excellent and exhaustive review of "broken-bond" type $\gamma(\vec{n})$ functions and variations on this theme can be found in Kern's chapter 2 in the series Morphology of Crystals (KERN 1987) This review includes modifications brought about in $\gamma(\vec{n})$ plots by phenomena such as surface reconstruction, adsorption etc.

It is not the purpose of this monograph on the Geometry of Crystal Growth to dig deep into detailed atomistic models of crystal surfaces and the $\gamma(\vec{n})$ functions corresponding with those models; this field has adequately been covered by Kern's and others' chapters in this series. Rather we want to point out, as a matter of principle, the way in which broken-PBC type functions fail to describe a number of elementary phenomena:

a: Surface energy.

The deficiency of broken-bond type $\gamma(\vec{n})$ expressions can easily be overlooked because they explain so much already. (e.g., size and orientation of facets on equilibrium crystals). For crystal shapes consisting of only facets which are all below the roughening transition (and many crystals fall in this category) broken-bond models are even to be considered as perfect, at least as far as the applications are restricted to equilibrium crystals.

However, the roughening transition is an experimentally observable fact and many cases of nonfaceted but nonspherical equilibrium forms are known (low-α factor surfaces; see, e.g., JACKSON, UHLMANN and HUNT 1967; LEAMY, GILMER and JACKSON 1975). It is the class of phenomena related with this transition which requires an essential amendment to broken-bond models. Ideally, the transition from a perfectly faceted surface, via a rounded but still nonspherical, to a "completely roughened" (i.e., spherical, liquid-like) surface would have to be described mathematically by a parametrized $\gamma(\vec{n})$ function where this transition would correspond with the roughening parameter varying between the values 0 and 1, for instance. In chapters 11 and 12 we will see that such parametrized functions can easily be constructed (even in an infinite variety of ways). The important point is, however, that such an approach requires each facet (i.e., each family of PBC's) to be treated individually. The broken-PBC type of expression does not give the possibility for describing the consequences of roughening, let alone for individual facet families.

b: Growth rate.

For analytic growth rate functions, it is much more obvious that a broken-PBC type function of the form of eq. (9.1) is deficient: it does not work at all. (e.g., perfect facets do not grow at low supersaturations). Still, there is some "broken-bond likeness" in the kink-motion and ledge-motion models introduced by BURTON, CABRERA and FRANK (1951). Maybe it is this seeming contradiction which has discouraged up to now attempts to the mathematical construction of physically reasonable analytic $R(\vec{n})$ functions.

To see where the problems are, it may be instructive to be aware of what a broken-PBC like $R(\vec{n})$ plot would mean physically: namely that all PBC's ending on the crystal surface would keep growing (= adding material to the crystal) with a constant velocity (each PBC type with its own characteristic velocity), *independently of the angle which the PBC vector* $\vec{p}$ *makes with the surface normal* $\vec{n}$. This is obviously not realistic for faceted crystals: PBC's hitting a facet under a large angle do not grow at all, whereas kink motion along a PBC-oriented ledge can be considered as the growth of one PBC (and only one at a time), which has to be located at a very specific position (at the front of a ledge on a facet). Explicit, practical construction of an R function on the basis of the kink velocity data involves the counting of only those PBC's (from all those ending on the crystal surface) which are able to grow, i.e. which have an "accessible end". To this purpose they have to hit the crystal under a small angle so that the accessible kink site in which it ends is a "repeatable step", where an atom is kept in place by sufficient binding energy to stabi-lize it against thermal desorption. Such a kink site is automatically surroun-ded by a sufficiently large planar area at the facet orientation, which can act as a receptor or collecting apron for new atoms from the parent phase.

Mathematically, this means (roughly) that the $|\vec{n}\cdot\vec{p}|$ factor has to be counted fully for low values ($\ll$1), but has to be suppressed at "high" values (approaching unity). Note that this kind of suppression is required in the R-function case only, not in the γ-function case. In 10.3 we will show that such a suppression can be brought about quite naturally by the mathematical trick of "assembling" of sub-functions representing different facets.

Apart from this, "roughening" of the ledge (i.e. nucleation of new kink sites) means advancement of the ledge *independent* of kink motion. In other words, in the very case for which broken-bond γ functions work best, a "broken-PBC" R function does not work at all.

Concluding: for R function modelling, much more than for γ-function modelling, it is essential to use an approach based on the peculiarities of facets (PBC *pairs*). Effects of roughening can then be introduced by the use of appropriate parameters as in the case of γ plots. The fact that the $R(\vec{n})$ function has a cusp going to zero *for each (non-roughened) facet direction* implies that its construction will require something like a *product* rather than a *sum* of PBC-related $|\vec{n}\cdot\vec{p}|$ terms.

c: The use of $\gamma(\vec{n})$ functions as an approximation to $R(\vec{n})$ functions

In many discussions, and even in textbooks on crystal geometry, the $\gamma(\vec{n})$ function is without further comment postulated as a valid point of departure to explain the shape of *macroscopic* crystals.*) (actually this validity is restricted to microscopic crystals: see 8.3 for validity range and transition) At this point it is of interest to explain why this assumption, although fundamentally and theoretically incorrect, can still *in practice* be a good approximation.

In the first place, the BCF theory about kink and ledge motion as growth mechanisms, explains why (cusped) minima in the $R(\vec{n})$ and $\gamma(\vec{n})$ functions coincide: the microscopic surroundings of kinks and ledges during (near-equilibrium) growth are equilibrium facets, i.e. occur at $\gamma(\vec{n})$-minimum orientations. So, facets on the $\vec{Q}R$ and $\vec{Q}\gamma$ shapes occur at the same orientations - but still the minima would be (almost) zero for the R function, in contrast to finite minima for the γ function.

However, in practice most technical and natural crystals are full of dislocations, which implies that the zero-$\vec{Q}R$ minima are "short- circuited" by the vector velocities of these dislocations, $\vec{\mathfrak{R}}^*_d$ (see ch. 6.4 on our definition and notation of velocity sources with vector velocity $\vec{\mathfrak{R}}^*$). So at first glance, the *macroscopic* crystal looks much like a $\vec{Q}R(\vec{n})$ shape derived from an $R(\vec{n})$ function with *positive* cusped minima.

Such a function is already more similar to the $\gamma(\vec{n})$ function than a "perfect-crystal" $R(\vec{n})$ function, defined here as an R function for a crystal without dislocations. On closer look, of course, the "facets" of the growing crystal consist of dislocation pyramids; i.e. of sets of "pseudofacets" of vicinal orientation. One facet may exhibit a forest of dislocation pyramids or just a few or even a single one. When dislocations with different Burgers vectors are present, their $\vec{\mathfrak{R}}^*_d$ vectors may differ so that ultimately the fastest of them overgrow the others.

Finally, it may be expected that adatom surface density on relatively high-energy facets (being related with the bonds extending from the facet) is higher than on lower-energy facets. Thus, the dislocation-pyramid velocity $\vec{\mathfrak{R}}^*_d$ might generally be higher for higher- free-energy facets - a further similarity between the "macroscopic" $R(\vec{n})$ and the $\gamma(\vec{n})$ function.

Altogether, for *faceted* crystals the similarity between the $\gamma(\vec{n})$ and the "macroscopic" $R(\vec{n})$ functions (defined by the dislocations' $\vec{\mathfrak{R}}^*_d$ vector velocities as pseudo "cusp minima") is no big surprise. However, kink and ledge advance rates are much higher than those corresponding with a real "sphere plot" with the same R minima. For *nonfaceted* crystals the similarity ends altogether, at least apart from the coincidence of orientations of minima and maxima.

*) HARTMAN and BENNEMA (1980) amended this postulate in the sense that the E_{slice} function rather than the γ function would have to be considered. Although the agreement may be closer, there is still no ground for an identification.

9.3. STRUCTURE OF CHAPTERS 10 -12

In order to follow a logical way to discuss the construction of a "complete" γ or R function, we will ultimately (in ch.12) subsequently discuss the following steps in this construction (resulting in functions of increasing complexity):

1) *2-D (two-dimensional) functions* *)
These describe the behaviour of (in general monatomic-height) ledges on a facet. The following items are of interest:

* rounding of cusps in connection with "roughening" and the corresponding analytical representations;
* construction of "combined" R functions representing more than one (pair of) facets;
* eventual cutoff of functions above a maximum imposed by any mechanism leading to surface-flux limitation.

2) *3-D functions applicable to a small orientation range near a certain facet. ("3D-SC" functions)*
* With the angle α representing the deviation from the exact facet orientation, the 3-D plots are essentially derivable from the 2-D plots discussed first, by multiplication with $|\sin\alpha|$ (for sufficiently small α).
* "Roughening" of the facet requires the possibility to "smooth" the cusp in the $|\sin\alpha|$ function (as in the 2-D case, but with a different smoothing parameter).

3) *"Complete" 3-D functions* representing not just a small orientation range but the full range of the unit sphere, with all facets occurring on the crystal. Depending on the required resolution, such complete functions can be:
* assembled from the 3-D-SF functions (see 2)) of all occurring facets;

*) Note that there are two types of 2-D plots: a) those representing the effectively 1-D object of monatomic-height ledges on a facet; b) those representing actually extended surfaces which are by some mechanism confined to be perpendicular to a certain fixed plane (i.e. cylindrical, in the mathematical sense). The latter situation occurs, for instance, in the case of the thin-layer growth experiments of JACKSON and HUNT (1965). For case b), the relevant 2-D plots are the shadow of the full 3-D plots cast onto the plane of interest. This is in general quite different from the case of the 2-D plot describing ledges on a facet. In the context of chapters 9-12, the term "2-D plot" is used exclusively in connection with monatomic ledges (case a).

* with less detail but greater simplicity: by various combination tricks of simpler functions, e.g.

$$\prod_i R_i \,|\, \vec{n} \wedge \vec{f}_i \,| \quad \text{or} \quad \left(\sum_i \frac{1}{\cos^2 \alpha_i} \right)^{-1/2}$$

if necessary with amendments for "cusp smoothing".

4) *"Upward-limited" ("series-resistance") functions*; i.e. complete 3-D functions describing growth determined by surface processes and maximized by a diffusion boundary layer or another flux-limiting mechanism.

5) *Functions including the effect of dislocations.*

Preceding the "construction manual" for R functions in ch. 12, we will give a general mathematical discussion of the assembling methods to be used as a tool, and of the effects which are generated by various assembling steps (in chapter 10).

In this mathematical discussion, we will illustrate that there is an infinite number of ways to construct "combined functions" from individual functions in such a way as to preserve essential features of the individual functions in the combined one (e.g., their behaviour in the vicinity of cusped minima). It has to be stressed, however, that this mathematical possibility, even if it results in functions which seem to simulate the actual situation reasonably, is no guarantee for the physical "soundness" of the function. ("any resemblance with real crystal growth phenomena is purely coincidental"). It is the task of theoretical analysis to investigate the physical credibility of a function, and to make a correct choice of the parameters by which it is defined.

In conjunction with the mathematical toolbox presented in ch.10, we will discuss some of the *physical aspects* of specific assembling operations in ch.11. The choice for a particular assembling operation or for a particular assembling exponent can be based on such physical arguments.

Chapter 10

NONLINEAR NETWORKS AND THE "ASSEMBLED FUNCTION" NOTATION

In science, each new point of view calls forth a revolution in nomenclature.

Friedrich Engels (1820-1895)

10.1 STARTING FUNCTIONS: NOTATION AND REPRESENTATION

10.1.1 Periodic Bond Chains and facets

For PBC directions, we will use normalized vectors $\vec{p}$ (eventually labeled with a suffix, $\vec{p}_i$) and for facet directions, normalized vectors $\vec{f}$ (labeled with two suffices: $\vec{f}_{ij}$) representing the facet normal

$$\vec{f}_{ij} = \frac{\vec{p}_i \wedge \vec{p}_j}{|\vec{p}_i \wedge \vec{p}_j|} \tag{10.1}$$

Note the necessity of keeping account of the *direction* of $\vec{p}$ and $\vec{f}$ vectors because the structure and properties of their head and tail ends can be asymmetric. Consequently, forward and backward growth velocities may be different and an $\vec{f}_{ij}$ direction may be a facet while $\vec{f}_{ji}$ is not a facet at all (non-centrosymmetric case).
(The general orientation variable, as always, is written as a unit vector $\vec{n}$).

As an alternative for vector notation, we can use angular variables. In fact we have then to discern between two types of such variables, and we will adopt a special notation habit to highlight the difference:
a) angular variables for the orientation of fixed vectors such as $\vec{p}_i$ or $\vec{f}_{ij}$ *in a certain reference frame*. For such angles we will use - for the $\vec{p}_i$ case- symbols φ_i (in 2D descriptions) and (φ_i, ϑ_i) (in 3D spherical-coordinate

descriptions), or φ_{ij} etc. for the $\vec{f}_{ij}$ case.
b) angular variables for the angle *between the orientation vector* $\vec{n}$ *and vectors* $\vec{p}_i$ *or* $\vec{f}_{ij}$. (and as such independent of a reference frame). For such angles we will use symbols α_i and α_{ij}, respectively. Accordingly, $|\vec{n}\cdot\vec{p}_i|$ products have the form $|\cos\alpha_i|$, whereas $|\vec{n}\wedge\vec{f}_{ij}|$ products have the form $|\sin\alpha_{ij}|$. *Generally speaking, the consequence of our notation habit is that angular variables occurring in trigonometric expressions for R or γ functions are invariably of the "α" type.* When we would also introduce an angular variable ν for the $\vec{n}$ position in the reference frame, then we would have (for the 2D case) $\alpha_i = \nu - \varphi_i$. Anyhow, variables of type α_i or α_{ij} are all varying with the surface orientation $\vec{n}$, whereas variables φ_i or φ_{ij} are not.

(Note: the use of angular variables can in some cases highlight the analogy between 2D and 3D-case functions, because trigonometric identities can be exploited to simplify the expressions. One example is the analogy between formulas (10.45) and (10.62).

When an explicit coordinate notation of vectors is required, we will in this chapter use either a Cartesian frame (h,k,ℓ) - as projected onto an $(\vec{e}_1,\vec{e}_2,\vec{e}_3)$ frame - (with $h^2+k^2+l^2=1$); a spherical $(\alpha,\beta,\gamma) = (\arccos h, \arccos k, \arccos l)$ or a cylindrical coordinate system $(\varphi,\vartheta) = (\arccos h, \arccos k/\sqrt{k^2+l^2})$.

(This applies to normalized vectors. When the "crystallographic" habit of integer numbering of directions is followed (e.g, (221)), then we have first to normalize by dividing all components by $\sqrt{h^2+k^2+l^2}$, i.e. by 3 in the example).

Surface bond chains; SBC's

For the sake of surface properties such as surface free energy and growth rate, it makes no difference whether a "bond chain" involves bonds which are present throughout the crystal lattice (i.e., PBC's) or only at the surface itself. At the surface, the possibility of forming bond chains is increased (or sometimes decreased) (GILING and VAN ENCKEVORT (1985)):
a) by dimerization, i.e. without interaction with other species;
b) by chains involving bonds with adsorbed lattice-strange atoms or molecules. (this class, of course, is highly sensitive for the adsorbate content of the parent phase).

The extra possibility at the surface concerns 2-dimensionally periodic chains with a translation vector tangent to the surface only. Such non-bulk periodic bond chains may be called *surface bond chains* (SBC's). Their rôle as participants in the "PBC game", i.e. in determining γ and R functions, is well known (facet and habit modifications induced by surface reconstruction and adsorbates). As their mathematical rôle in this context cannot be distinguished from bulk PBC's, we will adopt the convention that the $\vec{p}$ symbol can also represent a *surface* bond chain. When distinction is required, we will use the symbol $\vec{s}$.

There is, however, a *phenomenological* difference between PBC's and SBC's. The possibility of forming extra or "side" chains, i.e. to transform a regular PBC into an SBC, is orientation-dependent. Consequently, the "partial surface energy" γ_i in the broken-bond summation for $\gamma(\vec{n})$ is not just a constant as it would be for a "real" PBC, but it has a value which can change with orientation. The same argument applies to the partial volume increase rate which plays a rôle in $R(\vec{n})$ functions. In fact, "pure" PBC's without surface interaction at all may turn out to be exceptional, after all.

10.1.2 Starting functions.

The basic "raw material" from which to assemble the orientation-dependent part of all R and γ functions (together with a number of appropriate constants or affinity-dependent factors) consists of only one function

$$|\vec{n} \cdot \vec{p}_i| = |\cos \alpha_i| = |h| \tag{10.2}$$

(for $\vec{p}_i$ aligned with the $\vec{e}_1$ axis of a reference frame).
The plot representation of this function consists of a pair of unity-*diameter* spheres passing through the origin and symmetrical on the axes $\vec{p}$ and $-\vec{p}$. For all $\vec{n}$ perpendicular to $\vec{p}_i$, the function has a cusped zero-minimum. We will in this connection speak of the " $\vec{p}_i$ pole" and the " $\vec{p}_i$ equator" on the unit sphere (= sphere with *radius* 1 centered on the origin).

It is convenient to define a "broken-PBC counting function" $C(\vec{n},\text{i})$ representing *the number of PBC's $\vec{p}_i$ crossing a surface with orientation $\vec{n}$ per unit area.* This function (specific for each particular PBC $\vec{p}_i$) is obtained from eq. (10.3) by multiplication with a_i^{-1}, i.e. the number of ($\vec{p}_i$)-type PBC's crossing a unit surface element perpendicular to $\vec{p}_i$ ($\vec{n} = \vec{p}_i$) - so a_i is the area of a crystallographic unit cell in a cross-section perpendicular to $\vec{p}_i$. (warning: starting from section 10.2, the symbol a will also be used for the "assembling exponent" - but confusion is unlikely).

$$C(\vec{n},\text{i}) = (1/a_i)|\vec{n} \cdot \vec{p}_i| \tag{10.3}$$

In γ or R functions, the $C(\vec{n},\text{i})$ function is used in combination with some appropriate OID factor (which may depend on variables as affinity, temperature, chemical composition etc.) The $C(\vec{n},\text{i})$ counts the broken

PBC's, whereas the OID factor represents some quantity to be summed per broken PBC. (energy for γ, volume increase for R).

It may be appropriate here to remark that even the orientation independence of these factors can be relative:

a) as remarked above, the conditions for part of the bond chain of an SBC to be formed may not be present for the full circle of $\vec{n}$ orientations perpendicular to $\vec{s}$.

b) for asymmetric PBC ends, the constant for the $\vec{p}$ head-end may differ from the tail-end (so that const. $|\vec{n} \cdot \vec{p}_i|$ is represented by a set of unequal-diameter spheres).

"Asymmetric" PBC's

An alternative description of asymmetric PBC's makes use of functions counting the head-ends and the tail-ends separately. For the head-end, we have to define the function (including the $\vdots$ symbol)

$$C_{\text{head}}(\vec{n},\mathrm{i}) \equiv \frac{1}{\mathrm{a_i}} \vdots\, \vec{n} \cdot \vec{p}_i \,\vdots \equiv \frac{1}{\mathrm{a_i}} \frac{(\vec{n} \cdot \vec{p}_i) + |\vec{n} \cdot \vec{p}_i|}{2} \tag{10.4a}$$

which implies that the tail-end counting function is

$$C_{\text{tail}}(\vec{n},\mathrm{i}) = \frac{1}{\mathrm{a_i}} \frac{(\vec{n} \cdot \vec{p}_i) - |\vec{n} \cdot \vec{p}_i|}{2} = \frac{1}{\mathrm{a_i}} \{ \vdots\, \vec{n} \cdot \vec{p}_i \,\vdots - (\vec{n} \cdot \vec{p}_i) \} \tag{10.4b}$$

When the head-end and the tail-end multipliers, e.g. for a γ or R function, are $\mathrm{b_1}$ and $\mathrm{b_2}$, respectively (including the $(1/\mathrm{a_i})$ factor), then the complete function is represented by

$$\begin{aligned} \mathrm{b_1} \vdots\, \vec{n}_i \cdot \vec{p}_i \,\vdots + \mathrm{b_2} \{ \vdots\, \vec{n}_i \cdot \vec{p}_i \,\vdots - (\vec{n}_i \cdot \vec{p}_i) \} = \\ = \frac{\mathrm{b_1}+\mathrm{b_2}}{2} |\vec{n} \cdot \vec{p}_i| + \frac{\mathrm{b_1}-\mathrm{b_2}}{2} (\vec{n}_i \cdot \vec{p}_i) \end{aligned} \tag{10.5}$$

Another function which can play a role is $|\sin \alpha_{ij}|$, i.e. the factor relating the growth rate of a near-facet oriented surface with the advance rate of ledges on the facet:

$$|\sin \alpha_{ij}| = |\vec{n} \wedge \vec{f}_{ij}| = \tfrac{1}{2} |\vec{n} - \vec{f}_{ij}| |\vec{n} + \vec{f}_{ij}| = \sqrt{k^2 + \ell^2} \tag{10.6}$$

(again $\vec{f}_{ij}$ aligned with $\vec{e}_1$).

Note that in three dimensions, $|\sin \alpha_{ij}|$ is represented by a toroidal shape symmetrical around the axis of the cusped minima, in contrast with the double-sphere shape of $|\cos \alpha_{ij}|$. In two dimensions, both $|\sin \alpha_{ij}|$ and $|\cos \alpha_{ij}|$ are represented by a double-circle plot !)

Monopoles.

In the description of non-centrosymmetric facets (e.g., tetrahedron) it can be practical to use the function

$$2\left|\sin\tfrac{1}{2}\alpha_{\mathrm{ij}}\right| = \left|\vec{n}-\vec{f}_{\mathrm{ij}}\right| = \sqrt{k^2+\ell^2+(1-h^2)} = \sqrt{2-2h} \qquad (10.7)$$

This function approaches $|\sin\alpha_{\mathrm{ij}}|$ in the limit $\alpha_{\mathrm{ij}} \to 0$ but has no zero-cusp in the opposite direction, $\vec{n} = -\vec{f}_{\mathrm{ij}}$. For two-dimensional plots, this function can be used in assembling procedures (see below) to construct functions with an uneven number of zero-cusps.

In three-dimensional functions, the monopole expression can be used to construct simplified prototype functions for crystals with non-centrosymmetric facets.
(note that $|\sin\alpha|$ and $2\,|\sin\tfrac{1}{2}\alpha|$ differ already in the third-order term: i.e.

$$-\tfrac{1}{6}|\alpha|^3 \text{ for } |\sin\alpha|, \quad -\tfrac{1}{24}\,|\alpha|^3 \text{ for } 2\,|\sin\tfrac{1}{2}\alpha|.$$

A monopole expression which approximates $|\sin\alpha|$ to 5th order is derived in 10.4.4:

$$\frac{2\,|\sin\tfrac{1}{2}\alpha_{ij}|}{\sqrt{1+\sin^2\tfrac{1}{2}\alpha_{ij}}} = 2\sqrt{\frac{1-h}{3-h}} \qquad (10.8)$$

10.2 THE ELEMENTARY ASSEMBLING OPERATION

In this section we introduce the concept of "assembling" as a strictly defined mathematical operation which combines (more than one) functions and/or constants into a single "assembled" function. We have to stress, however, that this operation does not involve any new mathematical principle; rather it is to be considered as just a shorthand notation of elementary algebraic expressions. For a single assembling operation, our shorthand notation is hardly an improvement over conventional notation. The virtues of the assembling notation only become apparent when calculation of a physical quantity involves a *series* of several assemblings. In such a case the explicit algebraic expression in conventional notation soon becomes intractable. In numerical calculation procedures, the concept of sequenced assembling can directly be translated into a practical algorithm.

An obvious area of application is in the formulation of the impedance or admittance of multiply connected electrical networks. The assembling notation enables to express such an impedance or admittance in a compact formula, even for the generalized case of networks containing nonlinear elements. In chapter 11 we will explain that the growth rate function $R(\vec{n})$ can be considered as a case of such an "admittance function". Also surface-energy functions can be expressed in terms of an assembling procedure. Of course, in this monograph our attention will remain focused on the $R(\vec{n})$ and $\gamma(\vec{n})$ functions related with crystals.

The $\vec{n}$ parameter in these functions might, in a way, be compared with the *frequency* parameter in electrical networks with inductive, capacitive and resistive (i.e., "passive") elements. In this analogy, singularities in the overall impedance occurring at resonance frequencies of LC loops in the electrical network correspond with singularities in the inverse of $R(\vec{n})$ for facet $\vec{n}$ directions. Damping of the resonance by dissipative elements corresponds with smoothing of the cusps in the $R(\vec{n})$ function, i.e. with the "roughening" phenomenon.

Assembling is not one single operation; it is a *group* of operations which can all be defined mathematically as the combination of (more than one) elementary functions into one "assembled" function. (addition and multiplication of the elementary functions are just two of the many possibilities to obtain an assembled function).

The method of assembling can be chosen in accordance to what features of the elementary functions have to be found back in the assembled functions. We will give examples of different assembling steps and of the corresponding requirements for the features to be kept preserved in 10.3.

To introduce the basic assembling step as a general mathematical operation, we have to define two or more starting or input functions $F_1, F_2, \ldots F_i, \ldots F_p$ which are all functions of a variable $\vec{n}$ in a $\geqslant 1$ - dimensional parameter space. (in our context, of course, the relevant parameter is orientation). We want to construct one "assembled function" F_a of the same variable $\vec{n}$ which exhibits a combination of certain characteristic features of all starting functions.

We define a "distorting operator" D which is basically a mathematical operation (see examples below), which exaggerates the features to be combined or preserved, while levelling the rest. Associated with the operator D we can define a D-*eliminating* operator which we will write as $\mathsf{D}_{\mathrm{INV.}}$, i.e. $\mathsf{D}_{\mathrm{INV}}(\mathsf{D}\,F) = F$ (so that $\mathsf{D}_{\mathrm{INV}} . \mathsf{D} = \mathsf{D} . \mathsf{D}_{\mathrm{INV}} = \mathrm{I}$ (identity). For instance:

$$\mathsf{D}\, F = F^{a} \qquad \mathsf{D}_{INV}\, F = F^{1/a}$$

$$\mathsf{D}\, F = (bF)^{a} \qquad \mathsf{D}_{INV}\, F = (1/b)F^{1/a}$$

$$\mathsf{D}\, F = b\{(F+c)\}^{a} \qquad \mathsf{D}_{INV}\, F = \{(1/b)F^{1/a}\}-c$$

$$\mathsf{D}\, F = \log F \qquad \mathsf{D}_{INV}\, F = \exp F$$

$$\mathsf{D}\, F = \log (F+a) \qquad \mathsf{D}_{INV}\, F = (\exp F) - a$$

$$\mathsf{D}\, F = \tan F \qquad \mathsf{D}_{INV}\, F = \arctan F$$

etc.

Now the elementary assembling operation is

$$F_{ass.} \equiv \mathsf{D}_{INV}\left(\sum_{i=1}^{n} \mathsf{D}\, F_i \right) \tag{10.9}$$

or in a generalized notation

$$F_{ass.} \equiv \overset{n}{\underset{i=1}{\boxed{\mathsf{D}}\!\times}}\; F_i \;\times\!\!\circ \tag{10.10}$$

with D either written explicitly in the assembling-symbol box like $\boxed{\log}\!\times$ or $\boxed{\tan}\!\times$, or defined separately like $\boxed{\mathsf{D}}\!\times$; $\mathsf{D}\, F = b\{(F+c)\}^{a}$. The F_i (functions and/or constants) are the *assemblands.*

In the context of this chapter we will often use the *exponential* D operator ($\mathsf{D}\, F = F^{a}$). It is convenient to use the special notation

$$F_{ass.} = \overset{n}{\underset{i=1}{\boxed{\mathsf{D}}\!\times}}\; F_i \;\times\!\!\circ \quad ; \; \mathsf{D}\, F = F^{a}$$

$$\equiv \overset{n}{\underset{i=1}{\textcircled{a}\!\times}}\; F_i \;\times\!\!\circ \tag{10.11}$$

In the following we will see that the effect of "exponential assembling" changes fundamentally when the assembling exponent shifts from a positive to a negative value. In the discussion of the physical meaning of assembling procedures it is therefore practical to use a notation which reveals automatically whether the assembling exponent, in parameter form, is restricted to positive or negative values, or not (instead of having to add $a<0$ or $a>0$ statements explicitly). For this purpose we introduce the following habit:

* unspecified exponent: $(\mathrm{a}\!\times$

* $\mathrm{a} > 0$: $(\underline{\mathrm{a}}\!\times$

* $\mathrm{a} < 0$: $(\bar{\mathrm{a}}\!\times$ (10.11b)

"Inverted assembling"

Another useful extension of the definition formulas (10.9) - (10.11) is the possibility to *subtract* certain $\mathsf{D}\,F_i$ terms instead of just adding them as in (10.9) - (10.11). As a notation for this extension we will use

$$F_{ass} = \boxed{\mathsf{D}}\!\times F_1, F_2, \ldots, F_h, (-)F_j, (-)F_{j+1}, \ldots, (-)F_p \quad \text{(notation)}$$

$$\equiv \mathsf{D}_{\mathrm{INV}} \left\{ \sum_{i=1}^{h} \mathsf{D}\,F_i - \sum_{i=j}^{p} \mathsf{D}\,F_i \right\} \quad \text{(definition)} \qquad (10.12)$$

where the $F_j \ldots F_p$ can be called "inverted assemblands". The association with "inversion" is suggested by an "inversion" relation between the functions

$$G_+(F) = \boxed{\mathsf{D}}\!\times F, C \times\!\!\circ \quad \text{and} \quad G_-(F) = \boxed{\mathsf{D}}\!\times F, (-)C \times\!\!\circ$$

(see fig.10.2 a and b on page 263), where F is considered as a variable and C as a constant. The relation follows directly from the definition:

$$\mathsf{D}\,G_- = \mathsf{D}\,F - \mathsf{D}\,C \;\Rightarrow\; \mathsf{D}\,F = \mathsf{D}\,G_- + \mathsf{D}\,C, \text{ or } \; F = \boxed{\mathsf{D}}\!\times G_-, C \times\!\!\circ$$

When we confront this with $G_+ = \boxed{\mathsf{D}}\!\times F, C \times\!\!\circ$ we see that

$$G_- = \arg(G_+) \qquad (10.13)$$

i.e., the function $G_+(F)$, plotted as a function of F, is the 45-deg mirror image of the function $G_-(F)$. In fig. 10.2a we show the functions $(\underline{\mathrm{a}}\!\times \varepsilon, \kappa \times\!\!\circ$ and $(\bar{\mathrm{a}}\!\times \varepsilon, (-)\kappa \times\!\!\circ$ for a number of *positive* assembling exponents, in fig.10.2b the same for a number of *negative* assembling exponents.

An application of an inverted assembland is the variable-modulation principle used for an orientation-dependence parameter ε, to be tuned between 0 and 1:

$\textcircled{2}\!\times \ |\cos\alpha|\,,\ \varepsilon\,|\sin\alpha| \ \times\!\circ$ (in polar coordinates)

$\textcircled{2}\!\times \ h,\ \varepsilon,\ (-)\varepsilon h \ \times\!\circ$ (in cartesian coordinates)

with (h,k) = vector on unit circle $h^2 + k^2 = 1$. (10.14)

NOTE:

In a sense, the (–) notation (together with the whole concept of inverted assembling) might be considered as an artefact, resulting from the limitation to real positive assemblands. In a generalized version, inverted assemblands might be replaced by ordinary but complex assemblands.

Nevertheless, the (–) notation is practical in the context of the present application where complex values of physical quantities are difficult to visualize. In maintaining this notation, however, we have to be aware that the link with the complex-number alternative can pop up in certain cases where a uniform notation is preferred. For instance, in equation (10.56) we use a parametrized inverted assembland

$(-)\sqrt{\dfrac{12}{m^2-4}}$ in a $\textcircled{-2}\!\times$ assembling, in which the integer m can take all values from 1 upwards. For m > 3, the square-root value is real. For $m = 1$, however, it is imaginary: but for that case (and for $\textcircled{-2}\!\times$ assembling) we have to realize that

$$\textcircled{-2}\!\times \ldots, (-)\sqrt{\frac{12}{m^2-4}} \ \times\!\circ_{(m=1)} = \textcircled{-2}\!\times \ldots, \sqrt{\frac{12}{4-m^2}} \ \times\!\circ \qquad (10.15)$$

For m = 2 the square root becomes singular, corresponding with a shortcut in the $\textcircled{-2}\!\times$ assembling, so that we can omit the assembland altogether:

$$\textcircled{-2}\!\times \ldots, (-)\sqrt{\frac{12}{m^2-4}} \ \times\!\circ_{(m=2)} = \textcircled{-2}\!\times \ldots\ldots \times\!\circ \qquad (10.16)$$

< end of note >

This general procedure offers a wide versatility. It includes, for instance, elementary multiplication ($\mathbf{D}\ F = \log F$), addition ($\mathbf{D}\ F = F$) and "parallel resistance" (i.e., reciprocal) addition ($\mathbf{D}\ F = F^{-1}$). This versatility is again increased when we perform a number of assembling steps *in succession*, each with a different $\mathbf{D}$ operator. For two steps:

$$F_{ass.2} = \overset{n}{\underset{i=1}{\boxed{\mathbf{D}(2)}\!\times}} \ \overset{m}{\underset{j=1}{\boxed{\mathbf{D}(1)}\!\times}} \ F_{ij} \times\!\circ\times\!\circ = \mathbf{D}_{INV.2}\left\{\sum_{i=1}^{n} \mathbf{D}(2)\, F_{ass.1,i}\right\} \qquad (10.17)$$

etc. We will show later on that quite realistic γ or growth-rate functions can be built up by a succession of only a few (3 to 5) of such elementary assembling steps, each fulfilling a specific aim or function. The trick lies in finding suitable starting functions and assembling operators for all the steps, in agreement with a physically realistic model. We will show (in 11.4) that such models can symbolically be represented in the form of a (nonlinear) *network*. The elements in such a network represent the assemblands in the assembling steps; each "topologically" discernible branch in the network corresponds with a separate assembling step.

Two important corollaries of the definition of an assembled function will be frequently used:

1) In exponential assemblings, a common factor (k) of all assemblands can be split off from the assembling as such:

$$\textcircled{a}\!\times\; kF_1, kF_2, \ldots, kF_h, (-)kF_j, \ldots, (-)kF_p \;\times\!\circ \; =$$

$$= k.\,\textcircled{a}\!\times\; F_1, F_2, \ldots, F_h, (-)F_j, \ldots, (-)F_p \;\times\!\circ \qquad (10.18)$$

2) A sequence of assemblings with identical D operators can be written as a single one (i.e., exponential as well as general case):

$$\boxed{D}\!\times\; \boxed{D}\!\times\; F_1, F_2, \ldots, F_h \;\times\!\circ, F_j, \ldots, F_p \;\times\!\circ \; =$$

$$= \boxed{D}\!\times\; F_1, F_2, \ldots, F_h, \; F_j, \ldots, F_p \;\times\!\circ \qquad (10.19)$$

10.3 EXPONENTIAL ASSEMBLING

10.3.1 Inequalities

We will now discuss the effect of the special class of *exponential* assemblings $\textcircled{a}\!\times G, H, \ldots \times\!\circ$. The assembled functions $G, H, \ldots$ are chosen to be $\geqslant 0$; they can have zero-cusps. (In the practice of our plot-forging programme we will mostly be concerned with $|\sin \alpha|$ and $|\cos \alpha|$ functions, and constants). The effect of exponential assembling - with the assembling exponent as a parameter - is illustrated by figs. 10.1a and b.

The choice a = 1 represents the *sum operation* and is obvious. Cusps are preserved. The "classical" broken-bond function (eq. 10.1) is a typical example.

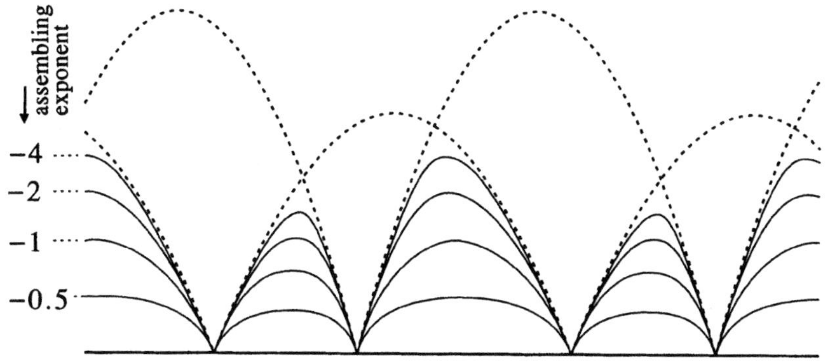

FIG. 10.1a Illustrating the effect of exponential assembling operation for *negative* exponent a. Assemblands are the two $|\sin \varphi+\alpha|$ functions (dotted lines); assembled functions are the full lines. Assembled functions combine the zero-cusps of all assemblands.

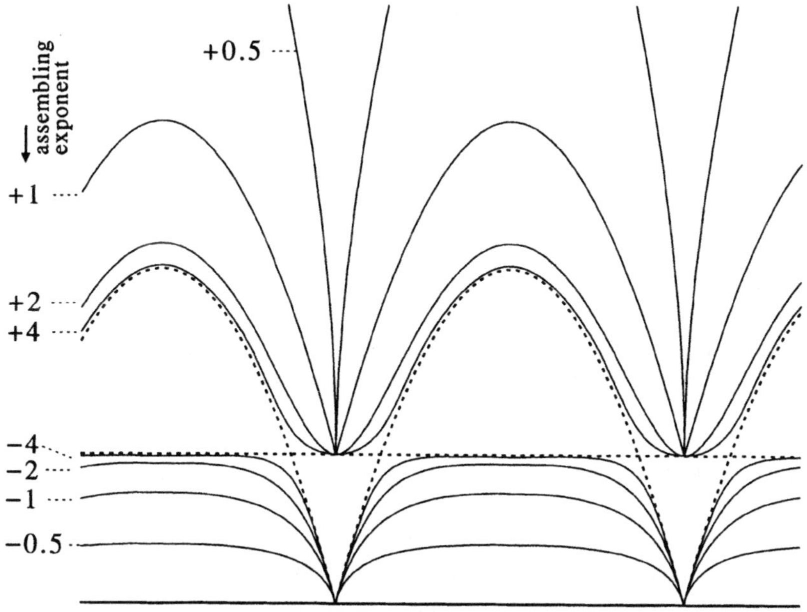

FIG. 10.1b Exponential assembling of a "zero-cusped" positive function and a positive constant, with positive and negative exponents. The "smoothing" of a cusp requires a > 1.

For $1 < a < +\infty$, we have:

$$\mathrm{Max}\,(G,H,\ldots) < \textcircled{\underline{a}}\!\times G,H,\ldots \rtimes\!\circ < \Sigma\,(G,H,\ldots) \qquad (10.20)$$

and the zero-cusps with finite slope in G,H etc. are smoothed to a positive minimum in the assemblage. (unless they coincide). Cusps in $G,H\ldots$ at a positive parameter value are preserved. The limits given above are approached for $a \to 1$ and $a \to +\infty$:

$$\lim_{a \searrow 1} \textcircled{\underline{a}}\!\times G,H,\ldots \rtimes\!\circ = \Sigma\,(G,H,\ldots) \qquad (10.21)$$

$$\lim_{a \nearrow \infty} \textcircled{\underline{a}}\!\times G,H,\ldots \rtimes\!\circ = \mathrm{Max}\,(G,H,\ldots) \qquad (10.22)$$

(i.e., Max = $\textcircled{\infty}\!\times \ldots \rtimes\!\circ$)

For $-\infty < a < 0$, we have (fig.10.1a ; $a < 0$ curves)

$$0 < \textcircled{\bar{a}}\!\times G,H,\ldots \rtimes\!\circ < \mathrm{Min}\,(G,H,\ldots) \qquad (10.23)$$

$$\lim_{\bar{a} \searrow -\infty} \textcircled{\bar{a}}\!\times G,H,\ldots \rtimes\!\circ = \mathrm{Min}\,(G,H,\ldots) \qquad (10.24)$$

(i.e., Min = $\textcircled{-\infty}\!\times \ldots \rtimes\!\circ$)

Zero-cusps in $G,H,\ldots$ are preserved.

The range $0 < a < 1$ yields (fig.10.1a ; $a = 1/2$ curve)

$$\textcircled{\underline{a}}\!\times G,H,\ldots \rtimes\!\circ > \Sigma\,(G,H,\ldots) \qquad (10.25)$$

and the zero-cusp is preserved in the assemblage. (This range of the a parameter is of no practical use for our present purposes).

Inverted assemblands:

Equations (10.20-10.25) refer to ordinary "positive" assemblands $G,H,\ldots$ The effect of inverted assemblands (see eq.(10.12) and fig. 10.2a/b) is to switch between the rôles of the assembled function and its argument. Writing ε' for the assembled function $\textcircled{a}\!\times \varepsilon, (-)\kappa \rtimes\!\circ$, we see that the effect of the *inverted* assembland $(-)\kappa$ is that the assembled function ε' is made ∞ for $\varepsilon = \kappa$ when $a < 0$ and is made $= 0$ when $a > 0$. In physical terms, $\varepsilon' = \infty$ means that the element is *short-circuited*, $\varepsilon' = 0$ that the path containing the element is *interrupted*.

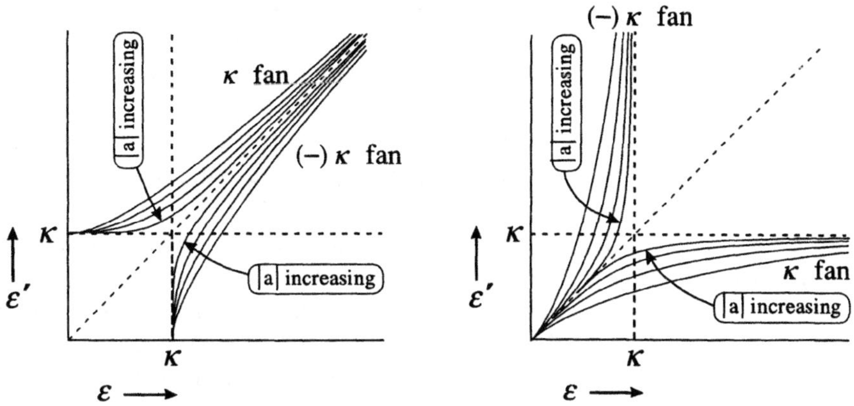

FIG.10.2 Illustrating the "mirror imaging" of inverted versus positive assemblands.
Case a) (left) represents positive assembling exponents:
$\varepsilon' = (\underline{a}) \times \varepsilon, \kappa \times\!\!\circ$ (κ fan) and $\varepsilon' = (\underline{a}) \times \varepsilon, (-)\kappa \times\!\!\circ$ ((−)κ fan);
case b) (right) negative ones:
$\varepsilon' = (\bar{a}) \times \varepsilon, \kappa \times\!\!\circ$ (κ fan) and $\varepsilon' = (\bar{a}) \times \varepsilon, (-)\kappa \times\!\!\circ$ ((−)κ fan)

10.3.2 Applications

By using appropriate assembling exponents we can, starting from a number of functions $G,H,\ldots$ and/or constants ε, L etc. (eventually as inverted assemblands), shape an assembled function with modified characteristics. In the following items 1) - 5) we will present elementary examples involving only one assembling step (thereby combining two or more assemblands). In section 10.5 we discuss more complicated examples involving two or more assembling steps, i.e. series of elementary assemblings. In the present ch.10, these examples are just considered as mathematical tricks designed to construct functions with certain properties. The physical aspects of these tricks will be discussed in chapter 11.

1) A *cusped zero*-minimum in a function G can be removed and replaced by a smoothed positive minimum ε. For this purpose, G has to be assembled with a function H which has the desired value ε at the place of the cusp in G. (or just with the constant ε itself).

The parameter a should be chosen > 1. For a = 2, $G = |\mathrm{b}x|$, and $H = \varepsilon$, the resulting assembled function is the hyperbola $\sqrt{\mathrm{b}^2x^2 + \varepsilon^2}$.

The "sharpness" of the minimum ($\mathrm{d}^2/\mathrm{d}x^2$ [$(a) \times G,H,\ldots \times\!\!\circ$] can be varied between zero and a delta function by varying a between ∞ and 1. In fig.10.1a, this smoothing is illustrated by the $1 < \mathrm{a} < \infty$ curves. ($G = |\cos\alpha|$; H = constant = 0.4)

In 11.1.4 we will argue that the assembling exponent a = 2 is to be preferred on physical grounds. For this value (and for this value only), $\mathbf{S}\,[\,(\underline{a})\!\!\times\; |\sin\alpha|,\ \varepsilon\,|\cos\alpha| \times\!\!\circ\,]$ has a single *flat* maximum, for any of the smoothed-function possibilities mentioned below (eqs (10.32a-e).

2) A function G can be *upward-limited* to a given maximum (either a constant or another function H). For this purpose we can assemble G and H exponentially with a < 0. Note that for a = −1, $(\bar{a})\!\!\times G,H,\ldots \times\!\!\circ$ represents the formula for the current or flux through linear resistances in series. The upward limitation is illustrated by the a < 0 curves in fig. 10.1a/b.

We can also have *nonlinear* flux-affinity or current-voltage relations. Writing, for the nonlinear "resistances": $i_1 = k_1 V^a$, $i_2 = k_2 V^a, \ldots$ (i.e., same exponent a for nonlinearity), then the *series* connection of the two resistances obeys the relation

$$i_{series} = k_{ass}\, V^a \quad \text{with } k_{ass.} = (-1/a)\!\!\times\; k_1, k_2, \ldots \times\!\!\circ \tag{10.26}$$

3) *Combination or "merging" of functions:*

Again for −∞ < a < 0, zero-cusps in either of the functions $G,H,\ldots$ are preserved. This preservation includes the shape of the cusp in the vicinity of the zero-minimum, i.e. the derivatives dF/dx. In other words, the assembled function $(\bar{a})\!\!\times G,H,\ldots \times\!\!\circ$ approaches each of the individual functions $G,H,\ldots$ in those ranges of $\vec{n}$ values where the *distorted* individual functions G^a, H^a etc. are much larger than the sum of the rest; i.e. when the functions $G,H\ldots$ *themselves* are much *smaller than the sum of the distorted rest.* The higher the value of $|a|$, the more closely the individual function is followed in the cusp region.

In a sense, the assembled function $(\bar{a})\!\!\times G,H,\ldots \times\!\!\circ$ can be considered as a "multiple exposure" picture of all the $G,H,\ldots$ functions as far as their minima is concerned, resulting in an assembled function *exhibiting all minima together.* (fig.10.1b, curves for a < 0).

3 bis) *Increasing the "fit" of combined functions:*

When a common maximum M can be determined for all assemblands $G,H\ldots$: $\quad \mathrm{Max}\,(G,H,\ldots) = M$

we can construct an even closer-fitting "multiple exposure" assembled function with the aid of the (−) option. For two assemblands, and −∞ < a < 0 :

$$(\bar{a})\!\!\times G,H \times\!\!\circ \;<\; (\bar{a})\!\!\times G,H,(-M) \times\!\!\circ \;<\; \mathrm{Min}\,(G,H) \tag{10.27}$$

because the (−)M term increases the $(\bar{a})\!\!\times G,H \times\!\!\circ$ assembling but not enough to surpass any of the individual assemblands.

This principle can be repeated for every additional assembland, so that for a total of n assemblands we can add $(n-1)$ $(-)$ M terms, i.e. effectively one $(-)$ $(n-1)^{1/a}$ M term:

$$(\bar{a})\!\times G_1, G_2, \ldots G_n \times\!\!\circ \; < \; (\bar{a})\!\times G_1, G_2, \ldots G_n, (-)\,(n-1)^{1/a} M \times\!\!\circ \; <$$
$$< \mathrm{Min}\,(G_1, G_2, \ldots G_n) \qquad (10.28)$$

In particular for a = −2, when the G_i are all $\sin \alpha$ functions of equal amplitude but different in phase, eq. (10.28) can be written as

$$(-2)\!\times |\sin\alpha_1|, |\sin\alpha_2|, \ldots |\sin\alpha_n| \times\!\!\circ < (-2)\!\times |\tan\alpha_1|, |\tan\alpha_2|, \ldots |\tan\alpha_n|, 1 \times\!\!\circ <$$
$$< \mathrm{Min}\,(|\sin\alpha_1|, |\sin\alpha_2|, \ldots |\sin\alpha_n|) \qquad (10.29)$$

We will use these inequalities in the construction of the multi-facet assembled functions to be discussed in 11.4.

Equations (10.28) and (10.29) are valid for *any* distribution of orientations α_i (i.e, cusp positions of G_i). For an *equidistant* distribution, however, an even closer approximation can be found, i.e. the compensating $(-)$ assembland can be increased above the $(n-1)^{-1/a}$ value of (10.28). (a 2D equidistant distribution corresponds with a regular polygon, a 3D one with one of the Platonic solids or a regular prism). We will discuss these cases in 10.4.4 and 10.4.5, respectively.

4) *"Distortion" of functions and parameters:*

In examples 1) and 2) above, assembling was used to modify an original function $F(\vec{n})$ in such a way that the modified function $F_{ass}(\vec{n})$ *accounts for some extra physical influence* (e.g, roughening in 1), limitation in 2)). The magnitude of this influence is expressed by a parameter occurring as an assembland, and mathematically this parameter can be considered to increase the dimension of the variable space. Whereas $F(\vec{n})$ has the variables of $\vec{n}$ only, the variables of F_{ass} are the $\vec{n}$ coordinates plus ε and/or M.

In example 3), there is no extra physical effect requiring an extra parameter, but rather two similar effects have to be combined and result in a function F_{ass} defined in the *same* variable space as F_1 and F_2 on their own.

Now we can distinguish another application of assembling in which not a physical interaction is described, but rather a function or a parameter is "distorted" by assembling with a numerical constant. Consider, for instance, the expressions

$$\textcircled{-2}\!\times \varepsilon, (-)1 \times\!\!\circ \quad = \frac{\varepsilon}{\sqrt{1-\varepsilon^2}} \quad = \varepsilon' \tag{10.30a}$$

$$\textcircled{-2}\!\times \varepsilon', 1 \times\!\!\circ \quad = \frac{\varepsilon'}{\sqrt{1+(\varepsilon')^2}} \quad = \varepsilon \tag{10.30b}$$

$$\textcircled{-2}\!\times \sin\alpha, (-)1 \times\!\!\circ \quad = \tan\alpha \tag{10.30c}$$

$$\textcircled{-2}\!\times \tan\alpha, 1 \times\!\!\circ \quad = \sin\alpha \tag{10.30d}$$

In (10.30a) the "sweep range" of the ε parameter is changed from $(0-1)$ to $(0-\infty)$. In this chapter we assume the habit to write this distorted parameter, ranging between 0 and ∞, as ε'. In the reciprocal relation (10.30b), a $(0-\infty)$ range is compressed again to $(0-1)$. Equations (10.30c) and (10.30d) show that these distortions correspond with the transition from $\sin\alpha$ to $\tan\alpha$ and vice versa (at least, for assembling exponent -2; but of course these statements are nor restricted to a particular choice of a). The distortions, for (+) and (−) assemblands and for positive and negative assembling exponents, are illustrated by fig.10.2a and b.

In other words, the assembling with a numerical constant (either positive or inverted) is identical with the choice of another, "distorted", function or parameter. When such a function or parameter is used as an assembland in a subsequent assembling step, we can write the latter assembland either as the result of an explicit assembling (left-hand terms of eq. (10.30 a-d) or just as an alternative function or parameter (right-hand terms). In this context, the algebraic functions used can be considered to result from "hidden assembling steps". For such hidden steps it is not the *dimension* of the variable space of F_{ass} which is affected by such an alternative, but the *sweep range of the variables* or of F_{ass} itself.

Such a distortion of parameters or of functions can be used for the following purposes (both examples are related with item 5) below):

1) to make a parameter correspond with some thermodynamic or physical quantity. An example is the preference for a roughening parameter ε to range between 0 and 1 and so to correspond with the value of $R/\mathbf{S}R$.

2) to bring the value of an assembled function, for the extreme values of some assembland parameter, to a fixed setpoint. An example is the choice for the $\rightarrow\infty$ limit for a "completely roughened" $|\vec{n}\cdot\vec{p}|$ function by equation (10.30c) or its cotan equivalent. By this distortion, we can accomplish that not only the $v_i|\vec{n}\cdot\vec{p}_i|$ term becomes OID for complete roughening, but that the v_i parameter is eliminated altogether from the end function in that case.

5) *Transition from a cusped OD function to an OID function as characterized by a roughening parameter.*

As mentioned in example 4), the concept of a roughening transition implies not just the phenomenon of roughening as such (to be described by the smoothing of cusps as discussed in item 1)), but also that this roughening can be "completed" - i.e., all orientation dependence annihilated - for a certain end value of the roughening parameter. In 4) we have shown that the end values (sweep range) of any parameter can be chosen at will, just by an extra assembling step dedicated to that purpose. Therefore, we can *choose* a range 0 - 1 without restriction of the possibilities.

Even then, several choices for such "transition-parameter functions" can be made, depending on the required limit functions and in-between behaviour. Examples are:

$$\textcircled{\underline{a}}\!\times\, G,\, (-)\varepsilon G,\, \varepsilon \,\rtimes\!\circ \qquad (10.31\text{a})$$

$$\textcircled{\underline{a}}\!\times\, G,\, (-)\varepsilon G,\, \varepsilon\kappa \,\rtimes\!\circ \qquad (10.31\text{b})$$

$$\textcircled{\underline{a}}\!\times\, (1-\varepsilon)G,\, \varepsilon \,\rtimes\!\circ \qquad (10.31\text{c})$$

$$\textcircled{\underline{a}}\!\times\, (1-\varepsilon)G,\, \varepsilon\kappa \,\rtimes\!\circ \qquad (10.31\text{d})$$

$$\textcircled{\underline{a}}\!\times\, G,\, \frac{\varepsilon}{\sqrt{1-\varepsilon^2}} = \textcircled{\underline{a}}\!\times\, G,\, \varepsilon' \,\rtimes\!\circ \qquad (10.31\text{e})$$

In all functions, ε is the transition parameter. Equations (10.31 a-e) correspond with different alternatives covered by item 4). In (10.31 a-d), the OID limit is reached by *pinching the cusped function G off*; in (10.31 e) *by short-circuiting* it with an ε' obtained from ε by a sweep-range modifying "hidden assembling step". Furthermore, in (10.31 a and b), the function G is assembled by a (–) G term using the same assembling exponent as used for smoothing; in (10.31 c and d) by the same (–)G term but with a (possibly) different assembling exponent = 1.

Finally, eqs (10.31 b and d) illustrate that the OID limit values can be chosen at will by adding an extra assembland for that purpose. Also, the OID limit value can be made ∞ such as in eq. (10.31e). Altogether, the examples show that the method can easily be adapted to any specific requirement, including the behaviour in-between the limit values.

We will discuss the possible alternatives in a more systematic way in connection with the "network" representation of sequenced assemblings in 10.5. The physical differences between the possible alternatives for such networks will be discussed in 11.4.2.

Conclusion:

For the special class of exponential assembling operations $\textcircled{a}\!\times \ldots \times\!\circ$ we have seen that, according to the range of the a parameter, we can establish the following effects on functions $G(x)$, $H(x)$ etc:

a) for a >1: smoothing of the zero-cusps of a function.
The remaining "sharpness" of the minimum depends on the value of a; decreasing for a > 1 to a $\rightarrow +\infty$. By special choice of the assembling operation and the smoothing parameter, the smoothing can be varied between no-smoothing-at-all and "complete" smoothing, i.e. elimination of all variation with $\vec{n}$.
In 11.1.4 we will argue that a = 2 is the most "natural" choice for the smoothing "assembling exponent".

b) for a < 0: limiting a function to a certain maximum.
This can be done either with a smooth or with a sharp transition, as a ranges from 0 to $-\infty$. Depending on the mechanism of limitation, "natural" choices for a are −1 and −2.

c) equally for a < 0:
constructing an assembled function exhibiting all features of the individual functions in those x-parameter ranges where these individual functions are smaller than the others.
For arguments given in 11.4, the value a = −2 seems to be the most natural choice for combination of $|\vec{n} \cdot \vec{p}_i|$ "assemblands" for the construction of growth rate functions.

d) equally with a < 0: constructing a continuous and differentiable "smooth function" having prescribed function values and derivatives at given x values (to be discussed in 10.3.3).

"Physically speaking" we can say that exponential assembling of terms with a > 0 applies to mutually reinforcing *or* supportive *mechanisms (or just additive for a = 1, such as fluxes through parallel resistances). In contrast, the a < 0 case applies to mutually* competitive *mechanisms or contributions (such as a flux through resistances in series, for which a = −1). We will use this analogy with parallel and series connection in the pictural representation to be explained in 10.5.*

In our plot-forging programme, the effects a, b and c will be used for the "physical" categories of "cusp-smoothing", "limitation" and "combination" of the elementary $|\vec{n} \cdot \vec{p}_i|$ building units. We have to stress that the assembling operation is not restricted to the special class $\textcircled{a}\!\times \ldots \times\!\circ$;

e.g. also $\boxed{\log}\!\times \ldots \times\!\!\circ$ can serve the application of "combinination". In chapter 11 we will consider these physical categories in more detail, if necessary in connection with physical models.

10.3.3 "Territory functions" and their application

Comparing a function with a single cusp pair (e.g., $F_i = v_i\,|\vec{n}\cdot\vec{p}_i|$ or $v_k\,|\vec{n}\wedge\vec{f}_k|$) with an assembled "multi-exposure" function (for instance $(-2)\!\times v_i\,|\vec{n}\cdot\vec{p}_i|\times\!\!\circ$ or $(-2)\!\times v_k\,|\vec{n}\wedge\vec{f}_k|\times\!\!\circ$ in which it is assembled with other functions, it is easy to determine the $\vec{n}$ range where the single function dominates in the assembled function. Consider the ratio

$$T_i(\vec{n},\mathrm{a}) = [(\text{single function})/(\text{assembled function})]^{\mathrm{a}}$$

$$= \left(\frac{F_i}{\underset{\text{all } k\in\{k\}}{(\mathrm{a})\!\times} \ldots F_k \ldots \times\!\!\circ}\right)^{\mathrm{a}} \quad (\leqslant 1) \qquad (10.32)$$

This ratio is $\simeq 1$ in the range near the "$\vec{p}_i$ equator" (for a $|\vec{n}\cdot\vec{p}_i|$ factor) or near the "$\vec{f}_k$ pole" (for a $|\vec{n}\wedge\vec{f}_k|$ factor) where this factor dominates the rest, and $\simeq 0$ beyond that range. Accordingly we will call $T_i(\vec{n})$ the *"territory function"* of this equator or pole. (note that this function also depends on the v_i multipliers connected with such cusps). It follows directly from the definition that

$$\sum_{\text{all } i\in\{k\}} T_i(\vec{n},\mathrm{a}) \equiv 1 \qquad (10.33)$$

so that the individual territory functions really complement each other. The sharpness of the transition from territory to nonterritory, i.e. 1 to 0 value, depends on the assembling exponent a (sharper transition for higher $|\mathrm{a}|$).

Territory functions can be combined to define "AND" and "OR" territory functions for the determination of sum and overlap of territories:

"OR" (sum of territories i and j) : $T_i(\vec{n}) + T_j(\vec{n})$ (10.34)
"AND" (overlap of territories) : $T_i(\vec{n}) \times T_j(\vec{n})$ (10.35)

The overlap of i and j territories ($i \neq j$) (defined on the basis of the same assembled function) is only appreciable in a border region. The width of this region tends to zero for $\mathrm{a} \to -\infty$. In other words, $T_i \times T_j$ can be considered as a sort of *interface or border function* between i and j territories; a "thick" border corresponds with a low assembling exponent.

Also, in agreement with eq. (10.34) we can define "non-territory" functions $(1-T_i(\vec{n},\mathrm{a}))$ which enable to extend the possibilities of the "Boolean" functions defined above. (Note that the "mechanism" of the territory functions is in a way analogous to the electronic trick in analog-to-digital conversion by the use of opamps).

It is of interest to remark that the territories, and the boundaries between them, can *experimentally* be made visible on an "orientation map". This can be done in a special version of a sphere-growth experiment, where such a map can be generated by Schlieren optics. The territory boundaries *and the territory centres* are shown as dark lines and points, respectively, in an otherwise bright field. (indicating that these are the only positions on the sphere where the orientation does not change as a result of growth). The method is discussed in ch. 13.7. The experimental Schlieren image can be compared with a computer-generated image based on an assembled $R(\vec{n})$ function. In this way the parameters of the $R(\vec{n})$ function can be fitted.

2) Other family-specific parameter functions

On the basis of the territory function concept, it is possible to define related functions which generate a certain *family-specific parameter value* (as a function of $\vec{n}$) rather than just the value 0 or 1 as given by the territory functions themselves:

$$\mathrm{Sm}_i(\vec{n},a) = \sum_{\text{all i}} \zeta_i \cdot T_i(\vec{n},a) \tag{10.36}$$

Such functions can be practical for the description of family-specific physical quantities; alternatively they can be used as a graphical tool in computerized rendering of crystal surfaces (e.g, using a "colour" parameter).

The symbol $\mathrm{Sm}_i(\vec{n},a)$ is used to symbolize a "smooth" function which is identical with the family-specific parameter ζ_i only for $\vec{n}$ located in the (pole or equator) cusp, and gradually changes to the neighbour-territory ζ_k value. (with a transition sharpness in accordance with that of the territory function's a parameter). The functions $\mathrm{Sm}_i(\vec{n},a)$ obey the following conditions:

For $\vec{n}$ coinciding with any $\vec{f}_{ik}$ out of a finite set of fixed orientations $\{\vec{f}_i\}$:
1) $\mathrm{Sm}_i(\vec{n}) = \zeta_i$ (ζ_i is a fixed parameter specific for the set $\{\vec{f}_i\}$)
2) $(d/d\vec{n})\ \mathrm{Sm}_i(\vec{n}) = 0$, and for $\vec{n}$ not coinciding with any $\vec{f}_{ik}$, Sm_i is a continuous and differentiable function of $\vec{n}$.

The sets $\{\vec{f}_i\}$ correspond with facet "families", each family having a limited number of members (e.g. 6 for the $\{100\}$ cube family).
The Sm function concept is only relevant when we have to express a parameter for a crystal with more than one family of facets.

10.4 *SOME SPECIAL IDENTITIES WITH ②✕..✕○ and (−2)✕..✕○*

The ②✕ . . . ✕○ assembling plays a special rôle in some trigonometric relations. This rôle is based on the Pythagorean identity

②✕ cartesian coordinates c_k of vector in D-dim space ✕○ =

$$= \| \vec{n} \| \text{ (= length of vector)} \tag{10.37}$$

so that we can write a ②✕ assembled function simply as a vector length. (see 10.4.1)

10.4.1 Four-dimensional representation of smoothed $|\sin \alpha_{ij}|$ *functions*

One special choice of a cusp-smoothing function, i.e.

$$②✕ \; |\sin \alpha_{ij}|, \varepsilon |\cos \alpha_{ij}| \; ✕○ \tag{10.38}$$

can be represented as a four-dimensional *unsmoothed* $|\sin \alpha^*_{ij}|$. For this purpose we define $\vec{f}^*_{ij}$ as a normalized vector in a 4-dim. hyperspace, writing it with an asterisk to discern it from the 3-dim. $\vec{f}_{ij}$.

When the real $\vec{f}_{ij}$ is aligned with $\vec{e}_i$ in a Cartesian frame, we add a fourth perpendicular axis $\vec{e}_4$ and define $\vec{f}^*_{ij}$ as the vector ($\sqrt{1-\varepsilon^2}$, 0, 0, ε). With $\vec{n} = (h, k, \ell, 0)$ in "ordinary" space it is easily verified that indeed

$$|\sin \alpha^*_{ij}| = \tfrac{1}{2} |\vec{n} - \vec{f}^*_{ij}| \, |\vec{n} + \vec{f}^*_{ij}| = \sqrt{\sin^2 \alpha_{ij} + \varepsilon^2 \cos^2 \alpha_{ij}} =$$

$$= \sqrt{k^2 + \ell^2 + \varepsilon^2 h^2} \tag{10.39}$$

(*Note:* Apart from the computational advantage of the a = 2 assembling exponent, this value is also the "natural" choice from a physical point of view. An argument for this statement is given in 11.1.4. Accordingly we will use the a = 2 exponent for smoothing.

Smoothed monopoles:

In analogy, we can extend the "monopole" function $2|\sin \tfrac{1}{2}\alpha_{ij}|$ to a smoothed function by writing it with a 4-dimensional α^*_{ij} :

$$2|\sin \tfrac{1}{2}\alpha^*_{ij}| = |\vec{n} - \vec{f}^*_{ij}| = \sqrt{k^2 + \ell^2 + (h^2 + 1 - 2h\sqrt{1-\varepsilon^2}\,)}$$

$$= \sqrt{2(1 - 2h\sqrt{1-\varepsilon^2}\,)} \tag{10.40}$$

10.4.2 Geometrical interpretation of $(2)\!\!\times$ *and* $(-2)\!\!\times$ *assembling*

As a consequence of eq. (10.37), most "smoothing" assemblings using a = 2, and "combination" assemblings using a = −2, can be represented as a geometrical construction. Two examples are given in figs. 10.3 and 10.4.

Fig.10.3 illustrates the function

$$(2)\!\!\times\, A|\sin\varphi|,\ (-)\varepsilon A|\sin\varphi|,\ \varepsilon\kappa\ \times\!\!\circ\ = \sqrt{(1-\varepsilon^2)A^2\sin^2\varphi+\varepsilon^2\kappa^2} \qquad (10.41)$$

Consider a cartesian (x,y,z) frame. A plane which contains the y axis makes an angle $\phi = \arcsin\varepsilon$ with the z axis. In this plane, a circle with radius A is drawn around the origin. A point P travels along this circle, its position being given by the angular variable φ. The projection of P on the z axis is T.

Furthermore, on the line defined as the intersection of the inclined plane with the (xz) plane, we draw a point Q at distance κ from the origin. The projection of Q onto the x axis is called U.

It is now easy to verify that the assembled function as defined above is equal to the length TU.

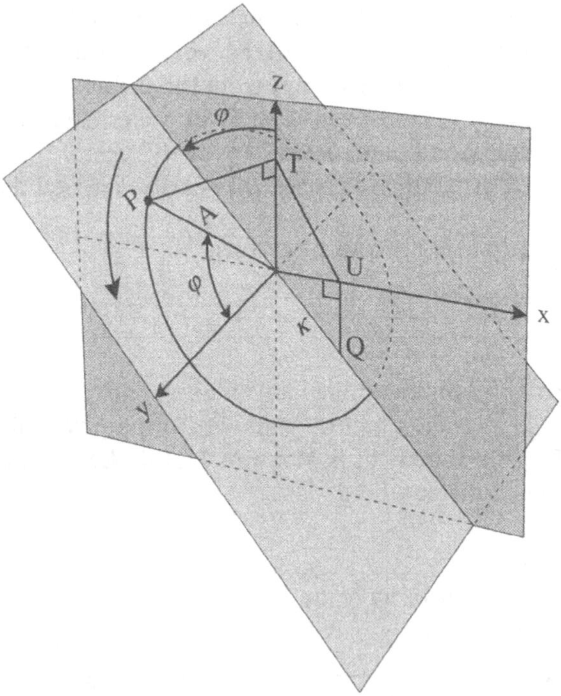

FIG. 10.3

Fig.10.4 illustrates the function

$$\ominus\!\!2\!\!\times\!\! A\,|\sin\varphi|,\, M \times\!\!\!\circ = \frac{M.A|\sin\varphi|}{\sqrt{M^2+A^2\sin^2\varphi}} \qquad (10.42)$$

Consider the cartesian frame in fig. 10.4. A circle with radius A is drawn around the origin in the (xy) plane; the point P travelling along it (position parameter φ) is projected onto the x axis (projection point Q).

In the (xz) plane, a circle with diameter M is drawn symmetrically with respect to the z axis and with poles in the origin and in $(z{=}M,\ x{=}0)$. The latter point is called M. The intersection of the latter circle with MQ is called T. The $\ominus\!2\!\times$ - assembled function defined above is identical with the length OT (O = origin).

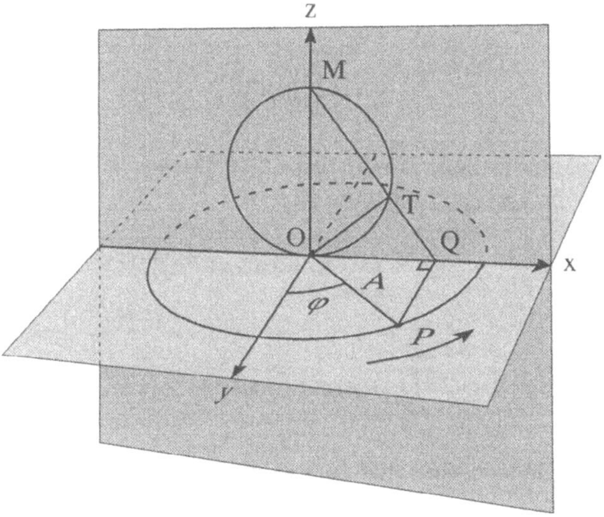

FIG. 10.4

10.4.3 Identities for regular 2D-distributions of identical facets

As we prefer to represent γ and R functions of symmetric crystals in as compact a shape as possible, the following identities are of interest.

a) *m- and 2n- polygons.*

We use the angular variable α as the surface-orientation variable measured from one of the facet orientations, in agreement with the habit introduced in 10.1.1. For a regular 2n- polygon (i.e., facet directions spaced by π/n) we have the two identities

$$\underset{k=0}{\overset{n-1}{(-2\!\times}}\; |\sin(\alpha+k\pi/n)| \times\!\circ = \frac{2^{n-1}}{n} \prod_{k=0}^{n-1} |\sin(\alpha+k\pi/n)| \tag{10.43}$$

$$= \frac{1}{n}|\sin n\alpha| \tag{10.44}$$

This equality can easily be extended to uneven-sided polygons, by starting from "monopole" functions

$$2\left|\sin\left(\frac{\alpha}{2}+\frac{k\pi}{m}\right)\right|$$

(m = uneven number of monopoles, equally distributed over full 2π range)

$$\underset{k=0}{\overset{m-1}{(-2\!\times}}\; 2\left|\sin\left(\frac{\alpha}{2}+\frac{k\pi}{m}\right)\right| \times\!\circ = \frac{1}{m} \prod_{k=0}^{m-1} 2\left|\sin\left(\frac{\alpha}{2}+\frac{k\pi}{m}\right)\right| \tag{10.45}$$

$$= \frac{1}{m}\left(2\left|\sin\frac{m\alpha}{2}\right|\right) \tag{10.46}$$

(the latter equations (10.45, 10.46) reduce to (10.43, 10.44) for even m by substituting $m = 2n$ and remembering that

$$\underset{k=0}{\overset{m-1}{(-2\!\times}}\; 2\left|\sin\left(\frac{\alpha}{2}+\frac{k\pi}{m}\right)\right| \times\!\circ = \underset{\ell=0}{\overset{n-1}{(-2\!\times}}\; \left|\sin\left(\alpha+\frac{\ell\pi}{n}\right)\right| \times\!\circ \tag{10.47}$$

and

$$\prod_{k=0}^{m-1} 2\left|\sin\left(\frac{\alpha}{2}+\frac{k\pi}{m}\right)\right| = 2^n \prod_{\ell=0}^{n-1} \left|\sin\left(\alpha+\frac{\ell\pi}{n}\right)\right| \tag{10.48}$$

b) It is of interest to notice that for assembling exponents different from -2 (but < -1),

$(\bar{a}\!\times \left|\sin\left(\alpha+\frac{k\pi}{n}\right)\right| \times\!\circ$ can *always* be written as $\prod_{k=0}^{n-1} \left|\sin\left(\alpha+\frac{k\pi}{n}\right)\right|$

multiplied by a "smooth function" with the same n-periodicity, which can be written as a sum of harmonics. Only for a $= -2$ this smooth function reduces to a constant. Nevertheless, for a $\neq -2$, a < -1, the "smooth function" has zero α-derivatives at the facet directions and does not seriously affect the behaviour of the function near the facet orientations. Therefore the Π or the $|\sin n\alpha|$ expressions can always be used, with adapted prefactors, instead of the $(\bar{a}\!\times \ldots \times\!\circ$ expression in those cases where only near-facet behaviour is relevant.

c) A remark similar to b) applies to facet distributions which are not arranged in a regular polygon and/or which have to be multiplied with non-identical prefactors. Here, too, a Π expression can be used, again multiplied by a "smooth function" to renormalize the individual functions to their original value in the vicinity of the zero-cusps. (the Π form has of course the same zeros as the $\textcircled{\bar{a}}\!\!\times \ldots \times\!\!\circ$ expression).

d) It is tempting to use the smoothed versions of the Π expressions (10.43, 10.45) and/or the $|\sin\frac{m\alpha}{2}|$ expressions instead of $\textcircled{-2}\!\!\times |\sin(\alpha^* + \frac{k\pi}{n})| \times\!\!\circ$, etc. Explicitly, these smoothed versions are written

$$\underset{k=0}{\overset{n-1}{\textcircled{-2}\!\!\times}} |\sin(\alpha^* + \tfrac{k\pi}{n})| \times\!\!\circ = \underset{k=0}{\overset{n-1}{\textcircled{-2}\!\!\times}} |\sin(\alpha + \tfrac{k\pi}{n})|, \varepsilon\, |\cos(\alpha + \tfrac{k\pi}{n})| \times\!\!\circ \qquad (10.49)$$

$$\prod_{k=0}^{n-1} |\sin(\alpha^* + \tfrac{k\pi}{n})| \times\!\!\circ = \prod_{k=0}^{n-1} \textcircled{2}\!\!\times |\sin(\alpha + \tfrac{k\pi}{n})|, \varepsilon\, |\cos(\alpha + \tfrac{k\pi}{n})| \times\!\!\circ$$

$$|\sin^* n\alpha| = \sqrt{\sin^2 n\alpha + \varepsilon^2 \cos^2 n\alpha} \qquad (10.50)$$

and similarly for the "monopole" version.

However, neither eq. (10.49) nor eq. (10.50) are appropriate representations for a family of identical facets which all behave like a single one with the same v and ε parameters. The physically more realistic representation of roughening for "regular-polygon" or "Platonic" families of facets will be derived in section 11.4.2 (see the "networks" of figs. 11.8).

e) For the case of regular periodic functions in a plane built up by assembling $|\sin\alpha|$ functions, it is just a trivial remark that the same identities apply to $|\cos\alpha|$ functions because they can be translated into $|\sin\alpha|$ by simple rotation. In 10.4.4 we will extend identities (10.43) and (10.44) to the 3-dim. case of regular polyhedra. Then the transition to cos functions cannot be made, as in 3 dimensions the $|\sin\alpha|$ function is not similar to the $|\cos\alpha|$ function.

10.4.4 Improving the fit to Min ($|\sin\alpha_i|$)
 (extension of eqs (10.28)-(10.29))

In the discussion of eqs. (10.28) and (10.29) we argued that a "multi-exposure" assembled function $\textcircled{\bar{a}}\!\!\times\, G, H, \ldots \times\!\!\circ$ can be pushed more tightly to the asymptotes Min($G,H,\ldots$) *by adding an extra "compensating" assembland* $(-)(n-1)^{1/a}M, M$ being a maximum common to all functions $G, H,\ldots$ We also anticipated that this extra assembland was only a worst-case choice, applicable to any *arbitrary* set of functions obeying $\mathrm{Max}(G,H,\ldots) = M;\ G,H,\ldots \geqslant 0$.

The *regular* distributions of similar $|\sin\, \alpha_{\mathrm{i}}|$ functions referred to in eqs (10.43 - 10.48) are *not* just an arbitrary set of functions. As a consequence of the symmetry of this case, the compensating assembland can be pushed further so as to improve the fit to Min ($|\sin\, \alpha_{\mathrm{i}}|$) above that for the "worst-case" $(-)(n-1)^{1/a}$ choice. We will present the calculation of this extreme assembland in some detail, the more because it provides an illustrative example of "assembling algebra".

a) Splitting up $\textcircled{-2}\!\!\times\, |\sin\alpha_1|, \ldots.,\, |\sin\alpha_n| \times\!\!\circ$ as

$$\textcircled{-2}\!\!\times\, |\sin\alpha_1|, \ldots.,\, |\sin\alpha_n| \times\!\!\circ\ =$$

$$\textcircled{-2}\!\!\times\, |\sin\alpha_{\mathrm{i}}|,\ \underset{k\neq \mathrm{i}}{\textcircled{-2}\!\!\times}\, |\sin\alpha_1|, \ldots, |\sin\alpha_k|, \ldots, |\sin\alpha_n| \times\!\!\circ \times\!\!\circ \tag{10.51}$$

we can see that the $|\sin\alpha_{\mathrm{i}}|$ assembland in the total (all i) assembled function is unnecessarily suppressed by the
$\underset{k\neq \mathrm{i}}{\textcircled{-2}\!\!\times}\, |\sin\alpha_1|, \ldots, |\sin\alpha_k|, \ldots, |\sin\alpha_n| \times\!\!\circ$ assembland, which is <1.

The requirement for the assembled function is only $F_{\mathrm{ass.}} \leqslant |\sin\, \alpha_{\mathrm{i}}|$. This requirement is still obeyed when we compensate the unnecessary suppression expressed by eq. (10.51); i.e. by adding an *inverted* assembland to the original assembling:

$$A_{\mathrm{comp}} = (-)\, \underset{\mathrm{all}\ \vec{n}}{\mathrm{Max}}\ \underset{k\neq \mathrm{i}}{\textcircled{-2}\!\!\times}\, |\sin\alpha_1|, \ldots, |\sin\alpha_k|, \ldots, |\sin\alpha_n| \times\!\!\circ \tag{10.52}$$

b) Now for the special case of a *regular* distribution of orientations α_{i} it is easily verified (just by symmetry considerations) that the maximum of $\textcircled{-2}\!\!\times |\sin\alpha_1|, \ldots, |\sin\alpha_k|, \ldots, |\sin\alpha_n| \times\!\!\circ$ is reached *at the cusp position of* α_i, i.e. $\alpha_k = (2\pi/n)(k-1)$. So we can write

$$-A_{\mathrm{comp}} = \underset{k\neq i}{\textcircled{-2}\!\!\times} \left|\sin\{\tfrac{2\pi}{n}(\mathrm{k}-\mathrm{i})\}\right| \times\!\!\circ\ =$$

$$= \lim_{\alpha\to 0}\ \textcircled{-2}\!\!\times\ \underset{all\ k}{\textcircled{-2}\!\!\times}\ \left|\sin\{\tfrac{2\pi}{n}(\mathrm{k}-\mathrm{i}) + \alpha\}\right| \times\!\!\circ, (-)\,|\sin\alpha| \times\!\!\circ =$$

$$= \lim_{\alpha\to 0}\ \textcircled{-2}\!\!\times\ \tfrac{1}{n}|\sin n\alpha|, (-)\,|\sin\alpha| \times\!\!\circ =$$

$$= \lim_{\alpha\to 0} \left(\frac{|\sin\alpha|\,.\,|\sin n\alpha|}{\sqrt{n^2\sin^2\alpha - \sin^2 n\alpha}} \right) \tag{10.53}$$

c) Expression (10.53) tends to $0/0$ as $\alpha \to 0$, but we can calculate its limit value by substituting the series expansions. Working out the algebra provides

$$-A_{\text{comp}} = \sqrt{\frac{3}{n^2-1}} \tag{10.54}$$

Altogether, combining eqs. (10.52) and (10.54), we obtain the expression for the "closest possible fit" for a $(-2)\!\!\times$ assembled regular 2D distribution of identical facets:

$$(-2)\!\!\times |\sin\alpha_1|, \ldots, |\sin\alpha_i|, \ldots, |\sin\alpha_n| \times\!\!\circ \leqslant$$

$$\leqslant (-2)\!\!\times \tfrac{1}{n}|\sin n\alpha|, (-)\frac{1}{\sqrt{n-1}} \times\!\!\circ \quad \text{(i.e., expression in (10.28)} \leqslant$$

$$\leqslant (-2)\!\!\times \tfrac{1}{n}|\sin n\alpha|, (-)\sqrt{\frac{3}{n^2-1}} \times\!\!\circ \leqslant \underset{\text{all i}}{\text{Min}} |\sin\alpha_{\text{i}}| \tag{10.55}$$

d) This result can be generalized to the uneven-sided regular polygon case. To this purpose, we just have to replace n by $m/2$ in (10.55):

$$(-2)\!\!\times |\sin\alpha_1|, \ldots, |\sin\alpha_i|, \ldots, |\sin\alpha_n| \times\!\!\circ \leqslant$$

$$\leqslant (-2)\!\!\times \tfrac{2}{m}|\sin\tfrac{m\alpha}{2}|, (-)\sqrt{\frac{12}{m^2-4}} \times\!\!\circ \leqslant \underset{\text{all i}}{\text{Min}} |\sin\alpha_{\text{i}}| \tag{10.56}$$

(m odd or even)

The first $\leqslant$ is an equality for $m = 2$, because the (−) term is ∞ and can be omitted. (negative assembling exponent).
The case $m = 1$ may require some comment because the compensating assembland is imaginary. However, in 10.2 we remarked that $(-2)\!\!\times \ldots, (-)k\mathrm{i} \times\!\!\circ$ can be written as $(-2)\!\!\times \ldots, k \times\!\!\circ$, so that for $m = 1$ the middle expression in (10.56) reduces to

$$(-2)\!\!\times\ 2\,|\sin\tfrac{1}{2}\alpha|, 2 \times\!\!\circ = \frac{2\,|\sin\frac{1}{2}\alpha|}{\sqrt{1+\sin^2\frac{1}{2}\alpha}} \tag{10.57}$$

(i.e., with maximum $\sqrt{2}$ for $\alpha = \pi$). In other words, the closest-fitting expression for the "monocusp" function is not simply $2\,|\sin\frac{1}{2}\alpha|$. We anticipated on this result in 10.2.

The assemblands $(-)\sqrt{3/(n^2-1)}$ and $(-)\sqrt{12/(m^2-4)}$ have the mathematical character of an "L type element" in the nonlinear network, in the sense to be explained in 11.4.3. In accordance, we may symbolize such an assembland as $(-)\,L^*$. In 11.4.3, "physical" surface-based processes tending to limitation of the growth rate are associated with assemblands which are labeled as L-type (Limiting) elements (and which in that case are positive). The occurrence of an *inverted* L^* assembland which is to be considered as part of the combination assembling step, implies that the "zero-point" of the *physical L* assembland is effectively shifted (see 11.4.3, eq. (11.30).

e) So far, the discussion on "closest-fit" approximations for regular polygons was restricted to the case a = −2 (assembling exponent). But of course, the principle can be generalized. Equation (10.28), as the variable-a version of eq. (10.29), already improves the fit for non-regular polygons by adding the assembland $(-)(n-1)^{-1/a}$.

By the same reasoning as followed in 10.4.4 a-e, we can find a tighter compensating assembland for regular polygons for *arbitrary* a values, as a generalisation of the $(-)\sqrt{12/(m^2-4)}$ assembland. (not worked out further here).

f) Another generalisation of the principle followed in 10.4.4 sub a-e can be made to the 3-D case of regular cusp distributions ("Platonic" solids and regular prisms, see section 10.4.5).

g) *Compensating assemblands for non-regular distributions:*
By now we have encountered the worst-case compensating assembland $(-)(m-1)^{-1/a}$ (for arbitrary distribution of $m\ |\sin\alpha_i|$ cusp positions) and the "maximum" compensating assembland $(-)\sqrt{12/(m^2-4)}$ (for perfectly regular distribution of these positions, and the special choice a =−2). Of course, for a nonregular but specified set of cusp positions the compensating assembland will be in-between these extremes. It can be written as

$$\operatorname*{Max}_{i}\ \operatorname*{Max}_{k\neq i}\ \textcircled{-2}\!\!\times\ |\sin\alpha_1|,\ldots,\ |\sin\alpha_k|\ ,\ldots,|\sin\alpha_n|\ \times\!\!\circ \tag{10.58}$$

(because the distribution is not symmetric, the second maximum is no longer in a cusp position and it depends on i. A similar approach can be followed when the $|\sin\alpha_i|$ terms have different prefactors v_i).

h) For a = −2, the worst-case compensating assembland $(-)(m-1)^{-1/a}$ can be combined with sin or cos OD functions in the alternative expression

$$\textcircled{-2}\!\!\times\ |\sin\alpha_1|,\ldots\ldots,|\sin\alpha_n|,(-)(m-1)^{-1/a}\ \times\!\!\circ\ =$$

$$=\ \textcircled{-2}\!\!\times\ |\tan\alpha_1|,\ldots\ldots,|\tan\alpha_n|,1\ \times\!\!\circ \tag{10.59}$$

(or the cos/cotan analog).

10.4.5 Three-dimensional regular distribution of identical facets

There is a limited number of 3-dimensional polyhedra with a perfect symmetry displaying identical facets: the “Platonic solids”. (regular tetrahedron, cube, octahedron, dodecahedron and icosahedron). The facet vectors $\vec{f}_i$ of these polyhedra can be represented in a cartesian frame and consequently the angles α_i and $\sin\alpha_i = \frac{1}{2}\,|\,\vec{n}-\vec{f}_i\,|\,|\,\vec{n}+\vec{f}_i\,|$ defined.

A 3-D equivalent for the 2-D expression $\sin n\alpha$ cannot be given, but the expressions $\textcircled{-2}\!\!\times\, |\sin\alpha_i|\, \times\!\!\circ$ and $\Pi\,|\sin\alpha_i|$ do have a 3-D meaning. (Below we write α_i instead of α_{ij} for the angle between $\vec{n}$ and the normal to the i-th facet of a “Platonic solid”)

In view of the identity given by eq. (10.43), between the $\textcircled{-2}\!\!\times$ and the Π expression (the latter with an appropriate normalizing factor F), we might expect a similar identity for the 3-D “Platonic” case, with F calculated as

$$F = \left[\ \prod_{i\neq k}\ \{\,|\sin\alpha_i|\,\}\right]^{-1} \tag{10.60}$$

so as to ensure that the $F\Pi$ expression approaches any individual $|\sin\alpha_i|$ term for $\vec{n}$ approaching its cusp position. (These F play the role of the factors $(2^{n-1})/n$ or $(2^m)/m$ in the 2D regular-polygon case).

However, it is easily verified that for this $F\Pi$ expression, an identity of the type of eq. (10.43) is not obeyed. Still we have the interesting *inequalities* between five comparands

$$\underset{i\in P}{\textcircled{-2}\!\!\times}\, |\sin\alpha_i|\, \times\!\!\circ \;<\; \underset{i\in P}{\textcircled{-2}\!\!\times}\, |\sin\alpha_i|\,,(-)\,1/\sqrt{n-1}\, \times\!\!\circ \;\leqslant\; F\prod_{i\in P}|\sin\alpha_i| \;\leqslant$$

$$\leqslant\; \underset{i\in P}{\textcircled{-2}\!\!\times}\; F\prod|\sin\alpha_i|\,,(-)L^*\, \times\!\!\circ \;\leqslant\; \underset{i\in P}{\mathrm{Min}}\,(|\sin\alpha_i|) \tag{10.61}$$

where P is the set of n pole axes of the Platonic solid under consideration. (for “half” axes, such as the tetrahedron, an adapted version applies: see table 6.2). The inequality between the third and fifth comparand in eq. (10.61) is based on the consideration that the product of all the “other” $|\sin\alpha_{i\neq k}|$ functions has a maximum at the position of the $|\sin\alpha_i|$ cusp. This is just a consequence of the symmetry of all cusps involved. Apart from the Platonic solids, also a regular “prismatic” distribution of facets (discussed in 12.4) obeys this condition and therefore obeys all inequalities of (10.61). The last inequality will be explained in connection with eq. (10.64).

Below we list the Π expressions, including the F factors, of the Platonic solids. Also the L^* assemblands are included in the list (these are discussed below). Both constants have been obtained by a numerical procedure, although in principle they might be represented as a trigonometric expression. The number n counts the facets.

TABLE 10.1

Prefactors and L assemblands for the Platonic 3D symmetries (a=2 case)*

tetrahedron ($n = 4$)	$L^* = 1.15476$ $F\Pi\|\sin\alpha_i\| = 3.674235 \prod_{i=1}^{4} \|\sin(\alpha_i/2)\|$	(10.62a)
cube ($n = 6$)	$L^* = 1.00000$ $F\Pi\|\sin\alpha_i\| = 1.000000 \prod_{i=1}^{3} \|\sin\alpha_i\|$	(10.62b)
octahedron ($n = 8$)	$L^* = 0.81646$ $F\Pi\|\sin\alpha_i\| = 1.193243 \prod_{i=1}^{4} \|\sin\alpha_i\|$	(10.62c)
dodecahedron ($n = 12$)	$L^* = 0.63245$ $F\Pi\|\sin\alpha_i\| = 1.746928 \prod_{i=1}^{6} \|\sin\alpha_i\|$	(10.62d)
icosahedron ($n = 20$)	$L^* = 0.47135$ $F\Pi\|\sin\alpha_i\| = 4.805420 \prod_{i=1}^{10} \|\sin\alpha_i\|$	(10.62e)
rhombic dodecahedron ($n = 12$)	$L^* = 0.72755$ $F\Pi\|\sin\alpha_i\| = 1.777777 \prod_{i=1}^{6} \|\sin\alpha_i\|$	(10.62f)

(The rhombic dodecahedron is not a Platonic solid in the exact sense but an "Archimedean" solid with almost-Platonic symmetry.)

The cartesian coordinates of the facet normals are listed in table 10.2. The $|\sin\alpha_i|$ can so be written, for any normal $\vec{n}$, in terms of distances to these facet normals.

TABLE 10.2: *Facet normals of regular polyhedra (not normalized)*

	normal directions	normalizing factor
tetrahedron	$(1\ 1\ 1)\ (1\ {-1}\ {-1})\ ({-1}\ {-1}\ 1)\ ({-1}\ 1\ {-1})$	$1/\sqrt{3} = 0.577350$
cube	$(\pm1\ 0\ 0)\ \ (0\ \pm1\ 0)\ \ (0\ 0\ \pm1)$	1
octahedron	$(\pm1\ \pm1\ \pm1)$	$1/\sqrt{3} = 0.577350$
dodecahedron	$(0\ \pm\tau\ \pm1)\ (\pm1\ 0\ \pm\tau)\ (\pm\tau\ \pm1\ 0)$	$1/\sqrt{1+\tau^2} = 0.525731$
(rhomb. dod.)	$(0\ \pm1\ \pm1)\ (\pm1\ 0\ \pm1)\ (\pm1\ \pm1\ 0)$	$1/\sqrt{2} = 0.707107$
icosahedron	$(\pm1\ \pm1\ \pm1)$ $(0\ \pm1/\tau\ \pm\tau)\ (\pm\tau\ \ 0\ \pm1/\tau)\ (\pm1/\tau\ \pm\tau\ 0)$	$1/\sqrt{3} = 0.577350$

with $\tau = (\sqrt{5}+1)/2 = 1.618034\ldots$ = golden section ratio.

Finally, we list the explicit $F\Pi\ |\sin\alpha_i|$ functions in terms of (h, k, ℓ) *normalized* $\vec{n}$ vectors $(h^2 + k^2 + \ell^2 = 1)$:

TABLE 10.3

tetrahedron:

$$F\Pi\,|\sin(\alpha_i/2)| = 3.674235.\ \tfrac{1}{4}\sqrt{\begin{array}{l}\{1-\tfrac{1}{3}\sqrt{3}\,(h+k+\ell)\}\,\{1-\tfrac{1}{3}\sqrt{3}\,(h-k-\ell)\}\\ \{1-\tfrac{1}{3}\sqrt{3}\,(-h+k-\ell)\}\{1-\tfrac{1}{3}\sqrt{3}\,(-h-k+\ell)\}\end{array}}$$

cube:

$$F\Pi\,|\sin\alpha_i| = \sqrt{(1-h^2)(1-k^2)(1-\ell^2)}$$

octahedron:

$$F\Pi\,|\sin\alpha_i| = 1.193243\sqrt{\begin{array}{l}\{1-\tfrac{1}{3}\,(h+k+\ell)^2\}\,\{1-\tfrac{1}{3}\,(h-k-\ell)^2\}\\ \{1-\tfrac{1}{3}\,(-h+k-\ell)^2\}\,\{1-\tfrac{1}{3}\,(-h-k+\ell)^2\}\end{array}}$$

dodecahedron:

$$F\Pi\,|\sin\alpha_i| = 1.746928\sqrt{\begin{array}{l}\{1-\tfrac{1}{1+\tau^2}\,(\ell+\tau k)^2\}\,\{1-\tfrac{1}{1+\tau^2}\,(h+\tau\,\ell)^2\}\\ \{1-\tfrac{1}{1+\tau^2}\,(k+\tau h)^2\}\,\{1-\tfrac{1}{1+\tau^2}\,(-\ell+\tau k)^2\}\\ \{1-\tfrac{1}{1+\tau^2}\,(h-\tau\ell)^2\}\,\{1-\tfrac{1}{1+\tau^2}\,(-k+\tau h)^2\}\end{array}}$$

icosahedron:

$$F\Pi\,|\sin\alpha_i| = 4.805421\sqrt{\begin{array}{l}\{1-\tfrac{1}{3}\,(h+k+\ell)^2\}\,\{1-\tfrac{1}{3}\,(h-k-\ell)^2\}\\ \{1-\tfrac{1}{3}\,(-h+k-\ell)^2\}\,\{1-\tfrac{1}{3}\,(-h-k+\ell)^2\}\\ \{1-\tfrac{1}{3}(\tfrac{h}{\tau}+\tau k)^2\}\,\{1-\tfrac{1}{3}(\tfrac{k}{\tau}+\tau\,\ell)^2\}\\ \{1-\tfrac{1}{3}(\tau h+\tfrac{\ell}{\tau})^2\}\,\{1-\tfrac{1}{3}(\tfrac{h}{\tau}-\tau\,k)^2\}\\ \{1-\tfrac{1}{3}(\tfrac{k}{\tau}-\tau\ell)^2\}\,\{1-\tfrac{1}{3}(\tau h-\tfrac{\ell}{\tau})^2\}\end{array}}$$

rh. dodecahedron:

$$F\Pi\,|\sin\alpha_i| = 1.777777\sqrt{\begin{array}{l}\{1-\tfrac{1}{2}\,(h+k)^2\}\,\{1-\tfrac{1}{2}\,(k+\ell)^2\}\\ \{1-\tfrac{1}{2}\,(h+\ell)^2\}\,\{1-\tfrac{1}{2}\,(h-k)^2\}\\ \{1-\tfrac{1}{2}\,(k-\ell)^2\}\,\{1-\tfrac{1}{2}\,(h-\ell)^2\}\end{array}}$$

Closest fit to Min ($|\sin\alpha_i|$) *by a compensating (−)L* assembland:*

We can construct a closer-fitting function than $F\Pi\,|\sin\alpha_i|$ by exactly the same procedure as followed for the 2D regular polygon case (10.4.4). In analogy with eq. (10.53), L^* is obtained as a limit:

$$L^* = \lim_{\alpha_i \to 0} \frac{|\sin\alpha_i| \,.\, F \prod_{k\neq i} |\sin\alpha_k|}{\sqrt{1 - F^2 \big(\prod_{k\neq i} |\sin\alpha_k|\big)^2}} \tag{10.63}$$

The "close-fitting" function involving this assembland is

$$(\!-2\!)\!\!\times\; F \prod |\sin\alpha_i| \,,\, (-)L^* \;\times\!\!\circ \tag{10.64}$$

Its position in the sequence of "inequation" (10.61) follows immediately from its derivation.

The regular polygons and polyhedra can be used as a basis for simple expressions for "prototype functions". Although the regular dodecahedron and icosahedron may seem rather exceptional models for real crystals, their high periodicity may be suitable for modelling purposes, e.g., for dendritic growth.

The smoothed versions of the Π expression will be used to construct symmetric "prototype" functions in section 12.4.

10.4.6 Three-dimensional prismatic distributions of facets based on two-dim. regular polygon distributions

The 2-D regular-polygon functions defined by eqs (10.43 through10.46) can be "embedded" in 3-D space so as to represent a regular-prism family of facets. However, the single-OD angular variable α or $n\alpha$ cannot be used for the case of facets with 2-D OID ledge behaviour. Instead we have to start from the (regularly distributed) cusp positions, which we choose to locate on the h-equator: $\varphi_j = 2\pi j/m$; m = number of cusps; j (= 1 to m) is the cusp identification number:

$$\begin{aligned} h_j &= 0 \\ k_j &= \sin\!\left(\tfrac{\alpha+\varphi_j}{2}\right) = \sin\!\left(\tfrac{\alpha}{2} + \tfrac{j\pi}{m}\right) \\ \ell_j &= \cos\!\left(\tfrac{\alpha+\varphi_j}{2}\right) = \cos\!\left(\tfrac{\alpha}{2} + \tfrac{j\pi}{m}\right) \end{aligned} \tag{10.65}$$

With these cusp directions we can construct either the $(\!-2\!)\!\!\times |\sin\alpha_i| \times\!\!\circ$ or the $(2/m)\Pi\,\{|\sin\alpha_i|\}$ expressions using $\{|\sin\alpha_i|\}^2 = 1 - |\vec{n}\cdot\vec{f}_i|^2 = \frac{1}{2}\,|\vec{n} - \vec{f}_i|\,|\vec{n} + \vec{f}_i|$. As remarked in 10.4.4, also the Π expression satisfies inequality (10.61) for this case). The factor $F = 2/m$ is in this case identical with the 2-D factor of eq. (10.45) because the 2-D function is to be copied in the h-equator plane. For the same reason, the L^* assemblands (to be used for closest-fitting assembled functions) are identical with those of the 2-D "parent case".

(In 11.4.2 and 11.4.3 we will explain how to add a pair of facets in the "c-axis" direction of such a prism or in other directions).

A different expression based on the $(1/n)\,|\sin(n\alpha)|$ function can be defined in a polar coordinate system (α, β): The 3-D function

$$\frac{2}{m}\left|\sin\frac{m\alpha}{2}\right| + \tfrac{1}{2}\left|\sin 2\beta\right| \tag{10.66}$$

has two "circular" facets at both c-axis poles and m "rectangular" facets at the equator. Alternatively, the 3-D function

$$\frac{2}{m}\left|\sin\frac{m\alpha}{2}\right| + \left|\cos\beta\right| \tag{10.67}$$

has no facets in the c-axis direction and m "rectangular" facets at the equator. Again, smoothing of these facets will be discussed in 11.4.2.

10.5 SUCCESSIVE ASSEMBLING STEPS

In 10.3 we have discussed how the assembling operation can be used for quite different purposes in the construction of γ and R functions, each purpose requiring a specific $\mathbf{D}$ operator or assembling exponent a. In general, we have to fulfill more than one of these purposes simultaneously: in other words, we have to apply a *succession* of assembling operations.

It is therefore appropriate to develop, as a last tool in the mathematical toolbox presented in this paragraph, an economical notation for functions which can be defined in terms of such an iteration of assemblings. In the following we propose several possible alternatives for the notation of such iterated assemblings:

1) In the form of a table, displaying all assemblings and parameters involved.
2) The assemblings in the table can formally be expressed in the form of an explicit formula, quite analogous to an expression with nested summations or integrations.
3) In addition to these formal representations, we can also draw a more pictural representation of the same iterative procedure in the form of

an "assembling tree".

4) We can represent the physical situation in a suggestive way by drawing the assembling tree in a different topological form, so that it assumes the shape of a *nonlinear network*. The advantage of the network representation is that the "physical" rôle of any assembland can be recognized more easily, so that it can naturally be used for an exhaustive and systematic exploration of all possibilities which are inherent in the sequenced-assembling approach as such.

All four representations are fully equivalent. Still, in the discussions on the physical aspects of sequenced assemblings to be given in chs. 11 and 12, we prefer to use the network representation for the reasons given above.

Below we work out the four alternatives on the basis of fig. 10.5a-d, where one and the same sequenced assembling is represented in the four equivalent modes. Although the example in fig. 10.5 is quite arbitrary and not related with R or γ functions in special, it has been chosen to show a number of complications which also occur in the R and γ functions considered later on.

Ad 1) Tabular and formula representation

A single exponential assembling is fully defined by the exponent a, the list of the functions and/or constants to be assembled, and the "range" or values of the assembling index over which to assemble. These can be summarized in a row of a table: e.g. if ${}^{(s)}\mathcal{A}$ is defined by

$${}^{(s)}\mathcal{A} = \textcircled{a(s)}\!\!\times\; c_i \cdot {}^{(s-1)}\mathcal{A}_i\,,\, M \rtimes\!\!\circ \quad \text{, by}$$

	exponent	assemblands	assembling indices or range
${}^{(s)}\mathcal{A}$	a(s)	$c_i \cdot {}^{(s-1)}\mathcal{A}_i\,,\, M$	$i \in$ set $\{k\}$

Here s is just a number indicating a position in a *sequence* of assemblings. A table consisting of more of such rows represents the whole sequenced procedure (fig. 10.5a, table 10.4). The assemblands have either to be specified separately or in the form of an example function with an index, together with a range for this index over which to assemble.

Ad 2) The formula representation (presented as "fig. 10.5b", because it just serves as an illustration) is the logical "contracted" form of the tabular representation; in fact it is analogous with a formula with nested summations or integrations. Note that the use of the assembling "closing symbol" $\rtimes\!\!\circ$ is essential to prevent ambiguity in the order of the assembling procedures.

Table 10.4

Tabular representation of successive assembling steps

assembling	exponent	assemblands	ass. range
${}^{(1)}\mathcal{A}$	a_1 (<0)	P_i	$i = 1 \ldots. k$
${}^{(2)}\mathcal{A}$	a_2 (>0)	Q_i	$i = 1 \ldots. l$
${}^{(3)}\mathcal{A}$	a_3 (<0)	${}^{(1)}\mathcal{A}$, ${}^{(2)}\mathcal{A}$	- - -
${}^{(4)}\mathcal{A}_i$	a_4 (>0)	R_{ij}	$j = 1 \ldots. m$
${}^{(5)}\mathcal{A}$	a_5 (<0)	${}^{(4)}\mathcal{A}_i$	$i = 1 \ldots. n$
${}^{(6)}\mathcal{A}$	a_6 (>0)	${}^{(3)}\mathcal{A}$, ${}^{(5)}\mathcal{A}$, S	- - -
${}^{(7)}\mathcal{A}$	a_7 (<0)	T_1, T_2, T_3	- - -
${}^{(8)}\mathcal{A}$	a_8 (>0)	${}^{(6)}\mathcal{A}$, ${}^{(7)}\mathcal{A}$	- - -

Formula representation:

$$\textcircled{a_8}\!\times\ \textcircled{a_6}\!\times\ \textcircled{a_3}\!\times\ \overset{k}{\underset{i=1}{\textcircled{a_1}\!\times}}\ P_i \succ\!\circ,\ \overset{l}{\underset{i=1}{\textcircled{a_2}\!\times}}\ Q_i \succ\!\circ \succ\!\circ,$$

$$\overset{n}{\underset{i=1}{\textcircled{a_5}\!\times}}\ \overset{m}{\underset{j=1}{\textcircled{a_4}\!\times}}\ R_{ij} \succ\!\circ \succ\!\circ,\ S \succ\!\circ,\ \textcircled{a_7}\!\times\ T_1, T_2, T_3 \succ\!\circ \succ\!\circ$$

FIG. 10.5 a and b

Ad 3) "Assembling tree" representation

A more visual reference (of sequence as well of assembling exponents) is given by the assembling-tree representation of fig. 10.5c. The relation with the tabular/formula representation is defined by the following keys:

1) The evolution of all subfunctions, "on their way" from elementary starting function towards participation in the end function (by successive amendments and merging with other amended subfunctions by assembling procedures), is represented by a branched-root like structure of *horizontal*

("path") lines. The elementary starting functions and the end function are represented in rectangles at the start and the finish of these path lines.

2) An assembling procedure is represented by a *vertical line* connected with the path line. The assembling exponent can be indicated in a circle at the end of this vertical line. As a means to distinguish positive-exponent and negative-exponent assembling procedures, these vertical lines can be drawn in opposite directions.

For indexed assemblands we can use an indication by dots (see P_i, Q_i and $R_{\ell i}$ in fig. 10.5c).

Assemblands can either be constants (and then they can be shown in rectangles connected to the corresponding vertical "assembling" lines) or functions which are themselves the result of an assembling procedure. In the latter case we have to do with more starting functions, each giving birth to a "root line" which at some stage merges with other root lines by a ("combination") assembling procedure. In such a case the different root lines, with all their attached assembling procedures, are in principle equivalent; the difference between their starting functions is symbolized by the index. Therefore, it is sufficient to draw only one of them in extenso so that its fellow assemblands can be considered to be copied from it up to the merge point. *Such copies are represented by (horizontal) dashed lines in an assembling tree*. So, each dashed line attached to a vertical "combination assembling" line symbolizes an assembled function which is an exact copy of the fully represented assembled function up to the merge point, apart from a variation symbolized by the assembling index. In other words, each combination assembling implying another bunch of dotted lines increases the "dimension" of the assembling procedure by one.
(Example: In fig.10.5 the assemblands of ${}^{(5)}\mathcal{A}$ are the ${}^{(4)}\mathcal{A}_i = \left(a(4)\right)\!\!\bigtimes_{j=1}^{m} R_{ij} \!\times\!\!\circ$.
In the tree, only ${}^{(4)}\mathcal{A}_1$ is drawn fully, whereas the other ${}^{(4)}\mathcal{A}_i$ are represented as dashed lines in ${}^{(5)}\mathcal{A}$).

In the context of our R and γ functions, indexed assemblings will be used in the tiered construction procedure of "single-family" and "multi-family" functions (see 11.4). In single-family functions, the index is related with a set of crystallographically similar facet orientations, in multi-family functions with the complete set of "facet families" (figs. 11.8 g/G and h/H)

Ad 4) Network representation of exponential assembling sequences

An even more illustrative representation of a series of assemblings can be made in the form of a symbolic "network" of conductive pathways. This is suggested by the analogy with a real network of linear resistances where

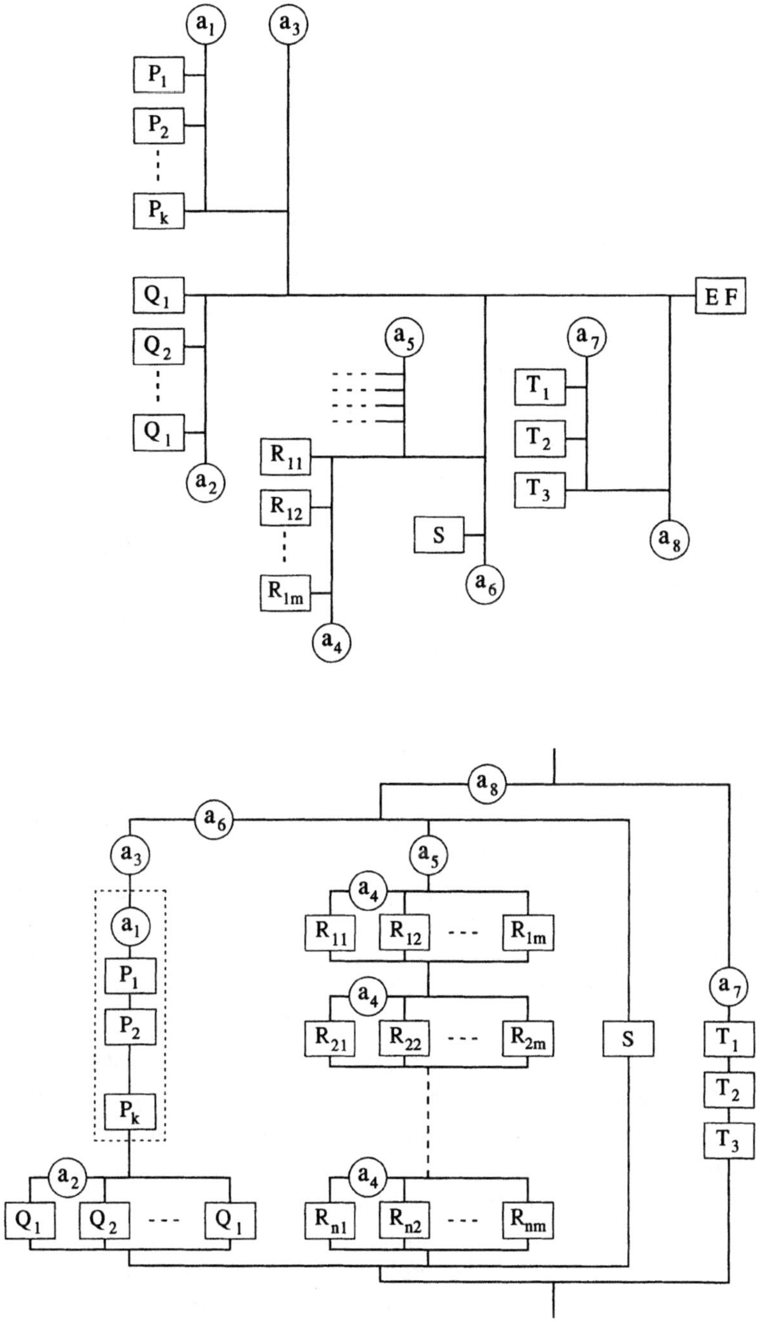

FIG. 10.5 c and d: Assembling tree (above) and network (below) representations of the example.

the effect of two or more resistances R_1, R_2,... can be written, in terms of their conductance $C_1 = R_1^{-1}$, $C_2 = R_2^{-1}$, . . . :

for resistances *in series* : $C_{\text{eff}} = (\!-1\!)\!\!\times\; C_1, C_2, \ldots \times\!\!\circ$

for resistances *in parallel* : $C_{\text{eff}} = (\!+1\!)\!\!\times\; C_1, C_2, \ldots \times\!\!\circ$

The analogy to be drawn for our pictural representation of exponential assembling sequences is that *all assemblings with positive exponent* are represented as a parallel arrangement of the assemblands, and *all assemblings with negative exponent* as a series arrangement.

We further introduce the habit to represent parallel arrangements as a horizontal row of elements, series arrangements as a vertical column. Notice, however, that for assembling exponents different from + and −1, the Kirchoff relations in such a network do not apply.

Our pictural representation just expresses that for the parallel arrangement, the effective flux $\geqslant$ maximum of individual fluxes; and for the series arrangement: effective flux $\leqslant$ minimum of the individual fluxes (all fluxes supposed $\geqslant 0$). In other words, the difference in physical character of competitive versus collaborative processes is correctly symbolized in the topology of such a network. Therefore this representation is suitable to obtain an intuitive understanding of the complex mechanism represented by a series of assemblings.

The value of the exponent a can be indicated in the network picture as shown in fig. 10.5d: positive a values in horizontal connections, negative a values in vertical ones.

(of course, the degree of freedom which is implicit in the assembling principle is much larger than can be represented in *only the structure* of a network: therefore the explicit rendering of the exponents is essential for a truly complete representation. Still we will omit the explicit rendering of the exponents in the networks to be considered in chs. 11 and 12, because all exponents are standardized to +2 and −2 so that no confusion can arise).

The key relation between the network and the tabular/assembling tree representation of an assembling sequence is that any closed loop in the network has to be written out as an assembling or a succession of assemblings before it can be integrated in a more extensive loop. This principle ensures a one-to-one relation between an assembling sequence as represented in tabular form and its "network". In connection with this, we have to remark that an essential requirement in the representation of successive assemblings with *different* negative exponents in a network (such as a_1 and a_3) is to distinguish the assemblands of both assemblings by the use of an enclosing box such as in $\overset{k}{\underset{i=1}{(a(1)\!\!\times}}\; P_i \times\!\!\circ$ in fig. 10.5d.

General remarks related with sequenced assemblings:

1) Assemblands can either be specified separately (such as in ${}^{(7)}\mathcal{A}$ in fig. 10.5a-d), or by one or more indices (such as ${}^{(1)}\mathcal{A}$ and ${}^{(2)}\mathcal{A}$ with one, and ${}^{(4)}\mathcal{A}$ with two indices). In the latter case, the index range over which to assemble is to be written in the assembling "opening symbol", in analogy with the convention in Σ and Π notations (fig. 10.5b). In trees and networks such indexed assemblands can be symbolized by a row of dots.

2) Assemblands can be assembled functions themselves, such as in ${}^{(3)}\mathcal{A}$, ${}^{(5)}\mathcal{A}$, ${}^{(6)}\mathcal{A}$ and ${}^{(8)}\mathcal{A}$. In those cases, the succession in which the assemblings have to be carried out is obligatory, i.e. insofar that ${}^{(1)}\mathcal{A}$ and ${}^{(2)}\mathcal{A}$ have to be carried out before ${}^{(3)}\mathcal{A}$, and all $j = 1 \dots m$ ${}^{(4)}\mathcal{A}$ before ${}^{(5)}\mathcal{A}$. In contrast, for "equi-levelled" assemblings such as ${}^{(1)}\mathcal{A}$ and ${}^{(2)}\mathcal{A}$ the order is irrelevant. In a complex succession of assemblings, we can indicate obligatory successions by bracket-and- arrow symbols $(\dots) \Rightarrow$, the bracket pairs embracing the assembling(s) to be worked out before the one following the arrow. The whole series of assemblings represented in fig.10.5 is then "topographically" characterized by

$$((({}^{(1)}\mathcal{A} , {}^{(2)}\mathcal{A}) \Rightarrow {}^{(3)}\mathcal{A} , ({}^{(4)}\mathcal{A}_{1\text{-}n}) \Rightarrow {}^{(5)}\mathcal{A}) \Rightarrow {}^{(6)}\mathcal{A} , {}^{(7)}\mathcal{A}) \Rightarrow {}^{(8)}\mathcal{A}$$

In connection with this, it makes sense to define the "dimension" of a sequenced assembling as the total number of bracket pairs required in such a topological structure, or in other words as the number of indices required to characterize the assemblands if represented in indexed form. Because (in a practical assembling procedure for R and γ functions) each assembling involves its own parameter, the dimension of the sequenced assembling is also the dimension of the parameter space.

3) Successive assembling steps with the same exponent cannot be distinguished from a single one with all assemblands combined (eq. (10.19)). When, in such a case, the set of assemblands represent more than one "physical" mechanism, it can be functional to represent two separate assemblings (one per mechanism, even if the assembling exponents are the same). In the first place this provides a visual means to distinguish between the different mechanisms at a glance; in the second place, we can modify the assembling exponent for the mechanisms separately in a process of parameter fitting.

An example can be found in the merging of several facet functions into a single-family function or of several single-family functions into a multi-family function (see fig. 11.8 g/G and h/H). In both cases we have to combine a number ($\geqslant 2$) of similar *functions* and an (optional) *limiting constant*.

Chapter 11

PHYSICAL ASPECTS OF ASSEMBLING OPERATIONS

Before I came here I was confused about this subject. Having listened to your lecture I am still confused. But on a higher level.

Enrico Fermi (1901-1954)

11.1 SMOOTHING OF CUSPS (DOWNWARD LIMITATION OF FUNCTIONS)

11.1.1 Cusp-smoothing as a consequence of interaction between different mechanisms.

Consider a "broken-PBC model" term

$$F_1 = (\text{const.}).\left|\vec{\boldsymbol{n}} \cdot \vec{\boldsymbol{p}}\right| \tag{11.1}$$

measuring the OD contribution of some PBC-related quantity F_1 (e.g., surface energy or growth rate).

Suppose we have a second contribution F_2 which is, for the time being, independent of orientation ($F_2 = b$). If there is no physical interaction between the mechanisms leading to F_1 and F_2, the result is obtained by mere addition (in "assembling" terms: $\mathbf{D}F = F$), so that

$$F_{\text{ass.}} = (\text{const.}).\left|\vec{\boldsymbol{n}} \cdot \vec{\boldsymbol{p}}\right| + b \tag{11.2}$$

i.e., the cusp at the minimum is maintained, the minimum itself is shifted from 0 to b.

However, there are many instances where a physical interaction does exist. Of particular relevance to growth-rate functions, for instance, is the interaction of ledge growth and the "birth-and-spread" (B&S) mechanism.

The latter generates ledges continuously by nucleation of 2-dimensional islands on the flat terraces in-between ledges; these nuclei grow out and so act as a source of new ledges (HILLIG 1966). For an originally perfectly-planar surface in the facet orientation (i.e., no ledges), the B&S mechanism will lead, after some time, to a (statistically) stationary situation with a certain average number of ledges, of all orientations, per cm^2. This statistical presence of ledges on a surface which is *macroscopically* oriented in a facet direction, i.e. a direction where no "orientation ledges" are present at all, can be considered as a *statistical vagueness* of this orientation itself. The macroscopic orientation, from this point of view, is just an average of a distribution of orientations on a smaller scale. We will refer to this distribution in 11.1.3 and show that this statistical vagueness corresponds to a smoothing of cusps of the original "delta" distribution of orientations.

When the macroscopic orientation is gradually tilted away from the facet direction, we introduce "orientation ledges" which also contribute to growth and thus interact with the advancing boundaries of nucleated B&S islands. The important point is that, up to a certain density of orientation ledges, the contribution to the growth rate by the orientation ledges is just compensated by the loss of boundary length of the nucleated islands. In other words, the derivative with respect to orientation becomes zero, and the cusp is removed from the growth-rate function.

This is illustrated by the 1-dim. model of fig. 11.1a. Here, a fresh B&S nucleus is sketched as a shaded island on a terrace amidst a train of orientation ledges. Remark that *independence* of the two mechanisms (i.e. a=1 assembling, eq. (11.2)) would imply that the left ($\ulcorner$) ledge of the B&S island would not simply *collide* with the advancing $\urcorner$ orientation ledge and so be annihilated, but rather that the two would *climb onto each other* and proceed further. This is obviously not realistic: rather we have to do with a kind of equilibrium statistics of $\ulcorner$ and $\urcorner$ ledges.

An exactly facet-oriented surface with B&S growth kinetics shows statistically equal densities of $\ulcorner$ and $\urcorner$ ledges or kinks: $n_{\ulcorner} = n_{\urcorner} = K$. Nucleation statistics maintains the product $(n_{\ulcorner} \times n_{\urcorner})$ at a fixed level K^2 (proof for this statement requires rather "reasonable" assumptions about birth statistics and some paperwork to evaluate the ($\ulcorner$ - $\urcorner$) annihilation - we omit it because of the limited value of 1-dim. models for this case). Tilting of the surface over an angle φ, away from the exact facet orientation, implies a statistical surplus

$$\Delta n = n_{\ulcorner} - n_{\urcorner} = \frac{1}{h} \tan \varphi \tag{11.3}$$

(h = ledge height). Now $n_{\ulcorner}$ and $n_{\urcorner}$ can be solved as a function of φ from eq. (11.3) by substituting $n_{\ulcorner} \times n_{\urcorner} = K^2$:

$$n_{\lrcorner,\llcorner} = \sqrt{K^2 + \varphi^2/2h^2 \pm \varphi/2h \sqrt{\varphi^2/h^2 + 4K^2}} \qquad (11.4)$$

to first order in φ. This can be worked out to find $R(\varphi)$:

$$R(\varphi) = v_{/\!/}\, h\ (n_{\lrcorner} + n_{\llcorner}) = v_{/\!/}\, h \sqrt{4K^2 + \varphi^2/h^2} =$$

$$= \textcircled{2}\!\!\times R_{or}, R_{B\&S} \times\!\!\circ \qquad (11.5)$$

where $R_{or.}$ is the pure "orientation-induced" growth rate corresponding with the orientation ledges alone, and $R_{B\&S}$ the nucleation-induced growth rate. (the ledge velocity $\vec{v}_{/\!/}$ is assumed to be independent of distance to other ledges).

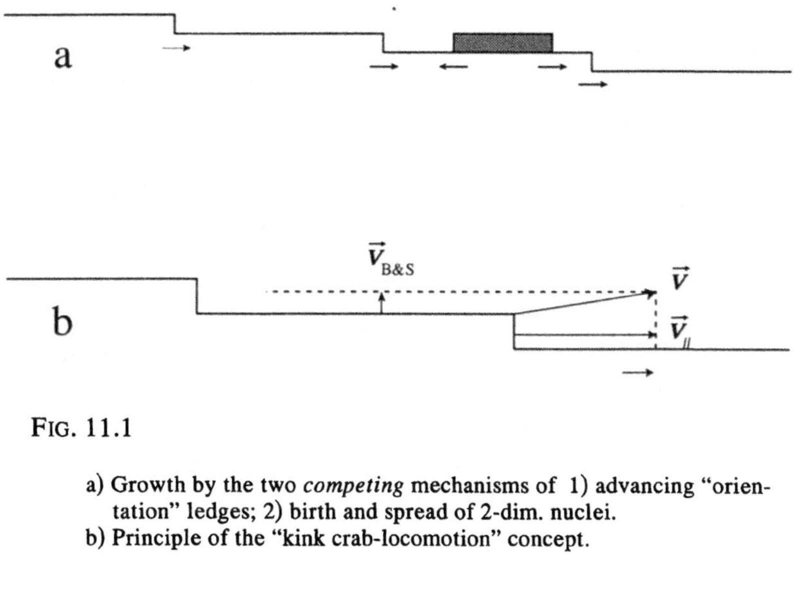

FIG. 11.1

a) Growth by the two *competing* mechanisms of 1) advancing "orientation" ledges; 2) birth and spread of 2-dim. nuclei.
b) Principle of the "kink crab-locomotion" concept.

This simple 1-dim. model illustrates that for an increasing density of orientation ledges, the B&S mechanism is annihilated (new nuclei have to start on the terraces in-between the ledges and have less time to grow out before the next ledge swallows them). This means that the F_1 function is an asymptote for the $F_{ass.}$ function.

Equation (11.5) can also be considered as an expression of the following "kink crab-motion" model. Basically, this is related with the fact that assembling with exponent a = 2 has the geometric significance of a vector length (eq. (10.37)).

In the classical kink motion model, a train of kinks or ledges (fig.11.1b) moves to the right; i.e. the vertical sections of the zigzag move (velocity $\vec{v}_{/\!/}$) whereas the horizontal ones remain stationary (until the next kink passes). In the crab-motion model equivalent with eq. (11.5), also the *horizontal* sections move *upwards* with the velocity $v_{\mathrm{B\&S}}$, so that the whole structure moves with a vector velocity $\vec{v}_{\mathrm{res}}$ obtained by adding both velocities as vectors. Obviously, both velocities are to be considered as *statistical* velocities, i.e. averaged over adsorption and desorption of atomic units and over nucleated monatomic-height islands on the terraces and at the kinks, respectively.

A theoretical derivation of the $R(\vec{n})$ function which takes exactly account of this interaction in the 2-dim. case is quite intricate and will not further be attempted here. The physics is complicated by the fact that for positive line energy, the birth of monatomic-height islands is a nucleated process involving the curvature of the ledges. This subject is typically related with atomistic models and thus falls beyond the scope of this monograph. Still it seems quite likely that the same assembling exponent a = 2 will be found to apply to the 2-dim. case. In terms of the kink crab-motion concept this would mean that in the $R(\vec{n}) = |\vec{n} \cdot \vec{v}_{\mathrm{res}}|$ expression, the "more complex physics" of the 3D- as compared with the 2D-case can just be expressed as a different (3D) value of the $v_{\mathrm{B\&S}}$ parameter. If so, eq. (11.5) with a = 2 is also valid in the 2D case.

(If not, we can still say that the mechanisms for "orientation-induced" and for "B&S-induced" growth are typically interactive, and that the interaction leads to elimination of the cusp from F_1 and to domination of the B&S contribution in a certain orientation range around the cusp orientation of F_1. All these features can adequately be represented by the choice DF = F^{a} with a > 1, $F_1 = v_{/\!/}\,|\vec{n} \cdot \vec{p}|$ and $F_2 = v_{\mathrm{B\&S}}$, i.e. independent of $\vec{n}$:

$$F_{\mathrm{ass}} = \textcircled{\scriptsize a>1}\!\!\times\; v_{/\!/}\,|\vec{n} \cdot \vec{p}|,\, v_{\mathrm{B\&S}} \;\times\!\!\circ \tag{11.6}$$

This quite general smoothing operation (which can also be applied to the surface-energy function, for the entropic contribution of ledges) has a complementary explanation in terms of orientation distribution functions, to be discussed in 11.1.2 and 11.1.3. The link is made by writing $v_{\mathrm{B\&S}}$ as $\varepsilon\, v_{/\!/}$, where the "roughening parameter" ε represents the average of $\sin|\phi^*|$ (ϕ^* = "microscopic" slope corresponding with roughening).

11.1.2 Cusp-smoothing considered as the effect of a statistical spread in PBC orientations

In this and in the next section, cusp smoothing is considered as the effect of a broadened distribution function. In the present section this distribution function regards the *PBC directions,* in 11.1.3 it regards the *surface orientations.* The difference, together with the whole concept of statistical broadening leading to cusp-smoothing, is illustrated by fig. 11.2 a, b and c.

Fig. 11.2a represents the straightforward example of a crystalline bulk material with unique PBC orientation $\vec{p}$ and a smooth (spherical) surface. The "PBC-counting function" $C(\vec{n})$ is

$$C(\vec{n}) = \frac{1}{a} \left| \vec{n} \cdot \vec{p} \right| \tag{10.3}$$

(if it is required to count head and tail ends separately, we can use the functions (10.4a) and (10.4b).

In fig. 11.2b, the complication is that the PBC orientations no longer have one fixed orientation but obey a distribution function centered around $\vec{p}$ - i.e., the spread occurs in the $\vec{p}$ of the $|\vec{n} \cdot \vec{p}|$ product. (for the moment we are not primarily interested in the physical meaning of such a spread , but rather in the mathematical phenomenology of the distinction between the b and c cases. Although PBC directions in a crystallographic sense are very sharply defined, a non-peaked distribution might be used, for instance, to describe eventual contributions by the free electron gas in metals).

In fig. 11.2c, the PBC orientations are fixed again but now the crystal *surface* is undulated; i.e. the statistical spread now occurs in the $\vec{n}$ of the $|\vec{n} \cdot \vec{p}|$ product. The concept of a statistical spread in "medres" orientations for one given "lores" (= macroscopic) orientation has been discussed in ch.1; 1.4 .

Comparing cases b and c, it is obvious that they are quite similar mathematically because of the symmetrical rôles of the $\vec{n}$ and $\vec{p}$ vectors in the $|\vec{n} \cdot \vec{p}|$ product. One phenomenological difference is that in the b case, the distribution function $F(\vec{p})$ for the PBC orientations is a property of the crystal bulk and therefore unique, whereas the distribution function $F'(\vec{n})$ of the "medres" surface orientations can depend on the macroscopic orientation $\vec{n}_0$ and should rather be written as $F'(\vec{n},\vec{n}_0)$.

Because of the possible pitfalls in the discussion of case c, we choose to discuss the more straightforward case b first.

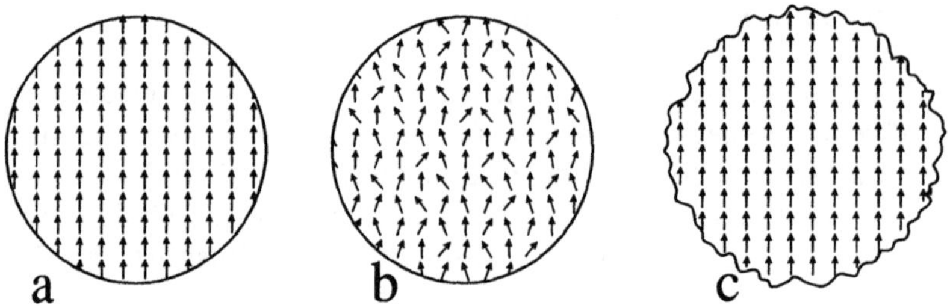

FIG. 11.2 Illustrating the correspondence between cusp smoothing and statistical spread in orientations.

a) sharp cusp: no spread in PBC vectors, no spread in surface orientation
b) smooth cusp: spread in PBC vectors, no spread in surface orientation
c) smooth cusp: no spread in PBC vectors, spread in surface orientation

PBC-orientation statistics and the $C(\vec{n})$ function

Considering fig. 11.2b as a 2-dimensional picture (i.e., a one-dim. orientation variable $\vec{p}$ for the PBC vectors, which are confined to a plane) we can define a PBC-orientation distribution function $F(\vec{p})$:

$$\int_0^{2\pi} F(\vec{p})\,d\vec{p} = 1 \tag{11.7}$$

and

$$\frac{\int_0^{2\pi} \vec{p}\,F(\vec{p})\,d\vec{p}}{\left|\int_0^{2\pi} \vec{p}\,F(\vec{p})\,d\vec{p}\right|} = \vec{p}_0 \tag{11.8}$$

(= "average" orientation, at least if not = 0/0)

We now *define* the function $C(\vec{n})$ as the function which counts the broken PBC's penetrating the surface per unit area, as a function of surface orientation $\vec{n}$. This function can now be written as

$$C(\vec{n}) = \frac{1}{a}\int_0^{2\pi} |\vec{n}\cdot\vec{p}|\;F(\vec{p})\,d\vec{p} \tag{11.9}$$

(i.e., a generalisation of (10.3))
Writing ψ for the $\vec{n}$ variable and φ for the $\vec{p}$ variable:

$$C(\psi) = \frac{1}{a}\int_0^{2\pi} \left|\cos(\varphi-\psi)\right| F(\varphi)\, d\varphi \tag{11.10a}$$

with the consequence that, for the radius of the $\vec{\mathbf{Q}}C$ curve, we can write

$$\mathbf{S}C(\psi) \equiv C(\psi) + \frac{d^2C(\psi)}{d\psi^2} =$$

$$= \frac{1}{a}\int_0^{2\pi} 2\delta(\varphi - \psi \pm \pi/2)\, F(\varphi)\, d\varphi = \frac{2}{a}\{\, F(\psi+\pi/2) + F(\psi-\pi/2)\,\} \tag{11.10b}$$

When we want to count only PBC heads or tails, the cos $(\varphi-\psi)$ in eq. (11.10a) has to be singly instead of doubly "rectified" - i.e., eq. (10.4a) applied - with the result that the factor 2/a in eq. (11.10b) reduces to 1/a.

Stated in words, eq.(11.10b) identifies the function $\mathbf{S}C(\psi)$ *with "images" of the* $F(\varphi)$ *distribution, shifted from the pole position of the original* $F(\varphi)$ *to the two equator positions at* $\varphi \pm \pi/2$. *The same can be said of* $\mathbf{S}\Phi(\psi)$ *for any physical quantity* $\Phi(\psi)$ *which is linearly related with* $C(\psi)$.

This reveals $\mathbf{S}C$ *as another distribution function, not of the PBC vectors themselves but of their tangent points to the circle.*

A direct consequence of (11.10b) is that

$$\int_0^{2\pi} \mathbf{S}C(\psi)\, d\psi = \frac{4}{a} \text{ (heads and tails together),} \tag{11.11}$$

$$\text{or} \quad = \frac{2}{a} \text{ (heads or tails separately)} \tag{11.12}$$

This equation is just another version of eq. (2.7).

These results can be generalized to the 3-D case. Now $\vec{p}$ is a locally 2-dim. variable; accordingly, in the definition equations (11.7) and (11.8), the line integrals over the full circle are now to be replaced by surface integrals over the full sphere, and the same applies to $C(\vec{n})$:

$$C(\vec{n}) = \frac{1}{a}\int_{\text{unit sphere}} \left|\vec{n}\cdot\vec{p}\right|\, F(\vec{p})\, d\vec{p} \tag{11.13}$$

but, in contrast with the 2-D ("cylindrical") case where all PBC vectors were confined to one plane, the projections of the penetrating PBC's can now have all directions in the $\vec{n}$-oriented plane.

For the elements of the $\vec{\vec{\mathbf{S}}}C$ tensor we have to choose a local orthogonal reference frame $d\vec{n}_1$, $d\vec{n}_2$ in the tangent plane at $\vec{n}$. Only $F(\vec{p})$ values on the $\vec{n}$ equator contribute to the $\vec{\vec{S}}C(\vec{n})$ tensor elements for the same reasons as in the 2-D case. They have to be integrated over the circumference of the equator, for which we define the angular variable α starting at the $d\vec{n}_2 = 0$ plane. (fig. 11.3). For this (arbitrary) choice of frame we have

$$\vec{\vec{S}}C(\vec{n}) = \frac{1}{a} \begin{pmatrix} \int\limits_{\vec{n}\ \text{equator}} F(\alpha)\sin^2\alpha\, d\alpha & \int\limits_{\vec{n}\ \text{equator}} F(\alpha)\tfrac{1}{2}\sin 2\alpha\, d\alpha \\ \int\limits_{\vec{n}\ \text{equator}} F(\alpha)\tfrac{1}{2}\sin 2\alpha\, d\alpha & \int\limits_{\vec{n}\ \text{equator}} F(\alpha)\cos^2\alpha\, d\alpha \end{pmatrix} \tag{11.14}$$

It follows directly that

$$\text{Trace}\,(\vec{\vec{S}}C(\vec{n})) = 1/a \int\limits_{\vec{n}\ \text{equator}} F(\alpha)\, d\alpha \tag{11.15}$$

When the position of the basis vectors $d\vec{n}_1$ and $d\vec{n}_2$ (i.e. the $\alpha = 0$ position on the $\vec{n}$ equator) is chosen such that the off-diagonal tensor elements vanish, then $d\vec{n}_1$ and $d\vec{n}_2$ are the principal directions of the tensor and the diagonal tensor elements can be considered as the densities of tangent PBC's to the surface, multiplied by their projections onto $d\vec{n}_1$ and $d\vec{n}_2$, respectively.

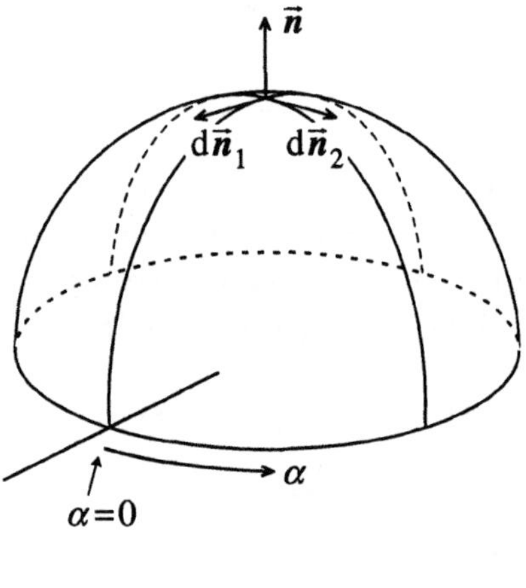

FIG. 11.3 Reference frame and definition of variables for integrations in eqs. (11.14) and (11.15).

11.1.3 Cusp-smoothing considered as the effect of statistical spread in surface orientations.

We started discussing cusp-smoothing in terms of distribution functions with the case of fig. 11.2b rather because its analysis is straightforward conceptually. However, the case of fig. 11.2c seems to be the more relevant from a physical point of view. At first sight, it might seem identical with case b because of the symmetry of $\vec{n}$ and $\vec{p}$ vectors in the $|\vec{n}\cdot\vec{p}|$ factor. Identity would imply that the cases could be brought to coincidence by merely transferring the scatter in medres $\vec{n}$ orientations (i.e., its distribution function $F'(\vec{n})$) to the $\vec{p}$ orientations. (in such a way that the medres orientation scattering at $\vec{n}$ is removed, and replaced by a scattered $\vec{p}$ so that the $(\vec{n}-\vec{p})$ distribution function is maintained). Note that we emphasize the difference between the $F'(\vec{n})$ and $F(\vec{p})$ functions by a distinction mark (′).

However, on closer examination a few amendments have to be made. The first is of a mathematical nature and was mentioned already in 11.1.2: because the "medres orientation distribution function" $F'(\vec{n})$ may depend on the macroscopic, lores orientation $\vec{n}_0$, such a change in statistics will affect the $\vec{\nabla}_n$ derivatives involved in the tensor $\vec{\vec{\vec{S}}}C$. Unless the $F(\vec{n})$ distribution is *independent* of $\vec{n}_0$, this makes the cases b and c mathematically dissimilar. Still, the rough conclusion that surface-orientation scattering can be related with cusp-smoothing remains intact; only the mathematical expressions become suspect.

The second amendment is of a more fundamental nature: indeed it affects the very basis of the continuum approach. Essentially, it is related with the incompatibility of the "corpuscular" structure of a crystal surface with a continuum representation. On a large (lores) scale, where we can neither see the distinct atoms nor the distinct ledges, this is not relevant. But we may begin to feel uneasy when looking at medres scale, in particular when we define "medres" such that we cannot see individual atoms, but just can discern the ledges. How then should we define an $F'(\vec{n})$ distribution function?

Consider a crystal surface below the roughening transition, with interledge distance $\gg$ ledge height (fig. 11.4a). For ease of discussion, we just consider a 1-dimensional case (only _|¯ and ¯|_ kinks).

On the question how to define an "orientation distribution function" for this surface, we soon recognize that the recipe is in a way arbitrary, namely dependent on the range or size over which to average the medres orientation. Now the wish to dispose of such a distribution function is based on applications describing physical quantities $\Phi(\vec{n}_{\mathrm{macroscopic}})$ by statistics according to

$$\Phi(\vec{n}_{\text{macroscopic}}) = \int F'(\vec{n}_{\text{local}}) \cdot \Phi(\vec{n}_{\text{local}}) \cdot \mathrm{d}\vec{n}_{\text{local}} \tag{11.16}$$

This brings us naturally to choose a scale on which the quantity Φ can be expected to have a "natural" relation with medres orientation: i.e. the interledge distance. The (still somewhat arbitrary) recipe reads as follows (for the 2D case; we write ψ instead of $\vec{n}_{\text{macroscopic}}$, ϕ instead of $\vec{n}_{\text{local}}$):

* Divide the surface in sections each counting *one* kink by making cuts exactly half-between any pair of kinks, whether _┌ _┌, _┌ ┐_, ┐_ _┌ or ┐_ ┐_ . (fig. 11.4b).
* Determine the microscopic "orientation" ϕ_i of each section (length ℓ_i) from $\arctan(h/\ell_i) = \phi_i$. ($h$ = ledge height).
* Determine the distribution function $F'(\phi)$, taking the total length of ϕ-oriented surface as statistical weight.

(A delta-peaked function corresponds with a "mathematically exact cut" through the crystal). This recipe leads to a slight problem for mixtures of _┌ and ┐_ kinks because the $F'(\phi)$ distribution function has a zero-cusp at the exact facet orientation when defined as above. (this can be remedied by choosing sections containing two or more kinks, but this makes the spread narrower). The definition dilemma becomes evident when we approach the roughening transition. Kink density becomes higher, and consequently the $F'(\phi)$ distribution wider - ultimately it will range from $\psi + \pi/2$ to $\psi - \pi/2$. However, for such a surface the local contribution per unit area to a physical quantity Φ can no longer be expected to be comparable to that of a much larger area of "mathematically exact cut" structure. (eq. (11.16) is based upon such a concept). At this stage, the whole reason for applying statistics has lost its grounds.

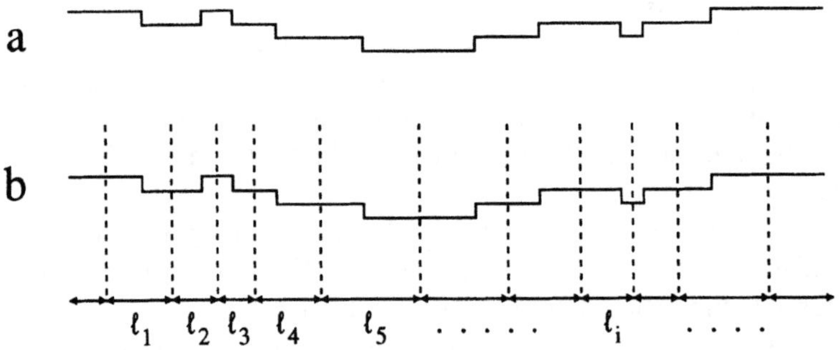

FIG. 11.4 a) Slightly roughened crystal surface, "1-dim. SOS model".
b) (arbitrary!) recipe for defining an orientation distribution function of the surface.

This remark illustrates that the explanation of cusp-smoothing as a result of medres-orientation spread inherently ends in vagueness, because an unambiguous definition for an orientation distribution function cannot be given. With this caution in mind, we can nevertheless proceed to translate the *congruence between* $\vec{\mathbf{S}}C(\vec{n})$ *and PBC-orientation spread* (case of section 11.1.2) into the alternative *congruence between* $\vec{\mathbf{S}}C(\vec{n})$ *and a spread in surface orientations* (present section). We discuss this translation specifically for the 2D case.

For this purpose we introduce an extra "orientation-scatter" parameter ϕ^*: whereas the macroscopic surface orientation parameter is ψ and the $\vec{p}$ orientation parameter is φ (just as in section 11.1.2), the local "microscopic" orientation ϕ is written as $\psi+\phi^*$. The surface-orientation distribution function can alternatively be written as $F'(\phi) = F'(\psi+\phi^*)$ (i.e., centered around ψ) or as $F'(\phi^*)$ (i.e., centered around 0). This function obeys definition equations like (11.7) and (11.8).

Instead of eq.(11.10a), we now have

$$C(\psi) = \frac{1}{a}\int_{-\pi}^{\pi} |\cos(\varphi-\psi-\phi^*)|\ F'(\phi^*)\,d\phi^* \qquad (11.17a)$$

For $\mathbf{S}\,C(\psi) = C(\psi) + d^2 C/d\psi^2$, we can still write down the analog of eq. (11.10b), now with ϕ^* taking the rôle of φ in eq.(11.10b):

$$\mathbf{S}C(\psi) = \frac{1}{a}\int_{-\pi}^{\pi} 2\delta(\varphi-\psi-\phi^* \pm \pi/2)F'(\phi^*)\,d\phi^* =$$

$$= \frac{2}{a}\{F'(\varphi-\psi+\pi/2) + F'(\varphi-\psi-\pi/2)\} \qquad (11.17b)$$

(Note: the PBC vectors are assumed here to be exactly aligned as in fig. 11.2c, so that φ has a fixed value. In principle it is also possible to describe the mixed case, i.e. with smeared-out $F(\varphi)$ *and* $F'(\phi^*)$ distribution functions. This requires a double integration:

$$C(\psi) = \frac{1}{a}\iint |\cos(\varphi-\psi-\phi^*)|\ F'(\phi^*)\,F(\varphi)\,d\phi^*\,d\varphi \qquad (11.18)$$

The result (11.17b) can be stated in words - compare with eq.(11.10b)! - as:

$\mathbf{S}C(\psi)$ *is still identified as an image of the* $\vec{p}$ *orientation, shifted from the pole position* φ *to the two equator positions* $\varphi \pm \pi/2$. *However, the image is "blurred": it is sampled with a scatter* $F'(\phi^*)$ *around* ψ *instead of measured at the exact* ψ *orientation. When* $F(\varphi)$ *itself is a delta function (as assumed in eq.(11.17b), then* $\mathbf{S}C(\psi)$ *represents the* $F'(\psi+\phi^*)$ *distribution function.*

This is analogous to the blurring of a narrow line (the $\vec{p}$ distribution function) when observed through corrugated glass. The width of the blurred image corresponds with the corrugation spectrum $F'(\vec{n})$ of the glass.

Also, we may quote an illustrative analog of the smoothed-cusp growth rate function which is provided by the transmission of light through a system consisting of polarizer and analyzer. Here too, the basic variable is an *orientation variable* (between polarizer and analyzer directions $\vec{p}$ and $\vec{a}$), and in the ideal case (perfect polarizers) the transmission is given by a cusped $|\vec{a}\cdot\vec{p}|$ function. When one of the polarizers is imperfect (i.e., ε = fraction of light transmitted in the "wrong" polarisation direction) then the transmission is given by $\textcircled{2}\langle\, |\vec{a}\cdot\vec{p}|\,, \varepsilon \,\rangle$.

We have to emphasize a difference between the cases of section 11.1.2 and eq.(11.10a/b), and that of section 11.1.3 / eq.(11.17a/b). For the $F(\varphi)$ distribution function of $\vec{p}$ vectors in eq. (11.10a), we have no a priori reason to exclude widely spread orientation distributions $F(\varphi)$. However, in the case of eq. (11.17a/b), we have to do with a *surface* orientation distribution $F'(\phi^*)$ for which we expect a restriction on intuitive grounds; e.g. $\phi^* = \pi$ corresponds with an "upside-down" local interface for which $F'(\psi+\phi^*)$ can be expected to be quite low.

More in particular, we can adapt equations (11.17a/b) *to the specific case of near-facet orientations.* The exact facet orientation ψ_{fac} is defined by $|\psi_{fac} - \varphi| = \pi/2$. We can measure all orientations from this ψ_{fac} as a new zero-point; this is expressed by the addition of a prime the angular variables ψ and φ. (i.e., $\psi'_{fac} = 0$, $\varphi' = \pi/2$; the variable ϕ^* = difference between local and macroscopic orientation remains unchanged). We consider the case that the facet is "not too badly roughened", more specific that the SOS restriction is applicable. This implies that $F'(\phi^*)$ is zero for all overhanging orientations: in other words, we can restrict the $(-\pi$ - $\pi)$ integration range in eq. (11.17a/b) to the range $(-\pi/2, +\pi/2)$ for ϕ^*, because the F' term for the overhanging orientations is zero anyhow. (when roughening is only slight, $F'(\phi^*)$ is concentrated in a narrow range around $\phi^* = 0$).

For SOS restriction and ψ' near facet orientation, i.e. $|\psi'| \ll \pi/2$:

$$C(\psi') = \frac{1}{a}\int_{-\pi/2}^{\pi/2} |\sin(\psi' - \phi^*)|\; F'(\phi^*)\, d\phi^* \qquad (11.19a)$$

and

$$\mathbf{S}C(\psi') = \frac{1}{a}\int_{-\pi/2}^{\pi/2} 2\delta(\psi' - \phi^*)\, F'(\phi^*)\, d\phi^*$$

$$= \frac{2}{a}\, F'(\phi^* = \psi') \qquad (11.19b)$$

(*note 1*: when $F'(\phi^*)$ depends on the variable ψ' too, eqs. (11.17b/11.19b) have to be supplemented with terms for $d^2/d(\psi')^2$ of $F'(\psi',\phi^*)$.
note 2: comparing with eq.(11.10a/b) or eq.(11.17a/b), we see that there is no "opposite" $F'(\psi')$ term in $\mathbf{S}C(\psi')$ any more).

Identification of the roughening parameter ε:

On the basis of eq. (11.19a), we can identify the geometrical meaning of the roughening parameter ε:

$$C(\psi^*=0) = \frac{1}{a}\int_{-\pi/2}^{\pi/2} \left|\sin(\phi^*)\right| \, F'(\phi^*)\, d\phi^*$$

$$= \frac{1}{a}\, \varepsilon_{\text{facet}} \qquad (11.20)$$

In other words, ε is the average value of $\left|\sin(\phi^*)\right|$, i.e. something as the *width of the scatter in microscopic orientation of the facet*. This identification is the logical consequence of equations (11.5) and (11.6):

* Consider specifically the smoothing of the cusps in the *growth rate function*. Here, _┌ and ┐_ ledges (if separated widely enough) essentially contribute to R in the same way, i.e. $R(\psi)$ is obtained simply by multiplying $C(\psi)$ with a.$v_{/\!/}$. (although the contributions of the _┌ and ┐_ ledges are simply additive, the *numbers* of _┌ and ┐_ ledges, as represented by the $F'(\phi)$ function, are not a simple addition of original "orientation" ledges and "B&S" ledges, but rather the result of an equilibrium reaction as explained in 11.1.1). In contrast, for the *surface free energy* $\gamma(\psi)$ we have, apart from the energy contributions of _┌ and ┐_ ledges, also to do with the entropy of the ledge system (which is not linear in $C(\psi)$).

* The identification of ε with a.$C(\psi_{\text{fac}})$ now follows directly from the comparison of eq. (11.5/6). ($R_{\text{facet}} = v_{\text{B\&S}}$) with eq.(11.20) ($\Rightarrow R_{\text{facet}} = \; < \sin\left|\phi^*\right| > v_{/\!/}$).

* Note that the argument for identification of ε by eq. (11.20) is applicable *only* to low values of ε ($\varepsilon \ll 1$): i.e. widely separated ledges, so that diffusion fields do not overlap. For ε approaching unity, as argued above, the surface-orientation spread concept becomes vague: moreover, the applicability of the $v_{/\!/}$ prefactor is lost because in such a wide orientation range other facets come within view.

Nevertheless, the argument for the low end of the ε range is sufficient for the identification. This also implies that the velocity parameter in the network elements of section 11.4 (fig. 11.8 a-f etc.) can only be $v_{/\!/}$ and not an independent parameter $v_{\perp}$ or so.

Equation (11.20) provides a common point of departure to describe the (simultaneous) smoothing of the cusps in the $R(\vec{n})$ and those in the $\gamma(\vec{n})$ functions: both are to be based on the same ε. However, only from the minimum in the $R(\vec{n})$ function (when low enough) we can directly determine the value of ε from eq. (11.21). (The value of ε, of course, is not the "physical" input parameter. Technically we have to do with parameters T, A etc, and thermodynamics has to predict how ε will adapt to them).

11.1.4 The special rôle of (2)✕ . . . ✕o *for cusp smoothing*

In 11.1.2 and 11.1.3 we have argued that any roughening, translated into cusp-smoothing in the γ and R functions, can be related with something as a statistical spread in orientation. More precisely, the *surface*-orientation distribution function itself can be related with **S***C* in the neighbourhood of a particular facet. We want to emphasize that this relation has nothing to do with any special way of representing R or γ functions - the discussion in the preceding subsections stands apart from the assembling techniques used to construct "tangible" functions. This means that, from a *physical* point of view, the quantity $R/\mathbf{S}R$ is the most natural candidate to be considered as a roughening parameter (variable in the range $0 < R/\mathbf{S}R < 1$ at cusp orientations). For the special *representation* of smoothed cusps by eq. (10.31e), we then find that

$$R/\mathbf{S}R = \varepsilon^2 \tag{11.21}$$

for the orientation of the original cusp and for the special choice a = 2 (which, as we will explain below, is indeed a preferred choice).

In the present subsection we will again consider the *assembling procedure for smoothing* in more detail. In 10.3 we have only remarked that a > 1 is a condition to be satisfied in order to obtain a *smoothed* cusp. However, now we have learnt to identify **S***C* with a distribution function, we have another viewpoint which can help to make a narrower choice for a, albeit on an intuitive argument. The argument is based on the *shape* of the distribution function, i.e. on **S***C* (or on **S***R*, etc.)
When we plot these functions for different a (starting from eq. (10.31e) with variable assembling exponent a) it turns out that *only for a = 2* we obtain a "natural-looking" bell-shaped curve with flat maximum. For 1 < a < 2 we obtain a *sharp* (upward-cusped) maximum, for 2 < a < 3, a curve with a central (downward) cusped relative depression, and finally for a > 3 a curve with a central *flat* relative depression; anyhow, for a > 2 the distribution function has two maxima surrounding a central relative minimum. Resuming, all situations *except that for a = 2* "look less natural" than the bell-shaped curve.

Of course, this argument becomes real only after such a distribution function has been calculated on the basis of thermodynamics - the subject we have deliberately left aside here. Until such proof is given, the present argument is intuitive - but still we might favour the choice a = 2 for the calculations in ch. 12. (Besides, as we have argued in 10.4, the choice a = 2 has some practical advantages when compact expressions are preferred).

Instead of working out our argument by starting with eq. (11.6) and investigating the distribution function **S***C*, we prefer, for convenience, to start with a smoothed $|x|$ cusp:

$$A = \textcircled{a}\!\!\times\ |x|, c \times\!\!\circ = (|x|^{a} + c^{a})^{1/a} \tag{11.22}$$

Because in this case, the second derivative of the $|x|$ function has not to be compensated (as in the $|\sin(x)|$ case), the "cusp distribution function" is just the second derivative itself:

$$F = d^2A/dx^2 = (a-1)\,c^{a}.\,(x^{a} + c^{a})^{(1/a\,-\,2)}.\,x^{a-2} \tag{11.23}$$

(we have only to consider the behaviour for $x > 0$ as the distribution function is symmetrical about $x = 0$). The shape of the F curve around $x = 0$ is imposed uniquely by the factor x^{a-2} . Indeed a smooth bell-shaped function corresponds uniquely with a = 2.

11.1.5 The effect of roughening: R functions versus γ functions

Consider fig. 11.5a and b. Here we have schematically represented the typical behaviour of γ (fig. 11.5a) and R (fig. 11.5b) functions upon transition from the "completely faceted" to the "completely roughened", i.e. OID situation. (corresponding with the transition of ε from 0 to 1, in the $\textcircled{a>1}\!\!\times \ldots \times\!\!\circ$ approach). For the sake of illustration, we have chosen a 4−symmetrical case with identical orientation of the minima. The transitions have been sketched intuitively so as to agree with our practical knowledge, both of actual crystal behaviour and of theoretical models:

1) In the γ case, roughening corresponds with smoothing of the cusps in the "raspberry plot" *and simultaneous decrease of the minima*, just for reasons of thermodynamic first principles. In the R case, on the other hand, roughening is characterized by smoothing of the cusp in the "lobe plot" *and simultaneous increase of the minima.* In the completely-smoothed, i.e. liquid-like limit, the OID R circle lies *beyond* all OD R plots, whereas the OID γ circle lies *within* all OD γ plots. We will have to formulate our mathematical assembling expressions so as to agree with this (a priori known) behaviour.

2a) In the "completely-roughened" limit (i.e., ε of all facets = 1) the functions are orientation-independent. This implies not only that individual (facet-related) contributions are OID on their own (this is ensured by the ε =1 substitution in equations like eq. (10.31 a-e), but also that the individual contributions *are not allowed to influence the end result* in the $\varepsilon = 1$ limit. The latter requirement is more stringent than the former (OID); it is a consequence of the fact that any R or γ function with a limited number of observable minima can be considered to have an *unlimited* number of higher-index "previous minima" which have already disappeared by complete

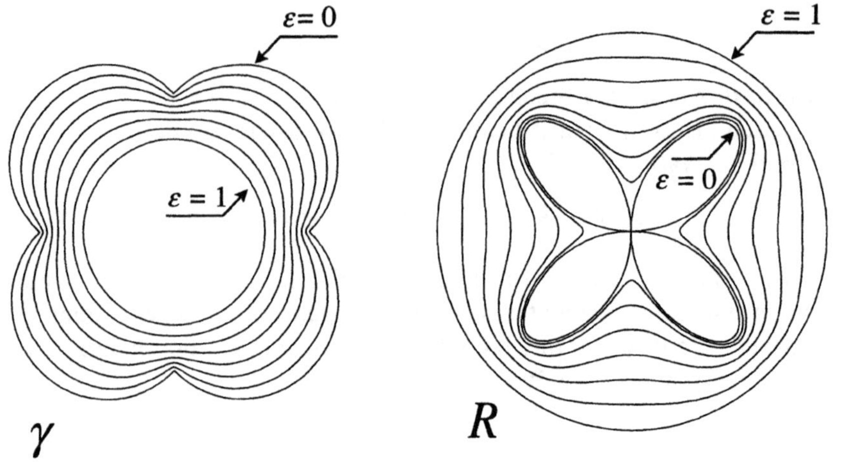

FIG. 11.5 Characteristics of the roughening transition: γ versus R functions

roughening. Consequently, in an expression which remains valid in the full ε range from 0 to 1, their influence should vanish in the $\varepsilon \to 1$ limit. In a "multi-family" function where two or more facet families, each with their own roughening parameter, are combined, complete roughening of one family should leave the other families unaffected.

2b) The *remaining* OID R or γ value in the completely-roughened limit is not zero (for γ) or ∞ (for R), as even the roughened, liquid-like surface has finite surface free energy and dissipative incorporation kinetics. We may write the OID limits as γ_0 and R_0, respectively.

2c) As a consequence of 2a) and 2b), we are led to require that a slight "de-roughening" of facets can be expressed as a correction to R_0 and γ_0:

$$(1-\varepsilon) = \delta \ll 1\ ;\quad \text{i} = \text{"family index"}\ ;$$

$$\gamma(\vec{n}) = \gamma_0 - \Sigma\, \Delta\gamma_i\,(\delta_{\gamma i}, \vec{n}) \tag{11.24a}$$

$$R(\vec{n}) = R_0 + \Sigma\, \Delta R_i\,(\delta_{Ri}, \vec{n}) \tag{11.24b}$$

in which the $\Delta\gamma_i$ and ΔR_i functions vanish for the corresponding δ approaching zero. The + sign for R and − sign for γ have been chosen in agreement with the direction of the change, as mentioned in 1).

2d) Whatever amendments we introduce in order to satisfy the requirements stated above, we will have to ensure that they do not affect the function behaviour at the other end of the scale, $\varepsilon = 0$. In particular for a γ

function, it should essentially reduce to a soapfroth plot ($\Sigma\ \gamma_i\ |\vec{n}\cdot\vec{p}_i|$) in the limit $\varepsilon_i \rightarrow 0$ for all ε_i .

3) Items 1) and 2 a,b,c,d) were discussed above as referring to a two-dimensional description. However, in ch.12 we will also construct functions which contain a) the 2-D information of the ledges on all individual facets, b) the 3-D information of the complete function. In such a case, the function contains $\varepsilon_{ij,k}$ parameters for the roughening of (k) ledges on an (ij) facet, and also ε_{ij} parameters for the roughening of the (ij) facet as a whole.

This implies that the complications as mentioned in 1) and in 2a, b,c,d) above (and the related remedies) actually arise at two levels: the 2-D level of ledges on facets *and* the 3-D level.

So far for the diagnosis. Now for the remedies, we have to be aware that the R function case and the γ-function case are fundamentally different and therefore deserve an individual approach, to be worked out in ch. 12.

For the R-function case, the use of the various smoothing operations represented by eqs. (10.31 a-e) seems appropriate because they all agree with simultaneous smoothing and *increase* of the minimum. Indeed, the discussions in 11.1.1 - 11.1.4 about the "compatibility" of the 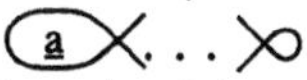 operation and the physical picture were almost exclusively related with R functions.

For the γ-function case, we have to find a physically reasonable way to account for the *decrease* of the minimum during smoothing. The physical origin of this decrease is the negative contribution by entropy terms. In the analytic expression of a γ function, this contribution requires an extra amendment as compared with the R-function case. This amendment is essential, but in the end result we have to do with the combined effect of *two* amendments. As there is a certain ambiguity - or "degree of freedom" - in the expression for the amendments separately, the final expression is subject to a "squared ambiguity" which makes the "γ" expressions given in ch.12.7 less satisfactory than their R-function counterparts. To give such expressions a solid physical basis (rather than the intuitive approach which we will follow in ch. 12.7) requires a more thorough analysis of the entropy terms (which depend strongly on the assumptions of the model used: SOS, etc.) In agreement with the general scope of this monograph, we made no attempts in this direction. Fortunately, in most physical and technical situations the emphasis is on the rôle of the R functions, whereas the γ functions occur in correction terms for capillarity effects.

Altogether, in view of this qualitative difference between the R-function and the γ-function cases, we will discuss the "remedies" separately.

11.2 UPWARD LIMITATION OF FUNCTIONS

(Upward limitation has special relevance for growth rate functions, rather than for free-energy functions).

11.2.1 Resistance mechanisms in series.

Strictly taken, the assembled-function approach for a series of resistance mechanisms gives an exact result only if all of the fluxes have the same exponential relation with affinity:

$$R_{\rm i} = \xi_{\rm i} A^n \Longrightarrow \text{then } R_{\text{series-combined}} = \xi_{\rm ass} A^n \quad \text{with}$$

$$\xi_{\rm ass} = (\!\!-1/n\!\!)\!\times \ \ldots \xi_{\rm i}, \ldots \times\!\!\circ \tag{11.25}$$

In practice, the mechanisms can be, for instance, diffusion through a boundary layer, combined diffusion processes on the interface, or chemical reaction on the interface (heterogeneous reaction). When not all processes are linear in A, we can choose for assembling with a = $-1/n$ with a "too high" n value: this results in an almost correct result in those orientation regions where one of the mechanisms is predominant over the others, wheras only the transition regions are rendered slightly in error.

11.2.2. Flux-imposed growth.

There are a few instances of crystal growth situations where the growth rate, for almost all orientations except narrow ranges around facets, is imposed by some flux-limiting mechanism: e.g. MBE, Verneuil and related techniques, growth from a low-pressure vapour at high supersaturation etc. In all such cases, growth particles hit the crystal surface with constant ("imposed") flux density and are sure to be attached into the lattice ("sticking coefficient = 1", except in facet ranges). Such cases can be dealt with in either of two ways:

a) by assembling the surface-defined growth rate (e.g., $R_{\rm s} = c_{\rm s} A\ |\vec{\boldsymbol{n}}\cdot\vec{\boldsymbol{p}}|$) with the constant imposed flux $R_{\rm i}$ by $(\bar{a})\!\times R_{\rm s}, R_{\rm i} \times\!\!\circ$, with a "large" value of $\bar{a}$. This is equivalent with describing the flux source as limited by a sublinear series resistance $R_{\rm i} = c_{\rm i}\, A^{-1/\bar{a}}$, but the competing surface resistance is then equally considered as sublinear with the same exponent and accordingly *the facet size is kept constant.*

b) by considering both resistances as linear (i.e., $R_{\rm s} = c_{\rm s} A\ |\vec{\boldsymbol{n}}\cdot\vec{\boldsymbol{p}}|$ and $R_{\rm i} = c_{\rm i} A$) but letting A increase to a high value while keeping $R_{\rm i} = c_{\rm i} A$

constant; i.e. $c_i = c_i^* A^{-1}$; c_i^* is a constant. Subsequently, both resistances are assembled by $\textcircled{-1}\!\times\ R_s, R_i \times\!\circ$. In this approach, the facet range (roughly defined by the crossover $R_s = R_i$) depends on A:

$$(\vec{n}_{\text{crossover}} \cdot \vec{p}) = c_i^* / c_s \,.\, A^{-1}.$$

Approach b) seems to be the more realistic from a physical point of view.

In both cases, in a sphere-growth experiment, the growth rate is limited to the imposed rate almost anywhere on the sphere surface except in narrow regions around equators and facets (where surface resistance imposes a lower limit than the external flux-limiting mechanism.) In terms of "sticking coefficient": this coefficient equals 1 except in the facet and equator regions where it is smaller than one.

This implies that in those regions a high surface gradient of the affinity exists which can drive a diffusion flux (either surface or bulk diffusion) of growth units from the equator and facet regions where they cannot be incorporated, towards the surroundings where the sticking coefficient is unity again. This diffusion flux results in the buildup of rims and corresponding concave regions around the facets (see fig. 13.10).

11.3 IN-BETWEEN THE MINIMA: SMOOTH OR CUSPED MAXIMA?

The choice for a small negative value of the assembling exponent corresponds with a *smooth* maximum in the multi-exposure function (10.23), below the value of the individual assemblands $v_i|\vec{n}\cdot\vec{p}_i|$ and $v_j|\vec{n}\cdot\vec{p}_j|$ at their intersection. However, certain models for interface properties at between-facet orientation (both γ and R) seem to suggest a *cusped* maximum. (note: a cusped maximum can be recognized experimentally by the occurrence of a facet in hollow-sphere growth experiments).

Consider the "square" 2-dim. Kossel crystal of fig. 11.6. Fig. 11.6a represents an exact (11) oriented surface: it counts 1/d re-entrant corners (REC) per unit length (d = diagonal of unit square). Fig. 11.6b has a tilt (angle φ) from the (11) direction and correspondingly shows 45° steps, i.e. positions with a missing REC. The fraction of missing REC's is 2 sin φ according to a similar goniometric relation which is known for ordinary steps, on a surface tilted away from the (10) orientation.

However, in contrast with these ordinary steps the contribution of the

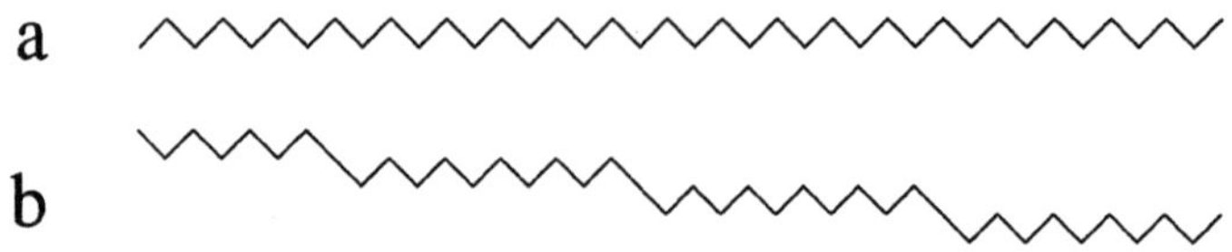

FIG. 11.6

a) Exactly (11) oriented surface of a 2-dim. "square" crystal with(10) and (01) facets. b) The same surface oriented slightly off-(11). $R(\bar{n})$ can have a cusped maximum in the (11) direction according to eq. (11.26).

45° steps is *negative*, as counted from the maximum corresponding with fig. 11.6a. (contribution : i.e. to broken-bond counting function, and consequently also to either γ or R functions).

In particular for growth rate functions, each REC can be a "fast" receptor site for atoms from the parent phase. When the maximum growth rate corresponding with fig. 11.6a is R_{max} and the 45° step positions are just inaccessible, then

$$R = R_{max}\,(1 - 2|\sin\varphi|) \tag{11.26}$$

i.e., a *cusped maximum*.

If, instead, atoms landing on 45° steps are directly transferred to the adjacent REC sites, thereby increasing their statistical chance of being filled, then just

$$R = R_{max} \tag{11.27}$$

independent of orientation. (when (10) is a facet orientation, then terraces between ledges on this facet act as a collecting apron for supply to the ledges, i.e. a terrace-to-REC transfer mechanism is active). Altogether, in a structure as sketched in fig. 11.6 we are likely to have something like

$$R = R_{max}\,(1 - \eta|\sin\varphi|) \qquad (\text{with } 0 \leqslant \eta \leqslant 1) \tag{11.28}$$

for φ counted from a maximum-R orientation, with the adaptable parameter η depending on the details of surface diffusion and REC-filling kinetics. The function (11.28) has a cusped maximum, but of course the cusp disappears when the regular structure of fig. 11.6b "roughens", analogous to the roughening of minimum-cusps.

(Apart from that, we have to be aware that any structure with a positive cusp *in the γ function* is formally unstable and has the tendency to break up into a zigzag structure of larger period than in fig. 11.6a. (see section 4.2).

Modelling the maxima in-between cusps

Disregarding the question whether an upward-cusped function - either for γ or for R - is physically realistic, we can consider how to *represent* an upward cusp in an assembling procedure:

a) For *positive* assembling exponent, an upward cusped maximum corresponds with an "inverted" assembland of the type $(-)\,|\vec{n}\cdot\vec{p}_i|$ or $(-)\,|\vec{n}\wedge\vec{f}_i|$. In particular, for the special case of summation we can just maintain the $\Sigma\,\gamma_i\,|\vec{n}\cdot\vec{p}_i|$ expression for the "soapfroth" γ plot but take the relevant γ_i factor negative where that particular PBC contributes a negative energy (KERN 1987). (Of course, the *total* surface energy has to remain positive for all orientations). Smoothing of the upward cusp requires previous $a > 1$ assembling of the $|\vec{n}\cdot\vec{p}_i|$ term with a positive constant.
b) For *negative* assembling exponent ($R(\vec{n})$ functions), we can consider the (upward) cusp orientation as the orientation of intersection of two $|\vec{n}\cdot\vec{p}|$ functions, $K = v_k\,|\vec{n}\cdot\vec{p}_k|$ and $M = v_m\,|\vec{n}\cdot\vec{p}_m|$:

$$\mathcal{A}_{res} = \mathrm{Min}\,(K,M) = \lim_{\bar{a}\searrow -\infty} (\bar{a})\!\times K,M \times\!\circ \qquad (11.29)$$

(we may remark here that for this particular case, the $(-\infty)\!\times$ assembling can be written alternatively as the *sum* ($(1)\!\times \;\ldots \times\!\circ$) including negative terms $-\,v_{cusp}\,|\vec{n}\cdot\vec{p}_{cusp}|$, because a sum of vector products $v_i\,|\vec{n}\cdot\vec{p}_i|$ with either positive or negative multipliers v_i can always be written as another vector product $v_k\,|\vec{n}\cdot\vec{p}_k|$. In such a summation, positive cusps correspond with negative v_{cusp} .

By taking a finite (negative) assembling exponent instead of $-\infty$ in (11.29), the positive cusp is smoothed.

Considering maxima in a "multi-exposure" function $(\bar{a})\!\times v_i\,|\vec{n}\cdot\vec{p}_i| \times\!\circ$ in general, we can apply three basically different methods in order to increase the sharpness of the maximum - the limit being the exact fit to $\mathrm{Min}\,(K,M)$:

1) Increase the absolute value of $\bar{a}$, as indicated in (11.29).
2) For fixed (finite) $\bar{a}$, we can add an inverted assembland $(-)L^*$ as discussed in 10.4.4. However, the effect is limited: a *sharp* cusp cannot be obtained in principle.
3) When more assemblands of the type $v_i\,|\vec{n}\cdot\vec{p}_i|$ are involved than the two intersecting functions K and M of interest, we can delete them

altogether. Essentially, this approach is comparable with a "nearest-neighbour approximation". It is explained in detail in 11.4.5 in connection with the "kaleidoscopic construction principle".

11.4 ARRANGEMENT OF FUNCTIONAL ELEMENTS IN MULTIPLY CONNECTED $R(\vec{n})$ NETWORKS

(systematic review of sequenced assembling operations for growth-rate functions)

In 11.1 through 11.3, we were mostly concerned with effects of (single) assembling procedures on their own. Such procedures can be represented by a simple network implying a limited number of assemblands in a series or a parallel arrangement. For "complete" R or γ functions we need sequenced assembling procedures, i.e. representations by multiply connected (eventually nonlinear) networks. In such networks, smoothing (ε), limiting (L or M) or range-deforming (see 10.3.1) elements can be placed in different positions, with different effects. The choice between these possibilities, as based on analysis of their (different) physical interpretation, is the subject of this subsection.

We have to emphasize that there is, as far as we can see, no Basic Principle which *a priori* imposes the choice for one particular network and rejects all others. Here we are finally confronted again with the ambiguity referred to in ch.1.4: a particular physical model corresponds with a function (network) representing that model; i.e. we can construct as many networks as models. Even the preferred function type may differ from crystal to crystal; maybe even in one and the same crystal, from one temperature range to another. As we have chosen to disregard specific atomistic models, we have to live with some ambiguity in the assembled end functions. Still, we can discuss the physical aspects of different choices of complex networks and compare the results. In doing so, we will show that the definition of a physically sensible set of limit conditions for the assembled functions reduces the ambiguity mentioned above to an acceptable degree, so that only two "realistic" construction principles remain.

(Of course, the physical basis of γ-function networks differs essentially from that of R-function networks. Therefore, γ functions are discussed separately in 12.7.)

As a network is more illustrative for a physical mechanism than the mere mathematical notation by an assembling tree or table, we will illustrate most of our discussions by networks.

11.4.1 Building up complexity, and alternatives for the OD-OID transition.

Terminology: single-cusp, single-family and multi-family functions

These terms are used to distinguish between assembled functions of increasing complexity:

1) *A single-cusp (SC) function or network* represents a single facet (2D or 3D) or "equator band". It is based on an orientation-dependent starting function of the type $v|\vec{n}\cdot\vec{p}|$ or $v|\vec{n}\wedge\vec{f}|$; in ch. 12.3 we will also use the more complex starting function $v(\vec{n}^{(2)})|\vec{n}^{(3)}\wedge\vec{f}|$ as a basis for a 3-D SC function. Apart from this OD function as an element, the network can contain other parameter elements to account for roughening and limitation.

2) *A single-family (SF) function or network* represents one family of facets (2D or 3D) which are identical from the point of view of crystallographic symmetry. They involve a set of more than one OD elements of the type as mentioned above for the SC functions, only differing in the $\vec{p}$ or $\vec{f}$ vectors (which are different in direction but identical in the sense of crystallographic symmetry). Consequently, the same parameters for roughening, limitation and ledge velocity apply to all the starting functions; i.e. the symmetry and the "cusp-multiplicity" of the assembled SF function just correspond with those of one particular family of identical facets of the crystal.

3) *Multi-family (MF) functions or networks* represent a collection of more than one facet families, implying different sets of "physical" parameters and of orientation ($\vec{p}$ or $\vec{f}$) vectors for each family.

The terms SC, SF and MF can be used at two levels: that of "two-dimensional" functions (representing ledge advance rate on a facet) and that of three-dimensional functions. We want to emphasize that all considerations on network constructions in this section 11.4 are of a fundamental kind and are thus applicable to the 2D and the 3D cases as well - therefore, we will hardly make the 2D-3D distinction here. Still we have to be aware that all networks which are going to be "derived" in this section can be considered to refer to 2D or 3D cases alternatively. Corresponding with this dimensional difference, the network elements themselves (considered as parameters of the end function) have a different physical explanation. For instance, $v_{\parallel}$ in the 2D network (eventually written as $v_{\parallel}^{(2)}$) represents kink velocity along a monatomic ledge, but in the 3D network $v_{\parallel}^{(3)}$ represents ledge advance velocity. Whereas $v_{\parallel}^{(2)}$ is one given parameter (possibly dependent on affinity, etc.), $v_{\parallel}^{(3)}$ can also be (ledge-) orientation dependent.

Only in ch. 12, which is a survey of the construction of "complete" 3-dimensional $R(\vec{n})$ functions, it is important to discriminate between the 2D and the 3D networks. There, neglect of the 2D networks of the facets speeds up the calculation considerably, but results in functions with orientation-independent ledge advance velocity on a facet.

4) *Dislocations*

In practice, even a 3-D MF function does not yet describe the growth rate of most *real* crystals, because dislocations act as velocity sources and obscure the cusps. An extra assembling step is required to amend the MF functions so as to account for their influence. This will be discussed in 12.6.

Parameter fields and conformity maps.

SC, SF and MF assembled functions are functions of an *orientation* variable $\vec{n}$ (one for the 2D, two for the 3D case) and also of a number of parameters v, M, ε etc. One of these parameters is the roughening parameter, extensively discussed in 11.1 etc. (for MF functions, we have as many roughening parameters as families). In the context of the present section, we choose to consider this roughening parameter (ε) as an additional variable of the assembled function, more or less equivalent with an orientation variable. Moreover, we choose it to range between the values 0 and 1. As explained in 10.3.1, sub 4, this is not a real limitation because such a limited range can be extended (even to infinity) by a suitable "hidden" assembling step. Our choice for the range 0 – 1 corresponds with a) the physical relation with the value of $R/\mathbf{S}R$ (see eq.(11.21); b) the mathematical identification of ε with sin $\alpha^{(4)}$ where $\alpha^{(4)}$ is a "tilt into the 4th dimension" as explained in 10.4.1.

As a result, we can visualize the variable space of the assembled function as a field with dimensions for the orientation *and* the ε variable(s). In fig. 11.7, we have represented just one $\vec{n}$ and one ε variable, such as required for a 2D SC function or a 2D SF function. The variable space is then the rectangle 2π (for $\vec{n}$, or rather $\vec{\varphi}$) $\times$ 1 (for ε). For more $\vec{n}$ or ε variables the variable space is extended accordingly: for 3D SC or SF functions we need one extra $\vec{n}$ dimension, for MF functions we need as many ε dimensions as families to be discerned.

In such multidimensional parameter fields, we can distinguish (lower-dimensional) loci where the networks (=functions) are essentially simpler than the general case, because certain elements disappear. In particular we will refer to the following loci or limit cases:

1) *the OD limit*: corresponds with the complete absence of any roughening, i.e. with $\varepsilon = 0$ (for a certain cusp or family of cusps)

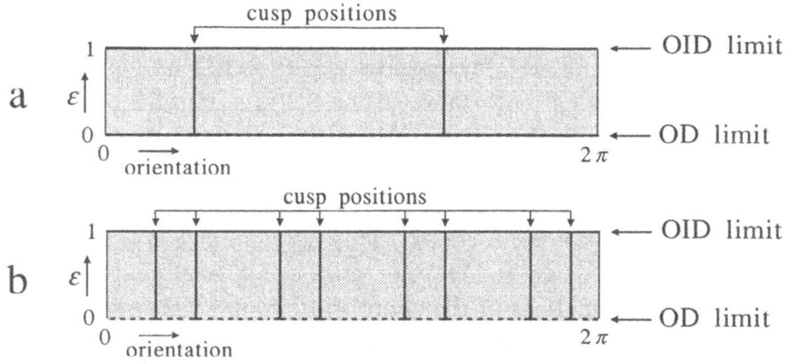

FIG. 11.7 Parameter fields / "conformity maps" for SC (case a) and SF functions (case b)

2) *the OID limit:* corresponds with complete roughening, i.e. with $\varepsilon = 1$ and with the disappearance of orientation dependence. (for a certain cusp or family of cusps to which this ε refers)
3) *the cusp positions:* correspond with those $\vec{n}$ or $\vec{\varphi}$ orientations where one of the OD_i starting functions (of a certain SC or SF) have a zero-cusp, i.e. $OD_i = 0$.

In fig. 11.7, the OD limit and the OID limit simply correspond with bottom and top boundaries of the $2\pi\times1$ rectangles. For 2D SC functions we count one ("monopole") or two (dipole; $|\vec{n}\cdot\vec{p}|$) cusp positions = vertical lines in fig. 11.7a. For SF and MF we can distinguish one or more families of cusp positions (fig. 11.7b).

We can investigate the behaviour of a particular network topology, by comparing the function corresponding with the network either with experimental data or with another, theoretical function which might be derived from an atomistic model. A logical approach for such a comparison is to start with particular choices for the parameters (= network elements) where the functions are inherently simpler: i.e. to consider both comparands on the loci referred to above (OD and OID limits, and cusp positions). Although a fair agreement between comparands on these loci does not guarantee agreement in the whole parameter field, an apparent *disagreement* may justify rejection of the corresponding network. In this context we might speak of "conformity maps" represented on the parameter fields of fig. 11.7. The "conformity" refers to the agreement between the network and either the experiment or the (independent) atomistic model. (Actual representation of this conformity requires some graphical convention: e.g. we can use drawn lines for fair, dotted for poor conformity)

A special conformity is the agreement between the simple SC function or network and the more complex SF or MF functions, describing more of the "single cusps" but at multiple positions. For the SF or MF functions we may require that the OID limit and the cusp functions are identical with those of the SC function; for the OD limit, of course, conformity is only possible within a certain range around the zero-cusps. We will see in 11.4.2 and 11.4.3 that we can make different choices for the topologies of SC and SF functions; these choices also differ in this "SC-SF conformity". This may be, already, an argument for preferring a particular topology, although a final choice cannot be made without an analysis of the conformity *between the network and an appropriate atomistic model.* Whereas the latter subject is out of scope in this monograph, the SC-SF conformity will be referred to in the discussion on SF and MF networks in 11.4.3 and 11.4.4.

Note: In the following subsections 11.4.2 and 11.4.3 we are going to sort out the possible network constructions by a systematic analysis of "all possible" arrangements. The elements of the network are the OD element(s), limiting (L or M) elements, the roughening parameter element $\varepsilon\, v_{\parallel}$ and eventual range-deforming elements.

All elements in a network have the same physical dimension; for a growth rate function, this is velocity. (also the limiting elements L and M have to include a velocity factor). Apart from this, the range in which we can freely choose a parameter has to be specified. The OD element is simply an $|\vec{n}\cdot\vec{p}|$ or $|\vec{n}\wedge\vec{f}|$ element, multiplied with a velocity factor $v_{\parallel}$. $(0 < \mathrm{OD} < 1)$.

The arguments involved in the analysis are of two kinds: a) the topological aspects: which network arrangements can in principle obey a conformity map as sketched in figs 11.7 (for conformity with an atomistic model or with experiment); b) what is the minimum, or the most economical, number of separate elements in a network required to account for the physical details to be represented.

The two alternative constructions for the OD-OID transition

In the OID limit ($\varepsilon = 1$), the OD element(s) has (have) to be "hidden from view" of the rest of the network. In principle, there are two alternative ways to accomplish this: i.e. either by *interrupting* the path in which it occurs or by *short-circuiting* it. Consequently we can distinguish two possibilities for the construction of R-function networks, which we will refer to as interrupt-type and short-circuit type. The difference is illustrated by the SC function case discussed below.

As a consequence of this difference, two different procedures (i.e., topologically different "network types") can be distinguished for the assembling of functions which represent one or more families of facets. Below we will discuss these differences. In 11.4.6 we will compare the two procedures and adstruct our preference for the "short-circuit" type.

Our arguments in the following subsections will be illustrated by figs 11.8 a-i. The successive-assembling procedures are presented as well in the "tree" representation (capital letters) and in the network representation (lower-case letters). Finally, the list of corresponding formulas is given in table 11.1.

11.4.2 THE SINGLE-CUSP (SC) CASE

Consider figs. 11.8a and 11.8b. For both cases, in the $\varepsilon = 0$ limit the circuits represent just the OD element. In the $\varepsilon = 1$ limit, fig. 11.8a represents an interrupted and fig. 11.8b a short-circuited connection. In both cases, the rest of a more extended network in which a circuit (one of figs 11.8a or b) is incorporated then does not see the OD element, but with different consequences.

(note that interruption as well as short-circuiting of the OD element involves the use of inverted assemblands. In the case of interruption, the OD element is made zero by the $(-)\ \varepsilon$. OD *parallel* element, in the case of short-circuiting it is shunted by a "zero resistance" = infinite-flux *series* element 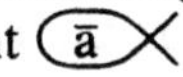

$\varepsilon\ v_{\parallel}$, $(-)\ v_{\parallel}$  . Of course, we might also interrupt a path including an OD element by an element containing $(1-\varepsilon)$ "in series", or by adding the factor $(1-\varepsilon)$ to the OD element itself. But this is, in fact, just a variation on the same theme, because $(1-\varepsilon)$ is just the parallel arrangement of 1 and $(-)\varepsilon$ with assembling exponent = 1).

Now this consideration affects only one of the requirements characterized by the map of fig. 11.7a: i.e. that orientation dependence be eliminated in the OID limit $\varepsilon \to 1$. However, there are other requirements which are not yet fulfilled. So the network of fig. 11.8a still has sharp zero-cusps for all values of $\varepsilon < 1$, whereas in the roughening concept we required smoothed cusps with height ε. In the case of fig. 11.8b, the growth rate goes to infinity for $\varepsilon \to 1$; but in most models we prefer to prescribe a finite maximum M.

These other requirements can be met by adding more network elements. Such extended networks (still for the SC case) are shown in fig. 11.8c (extension of 11.8a, i.e. the "interrupt" case) and in fig. 11.8d (extension of fig. 11.8b, i.e. the "short-circuit" case). At first sight, we have added even more extra elements than needed for a smoothed cusp and prescribed maximum. These extra elements have been inserted for two reasons: a) both networks (fig. 11.8c and 11.8d) now have identical OD and OID limits and cusp functions; b) the networks are more logical SC modules for constructing SF networks by series connection.

Remarks:

1) Although the functions corresponding with networks 11.8c and 11.8d are not identical in the whole parameter field, they do have identical OD limit, OID limit and cusp function. However, in the arrangement as shown the OD limit is not identical with the *OD function itself* (which has a maximum $v_{\parallel}$); instead the OD-limit function is distorted for higher values so as to reach a maximum M. The maximum can be restored to the original OD function by choosing $M = v_{\parallel}$.

The OID limit (for both networks) is M.

The cusp function is $\simeq \varepsilon\, v_{\parallel}$ for low ($\ll 1$) values of ε. For $\varepsilon \to 1$, the cusp functions reach the limit M, but the slope ∂(cusp function)/$\partial\varepsilon$ (in ε=1) is different for interrupt-type and short-circuit type network.

2) The OD elements have been left unspecified because they depend on the 2D or 3D character. (the functions are made explicit only from ch.12 onwards). They can be read as $v_{\parallel} \left|\vec{n} \cdot \vec{p}\right|$ or $v_{\parallel} \left|\vec{n} \wedge \vec{f}\right|$ where $v_{\parallel}$ is a kink or a ledge velocity. The limit M can be chosen between 0 and ∞ .

3) In 11.1.3 we pointed out that an unambiguous definition of the roughening parameter ε (according to eq. (11.20b) implies that the velocity factor in the cusp-smoothing elements, both in fig. 11.8c and 11.8d. is identical with $v_{\parallel}$. Both $v_{\parallel}$ and ε depend on parameters such as temperature and affinity. But whereas $v_{\parallel}$ is in general, to first approximation, a linear function of the growth affinity, the roughness "phase transition parameter" ε rather switches from $\simeq 0$ to $\simeq 1$ as a function of T, within a narrow temperature range near T_R . Also, even for $T \gg T_R$ but $T < T_M$ (melting temp. of bulk crystal), ε should remain slightly below the value 1 (see fig. 11.9).

4) Although a single "parallel" network is for convenience represented in fig. 11.8c, we might in principle use two different assembling exponents for the ${}^{(1)}\mathcal{A} = (\underline{a}_{(1)}) \!\!\times\! \text{OD}, (-)\varepsilon\,.\,\text{OD} \!\times\!\!\circ$ and for the ${}^{(2)}\mathcal{A} = (\underline{a}_{(2)}) \!\!\times\! {}^{(1)}\mathcal{A}\,,\, \varepsilon\, v_{\parallel} \!\times\!\!\circ$ assembling. The same remark applies to the three "series" assembling steps involved in fig. 11.8d.

5) Remember that a special case of fig. 11.8c, namely with (OD) = $\left|\sin\alpha\right|$, $M = 1$, assembling exponent = 2, can be written in terms of a 4-dimensional vector length (eq. (10.37).

6) In the short-circuit network 11.8d, the parallel path accounting for roughening relieves the indeterminacy of the site (in the network) for a series "resistance" limitation: it can be situated either in the parallel branch containing the OD term (and in such a position, we will use the symbol L for this limit) or in series with the result of the smoothing operation (in that position we use the symbol M). The effect is quite different, as can be verified when ε approaches 1 ($\varepsilon' \to \infty$): in that case L is short-circuited where-as M still imposes a limit. In the $\varepsilon \to 0$ limit, however, the distinction between L-type and M-type limitation vanishes. What is the physical relevance of this distinction?

As the M-type limitation is independent of the value of ε, we may identify it with a limit imposed by any mechanism *which has no relation with the surface structure*. In the first place it can be associated with a resistance situated in the bulk adjacent to the interface (e.g., 3-D diffusion in a boundary layer of the parent phase, thermal diffusion etc.) In the second place, even a surface-based resistance can have to be represented by an M-type limit, i.e. when it is associated with a residual resistance in the $\varepsilon \to 1$ limit. (the "growth resistance" of the completely roughened interface is not necessarily zero).

On the other hand, an L-type limitation is to be associated with a surface-based limitation anyhow, e.g. a surface-diffusion limitation or a limit related with adsorption-desorption kinetics, *at least insofar as such a limit vanishes on roughening of the facet*.

Altogether, L-type and M-type limits are discernible as well for their algebraic effect as for the kind of physical effect represented. Although the latter subject is related with the atomistic level of the models to be used and will not further be considered here, we have to be aware of the difference between both kinds in our general R-function cookery book.

So far, the L element has been considered as part of a SC network, where its rôle can be clearly recognized. In the next sections we will see that the extension to SF and MF networks implies a complication which more or less obscures the physical rôle of the L element (see item 3 below).

11.4.3 THE SINGLE-FAMILY (SF) CASE

In principle, we have to construct the network for a SF function by series connection of n (>1) SC modules ("multiple exposure principle"; see 10.3.2 sub3). The SC modules are identical apart from the cusp positions ($\vec{p}_i$ or $\vec{f}_i$ vectors) of the OD elements. We can do this either with the interrupt-type (fig. 11.8c) or with the short-circuit type network (fig. 11.8d) as a point of departure.

The M element has to be excluded from the n -fold repetition in the series; it should be included only once (otherwise, the OID limit would be $n^{1/a}.M$, instead of M as required for SC-SF conformity). With this amendment, we obtain fig. 11.18e for the interrupt-type and fig. 11.8f for the short-circuit type SF networks.

Remarks:

1) We can again compare the OD and OID limits and cusp functions of interrupt-type and short-circuit type networks:

* The OD limits are identical, = $(\bar{a} \times M, (-)\, n^{1/a}\, v_{/\!/}\,, \mathrm{OD}_1\,, \mathrm{OD}_2, \ldots \times\!\circ$ (at least when the L elements are omitted from the short-circuit modules). Just as in the SC function case, the maximum can be made equal to $v_{/\!/}$ by choosing $M = v_{/\!/}$.

* The OID limits are identical; = M.

* The cusp functions are nearly identical: both for fig. 11.8e and fig. 11.8f the i-cusp function can be written as

$$(\bar{a} \times M, (-)\, v_{/\!/}\,, \varepsilon\, v_{/\!/}\,, \underset{k \neq i}{(\bar{a} \times} \text{(other (k) modules)} \times\!\circ \times\!\circ$$

where only $\underset{k \neq i}{(\bar{a} \times}$ (other (k) modules)$\times\!\circ$ is (slightly) different for short-circuit or interrupt-type modules. However, the inclusion of $(\bar{a} \times$ (other (k) modules)$\times\!\circ$ in the straightforward series connection of fig. 11.8e and f affects the conformity between the SF and its prototype SC function.

2) This SC-SF conformity can be restored, i.e. the suppression by $(\bar{a} \times$ (other (k) modules)$\times\!\circ$ of the cusp functions relieved, by eliminating the horizontal connections in fig. 11.8f and using *only one* smoothing $(\bar{a} \times \varepsilon\, v_{/\!/}\,, (-)\, v_{/\!/} \times\!\circ$ element in the parallel branch. In doing so, we have again to exclude the L element from n -fold repetition, in order to have the OD limit function restricted by L rather than by $n^{1/a}.L$.

This results in the adapted short-circuit type SF network of fig. 11.8g. (note: in the drawing we have used the more general element $(-)L^*$ instead of $(-)n^{1/a}.\, v_{/\!/}$). The cusp function of this *short-circuit type* SF network is *identical* with that of the SC module. *But the possibility of assembling such a "reduced" network is restricted to the short-circuit case: this forms an essential difference between the interrupt and short-circuit principles.*

This fact, together with the possibility of distinguishing between L- and M-type limits, forms the argument for our preference for the short-circuit type network above the interrupt-type analog. (on the other hand, an

n-dependence of the smoothed minima in the cusp functions (as implicit in figs. 11.8e and f) may be physically realistic, because roughening is associated with loosening of PBC's - the more bonds, the less roughening).

3) *"Physical" and "mathematical" L elements*

In remark 6) on the SC function case we anticipated on a complication to L-type elements for SF and MF functions. This complication is related with the "compensating assembland" introduced in 10.4.4.

We have shown there that the fit of a series-assembled multicusp function (such as the series of OD elements in the left branch of fig. 11.8g) can be improved by adding a compensating assembland $(-)L^*$. Its addition implies that the maxima in-between cusps are somewhat lifted. (or lowered, for the special case $m = 1$ in eq. (10.56).

The mathematical rôle of the $(-)L^*$ element - i.e., its position in the network - is exactly similar to that of the "physical" L elements introduced in 11.4.2, sub 6). Indeed, also the latter affects the maxima in-between cusps. Of course, the $(-)L^*$ assembland is relevant only in the case of family functions where more than one OD functions are combined - therefore, the rôle of the "physical" L element can best be recognized in a single-cusp function. In a family function, we may take the "physical" L and the "mathematical" $(-)L^*$ together as

$$L_{\text{eff}} = (\bar{\mathrm{a}})\!\!\times\; L,\,(-)L^* \;\times\!\!\!\circ \qquad (11.30)$$

In other words, the effect of the $(-)L^*$ element can also be considered as an effective zero-shift for the L element. From a physical point of view this may mean that it is difficult to discern between any "physical limiting process" and the "competition between neighbouring facets" - or rather, that the latter kind of competition is the very kind of mechanism which should be listed under the category of physical limits. Nevertheless, in the networks of fig. 11.8 and the trees of fig. 11.9, we have represented the L and $(-)L^*$ elements separately in order to emphasize the distinction.

(*Note:* In the discussions of 10.4.4. and 10.4.5, the $(-)L^*$ was calculated as a numerical quantity, related to the amplitude (=1) of the $|\vec{n} \cdot \vec{p}|$ or $|\vec{n} \wedge \vec{f}|$ products. Therefore, in the network and tree representations of figs 11.8, 11.9 etc. we have to include the same $v_{\parallel}$ velocity factor in L^* which is also implicit in the OD and ε terms. The "physical" L (and M) elements are to be expressed as velocities anyhow.

For the magnitude of the $(-)L^*$ elements we refer to section 10.4.4. The special cases of symmetric families (2D and 3D) are represented at the "symmetric" level of fig. 12.1).

<text continued on page 328>

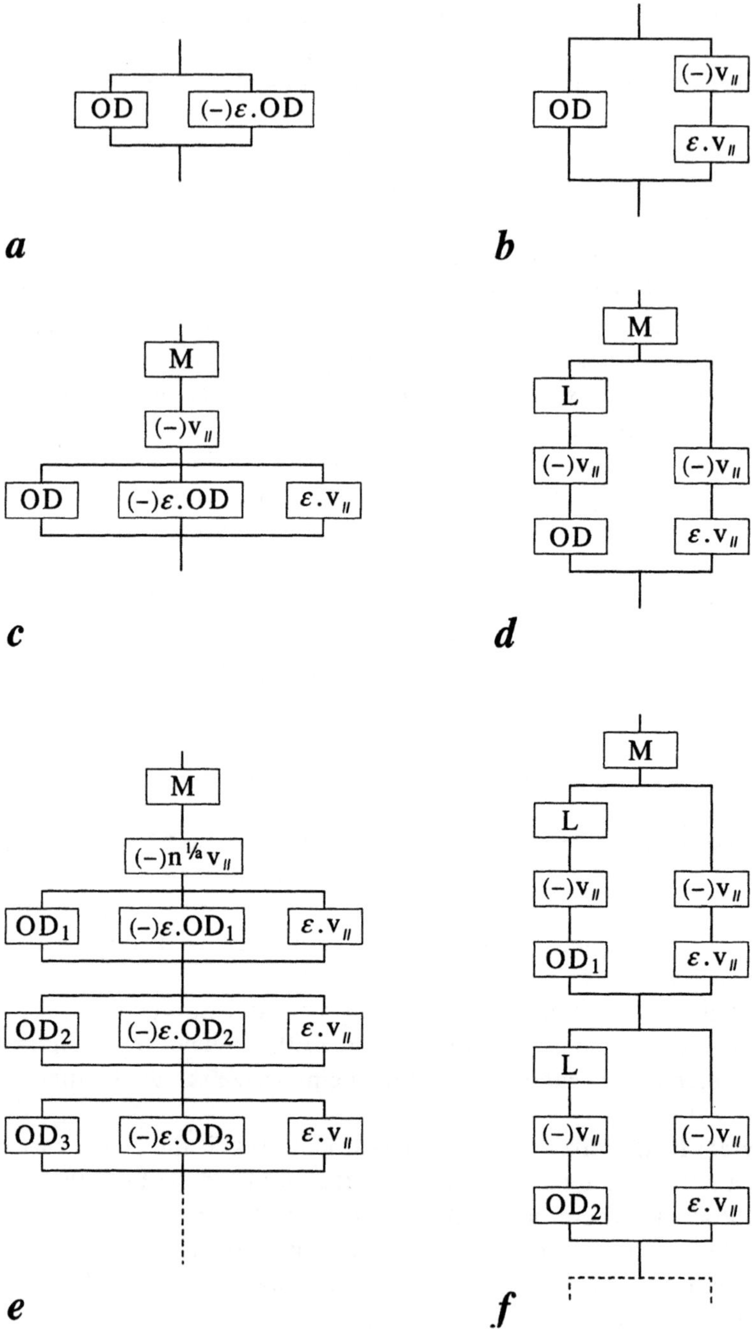

FIG. 11.8 a-f Network representations of SC and SF *R*-functions.

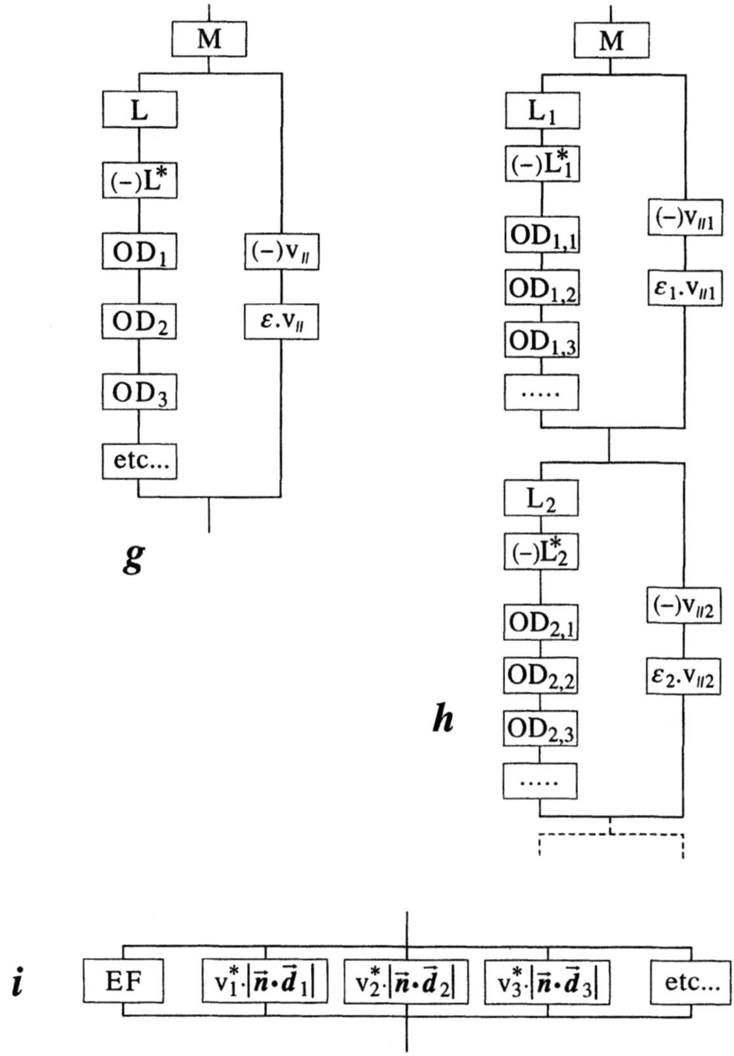

FIG. 11.8 g-i Network representations of SF and MF *R*-functions and the "dislocation amendment".

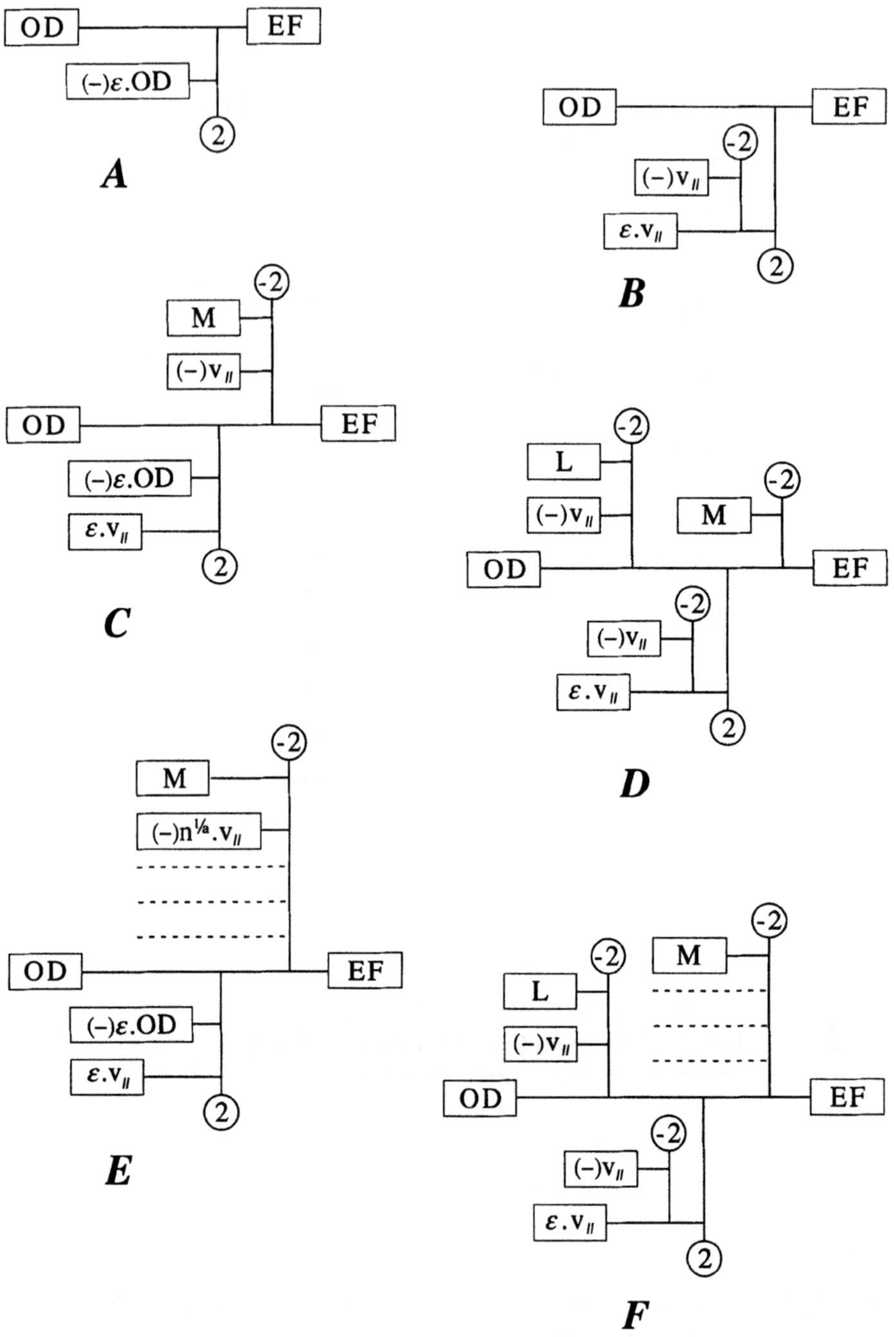

FIG. 11.8 A-F Assembling trees corresponding with FIG. 11.8 a-f

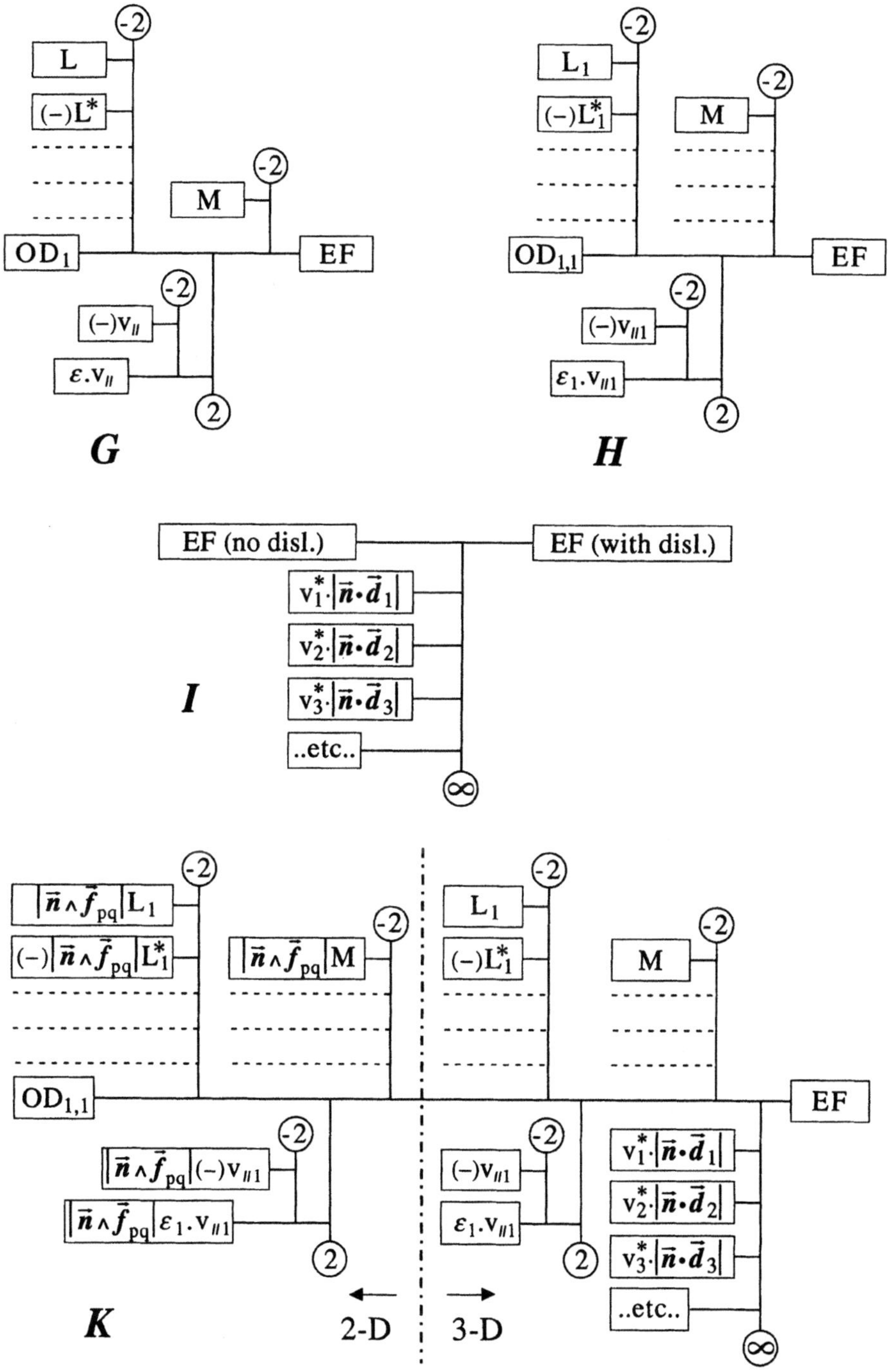

FIG. 11.8 G-K Assembling tree equivalents of Fig. 11.8 g-i. plus "2D embedding" (K)

The following table lists all assemblings occurring in the networks and trees of figs. 11.8 on the preceding pages.

TABLE 11.1

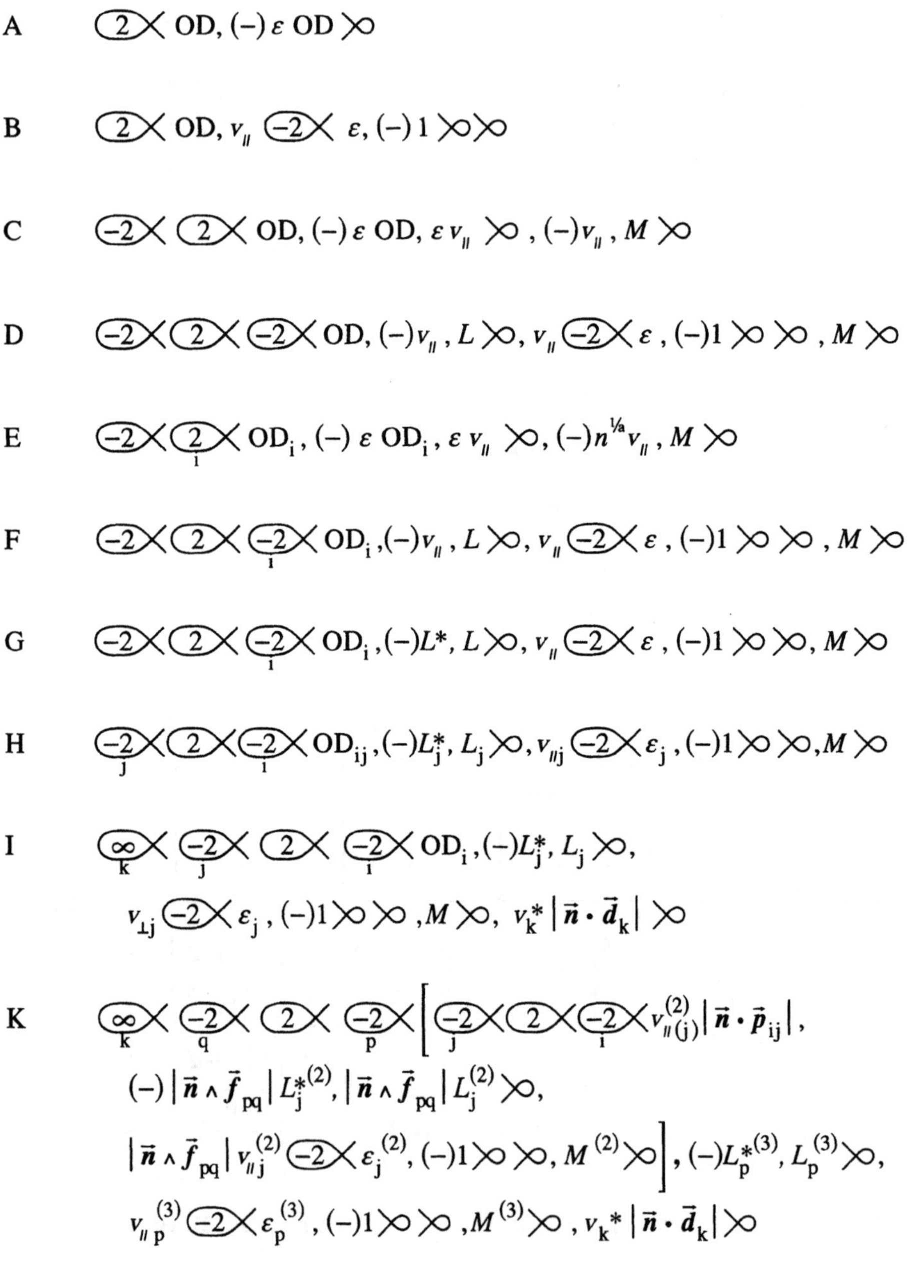

Comments to figs 11.8 and table 11.1:

1) Functions A through H represent 2D- as well as 3D- *R* functions. The suffix i counts SC functions within a family, the suffix j counts the families.

2) Function I (= function H with "dislocation amendment"; see ch. 12.6) is physically relevant for 3D *R* functions only.

3) For 2D *R* functions, the SC OD function should be read as $v_{\parallel} \, |\vec{n} \cdot \vec{p}|$ (with suffices i or ij if appropriate), for 3D functions as $v_{\parallel} \, |\vec{n} \wedge \vec{f}|$ (again with suffices i or ij if appropriate)

4) Function K represents a 3D function *with 2D detail* (see section 12.3), with the highest "detail level" possible; i.e. multi-family in 2D as well as in 3D, plus dislocation amendment. (fig. 11.8K). Accordingly, we now need *different* suffices to count the 2D SC functions (i) and families (j), and the 3D SC functions (p) and families (q). The embedding procedure explained in 12.3 is represented by the multiplication terms $|\vec{n} \wedge \vec{f}_{pq}|$ for the 2D parameters. In order to separate the 2D assembling stages (left of the dot/stripe lines in fig. 11.8K) visually from the 3D stages, they have been put between [ ] brackets in expression K (on page 326).

Note: in *figure* 11.8K (on page 325), the (2) and (3) superfixes have been omitted from the *L, M, ε* etc. symbols for reasons of space shortage.

5) Although we have written +2 and −2 for the positive and negative assembling exponents, these values are not really obligatory - actually the values of the assembling exponents are to be considered as parameters of the model under consideration.

Only for the "roughening" procedure, the value +2 seems the most appropriate (see 11.1.4).

For the SF assembling step in *symmetric* cases, the assembling exponent −2 has the advantage that the assembled expressions can be replaced by product expressions. (in the 2D case the expressions are identical, in the 3D case almost identical; see 10.4). However, this is not a *physical* argument.

11.4.4 THE MULTI-FAMILY (MF) CASE

Finally, for a MF network we have to combine the OD elements belonging to more than one family, in such a way that each OD element has to be smoothed by an ε_i parameter representative for its own family. In the case of the *interrupt network*, the extension from the SF to the MF case is straightforward - because all modules are just connected in series, there is no basic difference between the MF and the SF cases at all. (for this reason, the interrupt-type MF network is not represented explicitly in fig. 11.8). Still, we are confronted with the restriction that the M elements for all families have to be identical - but this is just the expression for the fact that the behaviour of the completely roughened surface is independent of the original family territory.

Of course, just as in the SF case, the height of the smoothed minima of one family is affected by the introduction of another family, or by the change in the ε parameter of another family.

The *short-circuit* version of the MF network is obtained by straightforward series connection of the SF networks, again with only one M element included. This, in fact, is the same method as we used for the SF interrupt-network case, and it has the same consequence, i.e. that smoothed minima of one family are affected by a change in another families' roughening parameter. This network is represented in fig. 11.8h.

A simplified "single-loop" MF network, in analogy with the transition from fig. 11.8f to fig. 11.8g, is not possible. However, we might in principle insert a more complex "ε branch" instead of the $(\bar{a}) \times \varepsilon v_{\parallel}, (-) v_{\parallel} \times\!\circ$ shunt in fig. 11.8g which generates a family-specific ε dependent on the facet-family "territory" (see 10.3.3 and equations 10.32 - 10.36).

11.4.5 Point group symmetry, the kaleidoscopic construction principle and reduction of the assembling set

In 11.4.2, when introducing the "facet family" concept, we only accounted for the *similarity of the cusps* in the assembled SF function (i.e. the agreement between the F_{ass} function in a certain range around each of the cusps and the original SC function). We did *not* use the (point-group) symmetry of the cusp *orientations* as such (apart from the calculation of L^* elements in certain highly symmetric cases). In the present subsection we argue that this symmetry may be exploited so as to obtain a faster, and maybe even a physically more realistic, assembling algorithm.

The point group symmetry of a crystal implies that *only one member* of each "family" has to be specified in detail, after which the other members of the same family are defined by the relevant symmetry (i.e., they are copies displaced by a combination of reflections and rotations). In the 2D case, the $\vec{n}$ unity circle can be subdivided into a number of sectors which are identical apart from reflection or rotation, so that just one of them can be considered as a prototype "elementary cell". In the 3D case, the total area of the $\vec{n}$ unity sphere can be subdivided into a number of either spherical triangles or "biangles" (meridian-bounded sectors) which are again identical by reflection and/or rotation. Only one of these triangles or "biangles" is in principle sufficient to cover the whole unity sphere, provided the rotation + reflection prescription is given. (this prescription is uniquely related with the symmetry group of the crystal under consideration).

When all symmetries involved are mirror symmetries, we can consider this as a "kaleidoscopic" construction. For instance, any cubic crystal is perfectly determined by only one "elementary (conical) kaleidoscopic cell" (EKC) bounded by the (100), (110) and (111) mirror planes (covering only 1/48 of the total 4π space angle). In an actual model of such a kaleidoscope we can see the complete crystal with every conceivable facet. When also rotational symmetries are involved, we might still speak of a "kaleidoscopic principle", but an actual working model of such a kaleidoscope requires either some sort of stroboscopic arrangement or a computer screen.

For each "single family" of cusps, this implies that the EKC *contains exactly one family member*. This has two consequences:

1) For the *starting functions* $|\vec{n}\cdot\vec{p}_i|$ or $|\vec{n}\wedge\vec{f}_{ij}|$, the EKC contains exactly one $\vec{p}_i$ or $\vec{f}_{ij}$. (more exactly: considering the spherical triangle or biangle of the EKC, we can find exactly one $\vec{f}_{ij}$ perpendicular to the surface and/or one $\vec{p}_i$ tangent to the surface, out of the whole set of symmetrically-identical $\vec{p}$ and $\vec{f}$ vectors). Once the point group of the crystal is given, it is sufficient to give the data for only the $\vec{p}_i$ *within* the EKC - or on its border - the others then can be found by rotation and reflection (just on the basis of the symmetry).

2) For the *end functions*, the kaleidoscopic principle implies that it is sufficient to calculate a $R(\vec{n})$ or a $\overline{Q}R(\vec{n})$ function for the orientations within the EKC only. For other orientations, the end function is identical apart from the proper rotation/reflection combination.

Reduction of the assembling set

Consequence 2) mentioned above does not imply in itself that *for the calculation of the end function* (according to the assembling networks of fig. 11.8 g or h) we would not need the $\vec{p}_i$ and/or $\vec{f}_{ij}$ positions of all the other family members (not in the EKC). However, comparing the *influence* of family members on the shape of the end function within the EKC, we can see that (OD_i) functions based on $\vec{p}_i$ or $\vec{f}_{ij}$ vectors *in direct neighbour images of the EKC* have the biggest influence.

It is therefore a logical step to take only these neighbour images into account for the assembling, when we want to economize on algebraic complexity or computer time. This step is quite comparable to the reduction to "nearest-neighbour interaction" in atomistic statistics. Of course, the simplification obtainable by this reduction depends on the crystal symmetry (for the lowest symmetries, there is no reduction at all).

In the 3D case, neighbouring "images" of the EKC can be subdivided into side-sharing and corner-sharing neighbours; including the latter implies, of course, a larger group. Because a common side is, generally speaking, a $\vec{\boldsymbol{p}}$ direction, the physical relevance of the side-sharing neighbours involves ledge competition, of the corner-sharing neighbours kink competition.

Reduction of the assembling set results in an assembled function which fits more closely to the limit Min (all remaining assemblands). In this sense, this restriction has a similar (not algebraically identical) effect as an increase of $|\bar{a}|$ or as a compensating (–) L^* element (see 11.3).

Further remarks:

1) Of course, reduction of the set of assembled functions in a "SF assembling step" to only neighbouring cells also affects the L^* parameter. To obtain a closest-fitting assembled function compatible with a given assembling exponent *and a reduced set of assembland functions,* the appropriate L^* value has to be worked out numerically along similar lines as followed in 10.4.4.
2) For the special case where all symmetries can be considered as *reflections*, the series-assembled functions of a family *with exponent* $-\infty$ (i.e., the assembling which just takes the minimum of all functions considered) *coincides with the elementary OD function within the EKC*. The neighbour images are thus, in this case, even completely irrelevant. This is not true for *rotational* symmetry, in general.
3) In order to represent the simplification in network, tree or formula representations, we have to make two things explicit: a) the range of OD functions is reduced to those related with $\vec{\boldsymbol{p}}_i$ or $\vec{\boldsymbol{f}}_{ij}$ in the EKC and in direct neighbour images of the EKC; b) With this reduced set, the resulting end function has not the required symmetry, because the nearest-neighbour assumption is only valid for the EKC we started with. Therefore, the *effective* end function should be constructed by *replicating* the assembled function in the EKC area and *pasting* it into the other $\vec{\boldsymbol{n}}$ ranges in accordance with the relevant point group. The symmetry of this point group automatically ensures that the "cut-and-pasted" end function has smooth derivatives (except for the $(-\infty)\times$ case).

4) The "nearest-neighbour" procedure is applicable as well to SF as to MF functions. Application to the MF case implies that a) the SF components on their own are assembled in the EKC only, over the reduced set of neighbour images; b) In the combination of the SF functions into the MF function, we have again to consider only the EKC + neighbour images, where all families are to be represented.

11.4.6 Short-circuit-type versus interrupt-type networks and a note on "de-roughening" (the $\varepsilon \to 1$ limit)

In the following we discuss, as a matter of principle or as a thought experiment, the hypothetical case that only one facet family is relevant. Moreover, the ε parameter is considered here purely as a phenomenological parameter expressing orientation dependence, without reference to any explanation in terms of surface roughness, etc. The roughening temperature T_R is supposed te be appreciably below the melting temperature T_M.

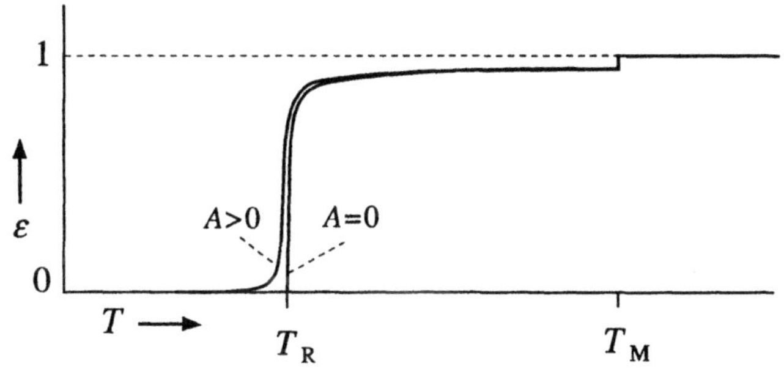

FIG. 11.9 Intuitive sketch of ε versus T behaviour; a) equilibrium ($A = 0$); b) growth ($A > 0$)

Considering the dependence of the ε parameter on temperature, we may expect two major features:

* a fairly sudden transition from 0 to $\simeq 1$ in a narrow temperature range near T_R. For $A = 0$, $\varepsilon = 0$ up to T_R; for $A > 0$ a "toe" may develop for $T < T_R$ (kinetic roughening).

* However, for the whole temperature range $T_R < T < T_M$, a slight orientation dependence should remain - i.e., $\delta = 1 - \varepsilon$ should keep a finite positive value reflecting the remaining influence of the crystalline bulk lattice. Only at $T = T_M$, the bulk melting transition, this δ (which may in practice be as low as 0.001) drops to zero in another sharp transition. These features are indicated schematically in fig. 11.9.

In 11.4.1, when discussing the possible alternatives for constructing networks to represent $R(\vec{n},A)$, we based our preference for the "short-ciecuit type" alternative on rather intuitive grounds. We have to emphasize that these arguments were related with the first deviations from the perfectly faceted crystal, i.e. with the $\varepsilon \ll 1$ side of the ε range; they are applicable only to the situation at $T \simeq T_R$. However, in this very $\varepsilon \ll 1$ range, the difference between the short-circuit and the interrupt-type networks is smallest (the OD limit for both networks being identical).

We may remark here that the other end of the ε range is relevant in the much wider temperature range $T_R < T < T_M$. The slight remaining orientation dependence is responsible for shape selection in cellular and dendritic growth patterns and therefore of considerable physical and technical interest. As remarked above, at this side of the ε range the argument for preference of the "short-circuit" network is hardly relevant, and the choice may be considered as an open question again. Indeed, the physical reality in the $T > T_R$ situation, and therefore the preferred model of the interface itself, may be rather different from that in the $T \simeq T_R$ situation, so that even a different network might be preferable at both ends of the ε scale.

Whereas the difference between short-circuit and interrupt-type networks is only slight for $\varepsilon \ll 1$, it is appreciable for $\delta = (1-\varepsilon) \ll 1$ (near the OID limit). More in particular, we may approximate $R(\vec{n},\delta)$ for the 2-n polygon case :

$$\begin{aligned} R(\vec{n},\delta) = R_o\,(\delta = 0) &+ a_1\,\delta &&+ a_2\,\delta^2 &&+ \dots \\ &+ b_1\,\delta \sin 2n\alpha &&+ b_2\,\delta^2 \sin 2n\alpha &&+ \dots \\ &+ c_1\,\delta \sin 4n\alpha &&+ c_2\,\delta^2 \sin 4n\alpha &&+ \dots \end{aligned} \qquad (11.31)$$

where R_0 is the OID limit;
a_1, a_2 etc. are the first-, second- etc. order in δ, OID shifts, b_1, b_2 etc. the the first-, second- etc. order in δ, second-harmonic OD shifts, etc.

The coefficients a_i, b_i etc. can be found by writing out the expressions explicitly, substituting $\varepsilon = \delta - 1$ and working out the coefficients. The difference between short-circuit type and interrupt-type functions can now be expressed in terms of the differences between the coefficients a_i, b_i, c_i etc. for both cases. As so many parameters are involved and no explicit models are evaluated in this monograph, the question as to the preferred choice has to be left open here.

Chapter 12

SYSTEMATIC ASSEMBLING OF $R(\vec{n})$ FUNCTIONS

One of the principal objects of theoretical research in my department of knowledge is to find the point of view from which the subject appears in its greatest simplicity.

J.W. Gibbs (1839-1903)

12.1 INTRODUCTION.

In 11.4 we have generated the networks corresponding with $R(\vec{n})$ functions from the physical requirements for such functions, including the possibility to describe a roughening transition for each facet family. We have introduced the concepts of single-cusp, single-family and multi-family functions, and shown how to construct the final function by assembling steps: single-cusp into facet-family, and facet families into an end function. (Consequently, we might call this the "facet procedure", as opposed to the "equator procedure" to be defined later in 12.5).

For the representation of the stepwise assembling procedure, we used, alternatively, the "network" or the "assembling-tree" representations which were explained in 10.5.

In the present chapter, we discuss a few finishing touches to complete our construction manual:

1) We can discern cases of increasing complexity in orientation dependence of starting functions (corresponding with crystal symmetry). Although this is not significant at all from a fundamental point of view, it does have practical consequences for the compactness of representation and the speed of computer calculations, and therefore it is worth while to work out the more symmetric cases separately into the "prototype" kind of function (see ch. 9.1).

2) The networks "derived" in 11.4 can be considered as complete (with the proper starting functions and parameters substituted) for 2-D functions, and for 3-D functions without 2-D detail (i.e., without the detail of the ledge-velocity orientation dependence on the facets). However, it is perfectly possible to define 3-D $R(\vec{n})$ functions *including* such detail. The step which is required for "embedding" the 2-D functions into the 3-D ones is explained in 12.3.
3) Under the condition that the 2-D roughening parameter of monatomic ledges on the facets is low enough, a growth rate function can be considered to be determined by *direct assembling of* $|\vec{n}\cdot\vec{p}|$ *vector dot products.* This implies that, in sphere-growth experiments, *equator bands* develop instead of discernible facets. The omission of the smoothing (a > 0) assembling steps simplifies the total assembling procedure considerably, as all resulting (a >0) steps can then be taken together. This simplified assembling procedure - i.e. without the roughening steps - can be characterized as the *equator procedure* and will be discussed in 12.5.
4) Finally, the effect of *dislocations* (acting as velocity sources) can be expressed as a final assembling amendment (12.6). (The networks as discussed up to now only concern "perfect" crystals, i.e. without dislocations).

12.2 A SURVEY OF ASSEMBLING PROCEDURES FOR 2D AND 3D $R(\vec{n})$ FUNCTIONS

Figs. 12.1 a, b and c have been designed as a concise and still complete summary of the main body of section 11.4 and ch. 12. They represent a 3-dimensional space (rendered in isometric perspective) with "vertical", "horizontal" (left-right) and "depth" dimensions. Items 1) through 5) below explain the details:

1) Assembling procedures are represented as assembling trees drawn on the "horizontal" planes. Starting and end functions are situated left and right, respectively, whereas the "assembling lines" are drawn in the depth direction.

2) In the vertical direction we discern three levels, corresponding with cases of different symmetry:
1) upper level: complete rotational symmetry; i.e. OID "OID"
2) middle level: *m*-symmetric "SYMM"
3) lower level: general case "GEN"

3) 2D and 3D functions: These three levels can be discerned for the 2D as well as for the 3D case. Consequently, fig. 12.1 has a 2D and a 3D version, with identical structure apart from the starting functions. When only 3D functions without 2D detail are considered, the 2D and the 3D versions can be considered as independent. But when 2D detail is to be incorporated into the 3D function, the end functions of the 2D version provide the starting functions for the 3D function. The "embedding procedure" involved is discussed in 12.3.

4) Assembling procedures evolve from left to right. On the way, the stations of SC, SF and MF functions are passed. If required, these stations can also be considered as end stations as indicated by the squares (2D) or boxes (3D) on the corresponding vertical bars.

(We remind the reader of the convention that dotted lines entering an assembling line symbolize functions which are identical with the fully represented one up to the merge point, apart from PBC vector directions and family-specific parameters. See 10.5).

The nature of the detail and of the symmetry of an assembled function can be characterized by the names of the end stations or "complexity labels". For 2D functions:

---	---	(2D - OID)
---	(2D - SF/SYMM)	(2D - MF/SYMM)
(2D - SC/GEN)	(2D - SF/GEN)	(2D - MF/GEN)

and for 3D functions, the corresponding 3D stations. (the OID "functions", of course, are trivial but are represented to complete the cataloguing picture).

Rendering of 2D detail in 3D functions is, of course, only possible on the 3D-SYMM or 3D-GEN levels. The complexity of such 2D-detailed 3D functions is to be characterized by a combination of (2D) and (3D) complexity labels.*)

5) In connection with the above, it is of interest to "count" the number of different parameters of one kind which is required to characterize a function. (when this number is >1, these parameters can be labeled in a systematic

*) Strictly taken, the distinction between the general and the symmetric case is nonexistent from a topological point of view: e.g. the symmetric $|\vec{n}\cdot\vec{p}|$ function describes already on its own a "regular polygon" with two sides (poles) and might therefore to be considered to belong to the "SYMM" category. Even the monopole might be situated there. The reason why we reserved a distinctive "symmetric" category is rather an algebraic one, namely that the corresponding functions have a special and simple representation (especially for assembling exponent = −2). By their simplicity of representation such functions can be considered as prototype functions in the sense of 9.1.

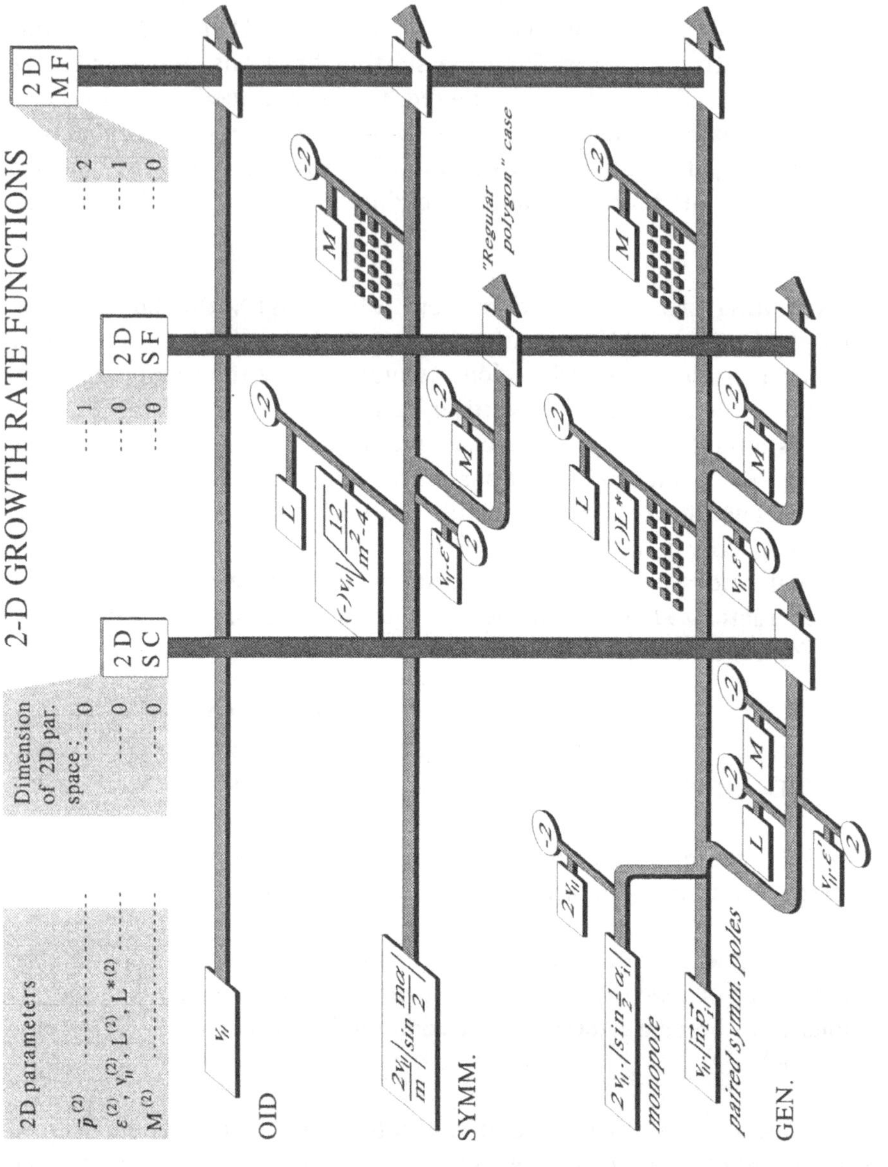

FIG. 12.1a (this page) and 12.1b (next page). For 3D R functions with 2D detail, the functions obtained by the procedures of fig. 12.1a have to be used as an input for the procedures of fig. 12.1b. Similarly, fig. 12.1c can be connected "in series" with the end functions of fig. 12.1b.

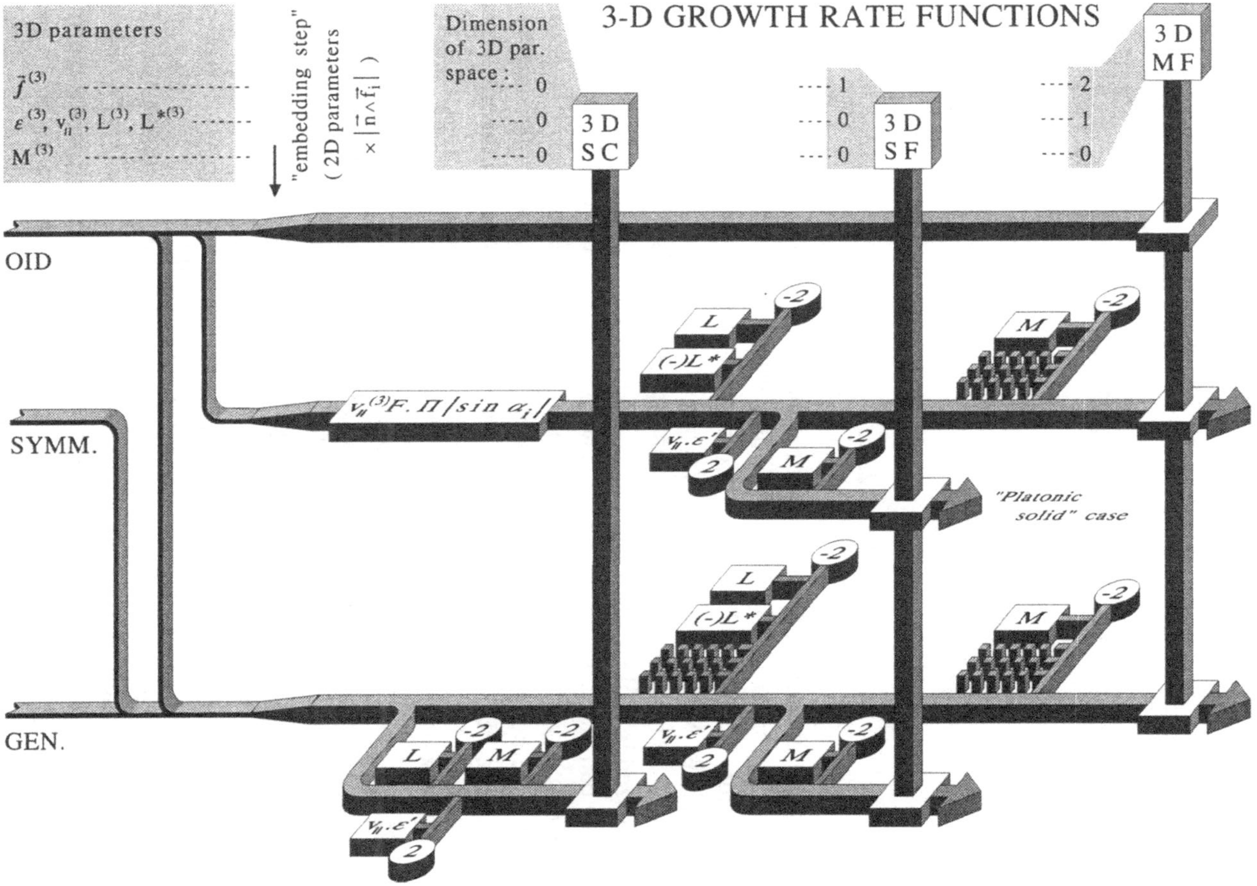
3-D GROWTH RATE FUNCTIONS
3D parameters
$\vec{f}^{(3)}$
$\varepsilon^{(3)}, v_{//}^{(3)}, L^{(3)}, L^{*(3)}$
$M^{(3)}$
"embedding step"
(2D parameters $\times |\vec{n} \wedge \vec{f}_i|$)
Dimension of 3D par. space :
0 0 0
1 0 0
2 1 0
3 D SC
3 D SF
3 D MF
OID
SYMM.
GEN.
$v_{//}^{(3)}F.\,\Pi/sin\,\alpha_i/$
L
(-)L*
M
-2
2
$v_{//}.\varepsilon'$
"Platonic solid" case

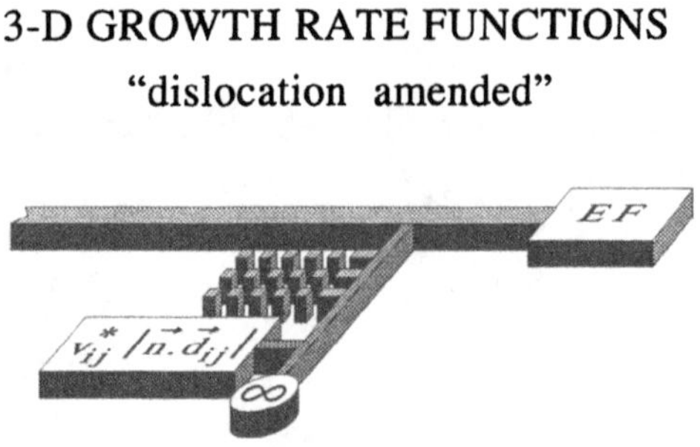

FIG. 12.1c

way by one or more indices). These parameters are indicated in the shaded areas at the top of figs 12.1a and b. They are subdivided into groups according to the "dimension" of their space, which can increase on going from SC to SF to MF functions. This dimension is represented in the shaded areas attached to the vertical function-type bars in fig. 12.1a-c, and again summarized in table 12.1.

We can discern *geometrical* parameters (i.e., the set of PBC vectors $\vec{p}_i$, and eventually the facet normals $\vec{f}_{ij}$) and "*physical*" parameters ($v_{\parallel}$, ε, L, L^*, M). Although their mathematical rôles cannot be distinguished, we consider the PBC vectors as fixed and given a priori (for a certain crystal).

Remarks:

a) SC and SF functions both have only one of the ($v_{\parallel}$, ε, L, L^*, M) parameters; the difference is in the number of *vector* parameters involved: one for SC, a "family" for SF; i.e. dimension = 0 for SC and =1 for SF. In agreement, the "single-family assembling" (in fig. 12.1a and b represented between the SC and SF vertical bars) is the "multiplication" of the single cusp by crystallographic point symmetry. This means that, in the computer algorithm, we only have to invoke the symmetry group of the crystal under consideration. ("kaleidoscopic principle"; see 11.4.5).

Also, this SF assembling step is the one in which the number of assemblands can eventually be reduced to "neighbour cells" in the kaleidoscopic construction principle (see 11.4.5, reduction of the assembling set). In that case, the end function represents the complete function only within the EKC; the really complete end function has to be constructed by "cutting and pasting" that end function to all other crystallographically identical positions.

The step from SF to MF functions involves an increase in the vector-parameter *and* in the ($v_{\parallel}$, ε, L, L^*) group dimension.

b) When 3D functions have to include 2D detail (see 12.3), then the 2D parameters also occur in the 3D functions. For a 3D SC function, we have a 1- or 2-dimensional $\vec{p}^{(2)}$ variable space (for representing SF of MF 2D detail, respectively), and accordingly 0- resp 1-dimensional $\{\varepsilon^{(2)}, v_{//}^{(2)}, L^{(2)}, L^{*(2)}\}$ variable space and 0-dimensional $M^{(2)}$ variable space. The step from 3D SF to 3D MF functions also involves an increase in all the relevant 2D-parameter variable spaces, including that of the PBC vectors and $M^{(2)}$.

In table 12.1, "imposed" (vector) parameters are marked by < > brackets, 2D-detail-in-3D-function parameters by ( ) brackets.

c) Labeling of parameters:
In a complete function, all parameters have to be labeled by a number of indices corresponding with the dimension represented in table 12.1. These indices are also the "assembling indices" in the relevant assembling steps represented in the trees. (In SF functions, we assemble over ($\vec{p}$ or $\vec{f}$) cusps, in MF functions over families).

6) The "dislocation amendment" to be discussed in 12.6 can be considered as a final assembling step. It is represented in fig. 12.1c which is to be considered as a last step connected to the arrows of the end functions in fig. 12.1b.

TABLE 12.1

Dimension of parameter space for 2D and 3D R($\vec{n}$) function parameters

PARAMETERS			2 D			3 D		
PBC vectors	physical dim.less	physical dimensional (velocity)	SC +SF symm	SF gen.	MF	SC +SF symm	SF gen.	MF
$\vec{p}^{(2)}$			<0>	<1>	<2>	<(2)>	<(3)>	<(4)>
	$\varepsilon^{(2)}$	$v_{//}^{(2)}, L^{(2)}, L^{*(2)}$	0	0	1	(1)	(2)	(3)
		$M^{(2)}$	0	0	0	(0)	(1)	(2)
$\vec{f}^{(3)}$						<0>	<1>	<2>
	$\varepsilon^{(3)}$	$v_{//}^{(3)}, L^{(3)}, L^{*(3)}$				0	0	1
		$M^{(3)}$				0	0	0

12.3 THE "EMBEDDING PROCEDURE": 2D DETAIL IN 3D R-FUNCTIONS

The sequenced assembling procedures symbolized by fig. 11.8a/A – i/I can be considered to represent a 3D growth rate function when the OD starting function is a simple explicit function of the normal vector $\vec{n}^{(3)}$ variable over the unit sphere. This is the case when OD can be written as $v|\vec{n} \wedge \vec{f}|$; i.e. when the ledge advance rate on the $\vec{f}$ facet is independent of the 2D orientation vector $\vec{n}^{(2)}$ in the facet plane.

Here we want to consider the case that $v = v(\vec{n}^{(2)})$ is a 2D-*orientation-dependent* function. Of course, we can still use figs 11.8 a/A - i/I as a valid point of departure; only we have the complication that we have to make a preceding calculation to find $v(\vec{n}^{(2)})$. Actually, $\vec{n}^{(2)}$ is the projection of $\vec{n}^{(3)}$ onto the facet, but renormalized:

$$\vec{n}^{(2)} = \frac{\vec{n}^{(3)} - |\vec{n}^{(3)}.\vec{f}|.\vec{f}}{|\vec{n}^{(3)} - |\vec{n}^{(3)}.\vec{f}|.\vec{f}\,|} \tag{12.1}$$

However, when $v(\vec{n}^{(2)})$ is calculated by a 2D assembling procedure starting from OD functions of vector dot-product type ($|\vec{n}^{(2)}.\vec{p}|$), then it is not necessary to determine $\vec{n}^{(2)}$ itself. This is because the first step in the 3D assembling procedure is a multiplication with $|\vec{n}^{(3)} \wedge \vec{f}|$ (= $|\sin\alpha|$, where α is the angle between $\vec{f}$ and $\vec{n}^{(3)}$). Multiplication of an assembled function with the factor $|\sin\alpha|$ is equivalent with multiplication of *all assemblands* with this factor. (this applies to the assemblands in a > 0 as well as in a < 0 assembling steps; i.e., ε , M, L and numerical constants, *plus the* $|\vec{n}^{(2)}.\vec{p}|$ *products*. Now the latter multiplication has the effect of replacing $\vec{n}^{(2)}$ by $\vec{n}^{(3)}$, so that $\vec{n}^{(2)}$ drops out:

$$|\vec{n}^{(2)}.\vec{p}_i|\,.\,|\vec{n}^{(3)} \wedge \vec{f}| = |\vec{n}^{(3)}.\vec{p}_i| \tag{12.2}$$

$$(\ |\vec{p}_i.\vec{f}| = 0\ )$$

So altogether, the 3D SC function can be calculated on the basis of the assembling procedure for the 2D $v(\vec{n}^{(2)})$ function, *by just replacing* $\vec{n}^{(2)}$ *by* $\vec{n}^{(3)}$ *in the OD element and adding a multiplication factor* $|\vec{n}^{(3)} \wedge \vec{f}|$ *to all other elements.*

(still, as this has to be done for all facets of all families individually, the calculation is complicated considerably - unless the assembling set is reduced according to the "kaleidoscopic principle" of 11.4.5!)

The "embedding procedure" is symbolized by the expansion of the path "ribbons" in fig. 12.1a into path "tubes" in fig. 12.1b.

12.4 *POLYGONAL, "PLATONIC" AND REGULAR-PRISM SYMMETRIES*

In 11.4 we derived the "short-circuit" version of a SF function (fig.11.8g). Both the 2D polygonal and the 3D "Platonic" families are examples of SF functions. As their special symmetry leads to compact expressions for $(\!-2\!\times$ (set of OD terms) $\times\!\circ$ (see 10.4.3 and -4), we will work out these SF functions explicitly.

1) 2D m-symmetrical families:

The assembling tree is represented in fig. 12.1a, "SYMM" level. The multiplicity integer *m* just represents the number of sides (cusps in the functions). The explicit expression is

$$(\!-2\!\times (\!2\!\times (\!-2\!\times\; v_{\parallel}\; \tfrac{2}{m}\left|\sin\tfrac{m\alpha}{2}\right|, (-)\;\; v_{\parallel}\sqrt{\tfrac{12}{m^2-4}}\;, L \times\!\circ,$$

$$v_{\parallel}\; (\!-2\!\times\; \varepsilon, (-)1 \times\!\circ \times\!\circ\;, M \times\!\circ \qquad (12.3)$$

2) The 3D "Platonic" families.

For the Platonic symmetries, as discussed in 10.4.4, we have the choice of using series-assembling expressions (in $(\!-2\!\times |\sin|$ or in $(\!-2\!\times |\tan|$ versions) or the product expressions eqs. (10.60). (see also the inequations (10.61). The latter choice is represented in the assembling tree of fig. 12.1b, middle ("symm") level. The *n* in the denominator of the OD function represents the number of "axes"; i.e. 3 for cube, 4 for octahedron, 6 for dodecahedron and 10 for icosahedron. The $(\!-2\!\times |\sin\alpha_i|$ version has slightly lower maxima.

The structure of the 3D Platonic-symmetry case is, of course, identical with the 2D regular-polygon case. The $(1/m).|\sin(m\alpha/2)|$ factor in the latter is replaced by $\mathrm{F}\Pi\, |\sin\alpha_i|$, by $(\!-2\!\times |\sin\alpha_i| \times\!\circ$ or by $(\!-2\!\times |\tan\alpha_i| \times\!\circ$ in the former. (the $\mathrm{F}\Pi\, |\sin\alpha_i|$ functions are tabulated in 10.4.4).

3) The 3D regular-prism families.

These again can be based:

a) either on the series-assembling expression of fig. 12.1a, with the facet orientations on a *h* equator substituted in eq. (10.65);

b) or on the "product" expression

$$v\,\frac{2}{m}\prod_{j=1}^{m}\sqrt{\left\{1-\left[k\sin\left(\frac{\alpha}{2}+\frac{j\pi}{m}\right)+\ell\cos\left(\frac{\alpha}{2}+\frac{j\pi}{m}\right)\right]^2\right\}} \quad (12.4)$$

Finally, the slightly artificial-looking functions in eqs. (10.66) and (10.67) can be used for a compact representation of 3D- regular prism families with "rectangular" 2D orientation dependence.

12.5 THE EQUATOR PROCEDURE AND ITS RELATION WITH STEPPED EQUATOR BANDS IN SPHERE-GROWTH EXPERIMENTS

All through chapters 11 and 12 up to now, we based the construction principles of our networks on a stepwise assembling procedure going from 2D ledge-motion $R^{(2)}(\vec{n}^{(2)})$ functions to 3D SC functions to SF and MF $R(\vec{n}^{(3)})$ functions. Reviewing all corresponding assembling trees, we can make the following fundamental observation:

The split-up into successive assembling steps is necessary because smoothing steps occur at different stages (2D and 3D): the associated positive-exponent assemblings have to be carried through *before* subsequent negative-exponent "combination" ones. *If no smoothing would have to be accounted for at any stage (neither 2D nor 3D), then all assemblings in the whole tree, from starting functions to end function, would be of "series" type.* If, moreover, all these assemblings can be made with the same (negative) exponent (say −2) *then the whole assembling procedure, including limitations and combinations, can be considered as one long series assembling.*

This implies that all the original starting functions (of type $|\vec{n}^{(3)}\cdot\vec{p}_i|$, each with their proper $v_{\parallel}$ factor) are directly visible in the end function (see below). These functions, therefore, can be called "equator" functions. They are relatively simple: e.g., the orientation dependence of the "cubic" equator function is just represented by $|h\,k\,l|$. (with $h^2+k^2+\ell^2=1$)

The consequence of an equator-type growth rate function can best be illustrated by considering a sphere-growth experiment. As a mental experiment, suppose that we have an $R(\vec{n})$ function based on just one single $|\vec{n}\cdot\vec{p}|$ term:

$$R(\vec{n}) = v\,|\vec{n}\cdot\vec{p}| = v\,|h| \quad (12.5)$$

The growth experiment will produce a "antibiotics-pill" shape (fig. 12.2a: consisting of two separating half-spheres of constant, original radius, with a cylindrical section in-between of length $2vt$). In other words, the sphere develops a cylindrical flat band or "equator band", not separate facets.

When more of such equator bands are present, the sphere expands in more directions: e.g. for the cubic symmetry we have

$$R(\vec{n}) = \textcircled{-2}\!\!\!\times v\left|\vec{n}\cdot\vec{p}_{\mathrm{i}}\right| \rtimes\!\!\circ = v\left|h\,k\,l\right| \qquad (12.6)$$

and the corresponding "sphere" will look as fig. 12.2 b. On the intersection of equators (corresponding with "facet" orientations), polygonal facets will develop, which are connected by cylindrical bands. In-between the cylindrical sections, spherical areas remain. (note that these types of surface correspond with F, S and K type in the terminology of HARTMANN and PERDOK (1955 a, b and c), and with $\overrightarrow{\overrightarrow{\mathbf{S}}}R$ tensors with two, one and no singular elements (ch. 2 of this monograph). Thermodynamically, an equator band might be characterized as an "S orientation below the roughening transition". This extends the concept of surface roughening to apply to cases where only one diagonal element of the $\overrightarrow{\overrightarrow{\mathbf{S}}}R$ tensor is singular).

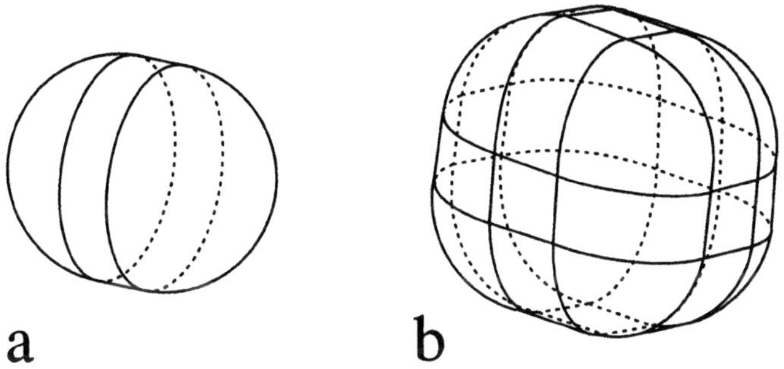

FIG. 12.2 : Equator bands corresponding with an $R(\vec{n})$ function assembled with negative assembling exponents exclusively ("equator procedure") a) one $\vec{p}$ vector; b) three orthogonal $\vec{p}$ vectors.

Such equator bands can actually be observed in sphere-growth experiments, although in practice the cylindrical areas are often stepped bands, composed of facet-oriented zigzags. Sometimes the cylindrical areas break up into a chain of satellite facets of relatively high-index orientations, related with a "family" of surface reconstructions (e.g. GARDENIERS c.s 1990, 1991). Gradual transitions from (partly) equator-type to "normal" facet-type behaviour can be observed for increasing temperature and/or affinity. In the light of $R(\vec{n})$ function-construction principles, such a transition can be considered as the consequence of a decrease of the roughening parameters involved.

The *initial* shape of a facet in a sphere-growth experiment reveals the character of the $R(\vec{n})$ function much better than the *final* shape which is likely to be polygonal anyhow just because neighbouring facet planes intersect in a straight edge. When this initial shape is circular this means that the 2D ledge $R(\vec{n}^{(2)})$ function is OID, i.e. that the "2D roughening is complete". Consequently, the equator-type character of the R function is completely lost for that orientation.

On the other hand, a polygonal initial shape of the facet reveals orientation dependence of the 2D $R(\vec{n}^{(2)})$ function; in other words incomplete 2D roughening or a partial equator-type character of the R function.

In connection with the equator procedure for R functions, we may also remark that the *surface-energy* "soapfroth" γ function is in principle also an "equator-procedure function", but with assembling exponent $a_1 = 1\ (>0)$

12.6 D-JUNCTION VELOCITY SOURCES AS LOW GROWTH-RATE LIMITS

In chapter 5, we have discussed how in crystals with dislocations, the original zero-cusps in the growth rate plot of the dislocation-less crystal are effectively bridged or cut short by an "effective plot" corresponding with the vector velocity of the D-junction(s). On a macroscopic scale (macroscopic as compared with the dimensions of individual dislocation pyramids) it looks as if the crystal were growing in accordance with a growth rate function consisting of the original plot, modified by the bridges. From this point of view, the situation is identical with that of a crystal in which the dislocations *were just another type of PBC's growing with a uniform speed,* to be accounted for by extra $v^*_{ij}\,|\vec{n}\cdot\vec{d}_{ij}|$ terms in the growth rate function. (here, the normalized $\vec{d}_{ij}$ vector is used to characterize the direction of a certain group of dislocations ending on the (ij) facet and imposing its velocity). All dislocations are supposed to be aligned with $\vec{d}_{ij}$ and to grow with the same vector velocity $v^*_{ij}\,\vec{d}_{ij}$, in contrast with real PBC's which only contribute near the $|\vec{n}\cdot\vec{p}|$ cusp. *In accordance, the assembling exponent is positive rather than negative.*

If, for instance for the sake of computer simulation of crystal growth, it is desirable to describe the crystal *with dislocations included* by a single growth rate function including the effect of the dislocations, it is perfectly possible to include the short-circuits in the R function:

a_1) In a very rough way, when $R^*_{\text{disloc.}} \ll v_{\text{ledge}}$ and when we are prepared to neglect the fact that the shape of the facet is actually composed of a number of pseudofacets (i.e. the slopes of the dislocation pyramids), we can just add R^* as a constant to the original function. This amounts to adding a sphere, centered around the origin, with radius R^*. Here we suppose that all facets on the crystal are identical and their dislocations grow ahead with the same velocity).

a_2) As an alternative, but similarly rough way we might simulate the dislocation-induced growth on the facets by a roughening-induced growth rate. This, of course, replaces the actual *sharp double-cusp* corresponding with the zigzag or pyramid structure by a *smoothed single cusp*. This can be permitted if the zigzag detail is not important. For this purpose, we just have to replace the term $v_{//}\varepsilon$ in the expressions or networks by $R^*_{\text{disloc.}}$.

b) More precisely, we can replace the short-circuited parts of the facet cusps by the effective-growth rate spheres - one sphere for each individual facet. (Note that in this case the sphere is passing through the origin, not centered around it; correspondingly the crystal develops vicinal pseudofacets instead of a single facet in the "exact" $\vec{f}_{ij}$ direction). This replacement can be expressed as an assembling step:

1) As the resulting plot can be considered as the *maximum* of the original "dislocationless" $R(\vec{n})$ function and the effective-plot v^* spheres, we can write

$$R(\vec{n}) = \text{Max}\,(R_{\text{orig.}},\ v^*_{ij}\,|\vec{n}\cdot\vec{d}_{ij}|\,) = \textcircled{\infty}\!\!\times R_{\text{orig.}},\ v^*_{ij}\,|\vec{n}\cdot\vec{d}_{ij}| \times\!\!\circ \qquad (12.7)$$

i.e., add a last assembling step of the form (12.7) to the trees in figs. 11.8. The connections between the $v^*_{ij}\ |\vec{n}\cdot\vec{d}_{ij}|$ spheres and the original plot can be made less sharp by choosing a finite assembling exponent instead of ∞. (dropping it to 2 wipes out the difference between dislocation-induced and roughening-induced growth on a facet orientation).
This procedure is illustrated in figs. 11.8 i/I + K and in fig. 12.1c.

2) Consider the hypothetical case of an original plot based on just one $\vec{p}$ vector, i.e. consisting of a pair of spheres: $R_{\text{orig.}} = v\ |\vec{n}\cdot\vec{p}|$. The bridged plot has cusped minima at the intersections between $v\ |\vec{n}\cdot\vec{p}|$ and the $v^*|\vec{n}\cdot\vec{d}|$ pair of spheres. Supposing $\vec{d}\ \cdot\ \vec{p}\ =\ 0$, this can be

rendered exactly by splitting up these sphere pairs into two or more sphere pairs (reduced size) with the minimum directions aligned with the cusps. The cusps correspond to an equation of the type

$$v^*|\vec{n}\cdot\vec{d}| = v\,|\vec{n}\cdot\vec{p}| \;\Rightarrow\; \vec{n}_{\text{cusp}} = \frac{v\vec{d} \pm v^*\vec{p}}{|v\vec{d} \pm v^*\vec{p}\,|}$$

and so, in terms of two new pairs of spheres we have

$$R(\vec{n}) = |\vec{n}\cdot(v\vec{p} + v^*\vec{d})| + |\vec{n}\cdot(v\vec{p} - v^*\vec{d})| \qquad (12.8)$$

Equation (12.8) is an alternative for (12.7) for the case $R_{\text{orig.}} = v\,|\vec{n}\cdot\vec{p}|$. For a more realistic function represented by an *assembling* of $v\,|\vec{n}\cdot\vec{p}|$ functions, we might replace all the $|\vec{n}\cdot\vec{p}|$ functions by sums of the type (12.8). The result is mathematically identical with (12.8).

3) Dislocation-induced growth is typically restricted to the 3D case. Still, 2D analogs of line faults acting as a velocity source can be imagined; also the cross-section through a 3D plot (in a plane through the $\vec{d}$ vectors) behaves as such.

4) The *effect of roughening* on the growth of dislocation pyramids can be considered as a case of competition: the original cusp behaves less sharp and elevated, and the pyramids become less pronounced. Finally, as $\varepsilon v_{\perp}$ reaches the value of v^*, the pyramids "drown" in the competing "B&S" growth and do not increase the growth rate any further.

12.7 ASSEMBLING γ FUNCTIONS

Again, we start discussing the 2-D case. By analogy with the R functions, we are led to "smooth" the cusps in the $|\cos\alpha|$ terms in eq. (9.1) but add a negative assembland:

$$\gamma_i \;\textcircled{a}\!\times\; |\cos\alpha_i|\,,\; \varepsilon_i\,,\; (-)\,\varepsilon_i\,|\cos\alpha_i| \;\times\!\!\circ \qquad (12.9)$$

The negative assembland cannot be γ_i similar to the R case, as this would make the total negative for all α values. However, a negative assembland $(-)\,\varepsilon_i\,\gamma_i$ pins the *minimum* of the (smoothed) function down at zero.

This is not enough. When the i cusp is the only one to roughen, the i minimum in

$$\sum_i [\, \gamma_i \, \textcircled{a}\!\times \; |\cos \alpha_i| \,,\, \varepsilon_i \,,\, (-)\, \varepsilon_i \, |\cos \alpha_i| \times\!\circ \; - \varepsilon_i \gamma_i \,] \qquad (12.10)$$

remains fixed for all ε_i values between 0 and 1, in contradiction with our "physical" concept of roughening as expressed in fig. 11.5 and the corresponding text. In fact, the basic mechanism leading to the simultaneous smoothing of the cusp and lowering of the minimum is based on the *interaction* between the "i" PBC ends (kinks) at the ledge and all other ($k \neq i$) PBC ends represented by the $k \neq i$ terms in the summation. In order to express such an interaction mathematically, we have to multiply these $k \neq i$ terms with a factor behaving as $(1-\varepsilon_i)$. The requirement to make this factor symmetrical for all terms, leads us to a factor

$$[\, 1 - \underset{k \neq i}{\textcircled{a}\!\times} \; (\textstyle\sum \varepsilon_k) \,,\, 1 \times\!\circ \,] \qquad (12.11)$$

(the assembling limitation to 1 is introduced because the factor should remain positive even when all ε's involved are = 1, but of course this expression is only one "guess" out of a large class of functions f obeying $0 \leqslant f \leqslant 1$; lim f = 0 ; lim f = 1. Altogether, the freedom - or rather uncertainty - in this term is much larger than in the *R*-function case).

The product of terms (12.9) and (12.10) would lead to zero value of γ in the limit $\varepsilon \to 1$ for all ε_i . This is not physically realistic; rather we should have a "liquid-surface limit" $\gamma_o > 0$. We might construct this by adding an orientation-independent term

$$\gamma_o \; \textcircled{a}\!\times \; (\Sigma\, \varepsilon_i) \,,\, 1 \times\!\circ \qquad (12.12)$$

ensuring that the "localized" depressions induced by the multiplication term eq. (12.11) are restituted as uniformly distributed OID terms. Again, this term is subject to ambiguity.

With all amendments, the final expession reads

$$\begin{aligned} \gamma(\vec{n}) = \sum [\, \gamma_i \, \textcircled{a}\!\times \; |\cos \alpha_i| \,,\, \varepsilon_i \,,\, (-)\, \varepsilon_i \, |\cos \alpha_i| \times\!\circ \; - \varepsilon_i \gamma_i \,] \times \\ \times \; [\, 1 - \textcircled{a}\!\times \; (\sum_{k \neq i} \varepsilon_k) \,,\, 1 \times\!\circ \,] + \\ + \gamma_o \; \textcircled{a}\!\times \; (\sum_i \varepsilon_i) \,,\, 1 \times\!\circ \end{aligned} \qquad (12.13)$$

Altogether the whole expression is less satisfactory than that for the *R* function. Still, it is better than the unamended expression because the $\varepsilon \to 1$ transition is rendered in a way which corresponds with actual physical behaviour at least in the $\varepsilon \to 0$ and $\varepsilon \to 1$ limits.

12.8 EPILOGUE

In retrospect, the message of chapters 9-12 is that $R(\vec{n},A)$ and $\gamma(\vec{n})$ functions can be represented as the admittance of a multiply connected passive "nonlinear" network.

On the one hand, this is a very general statement which allows for a wide choice of "physical" models. Details of such models have to be represented or translated by a) the configuration of the network itself, in a "topological" sense; b) the values of the parameters (network elements). Determination of both configuration and parameter values has to be the subject of a detailed "atomistic" consideration of surface equilibrium and kinetics which has been left out of our present monograph.

The need for such a thermodynamic extension is felt most painfully in the case of the $\gamma(\vec{n})$ function. In this chapter, the discussion of the γ function has remained underdeveloped for this reason - a realistic choice of the topological network structure and of the roughening assemblands has to be based on a statistical-thermodynamic approach of an atomistic model of the surface. In contrast, the construction of networks for the $R(\vec{n})$ function case is less dependent of thermodynamic aspects because much of the network structure can be understood without it. Only the meaning of the roughening parameter(s), and their dependence on T and A, is a subject for which thermodynamics is essential.

The "theory" presented in chapters 9 - 12 is *not* a theory on interface structure or kinetics. It is only an *analogy* between a mathematical construction and the Kossel model. Properties of the (abstract) mathematical construction show a striking similarity with those of the Kossel crystal and with actual crystals in growth experiments; details of the construction can be projected on details of the Kossel crystal. Still, physical arguments for network structures and assembling exponents can be given for many details, such as a = 2 for roughening and other arguments given in chapter 11.

The great advantage of the nonlinear-network approach is that it provides "assembled" $R(\vec{n})$ functions *defined for continuous* $\vec{n}$. (unlike most atomistic models). The dissection of the R and γ functions into networks with discrete elements provides a logical basis for the development of a continuum description of crystal growth. This includes the possibility of numerical simulation of growth experiments and a new theoretical approach to interface instability phenomena. The discrete elements in the network represent parameters such as kink and ledge velocities, limits imposed by diffusion, reaction kinetics or low pressure, and the roughening parameter. The dissection of the network allows these elements to be considered *apart from one another* in their dependence on temperature, affinity etc. Some of these dependences can be guessed intuitively; e.g., linear relation for $v_{\parallel}(A)$. The separate elements form an easier target for a thermodynamic approach than the analysis of a "tangled" network as a whole. Once the dependences are established for the network elements on their own, the behaviour of the complete network can be deduced from direct substitution. In this way, we may hope that, ultimately, physically realistic functions for $R(\vec{n},A)$ and $\gamma(\vec{n},A)$ can be constructed (i.e., including the explicit expression of dependence on affinity). Such functions would allow us to understand and simulate very "nonlinear" phenomena like cellular and dendritic growth.

But even without this extension, the parametrized functions can be used for direct computer simulation of 2D and 3D growth experiments such as sphere growth and growth in grooves and on mesa structures and masked substrates. Inversely, the parameters themselves can be "measured" by confronting the actual and the simulated experiments.

A striking result of the assembling approach to R and γ functions is the almost "cabbalistic" characterisation of the differences between a) R and γ functions, b) growth on facet-oriented surfaces induced by roughening and induced by dislocations.

For "R versus γ ", the difference is characterized by the exponent of the *combination assembling step* for all participating cusped functions. For the γ case, $a = +1$; for the R case $a = -2$ (or in any case $\leqslant -1$)

For "roughening versus dislocations", the difference is characterized by the exponent of the assembling step of *single cusped functions* and the *facet-orientation growth rate*. For the roughening case we have $a = +2$, for the dislocation-induced case, $a = +\infty$.

Extension of the theory to compositional aspects.

The present description, of theory as well as in the computer simulation, is restricted to unary systems, implying that only *geometrical* aspects of orientation dependence are considered. As we have roughly described in 8.5, it is possible to extend this to polynary systems, by defining "partial growth rates" for each species. This brings also the compositional aspects within view. (Orientation-dependent composition of epitaxial layers is of technical relevance in the growth of "3D" semiconductor devices such as laser and waveguide structures and 1- or 0-dim. quantum-well devices). Mathematically, this extension of the parameter space is to be characterized as a replacement of scalars by "species vectors" (see 8.5).

Chapter 13

NOTES ON SPHERE-GROWTH EXPERIMENTS

The first essential in chemistry is that thou shouldst perform practical work and conduct experiments, for he who performs not practical work nor makes experiments will never attain to the least degree of mastery. But thou, O my son, do thou experiment so that thou mayest acquire knowledge. Scientists delight not in abundance of material; they rejoice only in the excellence of their experimental methods.

Jabir ibn Hayyan (Geber) - 8th century

13.1 INTRODUCTION

Experimental determination of a growth rate function is most effectively done by growth on a spherical (convex or concave) substrate under carefully controlled growth conditions. The most spectacular feature on such a growth specimen is the eventual development of facets and stepped bands connecting some of the facets. For convex surfaces, they reveal faceted low-index orientations ("F" and "S" orientations, in the terminology of HARTMAN and PERDOK (1955 a,b,c), and sometimes the determination of these orientations is the only objective of the experiment: it provides the lowest "data level", i.e. of the *orientation* of those PBC's which are strong enough to play a role in the crystal's morphology. As the next data level we may consider the (minimum) values of the $R(\vec{n})$ function for these PBC directions: these determine which of the "visible" PBC's will generate a facet in the crystal's stationary morphology (the $\overline{\mathbf{Q}}R$ shape) and what will be the size and shape of this facet. In most papers reporting on sphere growth experiments up to now, investigations have been limited to these two data levels.

In this monograph we have repeatedly emphasized that the tensor quantity $\overline{\overline{\mathbf{S}}}R$ is at least as important as R itself. This can be taken both practically (e.g., in the external morphology of crystals and their kinetic behaviour) and theoretically (e.g., because its value undergoes a dramatic change in the vicinity of the roughening transition; much more than R itself). However, the experimental *determination* of $\overline{\overline{\mathbf{S}}}R$, involving derivatives, requires (at first sight!) the measurement of the $R(\vec{n})$ function *as a continuous function of orientation.* For the experimental practice, this implies a much denser grid of measuring points than is required for the "second level" of data, which may partly explain why this "third data level" has so scarcely been explorated up to now.

In other words, it seems that the present (1994) state-of-art in experimental $R(\vec{n})$ function measurement techniques is inadequate for a fruitful exchange between theory and experiment for at least a few fundamental aspects. The challenge to work out more sophisticated experimental techniques prompted me to "invent" an alternative experimental principle which may be called the "discotheque-sphere method". This optical method generates a map of the $\overline{\overline{\mathbf{S}}}R$ function directly, without the intermediate of thickness measurements.

Apart from this new principle, we will in this chapter emphasize the importance of four other aspects of sphere-growth experiments:

1) The possibility of growing on concave and convex substrates together, so that the data derived from these twin experiments can be used as a complement. (on a concave substrate, flat "facet" regions correspond with $R(\vec{n})$ *maxima* whereas the $R(\vec{n})$ minima are revealed as edges and vertices (whose position can also be determined quite accurately).
2) The importance of doing measurements *as a function of time* (preferably *in situ*, or else intermittently)
3) The principle of *orientation mapping* as an alternative for the (conventional) thickness mapping.
4) The importance of optimizing the duration of the growth (i.e., the thickness of the epilayer, as compared with the substrate sphere radius).

In this chapter we brought together several remarks on sphere-growth experiments. (by the way: in the context of this chapter, we will often - for shortness - use the word *sphere* when we actually refer to the non-spherical surface of the growing epitaxial layer *on* the sphere. A "map" of the sphere refers to a map representing either layer thickness or surface orientation, as a function of the spherical coordinates of the original spherical substrate.)

The structure of this chapter is as follows. In section 13.2 we present a catalogue of situations for small positive and negative crystal shapes- a subject which often leads to erroneous assumptions about the nature and the position of the facets. In 13.3 we discuss systematic errors and the resolution related with the method of thickness mapping; we will show that the radius of the initial surface is an important parameter which together with growth time has to be optimized for accuracy.

The alternative of orientation mapping is discussed in 13.4 -13.6. First we show in 13.4 that the values of the $\overline{\overline{\mathbf{S}}}R$ tensor elements, for the special case of a spherical surface, have a direct geometrical significance. In 13.5, orientation mapping and the information which can be derived from it is explained as a principle. In particular we will argue that mapping the orientation *as a function of time* provides a more direct approach to the determination of $\overline{\mathbf{Q}}R$ and $\overline{\overline{\mathbf{S}}}R$ data - and of the $R(\vec{n})$ function parameters related with these quantities - than thickness mapping. Moreover, the method is non-destructive and - once the experimental set-up is available - it is easier to obtain a dense grid of measuring points than by thickness mapping. The "discotheque-sphere method" referred to above is in principle a determination of the orientation distribution function of the sphere, which can - sufficient precautions provided - directly be translated into an $\overline{\overline{\mathbf{S}}}R$ map.

As another application of orientation mapping, we discuss the evolution of the size of facets in 13.6, together with "special effects" in connection with size evolution.

In 13.7, the experimental set-up for orientation mapping is considered in some detail. Basic requirement for this type of measurement is that the surface of the growing epilayer is used as part of the optical image-forming system - either as a lens or as a mirror. (As a consequence, in those cases where interface instabilities generate corrugations on the surface of typical wavelength $\lesssim \lambda$, the method is not applicable (at least not to the corrugated orientations, because diffraction effects affect the image formation.) The possibility to choose between a) convex or concave substrate, b) lens or mirror image formation, together result in four cases for which appropriate optics are sketched in 13.7.2. Furthermore, either optical or electronic image-processing tricks are required to obtain quantitative information on the "orientation" variable, as explained in 13.7.3.

Finally, in 13.7 we discuss the relevance of observing the evolution of facet size *in situ* as a function of growth time. This subject is relevant to the "thickness" as well as to the "orientation" version of mapping.

13.2 *POSITIVE AND NEGATIVE CRYSTALS IN GROWTH, EQUILIBRIUM AND ETCH SITUATIONS: WHERE ARE THE FACETS?*

The term "negative crystal" has become customary for a gas bubble or a liquid parent-phase droplet totally enclosed in a large single crystal. In agreement with this convention, we will also use it here for growth or etch experiments on hollow spheres or concave lenses, for which the radius is taken negative.

The contrasts positive/negative substrate radius, growth/etch, and (growth/etch)/equilibrium all together give rise to a rather confusing number of possibilities, so that it requires some mental gymnastics to be aware of the facet orientations to be expected. A survey of these orientations in the form of a map is presented in fig. 13.2. In order to emphasize the differences between the six cases distinguished in fig. 13.2, we have deliberately chosen a $\vec{\mathbf{P}}R(\vec{n})$ plot with (distinguishible) *noncentro-symmetrical* minima and maxima. This plot is drawn in fig. 13.1 (dashed curve), together with the resulting $\overleftarrow{\mathbf{Q}}R(\vec{n})$ shape (solid curve).

(Note: As our map in fig. 13.2 also involves equilibrium shapes, we would actually require a $\vec{\mathbf{P}}\gamma(\vec{n})$ plot in addition to determine that shape. For convenience, and because our only objective here is to summarize principles, we just consider the plot of fig. 13.1 also as a γ plot for this case, i.e. for the shapes numbered as 2 and 5 in fig. 13.2).

Variables in the map are *sphere radius* and *growth affinity*. Contrast between positive and negative curvature is represented vertically, between growth and etch conditions horizontally. More specific, the horizontal "axis" is intended to represent externally applied growth affinity, the middle corresponding with equilibrium *for a flat interface* - growth is right, etching left. As a consequence, we have to locate *a small positive crystal in equilibrium* a bit to the right of the middle and *a small negative crystal in equilibrium* a bit to the left, to agree with the curvature-induced shift *). In agreement with this, the effective growth affinity felt at the interface (= externally applied + GT shift, i.e. left-hand terms in eq.(1.8)) has to be measured from the GT-shifted position of the equilibrium crystal. (in fig. 13.1 all spheres are supposed to have the same radius initially).

Altogether we distinguish *six cases:* etch / equilibrium / growth (=3), for "positive" and "negative" radius (×2). In fig. 13.2 these cases correspond with six boxes, each containing a sketch of the situation plus the vector-algebraic expression of the shape.

*) The equilibrium can be either unstable (for a a (positive or negative) crystal in an infinite host phase) , or stable (in a finite host phase with rapid enough exchange between both phases). In the thought experiment to be discussed below we may assume that the equilibrium is stabilized, although this is not essential.

Because the occurrence and orientations of *facets* are the special topic highlighted in fig. 13.2, these facets (and their position) are emphasized in the sketches by *double-thick line sections.*

Consider the following thought experiment. By some ingenious trick we are able to start with six perfectly spherical starting geometries at $t = 0$. From $t = 0$ the affinity is active. (for the "equilibrium" case, this affinity originates from the difference between the spherical and the equilibrium shape, i.e. we have to do with capillarity-induced "ageing kinetics"). The question is: where can we expect facets to develop?

The vertical, i.e. concave/convex contrast corresponds with $\vec{n}$ versus $(-\vec{n})$ antisymmetry:

$$\gamma_{\ell s}(\vec{n}) = \gamma_{s\ell}(-\vec{n})\,; \qquad R_{\ell s}(\vec{n}) = R_{s\ell}(-\vec{n})$$

Centrosymmetric facet pairs therefore occur identically in + and – crystals, but in non-centrosymmetric facets (if they occur in the crystal) we see the difference $\vec{n} - (-\vec{n})$ - this is why we had to choose a non-centrosymmetric case to make this difference visible. This applies to equilibrium, growth and etch situations alike.

However, the growth/etch contrast has an inverse effect on $r>0$ and $r<0$ crystals. In increasing-radius situations the facets correspond to *minima* in the $R(\vec{n})$ function, i.e. with $(\vec{p}_i \wedge \vec{p}_j)$ orientations, in decreasing-radius situations to *maxima* (i.e., not necessarily low-index orientations). Assuming that $|R(\vec{n},A)|$ maxima (absolute values) for positive and negative A occur at identical orientations (a fairly safe assumption) - and the same assumption for $|R(\vec{n},A)|$ minima - we can expect to find corresponding maximum-facets in both "shrinking" spheres (numbers 1 and 6) and also corresponding minimum-facets in both expanding spheres (numbers 3 and 4) in opposite directions, $\vec{n}$ and $-\vec{n}$. Also in the equilibrium cases, where both the positive and the negative sphere develop facets in $\gamma(\vec{n})$-*minimum* directions, the facets occur in opposite directions. (of course, the equilibrium shape $\bar{\mathrm{Q}}\gamma$, towards which the original spherical shape is driven by the GT shift, is reached only for $t \to \infty$).

However, the minimum-R_{sl} directions for the growing positive sphere do not necessarily correspond with the minimum-R_{ls} directions for the etching negative sphere: this is only the case for centrosymmetric facet pairs, not for "unpaired" facets. The correspondence between these cases should therefore be characterized as *antisymmetrical.*

Of course, a similar antisymmetric-$\vec{n}$ relation exists for the maximum-$R(\vec{n})$ facets occurring on the etched positive and on the "growing" hollow sphere (decreasing in absolute value of the radius).

Summarizing, for the general case of non-centrosymmetrical minima and maxima in $\gamma(\vec{n})$ and $R(\vec{n},A)$ functions, we can expect:

1) Identical facet orientations in cases 2 and 3, and in cases 4 and 5 (but cases 4 and 5 are antisymmetric with cases 2 and 3)
1b) For centrosymmetric crystals, all cases 2, 3, 4 and 5 yield identical facets (in ($\vec{p}_i \wedge \vec{p}_j$), i.e. low-index orientations)
2) Maximum-rate facet orientations in cases 1 and 6 (antisymmetric, i.e. only coinciding for centrosymmetric crystals)

As a consequence, we might observe *transitions* and/or mixed forms involving *combinations of* $\gamma(\vec{n})$ *minima and* $R_{sl}(\vec{n},A)$ *maxima* by oscillating the affinity for a stabilized small crystal (i.e. a small crystal in equilibrium with a small volume of its parent phase). We should be alert for such mixed forms in observations on very small crystals.

Where are the edges?

The "map" of fig. 13.2 can be transformed into a map "where are the edges"? by just interchanging the words "maxima" and "minima". Strictly taken, the edges - i.e. self-intersections of the curve constructed for the growing surface - need not correspond exactly with $R(\vec{n})$ maxima (in cases 2-5), but their positions are likely to be quite near the positions of the $R(\vec{n})$ maxima orientations on the original sphere.

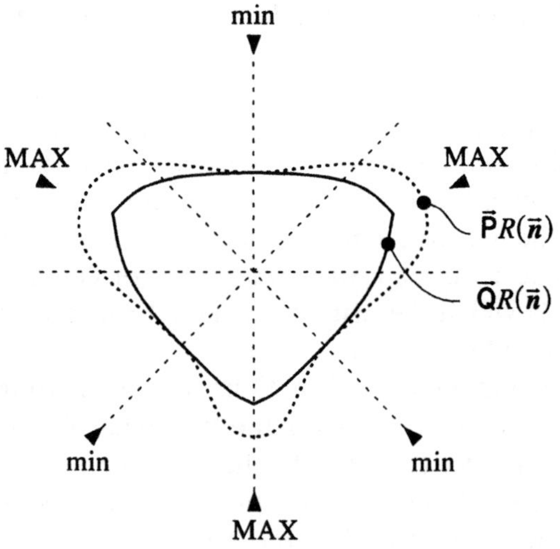

FIG. 13.1

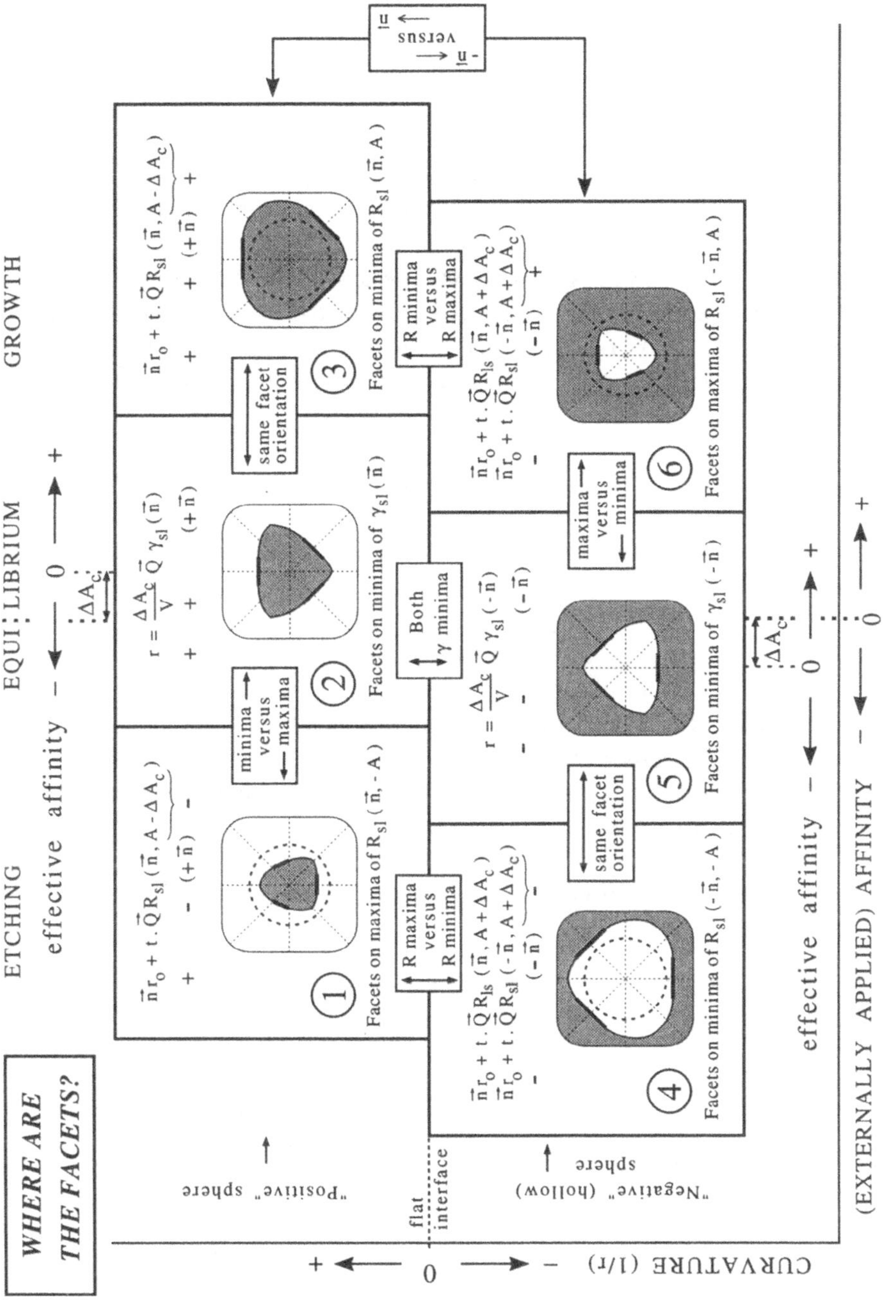
WHERE ARE THE FACETS?
ETCHING
EQUILIBRIUM
GROWTH
effective affinity
same facet orientation
minima versus maxima
R maxima versus R minima
Both γ minima
R minima versus R maxima
maxima versus minima
Facets on maxima of R
Facets on minima of γ
Facets on minima of R
Facets on maxima of R
"Positive" sphere
"Negative" (hollow) sphere
flat interface
CURVATURE (1/r)
(EXTERNALLY APPLIED) AFFINITY

FIG. 13.2

13.3 THE SYSTEMATIC ERROR AND RESOLUTION IN THICKNESS-MAPPING EXPERIMENTS

13.3.1 Systematic error

When the $R(\vec{n})$ function is determined by direct epilayer thickness measurement in a sphere-evolution experiment, care should be taken to compensate for the systematic error introduced by this method. The error is illustrated in fig. 13.3, which shows the spherical starting geometry, the growth rate vectors, and the surface after a time t in a two-dimensional figure. The thickness of the grown layer $d(\vec{n}_0)$, measured perpendicular to the *original* surface and at the location where the original surface was oriented at $\vec{n}_0$, is

$$d(\vec{n}_0,t) = \int_0^t R\{\vec{n}(t)\}\,\mathrm{d}t \qquad (13.1)$$

However, during growth the orientation of the surface at this point changes:

$$\vec{n}(t) = \vec{n}_0 + \frac{t}{r}\left(\frac{\mathrm{d}R}{\mathrm{d}\vec{n}}\right)_{\vec{n}_0} \qquad (13.2)$$

(to first order, and unless a facet orientation is hit).

We can compensate for this systematic error in either of two ways:

1) shift the *orientation* from $\vec{n}_0$ to the (first-order) average position $\vec{n}_0 + (t/2r)(d/d\vec{n})R$, or alternatively
2) keep the orientation fixed but apply a correction to $R(\vec{n}_0)$. (the two alternatives are, of course, mathematically identical to first order).

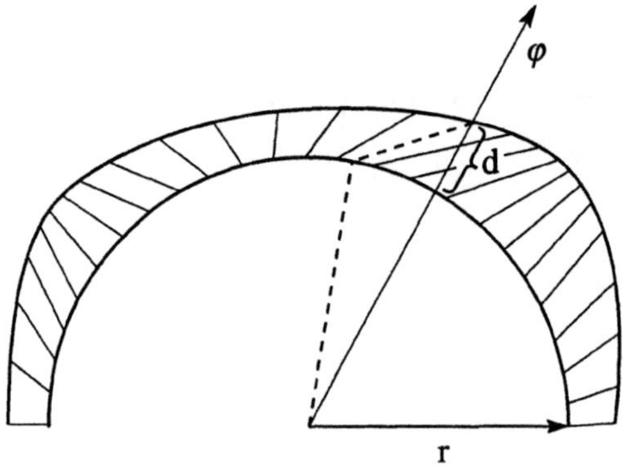

FIG. 13.3

Restricting the correction to first order and supposing that neither facets nor edges are hit, we have:

$$d(\vec{n}_o, t) = \int_o^t R\left\{\vec{n}_o + \frac{t}{r}\left(\frac{\mathrm{d}R}{\mathrm{d}\vec{n}}\right)_{\vec{n}_o}\right\}\mathrm{d}t$$

$$= \int_o^t \left\{R(\vec{n}_o) + \frac{t}{r}\left(\frac{\mathrm{d}R}{\mathrm{d}\vec{n}}\right)^2_{\vec{n}_o}\right\}\mathrm{d}t$$

$$= t\,R(\vec{n}_o) + \frac{t^2}{2r}\left(\frac{\mathrm{d}R}{\mathrm{d}\vec{n}}\right)^2_{\vec{n}_o}\Bigg\}\mathrm{d}t \quad \text{(+ higher-order terms)} \tag{13.3}$$

So, dividing by t to obtain $R(\vec{n}_o)$, the experimental result is in general *too high:* the first-order correction is

$$R(\vec{n}_o)_{corrected} = R(\vec{n}_o)_{exp.} - \frac{t}{2r}\left(\frac{\mathrm{d}R}{\mathrm{d}\vec{n}}\right)^2_{\vec{n}_o} \tag{13.4}$$

The only directions for which the measurement gives an exact result are those where $R(\vec{n})$ has an extremum: facets and maximum-growth rate directions. As $\mathrm{d}R/\mathrm{d}\vec{n}$ is, for "physical" arguments, always finite, the correction can be kept small by taking t small enough:

$$\text{correction} \ll \text{measured } R(\vec{n}) \text{ when } t \ll \frac{2R}{(\mathrm{d}R/\mathrm{d}\vec{n})^2}$$

$$\text{or } d \ll 2r\,\frac{R^2}{(\mathrm{d}R/\mathrm{d}\vec{n})^2} \tag{13.5}$$

This illustrates the importance of choosing a large enough radius of the spherical substrate. (the accuracy of the d measurement increases with t).

When the correction is essential, we have first to determine $\mathrm{d}R/\mathrm{d}\vec{n}$ from the uncorrected $R_{\mathrm{exp}}(\vec{n})$ data. In fact this procedure yields

$$\frac{\mathrm{d}R_{\mathrm{exp.}}}{\mathrm{d}\vec{n}} = \frac{\mathrm{d}R}{\mathrm{d}\vec{n}} + \frac{t}{r}\,\frac{\mathrm{d}R}{\mathrm{d}\vec{n}}\,\frac{\mathrm{d}^2R}{\mathrm{d}\vec{n}^2} \quad \text{(+ higher-order terms)} \tag{13.6}$$

So by compensation with this value, we are left with a residual error which is quadratic in t/r:

$$R_{\mathrm{corrected}}(\vec{n}) = R_{\mathrm{exp.}}(\vec{n}) - \frac{t}{2r}\left(\frac{\mathrm{d}R_{\mathrm{exp.}}}{\mathrm{d}\vec{n}}\right)^2$$

$$= R - \frac{t^2}{r^2}\left(\frac{\mathrm{d}R}{\mathrm{d}\vec{n}}\right)^2\left[\frac{\mathrm{d}^2R}{\mathrm{d}\vec{n}^2} + \frac{t}{2r}\left(\frac{\mathrm{d}R}{\mathrm{d}\vec{n}}\right)^2\right] \tag{13.7}$$

If necessary, an improved correction can be found by an iterative procedure, determining $\mathrm{d}R/\mathrm{d}\vec{n}$ again from the corrected graph.

An important conclusion to be drawn from eq. (13.4) is that, when the layer is so thick that correction is essential, it is not sufficient to measure $R_{\mathrm{exp}}(\vec{n})$ as a function of only one orientation parameter, e.g. along a great circle. The correction actually involves the derivatives of $R_{\mathrm{exp}}(\vec{n})$ in *two* directions:

$$\frac{t}{2r}\left(\frac{\mathrm{d}R_{\mathrm{exp.}}}{\mathrm{d}\vec{n}}\right)^2 = \frac{t}{2r}\left[\left(\frac{\mathrm{d}R_{\mathrm{exp.}}}{\mathrm{d}\varphi}\right)^2 + \left(\frac{\mathrm{d}R_{\mathrm{exp.}}}{\mathrm{d}\vartheta}\right)^2\right] \tag{13.8}$$

where φ,ϑ are local tilting angles about perpendicular axes in the surface. The layer thickness chosen for a measurement will in general be a compromise, the accuracy increasing linearly with thickness but the systematic error also increasing, quadratically or more (after correction). The only way of measuring $R(\vec{n})$ free from errors is choosing $r = \infty$, i.e. a flat substrate. The thickness measurement then has to be done at a distance

$$\ell \geqslant \frac{\left|\dfrac{\mathrm{d}R(\vec{n})}{\mathrm{d}\vec{n}}\right|}{R(\vec{n})}\,d$$

from the edges of the substrate, and the thickness grown has to be limited so as to allow for this condition (which can only be verified after a series of measurements giving R as a function of $\vec{n}$).

13.3.2 Resolution

Apart from the systematic error mentioned above, we have to be alert for two other complications in sphere-evolution experiments:

1) *"Hidden orientations"*:

The orientation $\vec{n}$ disappears altogether after some time when at least one of the elements of the $\vec{\vec{S}}R$ tensor, in diagonal form, is negative:

$$[\vec{\vec{K}}^{-1}]_{\mathrm{kwt}}(t) = r_o\,\vec{\vec{I}} + t\,\vec{\vec{S}}R \tag{13.9}$$

The corresponding curvature radius becomes zero (i.e., the surface develops an edge) as soon as $t \geqslant r_o.\vec{\vec{S}}_{11}R$ (or $d \geqslant R.\vec{\vec{S}}_{11}R$), assuming that $\vec{\vec{S}}_{11}$ is the lowest negative element of the $\vec{\vec{S}}R$ tensor. (however, we will find that the first-order correction validity range is by far exceeded before this happens).

2) *Resolution*:

The resolution of the $\vec{n}$ parameter decreases as a function of thickness, because the measurement, after correction, yields an average of $R(\vec{n})$ in the range from $\vec{n}_o$ to $\vec{n}_o + (t/r)(dR/d\vec{n})$; all details within this range are lost. In particular, this is relevant when we want to determine $R(\vec{n})$ in the direct neighbourhood of a facet direction in order to investigate deviations from the "ideal" sin α dependence and/or to calculate the parameter $\vec{\mathbf{S}}R$. The dimensions of a facet area start to increase linearly with t, with a growth rate $(dR/d\vec{n})_{\vec{n}\,\text{facet}}$, so that the facet boundary arrives above the nearby orientation $\vec{n}_f + d\vec{n}$ after $dt = r(dR/d\vec{n})^{-1}_{\vec{n}\text{fac.}}$ Deviations from a linear $R(\vec{n})$ dependence cannot be determined within the range $\vec{n}_f \rightarrow \vec{n}_f + d\vec{n}$.

Notation in terms of $\vec{\mathbf{T}}$ *operator.*

Both the value of the first-order correction and the accuracy of the layer-thickness measurement can adequately be expressed in terms of the vector operator $\vec{\mathbf{T}} = d/d\vec{n}$ ln. For the *fractional* value of the 1st order correction $\varepsilon =$ correction $/R(\vec{n})$, we have

$$\varepsilon = \frac{d}{2r_o}\vec{\mathbf{T}}^2 R \tag{13.10}$$

and for the resolution

$$\Delta\vec{n} = \frac{d}{r_o}\vec{\mathbf{T}}R \tag{13.11}$$

where d is the thickness of the grown layer, and r_o is the radius of the initial sphere.

A typical value of $\vec{\mathbf{T}}R$ in the experiments by HOLLAN (1972) on GaAs epitaxy in the GaAs/$AsCl_3$/H_2 system is $\vec{\mathbf{T}}R \simeq 2$. For a 1 cm. diameter sphere, this implies that the uncertainty in the $\vec{n}$ variable increases with about 1 degree for every 40 micron of epilayer thickness.

13.3.3 Alternative experimental approaches to thickness measurement

The conventional method to determine thickness of the layer on the sphere is by cross-sectioning and measuring the thickness microscopically, the substrate/epilayer interface being revealed by etching or by a thin marker layer. Being destructive and time-consuming, the method is discouraging in determining a dense grid of measuring points. An alternative, faster and easier method would therefore be worth while to develop, even at the cost of some investment in new instrumentation.

One obvious principle is the use of optical interference techniques. Distortion of Newton interference rings generated in the ring wedge between a flat glass plate and a sphere can in principle be used to measure the distortion of the sphere, but is hardly a practical way to generate a thickness map. Rather we would like to use interference between the epilayer surface and the *original substrate*. This can be done with hetero-epitaxial layers when the epilayer's refraction index differs sufficiently from that of the substrate; different absorption could also be used. But this is a rare occasion. Rather we might resort to holographic techniques - requiring to take a hologram in an optical set-up with the virgin sphere and then generate interference patterns in the same set-up with the growing sphere. (The optical set-ups as described in 13.6.1, which are primarily intended for Schlieren-type imaging, might in principle be adapted for such holographic techniques as well).

An additional virtue of holographic methods is that they work equally well for etching as for growth experiments. Microscopic determination of etch-depth profiles is not so easy, unless periodic marker layers are embedded in the substrate (difficult for *spherical* substrates).

Finally, a rather crude method might be applied to hetero-epitaxial systems differing from the substrate in chemical reactivity - such layers might be peeled off by just dissolving the substrate. This could be applied, for instance, to some cases of III-V heteroepitaxial systems. The thickness of the peels, if strong enough to be handled, can be measured by interference microscopy.

13.4 THE **S***R VALUE AND THE "BLOWUP PROJECTION CENTRE" OF THE SPHERE*

We have seen in ch.2 that the **S**R extrema can be identified with edges and facets, and **S**$R = 0$ with "parallel displacement" of the curved interface. What is the geometrical interpretation of **S**R = constant ($\neq 0$, and $\neq \infty$) ?

Consider a section of the circle between orientations φ_1 and φ_2 (fig. 13.4), and suppose **S**R to have a constant value there. This means that after a growth time t, the radius between φ_1 and φ_2 is still constant, i.e. r + **S**$R.t$. The geometrical similarity implies that the $\vec{\mathbf{Q}}R$ vectors all have a common point of intersection - a "projection centre". Where is its position?

In fig. 13.4, we have drawn the circles for substrate and epilayer, for the case that R has a minimum and **S**R has a maximum at orientation $\psi(R_{min})$ *) We suppose that **S**R has a constant (finite) value in a certain φ range around this minimum.

*) Strictly spoken, the mathematical condition for minimum R is different from that for maximum **S**R, the latter involving also the third derivative. However, the physical conditions to be invoked for non-coinciding (R-minimum, **S**R-maximum) seem rather exotic.

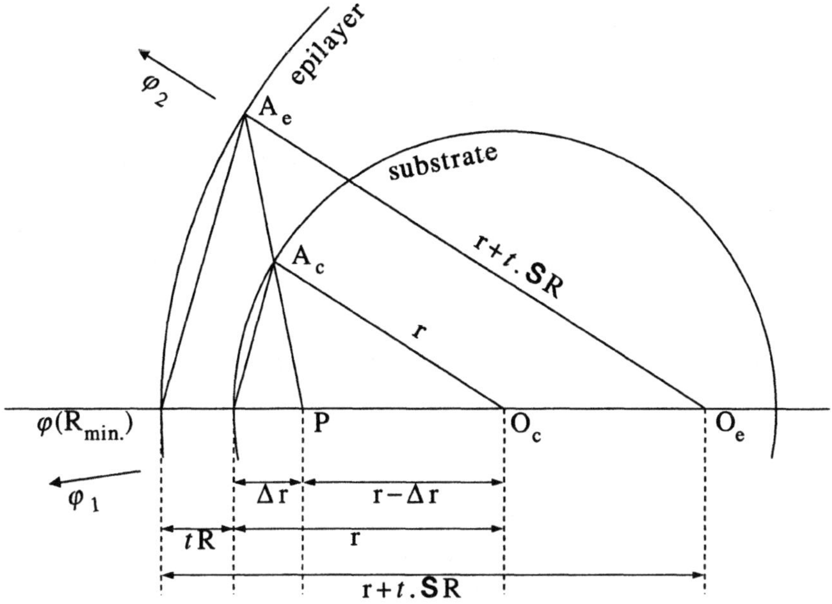

FIG. 13.4

As the $\vec{\mathbf{Q}}R$ vector at $\psi(R_{min})$ is in the normal direction, the projection centre P has to be situated on the radial line in the $\psi(R_{max})$ direction. Writing Δr for the "depth" of P below the substrate's surface, we find from the similarity of the triangles O_cA_cP and O_eA_eP:

$$\frac{\Delta r}{r} = \frac{\Delta r + tR}{r + t.\mathbf{S}R} = \frac{R}{\mathbf{S}R} \quad (= \varepsilon^2 \text{, see eq. (11.23)} \qquad (13.12)$$

For R = constant, $\Delta r = r$, i.e. the projection centre is in the sphere's centre; an obvious result. $\mathbf{S}R = 0$ corresponds with $\Delta r = \infty$ or "parallel translation" of the surface. For negative $\mathbf{S}R$ and positive r, Δr is negative, i.e. P is above the substrate, implying that the epilayer develops a sharp edge $(r=0)$ for the first time after a growth time $t_{(edge)} = \Delta r/R = r/\mathbf{S}R$. This provides a key for a direct experimental determination of $\mathbf{S}R$ (see 13.6).

Note that eq.(13.12) applies for *all* orientations, not only for R_{min} orientations. For other orientations, where $dR/d\psi \neq 0$, P has not only a depth Δr below the position under examination but also a tangential shift:

$$\frac{\Delta s}{r} = \arctan \frac{\vec{\mathbf{Q}}R(\psi)}{|\mathbf{Q}R(\psi)|} \frac{\Delta r}{r} = \frac{dR/d\psi}{R} \frac{\Delta r}{r} = \frac{dR/d\psi}{\mathbf{S}R} \qquad (13.13)$$

For the general case where $\mathbf{S}R(s)$ is a continuous variable along the surface, we may think of the projection point P as a *local* projection point = intersection point of the $\vec{\mathbf{Q}}R$ vectors in the vicinity of s. For a point travelling along the surface, the corresponding projection point follows a trajectory as sketched in fig. 13.5. The trajectory is divided in "external" (corresponding with $\mathbf{S}R < 0$ regions) and "internal" parts (corresponding with $\mathbf{S}R > 0$). The transitions ($\mathbf{S}R = 0$) correspond with asymptotes in maximum-$\mathrm{d}R/\mathrm{d}\psi$ directions. For the extreme-$\mathbf{S}R$ orientations, P is situated on the normal at position s, and the sharp edge occurs there *first*; i.e., this special point is the *starting point of the edge trajectory* corresponding with the $\mathbf{S}R$ extremum (this edge trajectory should not be confused with the trajectory of the projection centre itself, although it starts on it).

For all non-extreme values of $\mathbf{S}R$, $\vec{\mathbf{Q}}R$ vectors are annihilated by an already existing edge.

Two-dimensional case:

The above discussion, being concerned with principles in the first place, dealt only with one orientation parameter. In the general (2-dim) case, the $\vec{\vec{\mathbf{S}}}R$ tensor, when brought in diagonal form, can have different elements. In view of the correspondence between $\vec{\vec{\mathbf{S}}}R$ value and projection centre, this means that in the 2-D case we can have two projection "axes", for the curvature around each of these two (perpendicular) axes. After the start of growth, the originally spherical surface becomes locally ellipsoid.

A special case is an $\vec{\vec{\mathbf{S}}}R$ tensor with one peaked and one finite element. (which can occur, for instance, on an equator orientation of a "strong" PBC). This means that the radius in only one direction increases "infinitely" fast. Morphologically we see a cylindrical ribbon developing on the sphere, often as a sort of connection between facets (see 12.5). In the terminology of HARTMAN and PERDOK (1955 a,b,c) this is called an S orientation.

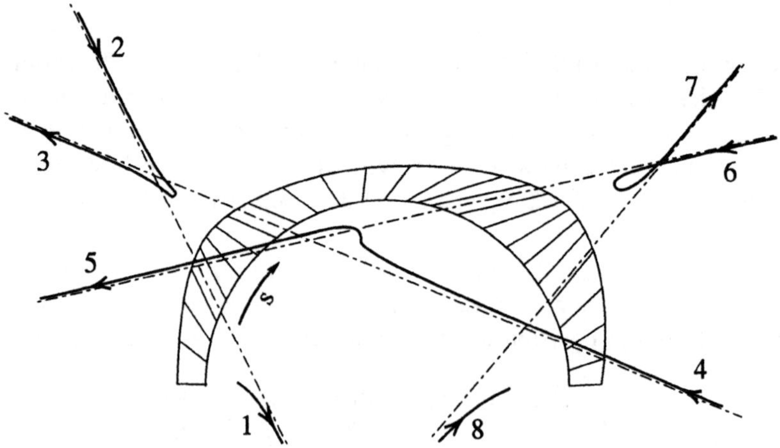

FIG. 13.5

13.5 THE ALTERNATIVE APPROACH : ORIENTATION MAPPING.

13.5.1 Principles.

The conventional way to extract data from a sphere-growth experiment consists of the combination of two types of measurements:

1) Using the facets as mirrors, their *orientations* are determined optically with a goniometer. Subsequent measurement of the facet *sizes* with the microscope can reveal their relative importance.

2) More directly, the *growth rate function* can be determined directly by measuring the layer thickness microscopically on a cross-sectioned sphere, as a function of orientation of the original spherical substrate. To increase accuracy and to minimize systematic errors, a large sphere radius and a small epilayer thickness/radius ratio are favourable (see 13.3) Such measurements provide the $R(\vec{n})$ function directly; $\vec{\mathbf{Q}}R$ and $\overline{\overline{\overline{\mathbf{S}}}}R$ data can be derived from this function.

In the present section we discuss an alternative principle in which the first and second *derivatives* of the $R(\vec{n})$ function are determined rather than $R(\vec{n})$ itself. Although the determination of the R function value itself requires additional, and independent, thickness measurements, the method provides a more direct and accurate way to measure the parameters of the R function as defined in chapters 9 - 12. An additional virtue of these orientation-mapping methods is that they are directly applicable to sphere-*etch* experiments as well, in contrast to most thickness-mapping methods.

For convenience, we consider the "cylindrical" (one orientation variable) case first. (two-variable complications will be discussed in 13.5.4). Convex and concave substrates, with radius $+r$ and $-r$, are considered in pairs as the information revealed by them is complementary. Fig. 13.6a and c shows these substrates and the epilayers grown on them.

We assume that an experimental device is available which permits the orientation ψ to be measured as a function of the location on the substrate. For the sake of explaining principles, we might think of a narrow laser beam of fixed orientation, scanning the cylindrical surface. The deflection of the mirrored beam 2ψ can be measured, so as to determine the orientation ψ of the surface.

Here and in the rest of this chapter, the symbol ψ is used as an orientation angle parameter. (in the 2-dim. case we will use ψ_1 and ψ_2 as a locally orthogonal set of directions). Accordingly, $\mathbf{S}R$ can be written as $R + d^2R/d\psi^2$ etc.

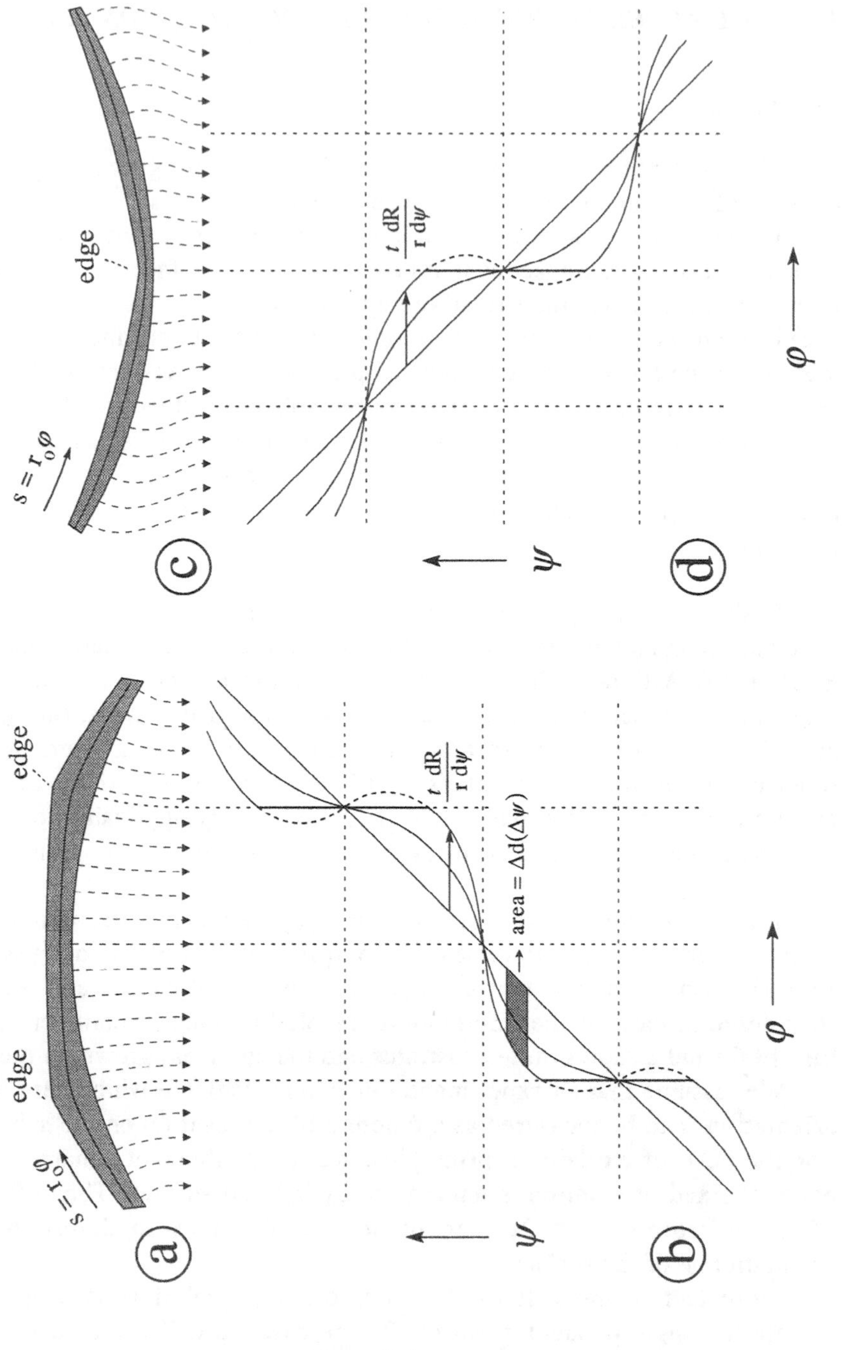

FIG. 13.6

Note on the representation of surface coordinates:

In the following we are going to plot the measured orientation as a function of the position on the surface. For the representation of this position - i.e. for the location of measuring points on the horizontal scale - we can use two alternatives: a) the "exact" choice of a real surface coordinate s (or u,v in two dimensions) and b) the "approached" choice of the surface coordinate of the *original surface, before growth* = $\varphi.r_0$. (here we introduce the angular variable φ as a reduced length variable, well to be distinguished from the real angular variable ψ). Choice of the "reduced" coordinate implies that the *original* substrate is represented by a 45° line. Note that choice b) is more in agreement with experimental practice: it is the same choice as is customary in the plotting of a thickness map - either in a sphere-growth experiment or in the representation of the Himalayas on a map of Asia. In contrast, choice a) implies the impractical consequence that the total width of the horizontal axis expands gradually during a growth experiment, because the size of the sphere increases.

However, it is important to be aware that the local curvature $d\psi/ds$ of the surface, in case of choice b), is not *exactly* identical with $1/r_0\, d\psi/d\varphi$ as determined from the $\psi(\varphi)$ map (the difference being first order in d/r_0, where d = layer thickness).

In the discussions to follow we will sometimes use choice b), especially when reference is made to actual maps of the sphere surface, and sometimes choice a) (in particular when discussing "maps" of the orientation distribution function $F(\psi)$, which is a function of the ψ variable and can be considered independently of the surface-map). When φ is used to represent location on the sphere, we will account for the errors made. In particular, the error can turn up in the form that $d^2R/d\varphi^2$ is found in the φ representation instead of $\mathbf{S}R$ in the s representation.

It is to be emphasized that the use of one or two angular variables instead of an ordinary surface variable is appropriate only in the case of growth on a cylindrical or spherical substrate and in connection with the experimental approach of "orientation mapping". Moreover, the method makes only sense during the initial stage of growth ($d \ll r_0$). This is the reason why this representation is used specifically in this chapter and not in the general discussion of ch. 5. In 13.5.2, some of the general statements of ch. 5 will be reformulated in terms of the $\psi(\varphi)$ representation.

Evolution of orientation maps

In fig. 13.6b and d, we plot the surface orientation ψ as a function of the "reduced surface coordinate" φ. ("choice b" as defined in the note above -

13.6b for the convex, 13.6d for the concave surface). The *original* substrate's surface, being perfectly cylindrical, is represented by a straight line $d\psi/d\varphi = 1/r_0$ = const. (note: because the reduced coordinate φ actually replaces s which is supposed to move "from left to right", we have to take account of the *sign* of r_0 to have $\varphi = s/r_0$ moving in accordance. Therefore, for the convex case c), $d\psi/d\varphi = 1$, whereas it is $= -1$ for the concave case).

When the epilayer starts to grow, the $\psi(\varphi)$ line is deformed. Points of the 45° line move horizontally with velocity $dR/r.d\psi$ (the horizontal motion, ψ = const., agrees with motion of the locus of fixed orientation along the kinematic wave trajectory). This implies that the deformed line keeps crossing the original 45° diagonal at ψ (or φ) positions corresponding with *R minima or maxima;* i.e. $d^2R/d\varphi^2 = 0$. The rotation of the $\psi(\varphi)$ curve around these fixed points (locally) is clockwise for R minima, anticlockwise for R maxima. Because the horizontal displacement of the $\psi(\varphi)$ curve is $t.\ dR/r.d\psi$, the area between the original and the shifted $\psi(\varphi)$ curves (shaded area) represents $(t/r)\ \Delta R(\Delta\psi) = (1/r)\Delta d(\Delta\psi)$; i.e. difference of the thickness of the epilayer as dependent on orientation. Only thickness *variations* are observable in this way, not the thickness itself.

Rotation of the $\psi(\varphi)$ curve around the fixed $R(\psi)$ min. or max. points can proceed until the $\psi(\varphi)$ curve is either horizontal (i.e., $R = d\varphi/ds = \infty$, a facet) or vertical (i.e. $R = 0$, an edge). Comparing the convex and the concave cases a and b, we see that an R minimum, i.e. clockwise rotation., generates a facet on the convex, and an edge on the concave substrate at the corresponding orientation. Mutatis mutandis, an R maximum generates an edge on the convex and a facet on the concave substrate.

In-between the $R(\psi)$-minimum and maximum orientations where the $\psi(\varphi)$ curve remains attached to the original diagonal, the orientations with $\mathbf{S}R = 0$ are characterized by $d\psi/d\varphi = 1$; i.e. those orientations where the $\psi(\varphi)$ curve is parallel with the original curve.

Note that the *sign* of the $d\psi/d\varphi$ slope can never change from that of the original starting value + or −1 in the case of constant affinity (straight kinematic-wave trajectories). A change of sign can only be induced by a) a velocity source such as a D-junction or a stabilized RE; b) in the case of inhomogeneous affinity, by the generation of inflection points (trifurcation of kwt's, see 5.2).

Altogether we can determine both the ψ *orientations* where the $R(\psi)$ minima and maxima occur, and the corresponding $d^2R/d\psi^2$ values, directly from the evolution of the $\psi(\varphi)$ curves with time - both on a concave and on a convex substrate. (Experimentally the observation of an "edge formation" could be an accurate method to determine the $\mathbf{S}R$ value of an $R(\vec{n})$ minimum, i.e. by measuring the incubation time of the first occurrence of an edge on a concave substrate).

It is important to be aware that both the φ-position and the orientation of "horizontal" crossings of the $\psi(\varphi)$ line with the original remain fixed, whereas the vertical crossings (once the edge has formed) can move in the φ dimension. We can draw the positions of such shifting edges at successive times in the *cross-sectional* drawing of the epilayer (fig. 13.7a) so as to reveal the *edge trajectories* corresponding with this motion of edges. Because in a sphere-growth experiment, especially in the initial stages, the edge-adjacent orientation can change continuously with time, these trajectories can be curved. Compare the case of ch. 6.5 (evolution of polylines), where the edge-adjacent orientations were fixed and therefore the edge trajectories were straight lines. On a convex surface, edges correspond with negative $\mathbf{S}R$ values; in-between each adjacent pair of edges is a stretch of surface with positive $\mathbf{S}R$ which, if $\mathbf{S}R \gg R$, may be called a "facet". We will call the area between the curvature maxima or the edges the *territory* of this facet. The extension of the territory concept to the 2-dim. case is trivial. (the territory concept is defined mathematically in 10.3.3). On a concave surface, of course, the rôle of edges and facets in relation with the sign of $\mathbf{S}R$ is inverted.

We will discuss edge trajectories in more detail in 13.6 in connection with the evolution of facet size as a function of time.

13.5.2 Special features in $\psi(\varphi)$ map evolution.

a) Dislocation-pyramid areas.

In areas where D-junction vector velocities exceed the original R function, the growing pyramids are directly translated into $\psi(\varphi)$ curve features:

The D-junction apex immediately generates a *fixed* vertical line at the φ-position of the D-junction, between the ψ orientations corresponding with the pyramid sides, which are obtained as the intersection of the $\vec{\mathfrak{R}}_d^*$ effective growth-rate circle with the $\vec{\mathbf{P}}R$ plot (fig. 6.31). The re-entrant edges at the pyramid base generate moving vertical lines in the $\psi(\varphi)$ plots. Their horizontal velocities are $(\mathrm{d}R/\mathrm{d}\varphi)_{\mathrm{p\,left}}$ and $(\mathrm{d}R/\mathrm{d}\varphi)_{\mathrm{p\,right}}$.
Note that the absolute value of $\vec{\mathfrak{R}}_d^*$ of the D-junction cannot be derived from the $\psi(\varphi)$ curve. Only the *difference* with the growth rate of the original substrate can be derived from φ_1, φ_2 and the displacement rates of the re-entrant edges.

As growth proceeds, the moving baseline-RE's of eventual neighbouring pyramids will collide and merge into stationary RE's. Sooner or later the whole region between $\varphi_{\mathrm{p\,left}}$ and $\varphi_{\mathrm{p\,right}}$ will be overgrown with

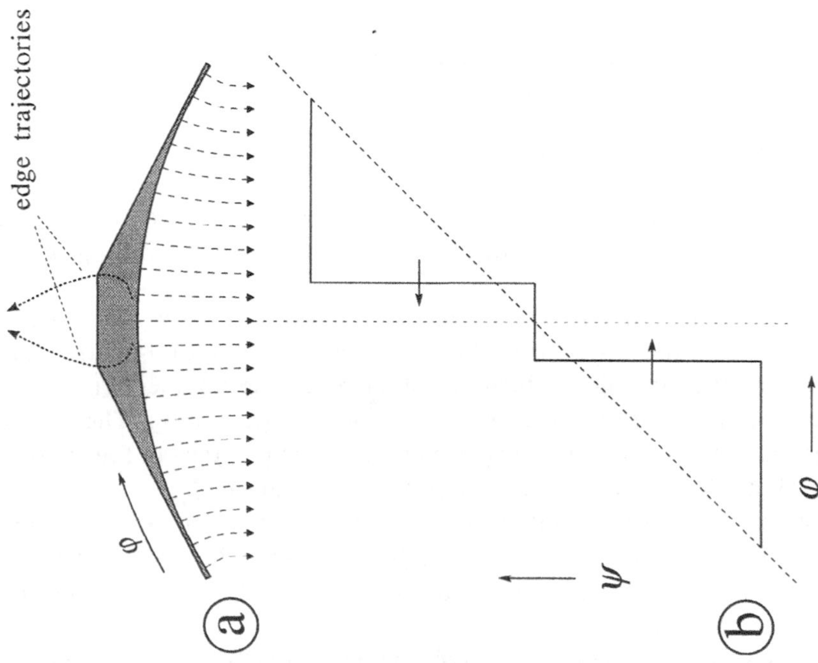

FIG. 13.7

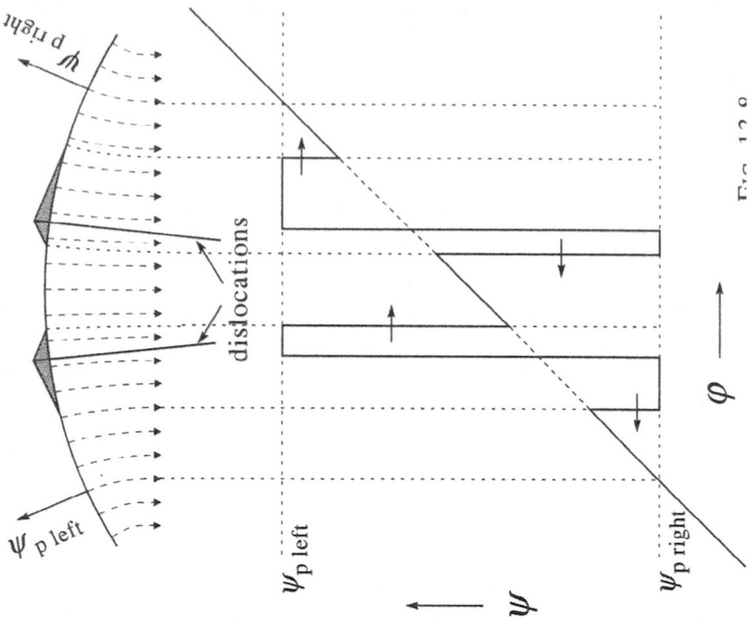

FIG. 13.8

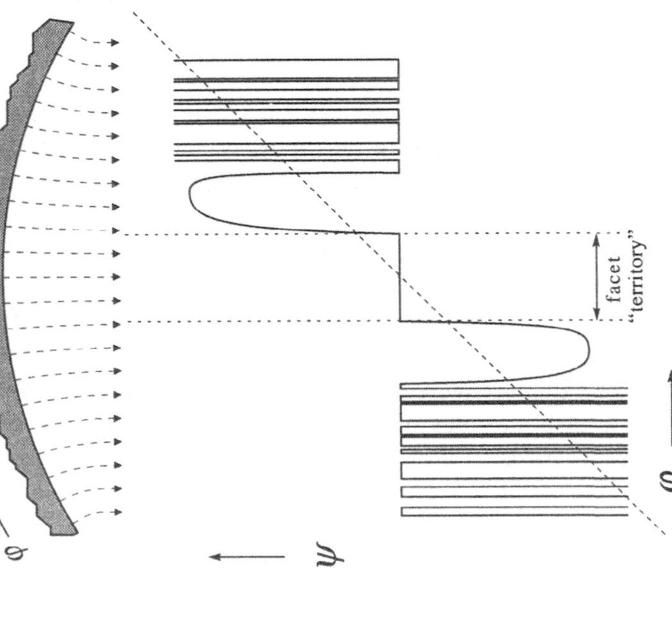

FIG. 13.9

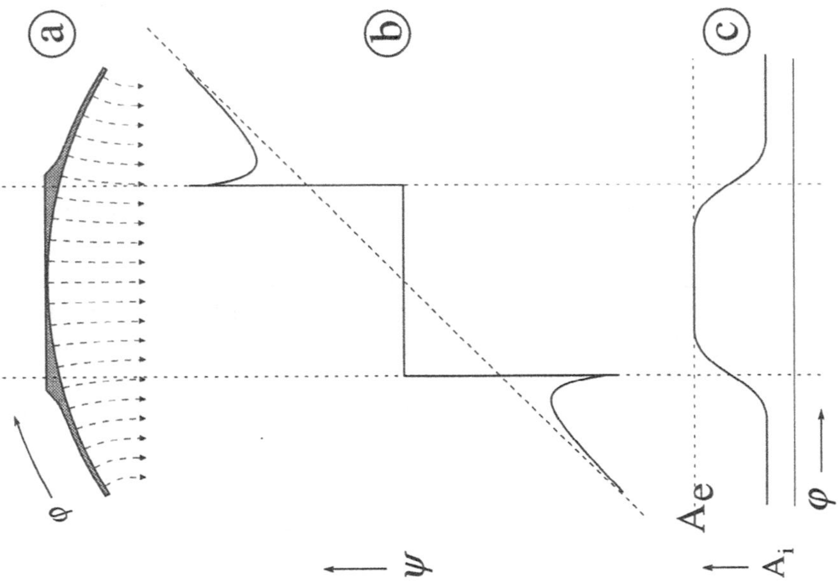

FIG. 13.10

pyramid(s) and surface orientations in-between will have vanished. The outermost RE's will merge asymptotically with the $\psi(\varphi)$ curve at $\varphi_{\mathrm{p\,left}}$ or $\varphi_{\mathrm{p\,right}}$ (for short growth times and/or low dislocation densities, the pseudofacets are surrounded by a RE).

When D-junctions with different Burgers vector occur, their $\vec{\mathfrak{R}}^*$ velocity and consequently their pyramid-side orientations will in general be different and the developing $\psi(\varphi)$ curve will accordingly look somewhat noise-like. Ultimately, the *fastest* D-junctions will overgrow the others.

Altogether, on the $\psi(s)$ or $\psi(\varphi)$ curve, a "clean" facet can be readily distinguished from a dislocation-induced "pseudofacet" consisting of vicinal orientations. Whereas the former shows a single $\mathrm{d}s/\mathrm{d}\psi$ peak at one orientation, the latter has a split-up peak with (at least) two (pyramid-side) orientations. (In 13.7 we will see that the $\mathrm{d}s/\mathrm{d}\psi$ peaks can readily be integrated by suitable experimental means so as to determine their orientations).

b) Zigzag or pyramid areas induced by instabilities:

Zigzag and pyramid-field corrugations can also develop on certain orientation sectors on a sphere as a result of any form of instability (constitutional supercooling; negative $\mathbf{S}\gamma$ or $\mathbf{S}R$). These instabilities have been discussed in 6.3.4 (see also fig. 6.30). In the $\psi(\varphi)$ graph of fig. 13.9, such corrugations look, of course, similar to the dislocation-pyramid features. The optical visibility depends on the size of the microfeatures involved.

c) Concave rims around facets:

We have repeatedly stressed the fact that *in the case of constant surface affinity,* a seed substrate with positive curvature everywhere and without "velocity sources" can never develop any concave area during growth.

In sphere-growth experiments, we are sometimes confronted with a phenomenon which seems to contradict this: when the faceted areas are limited in size and surrounded by "smooth" curved areas*), the facets are as it were protruding from the sphere, i.e. they are surrounded by a concave rim (fig. 13.10a). In the $\psi(\varphi)$ curve, the concave rim forms an "overshoot" of the ψ variable at both sides of the horizontal section corresponding with the facet. (fig. 13.10b).

*) This can either be a transient situation (during the initial stages of growth and before the facets have grown together) or a permanent one (due to the boundary-layer series resistance mechanism described in 8.2)

This phenomenon can be explained on the basis of a *surface gradient of the affinity*. Such a gradient can develop if crystal units have to pass a series resistance before they can settle into the lattice. The series resistance can be brought about by bulk diffusion in a boundary layer or by surface diffusion in an adsorbed layer. In fig. 13.10c, the affinity field is sketched as felt at the interface, when a constant affinity $A_{ext.}$ is applied "at the other side" of the boundary layer (dashed line at distance δ in fig.13.10a). In the facet area where the flux is very low, $A_{ext.}$ penetrates almost without loss to the interface. Around the facet where interface resistance is much lower and consequently the flux higher, an appreciable affinity drop occurs across the boundary layer. The resulting tangential gradients at the facet edge set up a tangential diffusion flux smoothing out the affinity gradient over a distance $\sim \delta$. This gradient deforms the edge of the facet into a concave rim.

As seen from the spherical substrate, we are no longer in the case of constant affinity. As seen macroscopically, we can consider the system (sphere + boundary layer) together and describe it with an "overall" growth-rate function which takes the series resistance into account (see 8.2). But on that larger scale, the detail of the concave rim cannot be discerned.

13.5.3 The "discotheque-sphere method"

In this section we will show that the exact value of $\mathbf{S}R$ can be found directly from two successive optical measurements, without resort to thickness measurements at all.

From the identity of $\mathbf{S}R$ and the "radius-increase velocity", it is obvious that we only have to measure the radius $ds/d\psi$ at two successive times, with a sufficient time interval. Now the derivative $ds/d\psi$ can also be considered as an *orientation distribution function,* at least under certain conditions. This orientation distribution function can be measured optically with the "discotheque-sphere method" (to be explained below). The *resolution* of this method is limited by the aperture of the illumination and by diffraction effects, i.e. the limit depends on the optical set-up used (and on the sphere radius, the wavelength etc.) We will now first investigate the conditions under which the sphere radius can be identified with the *measured* orientation distribution function, and afterwards explain how this function can be determined experimentally.

The orientation distribution function can only be identified with the slope $ds/d\psi$ of the ψ versus s plot (rotated over 90°) when two conditions are satisfied: (note that we use the s representation for the surface here)

1) The first condition is of a mathematical nature: a requirement is that the same ψ orientation does not occur more than once in disconnected "s" regions. This condition implies that no re-entrant corners, or in general concave areas occur on the "sphere". Planar regions are permitted and

correspond with delta spikes in the orientation distribution function.

In fig. 13.9, we have drawn a sphere with a facet *and* a region in-between two facet regions where the surface has been broken up in a zigzag instability. The "zig" sections have the facet's orientation; i.e. they contribute to the distribution function at this orientation. We can formally distinguish the different contributions by referring to the concept of facet "territories" (see 10.3.3 and 13.5.1). The actual facet territory (see fig. 13.9) contains only the facet, and it is the contribution from this territory which corresponds with the relevant term ($r_0 + t.\mathbf{S}R$) from which $\mathbf{S}R$ has to be derived. The contribution from the zigzags, i.e. from the adjacent unstable, negative-$\mathbf{S}R$ territories, might be said to have "tunneled" through the territory boundaries (edges). Experimentally we can shield it off by suitable screening (see 13.7.2.)

1bis) Narrowly related with the above is the case of dislocation-pyramid areas as discussed in 13.5.2, a. This may be considered as the disproportionation of one peak in the orientation-distribution function into two or more peaks, *within one facet territory*. The resulting adjacent "pseudofacets", in conjunction, can be considered to replace the original facet.

2) The second condition which has to be satisfied for the identification of curvature $d\psi/ds$ and the *measured* orientation distribution function is rather of a physical kind. It becomes obvious when we consider the case of a facet which is losing territory to agressive neighbour facet territories, i.e. whose territory-boundary edge trajectories are convergent (fig. 13.7). Mathematically speaking, there is nothing wrong with the slope/distribution function identification. But as time proceeds, the facet's orientation distribution function develops into a sharpening peak. When the territory size is constant, the integral over the peak remains constant (whereas the peak height keeps increasing in accordance with the radius). But in the case of a shrinking territory, the integral declines.

The difficulty is that it is experimentally impossible to measure the height of the peak, once its width has fallen below a certain aperture. Therefore, for practical reasons we have to conclude that the *observed* orientation distribution function does not reflect the surface's curvature distribution function any more when the imaging device which is used to map the $F(\psi)$ function can no longer resolve the peak. This has two consequences:

a) for a stationary-size territory, the *integrated* distribution- function peak remains constant, i.e. the facet remains visible in the $F(\psi)$ function but we can not follow the dr/dt from it any more.

b) for a shrinking territory, the integrated distribution function peak declines (and ultimately can disappear). This should not be misinterpreted as a declining curvature radius: it is due to the overgrowth of the facet. Rather we should say that the radius is measured *as averaged over a certain area* on the sphere. In view of the above we have to be aware that the determination of $\mathbf{S}R$ on the basis of eq. (13.13) has to be finished within a certain time window. (which can be really short for high-$\mathbf{S}R$ facets).

Now the relation between the evolution of curvature radius and of orientation distribution function has been established in detail, let us consider how this distribution function might be measured experimentally:

When the surface of a sphere is illuminated by a parallel light beam, the reflected light can be projected onto a far-away screen. For a perfect sphere, the light intensity on the screen is evenly distributed (apart from a "slow" cos 2ψ dependence accounting for the interception factors involved, at least when the screen is flat and perpendicular to the incident beam).

However, when facets are present on the sphere, they mirror the original intensity onto a spot corresponding with the facet's orientation (multiplied with the same cos 2ψ factor, and smoothed out by diffraction effects which are left out of consideration here). Conversely, edges (corresponding with missing orientation ranges) give rise to dark regions on the screen. We might adequately characterize this method of imaging the orientation distribution function as the *discotheque-sphere imaging method;* it is in fact a generalized goniometer method and an optical analog of X-ray methods.

To be more precise, for the real distribution function we have to focus the light of a given orientation in one point on a screen (with the parallel unfocused beam, it is distributed over an extended spot so that the ψ information is mixed with the φ information). Also, we might prefer to get rid of the cos 2ψ factor. The latter can be realized when the incident light is directed perpendicular to the sphere surface. However, in this case the change in intensity corresponds with $t.\mathrm{d}^2R/\mathrm{d}\psi^2$ again rather than with $t.\mathbf{S}R$.

A practical system, for each of the combinations convex/concave, reflection/refraction, is sketched in figs. 13.11a-d and discussed in 13.7.1. We will see there that such an optical system generates a "φ image" (an image of the sphere's surface) as well as a "ψ image", at a different position. The ψ image corresponds with the orientation distribution function. The φ image is a map of the surface orientation, but in order to *represent* orientation or orientation deviation we have to implement either a Schlieren optics or a grid illumination (section 13.7.2)

In order to determine a map of $\mathbf{S}R$ as a function of ψ, we should dispose of maps of the $F(\psi)$ distribution at two successive times. (the experimental form in which we measure $F(\psi)$ is as light intensity, $I(\psi)$) The first can be taken at $t = 0$ when no epilayer has grown yet; i.e. the radius is constant and the ψ image is uniformly illuminated, corresponding with the uniform radius. In order to convert I to r, we simply have to identify this uniform level $I(0)$ with r_o, i.e. multiply with the factor $r_o/I(0)$.

At a later time t (provided the conditions 1) and 2) mentioned above have been satisfied) we obtain an $I(\psi) \sim r(\psi)$ distribution. The difference $(I(t) - I(0))/t$ can be related with $\mathbf{S}R$:

$$\mathbf{S}R(\psi) = \frac{r_o}{I(0)} \frac{I(t,\psi) - I(0,\psi)}{t} \tag{13.14}$$

Any inhomogeneities due to stray light effects etc. are compensated by the difference method. (this is why $I(0,\psi)$ is considered to depend on ψ in eq. (13.14); $I(0)$ is just an average to represent the original radius).

Complications due to the 2-D nature of an actual surface are discussed in 13.7.2, together with techniques for obtaining the $\mathbf{S}R$ tensor elements separately.

Note that, at least in this version of the discotheque-sphere method, a bright-field illumination (of small aperture) is used - i.e., no Schlieren-stop screen. Combined versions of Schlieren-type optics and the discotheque-sphere methods seem possible to reveal the elements of the $\mathbf{S}R$ tensor separately or to amplify them, although in a nonlinear fashion.

13.5.4 Curvature in two directions: evolution of the curvature tensor.

Up to now, we have discussed the single-variable case of cylindrical substrates - or we have illustrated phenomena on spherical substrates by referring to just one orientation variable. The general case is to be described by a curvature tensor with four elements; in fact by three independent variables: the two principal radii and the orientation of their axes.

So, the amendment to be made to the discussion in 13.5.1 is that both the *substrate* will have to be scanned over a two-dim. area with variables φ_1, φ_2, and that the *deflection* will have to be measured in two directions ψ_1, ψ_2. From the results we should be able to determine the derivatives

$$\frac{d\psi_1}{r_o d\varphi_1} \qquad \frac{d\psi_2}{r_o d\varphi_1}$$

$$\frac{d\psi_1}{r_o d\varphi_2} \qquad \frac{d\psi_2}{r_o d\varphi_2}$$

together composing the second fundamental tensor $\vec{\vec{K}}$ of the epilayer surface growing on the spherical substrate. (for the moment we neglect the question how the measurement of the ψ deflections and the evaluation of the derivatives is to be arranged experimentally; we return to this in 13.7).

When $\vec{\vec{K}}$ is measured *as a function of time*, we have directly determined the $\vec{\vec{\mathbf{S}}}R$ tensor, in agreement with eqs. (1.9) and (5.2).

For the R maxima and minima, the $\vec{\vec{\mathbf{S}}}R$ tensor elements can be most accurately measured by determining the first occurrence of an edge on the appropriate (concave or convex) substrate.

Graphs versus maps

Although the above formulation of the 2-dim. case is correct in an abstract mathematical sense, we are confronted with something as a representation problem when it comes to the extension of 13.5.1 - 13.5.3 to the 2-dim. case. In the cited sections, $\psi(\varphi)$ correlations were discussed as represented in *graphs*. But what we obtain from experiments (and store in photographs or videotapes) is "map-like information" - either "direct" maps of the surface ($\psi(\varphi)$ maps) or distribution-function maps ($F(\psi)$ maps).

One-dimensional "maps": The $\psi(\varphi)$ graphical representation of the 1-D case in 13.5.1 has the advantage that the derivative is visualized as the slope of a line; this facilitated the discussions on the meaning of distortions occurring in this graph. However, the graph is *not* an "image" of the (1-dim.) cylindrical surface. We might construct a real image as follows: The "field" of the 1-dim. φ variable is a line segment. In order to represent orientation (ψ) *on this line*, we should dispose of a variable such as grey scale or colour to draw this line with: this is the 1-dim. version of the colour image as discussed below for the 2-dim. case, experimentally to be realized by the colour Schlieren method of 13.7.2.3. An alternative is to represent the ψ variable not continuously but in discrete steps (dots) at fixed intervals. The 2-dim. extension of this alternative is the "orientation-grid" visualisation as described in 13.7.2.5.

Two-dimensional maps and "graphs". Whereas the 1-D case involves a relation (between the two variables ψ and φ) which can be represented in a 2-dim. graph, the 2-D case involves a tensorial relation between two *two-dimensional vectors* (ψ_1,ψ_2) and (φ_1,φ_2). The representation of such a relation requires a 4-dimensional representation space. The information contained in a certain area of the epilayer surface can be characterized in "pixel" terms. When the area image is dissolved into n x m pixels, we have to collect and store the two orientation variables for each pixel; i.e. this amounts to two images of the area, one for the ψ_1 and one for the ψ_2 variable. (the "pixel depth" for each image corresponds with the required resolution). Technically we may combine these two (ψ_1,ψ_2) images again into one "colour" image because a colour field can be defined having two dimensions.

Either a set of 2 grey-scale images or one colour image have all the characteristics of a real image but little of the 1-D graph representation of

13.5.1 (although each of the two grey images can be represented in a 3-D frame and slopes be defined accordingly). Still, the principles illustrated for the 1-D case in 13.5.1 can be generalized to the 2-D case, and computerized image-processing techniques can be used to generate real maps of curvature, orientation or d/dt(curvature). (the latter requires comparison of images taken at different times). By such methods, a map of the separate $\overline{\overline{\mathbf{S}}}R$ tensor elements, or their trace, can be constructed from a video record of the appropriate images.

13.6 EVOLUTION OF THE SIZE OF FACETS

13.6.1 Facets, facet territories and edge trajectories

The development of facets is a spectacular feature of sphere-growth experiments. Still we might say that from a mathematical point of view "facets do not exist".

A suitable medium for expressing sizes and their evolution in a sphere-growth experiment is the $\psi(\varphi)$ orientation-map representation introduced in 13.5.2 (figs. 13.6a-d). For a positive ($r_0 > 0$) starting sphere, φ ranges with $\mathbf{S}R(\varphi) > 0$ behaviour correspond with decreasing $\psi(\varphi)$ slope, φ ranges with $\mathbf{S}R(\varphi) < 0$ behaviour with increasing $\psi(\varphi)$ slope. As long as differences in $\psi(\varphi)$ slope are gradual, there is no hard criterion to base the definition of a "facet" on. (orientation dependence of R does not necessarily imply $\mathbf{S}R(\varphi) < 0$ regions at all).

The first "event" which can be considered as the birth of a facet in a sphere-growth experiment is the *formation of an edge*, i.e. the starting point of an edge trajectory as defined in 13.5.1. (fig. 13.7) The incubation time for this formation is (see 13.4, eq. (13.12))

$$t_{\text{inc.}} = \frac{r_0}{|\mathbf{S}R|} \tag{13.15}$$

i.e., edge trajectories are only observable from the very beginning of the growth experiment in the exceptional case of a real zero-cusp, i.e. roughening parameter = 0. The surfaces meeting in the edge keep increasing in orientation difference, from zero at the start to the full span of the $\mathbf{S}R < 0$ region, asymptotically for $t \to \infty$. As long as these orientations keep changing, the edge follows (in general) a curved trajectory. We might say that all $\mathbf{S}R < 0$ orientations ultimately "condense" into edge trajectories, whereas $\mathbf{S}R > 0$ orientations expand to fill all space between them. Only in the asymptotic limit, a whole territory enclosed by edges has the maximum-$\mathbf{S}R$ orientation, corresponding with the conventional definition of a facet.

(of course, for a hollow-sphere starting geometry, $r_0 < 0$, the roles of $\mathbf{S}R > 0$ and $\mathbf{S}R < 0$ regions in connection with edges and facets are interchanged).

The above argument illustrates that the edge trajectories are the most logical point of departure for the rough description of the shape evolution in a sphere-growth experiment.

13.6.2 Dimensionless representation of size evolution

Features occurring on growing "spheres" can be recorded as "size versus time" data, preferentially in a dimensionless form so that experiments on spheres with different starting size can be compared directly.

For the "size" variable (d), when measured as an actual distance between edges or between loci of fixed orientation, this means that it has to be represented as d/r_0 (r_0 = initial sphere radius). Still, this variable keeps growing with t. A more logical choice for the size is the φ variable of 13.5.2. The total φ range is always fixed, and $t \rightarrow \infty$ asymptotes can directly be correlated with the edge orientations of the $\vec{\mathbf{Q}}R$ shape.

The "time" (t) variable has to be represented as

$$t^* = \frac{t \, . <R>}{r_0} \tag{13.16}$$

where $<R>$ is a proportionality factor (with the dimension of velocity) representing the overall size of the $R(\vec{n})$ plot. For instance, for a 2D $R(\vec{n})$ function we may choose

$$\frac{1}{2\pi} \int R(\varphi) \mathrm{d}\varphi$$

Different choices for $<R>$ just expand the time scale by some constant factor and are not essential. In particular cases we may choose other R-function parameters such as M or $v_{\parallel}$ (see ch.12), in particular for the purpose of investigating the effect of variation of *other* parameters while keeping the first-mentioned fixed.

13.6.3 Facet size evolution for M-limited R functions.

The trajectory of the edges depends on the details of the $R(\vec{n})$ function over the full $\vec{n}$ range - in particular in the $\vec{n}$ regions neighbouring that part of the ($\mathbf{S}R < 0$) range which has already been eliminated by the edge. Therefore, general statements on the *initial development* of territory size cannot be made. (the final (asymptotic) stage of the development, of course, agrees with the facet sizes in the $\vec{\mathbf{Q}}R$ shape).

The special case of *M-limited growth rate functions* forms an exception

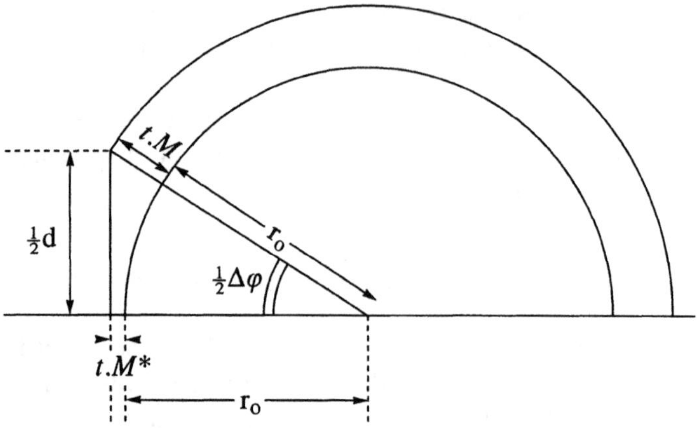

FIG. 13.11

to the above statement in the sense that the M parameter has an explicit effect on facet size evolution. Such a function can be written as resulting from an "un-limited" R function $R_{\text{unl}}(\vec{n})$ and a "series" limitation M:

$$R(\vec{n}) = \textcircled{\bar{a}}\!\left\langle R_{\text{unl}}(\vec{n}),\, M \right\rangle \tag{13.17}$$

with $\bar{a} = -1$ or -2. (see chs. 10-12). When M is $\ll$ the original maxima of R_{unl}, the end function $R(\vec{n})$ closely approaches the value M in the original high-growthrate sectors (the M limit can be imposed by series resistance mechanisms such as chemical reaction, adsorption kinetics or boundary-layer diffusion). In this special case, we can derive simple "growth laws" for facet size in a few characteristic cases.

Suppose that a minimum in the original function R_{unl} has the value $\varepsilon' v_{\parallel}$ (we adhere to the parameter nomenclature used in chs 10-12). This value is suppressed by the series limitation to

$$\textcircled{\bar{a}}\!\left\langle \varepsilon' v_{\parallel},\, M \right\rangle = M^* \tag{13.18}$$

In the following we are going to consider the cases $M^* = 0$ (zero-cusp, also in R_{unl}), $M^*/M < 1$, and $(M - M^*) \ll M$ (the latter corresponding with $\varepsilon' v_{\perp} \gg M$). The evolution of the facet/territory size can be roughly described by the simple geometrical construction of fig. 13.11:

From the Pythagoras equality it follows for the facet size d:

$$d^2 = 8\, r_0 t\, (M - M^*) + 4 t^2 \{ M^2 - (M^*)^2 \} \tag{13.19}$$

which suggests to represent the time variable in the dimensionless form $t^* = tM/r_0$. Moreover, we prefer to use the φ variable instead of d or d/r_0.

So eq. (13.19) can be rewritten as

$$\sin \tfrac{1}{2}\Delta\varphi = \frac{\tfrac{1}{2}d}{r_o + tM} = \frac{\sqrt{2t^*(1-M^*/M) + (t^*)^2.(1-(M)^2/M)}}{1+t^*} \quad (13.20)$$

In the initial stage of evolution, corresponding with $t^* \ll 1$, we have

$$\Delta\varphi = 2 \arcsin \sqrt{2t^*\,(1-M^*/M)} \simeq 2\sqrt{2t^*\,(1-M^*/M)} \quad (13.21)$$

i.e., a $\sqrt{t^*}$ behaviour with the extra factor $\sqrt{2\,(1-M^*/M)}$ (the higher the minimum M^*, the smaller the facet and the slower the growth of its size). The validity range of eq.(13.21) is restricted by the $t^* \ll 1$ condition; for large t^* the angular size $\Delta\varphi$ cannot exceed the asymptotic value of eq. (13.22).

In the final stage ($t^* >> 1$), $\Delta\varphi$ approaches the asymptotic value

$$\Delta\varphi_\infty = 2 \arcsin \sqrt{1-(M^*/M)^2} \quad (13.22)$$

Considering M as a parameter, we see that for *decreasing* M, (the parameters of the R_{unl} function being kept fixed), the ratio M^*/M increases towards 1 so that the asymptotic (angular) facet size decreases. This can be made more explicit for the case $\varepsilon' v_{//} \gg M$. The result depends, of course, on the assembling exponent a. For the two cases mentioned we obtain, writing $(M-M^*) = \delta$ so that $2 \arccos (M^*/M) = \sqrt{2\delta/M}$:

$$\text{for a} = -1 \text{ (and } \varepsilon' v_{//} \gg M): \quad \Delta\varphi_\infty = \sqrt{2\delta/M} = \sqrt{2M/\varepsilon' v_{//}} \quad (13.23a)$$

$$\text{for a} = -2 \text{ (and } \varepsilon' v_{//} \gg M): \quad \Delta\varphi_\infty = \sqrt{2\delta/M} = M/\varepsilon' v_{//} \quad (13.23b)$$

The behaviour of angular facet size versus t^* for these cases is sketched in fig. 13.12.

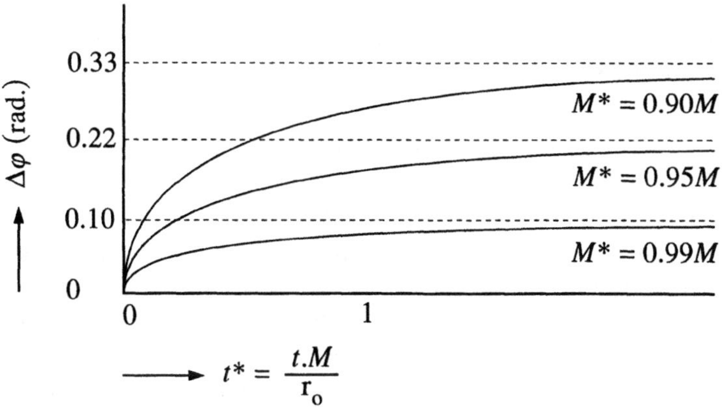

FIG. 13.12

Equations (13.23) have an important consequence for the persistence of facets (and thereby for their observability). The M limit makes the asymptotic angular size of *all* facets shrink. The asymptotic $\overline{\mathbf{Q}}R$ shape becomes a sphere with facets of limited angular size. As a result, those facets which would be of the "temporary" type without the M limitation - and therefore would be overgrown by "stronger" neighbour facets soon after the start of the experiment - are now becoming permanent because the *originally* permanent facets can no longer reach them. So we can increase the number of facet families which can be observed in a sphere-growth experiment by just introducing any "series-resistance" growth limitation.

If such a limitation has its origin in a *boundary layer* of a certain thickness, then we may expect the facets to increase slightly in size, and become more apparent because of "neck formation", as explained in 13.5.2c.

13.6.4 A "fundamental restriction" in the observability of low ε values

In connection with the size of facets, we may remark that there is a downward limitation in the *measurability of facet curvature* which is associated with the "quantisation" of the faceted-interface position in monolayer units. So in a sense, this may be considered as another fundamental restriction of the continuum approach, complementary to the FR at the *high* side of the curvature range which was discussed in 1.3.

A crystal with a slightly smoothed cusp in its $\gamma(\vec{n})$ and $R(\vec{n})$ function has a corresponding slightly curved facet. For a facet of width d, the smallest observable curvature is associated with a single monolayer island (height a) in the middle; with some phantasy the stepped surface can be considered as a curved surface with $r \simeq d^2/8a$. In other words, the curvature is "quantized" in units of $8a/d^2$: $1/r = n.8a/d^2$ $(n = 0, 1, 2, 3, \dots)$

Mutatis mutandis, this downward limitation restricts the possibility of measuring very low ε values on *small* crystals. In the first place this applies to the measurability of small $\gamma/\mathbf{S}\gamma$ values on small equilibrium crystals. A rough approximation shows that the downward limit is $(\gamma/\mathbf{S}\gamma)_{\text{obs. limit}} \geqslant 16a/\pi d$, i.e. for $d = 1\mu$, of the order of 10^{-3}.

For the determination of ε by a *sphere growth* experiment, we have to do with two limits. In the first place, to observe a difference between a zero-cusp ($R = 0$) and a smoothed cusp ($R_{\text{min}} = \varepsilon\, v_{/\!/}$) anyhow, we need a minimum growth time $t_{\text{min}} = a / \varepsilon\, v_{/\!/}$. During this time, the radius of a spherical starting substrate increases from r_o to $r_o + t_{\text{min}}.\mathbf{S}R = r_o + a/\varepsilon^2$. (by application of eq. (2.7). The corresponding (minimum observable) ε value is

$$\varepsilon_{\text{min.obs}} \simeq \sqrt{r_o/a}$$

Of course, these observability limits, being related with the atomic structure of the interface, stand apart from any *experimental* limits which may be connected with the measuring technique.

13.6.5 Conclusions

For the evaluation of sphere growth experiments and comparison of the results measured on unequally-sized spheres, it is appropriate to represent the data, in dimensionless form, as angular size versus dim.less growth time. For the latter we can use $t^* = t.<R>/r_0$ (eq. (13.16); see 13.6.2 for definitions). In such a plot, we can observe the following features:

1) *"Incubation time" of edges:*

By measuring the time t^*_{edge} we can determine the extreme values of **S**R directly as

$$|\mathbf{S}R|_{\mathrm{extr.}} = <R>/t^*_{\mathrm{edge}}$$

(the minima on a positive sphere, the maxima on a hollow one).

2) *Temporary versus permanent facets:*

Facets which are present in the $\vec{\mathbf{Q}}R$ shape will, of course, also be manifest in a sphere growth experiment. Other relative minima in the $R(\vec{n})$ function, when embedded by $(\mathbf{S}R)<0$ orientations, will also initially develop a facet territory on a sphere. The difference with the first case is that its bordering edge trajectories are converging (as in the case drawn in fig. 13.7) and ultimately merge, so that the territory disappears. We may call such territories or facets "temporary facets".

3) *Facet size reduction and suppression of territory competition by a series limit:*

In 13.6.3 we have shown that an "M-type" series limitation of a $R(\vec{n})$ function generally results in a decrease of the size of facet territories, and consequently in a decrease of the competition between neighbouring territories. Facet families which are "temporary" without the M limit may therefore be stabilized by such a limit. Generally speaking, an M-type limit results in a more "spherical" end shape with more and smaller facets; when the limitation is brought about by a diffusion resistance in a boundary layer, these small facets may be surrounded by a concave rim.

4) *Pseudofacets (dislocation-pyramid areas)*

The conclusion of 13.5.2a may again be referred to in relation with facet-size evolution. As shown in fig. 13.8, the presence of D-junction velocity sources results in the immediate emergence of a mixture of vicinal orientations (the pyramid slopes) over a finite φ-range simultaneously, finally covering the whole φ-range. It depends on the experimental observation technique whether this is resolved into a microscopic zigzag structure or rather measured as a finite φ-range of a facet-territory type. In the latter case, it is difficult to discern it from a facet related with a finite-M^* minimum as discussed in 13.6.2.

13.7 EXPERIMENTAL TECHNIQUES FOR ORIENTATION MAPPING

In 13.5.4 we defined the information content of an orientation map of the sphere's surface in terms of pixel densities and pixel depths - corresponding with resolution of the φ and ψ fields, respectively. We will now proceed to discuss the experimental techniques to obtain such maps. (at this place, a warning may be in order that the whole concept of orientation mapping, including the following techniques, are paper inventions of the author whose practical virtues still have to be proven).

All imaging techniques intended to generate a direct map of the growing sphere's surface in terms of orientation, curvature etc. *will have to use the growing epilayer's surface as part of the image-forming optical system*. The surface can be used either as a reflecting (mirror) or as a refracting (lens) surface; the choice will in general depend on the optical characteristics of the crystal bulk phase, the parent-phase bulk phase and the interface, and such considerations as the optical approachability of the growth cell. These factors will differ from case to case and will be left aside here. We will only discuss design criteria for the optical system (aperture, magnification etc) in relation with sphere size, wavelength and required accuracy.

13.7.1 The optical set-up

The four possible situations are concave mirror, convex mirror, concave lens surface and convex lens surface. The optical systems for these cases are sketched in figs 13.13 a-d (to be considered as a sketch for the purpose of illustrating principles; the choice of magnification, aperture etc. depends on substrate radius, growth time and required resolution). In practice, one flexible system might be designed allowing to be adapted to all four cases by minor changes.

Let us consider the image planes which are relevant for our purposes. Only the real image planes are of interest, as only these can be used to detect an image or to place a stopping shield for dark-field or Schlieren-type illumination.

1) The image plane(s) of the *spherical substrate surface* (the "φ planes"). This is where the "maps" are projected.

2) The (real) image planes of the *light source; in the case that the substrate is spherical*. In this image plane, we can intercept the light by an optical stop which corresponds exactly with the light source aperture. This is "Schlieren optics", or a refined version of dark-field illumination: the map of the substrate remains dark where the surface is exactly spherical, but those parts deviating from sphericity in orientation generate beams which can pass the optical stop and therefore are illuminated in the map. As such, a Schlieren picture of the substrate just reveals the "absolute" misorientation; in general in a nonlinear way. For many of our purposes we need more precise information including the direction of misorientation.In 13.7.2 we will see that variations on the Schlieren-optics theme can be designed which provide this type of information.

3) The (real) image planes of the light source *for those parts of the surface which are perfectly flat* (note the distinction with 2). Remember the introduction of the "discotheque-sphere imaging method" in 13.5.3: Different mirror orientations (ψ) on the sphere correspond with different light-source image positions in this plane, irrespective of the φ-position of the mirror. Therefore, we can characterize this plane as the ψ-plane; as explained in 13.5.3 the ψ-plane image represents the *orientation distribution function.*

Concentric arrangement of light source and sphere surface:

In all four cases of figs 13.13 a-d, we have deliberately chosen the light source image L (assumed here to be a point source) to be concentric with the sphere's surface (virtual light source image L_1 for convex mirror and concave lens, real L image for concave mirror and convex lens). This concentricity is not essential, but it has some practical advantages such as the annihilation of a correction term for off-axis beams (see 13.5.3) and the optimum use of aperture in a metallurgical microscope. On the other hand, the measurement will produce d^2R/ψ^2 rather than $\mathbf{S}R$ as mentioned in 13.5.1, and eq. (13.14) has to be adapted accordingly.

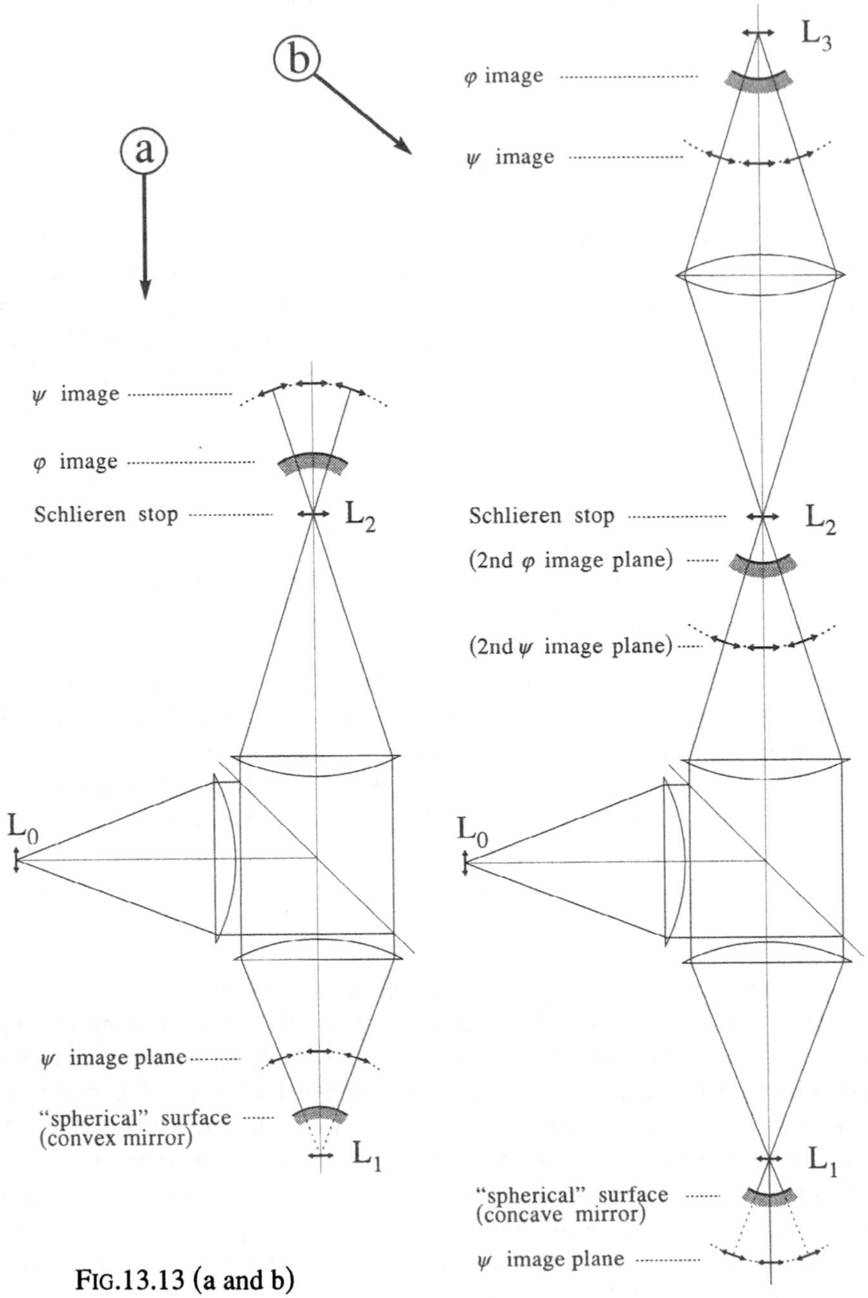

FIG.13.13 (a and b)

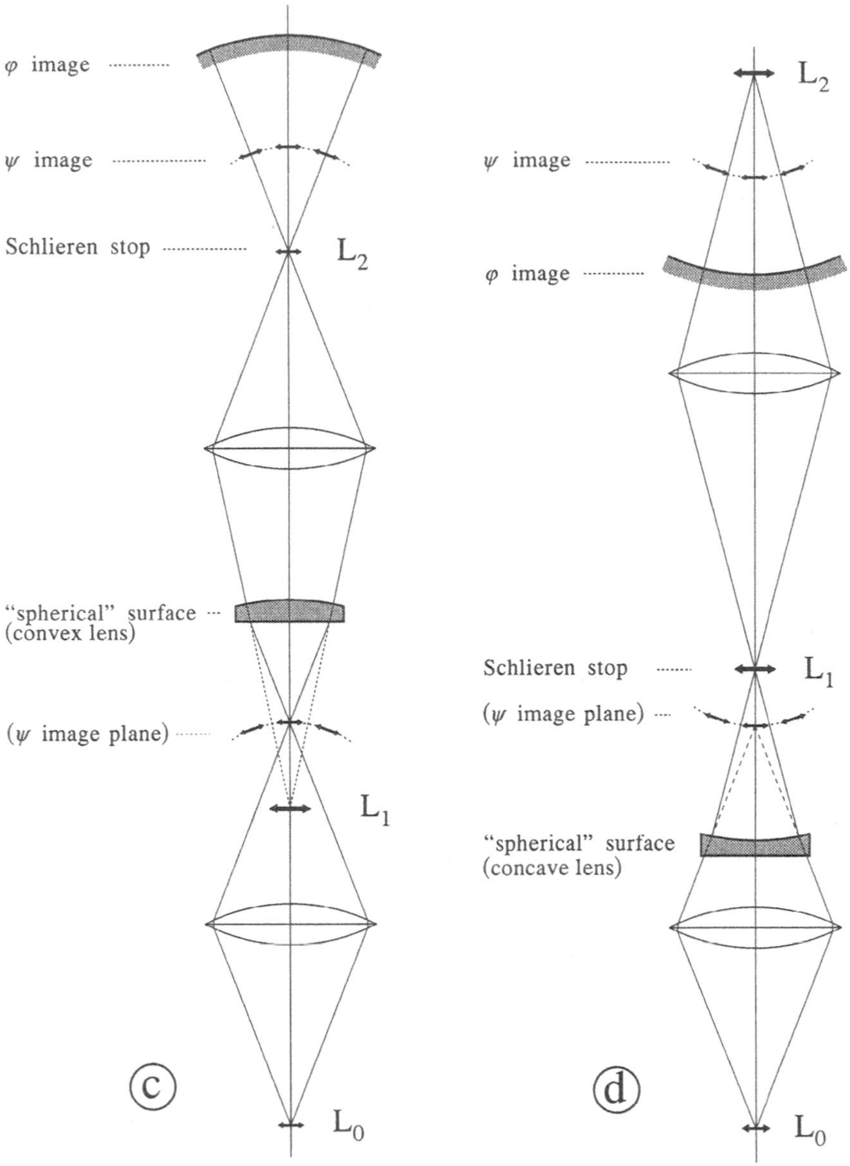

FIG.13.13 (c and d)

Aperture range of φ and ψ fields:

Both in the case of the φ and the ψ image, we have to do with a spherical object which cannot be imaged uniformly in focus onto an (in general) planar image receptor (photographic plate or TV camera). Although it might be partly remedied by deliberately choosing a lens system with the appropriate geometric aberration, this will in general limit the ψ and φ ranges in which a good image can be formed.

In order to reach *any* position on the sphere or half-sphere for aberration-free imaging sector by sector, we would have to mount the sphere on a goniometer head device so that any required spot could be positioned on-axis, remaining concentrical.

A simpler solution which increases the useful aperture with a factor of 2 as compared with a fixed sphere, is to make the sphere rotatable on a fixed axis making an angle α with the optical axis, when α is the maximum deviation which can be imaged sufficiently.

13.7.2 How to generate an orientation map

13.7.2.1. Double-scanning methods

In 13.5.1 we explained the measurement principle by the technique of a narrow laser beam scanning the φ field, combined with a detection device for measuring the deflection angle 2ψ. In fact, this illustration was not intended as a practical method, at least not for the 2-D case. When the principle would have actually to be applied to collect the data required for a 300 x 300 pixel image using a conventional TV camera, it would be utterly uneconomical. Suppose that the laser beam could be directed at precisely one pixel. (this implies a lower limit to the sphere radius in relation with the required angular resolution of the ψ variable). To measure the (ψ_1,ψ_2) position of the beam deflected from this single pixel, we would have to record a complete (ψ_1,ψ_2) picture, just to find the position of one spot. This takes something as 40 milliseconds - apart from that, the position data have to be analyzed from the video signal by apropriate electronic tricks. For all the 300 x 300 pixel points together we would need about 1 hour to collect the data.

A factor of 150 can be gained when the scanning of the area is not done pixel-by-pixel but line-by-line, both horizontally and vertically. Scanning the whole field then takes 600 pictures, requiring about 25 seconds. To find the (ψ_1, ψ_2) data for a given pixel at (φ_1,φ_2) position requires a cross-correlation of the (deformed) lines in the pictures corresponding to the φ_1 and φ_2 shots. This has to be computerized, but will still be time-consuming.

The ideal method would use, instead of a conventional TV camera, an image-analyzing device producing output signals for the ψ_1 and ψ_2 positions directly, i.e. without scanning the image, so that these signals could be stored synchronously with the scanning laser beam. The orientation map could then be taken and stored directly in a single 2-D field scan, i.e. in a fraction of a second.

Instead of scanning the object with an illuminating beam of one exact orientation and recording the transmitted orientation in the φ plane, we can, alternatively, illuminate the object with light of finite aperture and detect a Schlieren image in the ψ plane. This is a valid alternative for the double-scanning method, provided that we can manipulate the Schlieren image in such a way that it represents the orientation deviation. Possibilities for this are discussed in 13.7.2.2. - 13.7.2.4.

13.7.2.2 "Single-variable" Schlieren method.

In the heading of this subsection, the term single-variable is meant to apply to the representation of the ψ-space, not to the φ-space. Here we consider the conventional, dark/light version of Schlieren image formation, in dependence of the angular aperture of the light source.

When L is really a point source and when a "point stop" is used in L_2 of fig. 13.13a (a "nonphysical" possibility, because of diffraction effects) then a point in the image S_2 of the surface is either light (when the orientation deviates from the exact spherical orientation, even by the slightest amount) or dark (for zero deviation). When we prefer a more quantitative information, we can make the light source and the corresponding stop dimension finite. When ℓ_1 is the size of the L_1 source perpendicular to the optical axis (so that its aperture, as seen from S_1, is ℓ_1/r), then the light is only partly blocked by the stop as long as the condition $|(\psi - \varphi)| < \ell_1/r$ is fulfilled, and a corresponding fraction is passing. In fig. 13.14 the image intensity is sketched as a function of light-source aperture (provided the stop is dimensioned in accordance). When a circular diaphragm/stop combination is used, the linear behaviour is smoothed as in the dashed line. In other words: by choosing the light-source aperture and shape, we can adapt the "hardness" of the $\psi(\varphi)$ map image, or the maximum $(\varphi - \psi)$ deviation which can be translated back from the grey-scale of the S_2 image.

On a sphere where facets are beginning to appear, they will show up in the φ image as regions with a dark centre (because that centre is still in the original orientation) and light boundaries (where orientation deviation is highest). The edges between adjacent facet regions again will show up as dark lines. (note the contrast with the ψ-plane or discotheque-sphere image, discussed in 13.5.3).

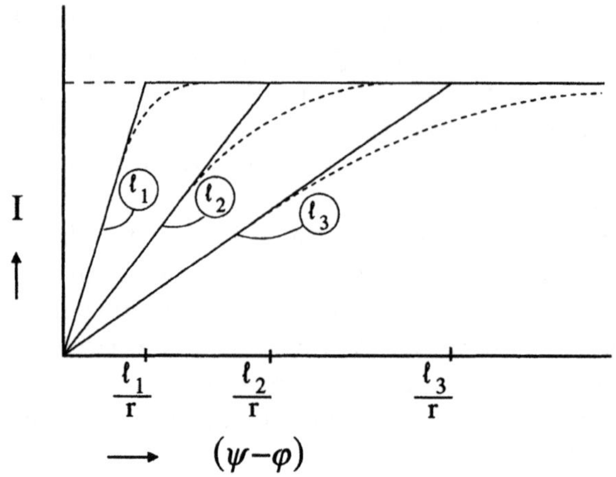

FIG. 13.14

13.7.2.3 Two-dimensional Schlieren method I: Colour Schlieren

Although the single-variable picture as described above is quite illustrative, it is not sufficient to reveal the *2-dimensional character* of orientation difference. The latter can be remedied by either of two tricks: colour and polarisation Schlieren.

In colour Schlieren we use, instead of a finite-area L_1 source emitting white light homogeneously, an equally finite-area source with an inhomogeneous RGB filter filling this area, for instance as sketched in fig. 13.15a. The range where (ψ_1,ψ_2) is represented is still dependent of the aperture according to fig. 13.14, but the *direction* of the deviation $(\psi_1-\varphi_1,\psi_2-\varphi_2) = (\Delta\psi_1,\Delta\psi_2)$, considered as a two-dimensional vector, is now translated in a unique RGB colour combination. (if required, both total intensity and RGB composition can be translated back into $\Delta\psi_1,\Delta\psi_2$ values). The generated φ images now have *coloured* maps.

Instead of the dark-field illumination technique we may use a bright-field technique by shielding off the deviating beam instead of the main beam; i.e. using a positive instead of a negative stop at L_2. This generates a white image with colour-bordered facets.

13.7.2.4 Two-dimensional Schlieren method II: Polarisation Schlieren

In analogy with the colour-Schlieren method, we can do a similar trick with polarizing filters. Here we can even arrange the filter geometry so that four maps are produced, representing $(\psi-\varphi)$ tilt maps in the $+\psi_1$, $-\psi_1$, $+\psi_2$ and $-\psi_2$ directions. The principle is as follows: (fig. 13.15b)

The *emitting* light source is a finite-area square which is for the left

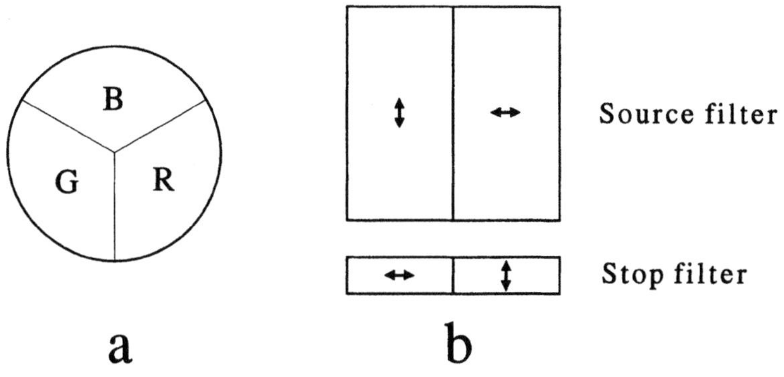

FIG. 13.15

half covered with a vertically polarizing, and for the right half with a horizontally polarizing sheet. The stop at L_2 consists of a horizontal slit opening, also covered with polarizers but with directions interchanged. When the light source image is perfectly aligned with the stop, the polarizers are crossed both left and right and no light passes through - so it is really a stop. When the image is misaligned to the left (i.e., by a negative ψ_1 deflection), then a certain width of the *vertical* polarizers overlap and light of vertical polarization comes through. Misalignment to the right (i.e., positive ψ_1 deflection) produces an opening transparent for horizontally polarized light. So when the image S_2 is observed with a vertical analyzer we obtain a map of *negative* ψ_1 deflection, when observed with a horizontal analyzer a map of *positive* ψ_1 deflection. The choice of a *slit-shaped* stop (the rest of the stop filter is black) guarantees that the image represents the ψ_1 deflection independently of the ψ_2 deflection, so that we get really a linear response for $\Delta\psi_1$ (linear up to the limit set by the aperture of the light source). Of course, for small details in the surface, the signal is blurred by diffraction. When we use a square stop instead of a slit, the ψ_1 and ψ_2 signals are mixed.

To obtain the two maps for the other direction of misorientation $\Delta\psi_2$, the illuminating square filter and the Schlieren stop filter both have to be rotated by 90°. The four maps can be combined in pairs so as to obtain two maps, for $\Delta\psi_1$ and $\Delta\psi_2$ tilt, respectively, but now in a representation which allows to record the *sign* of the signal.

Note 1: Such linear ($\Delta\psi_1$, $\Delta\psi_2$) "Schlieren" maps can easily be processed into (ψ_1, ψ_2) "point-source illumination" maps, i.e. by just adding the appropriate (φ_1, φ_2) values. This works as long as the real orientations have no larger deviations than the $\Delta\psi$ aperture of the polarized light source, i.e. as long as the image is a real linear representation of misorientation. The resulting processed image can be considered as a sort of bright-field image and has the appearance of a diffusively reflecting sphere illuminated by one or more light sources (i.e. when the RGB colour-coding method is used). It is similar to a diffuse computer-simulated surface when rendered with ray-tracing methods. In fact, actual computer-simulated sphere-growth experiments based on the $R(\vec{n})$ model functions as described in chs. 9-12 can be visualized in the same way. This provides a very direct comparison facility of experiment and $R(\vec{n})$ function parameters.

Note 2: Complications may arise when the crystal under investigation is birefringent. Refracted as well as reflected polarized light may then become elliptically polarized, affecting the Schlieren principle. Careful alignment of the optical axes with the polarizer directions is essential in this case.

13.7.2.5 Orientation-grid imaging techniques.

As a "simple" alternative to the methods mentioned above we can consider the "orientation-grid" technique.

A mirror sphere, as seen by the naked eye in a room with paned windows, reflects the window grid on its surface. We see the surface together with the deformed image of the grid (although strictly spoken the surface and the grid cannot be in focus simultaneously). Distortions in the sphere are revealed by distortions in the regularity of the grid. In other words: the grid reflections on the sphere can be used to make an orientation map of the sphere. The grid corresponds with a "ψ grid" in the sense of section 13.5.

Experimental conditions for this method are:

1) the aperture *of the ψ grid* should be large, i.e. at least as large as the ψ range of interest. (this condition agrees with the illumination aperture conditions of the other methods)

2) the aperture of the *image-forming objective* of the sphere should be small, i.e. the image should be taken "telescopically" rather than "microscopically". This aperture determines the resolution of the φ variable. The period of the grid, and the width of the grid lines (both in angular dimension) should be chosen in agreement with this aperture.

Instead of an actual grid, we might also use a field of variable colour (RGB field) or intensity ("shadow field") The latter responds to tilt in one direction only. The grid can be placed in an "L" real-image plane.

The optical configurations as sketched in fig. 13.13a-d can be used for this method, provided the image-forming aperture is decreased so as to accomodate for the grid period.

Chapter 14

SUMMARY: CONCLUSIONS AND OUTSTANDING QUESTIONS

The outcome of any serious research can only be to make two questions grow where only one grew before.

Thorstein Veblen (1857-1929)

The relation between crystal morphology and the orientation dependence of both surface free energy and growth rate or "interface mobility", is a classical as well as a neglected subject. It seems to me that the theory of the equilibrium shape (based on the $\gamma(\vec{n})$ function) and that of the growth shape (based on the $R(\vec{n})$ function) have always led a rather separate life. Both are "classical" subjects, but restricted in their applicability because, on their own, they only consider simple and exceptional extreme cases. In order to be applicable to situations and questions which are relevant in actual crystal-growth systems, both subjects have to be combined and generalized.

The present monograph is an attempt to accomplish such a generalisation and to apply the results to a number of problems which, in my opinion, are still poorly described or improperly recognized. Many of them are so complicated that a thorough analysis is beyond the scope of this book; in fact more new questions may have been raised than old questions solved. In such cases this book suggests at best a physically and mathematically more satisfactory statement of problems or an "attack strategy", rather than a solution.

Four main themes can be followed throughout this work:
The first is the formulation of *generalized equations of motion* for crystal interfaces (representing the effects of orientation-dependent surface free

energy and growth mobility) and their application to several morphological problems in crystal growth theory. Starting point for these equation are the functions $\gamma(\vec{n})$ and $R(\vec{n},A)$, to be considered as properties of a "mathematical" surface between continuous media and as given data. In this sense, this is a "phenomenological" theory; the determination of these γ and R functions from atomistic models of the crystal surface and its growth kinetics requires a complementary theory (see, e.g., KERN 1987).

The second theme, narrowly related with the first, is the "construction" of *explicit expressions for* $\gamma(\vec{n})$ *and* $R(\vec{n},A)$ *functions.* Such expressions are essential when we actually want to apply the EOM to practical situations. In this monograph we present a mathematical (chs. 9-12; item 11 of this summary) as well as an experimental approach (ch.13; item 6 of this summary).

The third main theme is the relevance of the behaviour of *junctions* for several crystal growth situations. Junctions are defined as meeting lines or points between interfaces mutually or between interfaces and dislocations. We discern D-junctions (dislocation/interface), 3-junctions (meeting line of 3 interfaces) and 4-junctions (meeting point of 6 interfaces and four 3-junctions). Several problems concerning the equilibrium and growth shapes of junctions are discussed in detail and the rôle of junctions as "velocity source" is emphasized.

The fourth theme is the rôle of *curvature effects,* both in relation with the stability of zigzag structures and in relation with surface-instability theory in general (ledges, terrace structures, cellular structures and superstructures)

In this summary the conclusions are grouped into

a) "classical" theory of equilibrium and growth shape,
b) extended formalism and equations of motion, and
c) nonlinear aspects and instability theory. (These groups do only partly correspond with the themes mentioned above).

GROUP 1: CLASSICAL THEORIES OF EQUILIBRIUM AND GROWTH SHAPE

1. Mathematical tools (ch.2)

Point of departure for the phenomenological description of interface behaviour are the functions $\gamma(\vec{n})$ and $R(\vec{n},A)$ in which the interface is considered as a mathematical surface between continuous media. For the description of such an interface and its evolution, differential geometry is the natural language; consequently we introduce the vector operator $\vec{\mathrm{Q}} = \vec{n} + \mathrm{d}/\mathrm{d}\vec{n}$ and the tensor operator $\vec{\vec{\mathrm{S}}} = \mathrm{d}\vec{\mathrm{Q}}/\mathrm{d}\vec{n}$, working on any function of $\vec{n}$ (normal vector characterizing interface orientation). With these operators, the classical theories of the equilibrium form and that of the growth form (kinematic wave theory) can be represented in a compact notation.

In particular, the application of these operators to the γ function leads to expressions for the force equilibrium in an interface and for the curvature shift in the growth affinity, and application to the R function to the concepts of vector velocity and "spreading rate" = rate of change of the radius of curvature. In this way, the mathematical similarity between the theories of equilibrium shape and that of growth shape is made evident (the results are, of course, identical with the classical formulation). For the equilibrium shape these have been developed, e.g., by GIBBS (1867), HERRING (1951, 1952), JOHNSON (1965) and HOFFMANN and CAHN (1972); for the growth shape by FRANK (1958, 1963) and CHERNOV (1963 a,b). The basic identity which shows this similarity is that between the plot of the vectors $\vec{\mathrm{Q}}\gamma(\vec{n})$ and $\vec{\mathrm{Q}}R(\vec{n})$ on the one hand and the shape obtained by the well-known Gibbs-Wulff construction from the γ and R plots on the other hand. As we do not a priori assume similarity between the γ and R functions, the equilibrium and the growth shapes can be different.

2. Junctions: equilibrium and growth shape (4.1 and 6.2, respectively)

Extensive consideration has been given, throughout the monograph, to junctions. Topologically, junctions are the connective elements in structures involving more than one interface and/or dislocations; mathematically they impose certain boundary conditions on the equations of motion for the interfaces involved. We discuss D-junctions, 3-junctions and 4-junctions as defined in the introduction to these conclusions.

Just as for interfaces without junctions, we can discern between the equilibrium configuration of junctions (where all interfaces are immobile) and the growth configuration where, in the simplest version, surface-energy effects are neglected, just as in the classical kinematic-wave description of the growth shape. The mathematical expressions for these configurations are dissimilar, and so are their solutions. (This difference affects the microscopic structure of moving junctions; see conclusion 13).

3. Wetting of junctions (4.1.2)

Depending on the solubility of the junction-equilibrium equation, junctions can be "wetted" or not. In a wetted 3-junction, the original single 3-junction has opened into a triangular pipe with stabilized cross-section. In a wetted 4-junction the original junction is replaced by a tetrahedron. A wetted D-junction is equivalent with a hollow-core dislocation.
In the case of *orientation-dependent* and *affinity-dependent* γ, respectively, the character of a junction can change between wetted and non-wetted by a *change in orientation* and by *motion*. Wetting (especially by liquid phases) is important in the kinetics of sintering.

4. The rôle of velocity sources in the growth shape (6.2.2; 6.4)

The presence of junctions in a growing crystal (dislocations and/or grain boundaries) affects their shape. Hence it is important to distinguish between the perfect-crystal growth rate function and derived growth shape, and the effects of the junctions.

We introduced the concept of "velocity source" in combination with a growth rate plot; D-junctions and 3-junctions can act as such and their vector velocity can, in some cases, be calculated by detailed analysis (see conclusion 13). This velocity is in general a complicated function of the affinity (and for D-junctions, of the Burgers vector). The presence of velocity sources can completely hide the perfect-crystal growth shape plot; in such cases an experimentally determined $R(\vec{n})$ function is representative for junction behaviour and for the statistical behaviour of the vector velocities of junctions, rather than for the surface of the perfect crystal.

When junction activity enhances growth in only one or two directions, whisker- and platelet growth results.

5. Insufficiency of the $R(\vec{n})$ function (6.3)

An important (and not generally recognized) conclusion of the present analysis is that even for a perfect, i.e. junction-free interface, the "macroscopic" growth rate plot is not always unambiguous; an interface may switch between two metastable structures, corresponding with metastable $R(\vec{n})$ functions. This is because another type of velocity source can occur, in particular when the $\vec{\vec{\mathbf{S}}}\gamma$ and/or the $\vec{\vec{\mathbf{S}}}R$ tensor has one or two negative eigenvalues (implying "strong" orientation dependence of the γ and R functions). This "homogeneous" velocity source is the stabilized edge or vertex. Its formation can be considered as a condensation phenomenon in a morphological phase transition and implies a nucleation barrier; therefore, the interface can grow according to the original R function, without the velocity source. Nucleation/annihilation kinetics and and the stability of the velocity source are affected by the perpendicular affinity gradient $\vec{\nabla} A \cdot \vec{n}$. From a mathematical viewpoint, a homogeneous velocity source is a soliton solution of the interface equation of motion. (see also conclusions 18 and 21).

6. Experimental determination of the $R(\vec{n})$ plot (13.5, 6 and 7)

The fact that the quantity $\mathbf{S}R$ is, in equations, physically more relevant than R itself suggests to tailor experimental measurement techniques in such a fashion as to determine this quantity directly.

This idea has been worked out (in ch.13) into an alternative evaluation principle for sphere-growth experiments. With an appropriate optical set-up (in which the surface of the growing sphere is essentially one of the image-forming surfaces), the elements of the $\vec{\vec{\mathbf{S}}}R$ tensor can be measured directly, in a nondestructive way, by following the evolution of an optical image as a function of time during the initial stage of growth.

GROUP 2: EXTENSION OF THE CLASSICAL THEORIES AND EQUATIONS OF MOTION

7. The general case: inhomogeneous affinity fields (5.3, 5.4)

The cases described by the "classical" textbook discussions on equilibrium and growth shapes are characterized by a constant growth affinity A (and temperature, chemical composition etc) all over the surface. In this case, orientation $\vec{n}$ is the only variable of the γ and R functions and shape evolution can be described by the classical version of Frank's kinematic-wave theory; i.e. the version with straight kinematic-wave trajectories.

In the general case of inhomogeneous affinity, surface gradients of A induce the kinematic wave trajectories to become curved. We can still describe the evolution of a surface by a "vector velocity" $\vec{\mathfrak{R}}$ and a "curvature evolution rate" $\vec{\vec{\mathfrak{S}}}$, but, in comparison with the constant-affinity case, the equations for these vector resp. tensor quantities involve the surface gradient of A and the surface curvature tensor. (plus of course the explicit dependence of γ and R on A, in the form of first and second derivatives).

Explicit expressions of the general EOM are given in ch. 5. They include terms for the effect of capillarity, which is accounted for by compensating the externally applied affinity $A_{|}$ with the expression for the curvature-induced shift in the affinity, ΔA_c, in the A argument of the $R(\vec{n}, A)$ function.

The expression $\overset{\rightrightarrows}{\mathfrak{S}} = 0$ can be used as an necessary and sufficient condition for a stationary structure (not for *stability* of the structure).

8. Equilibrium shape in an inhomogeneous affinity field and the Marangoni effect (3.3)

As a side-result of the extension mentioned in conclusion 7, we can also apply the extended operators to a surface free energy which depends on a second parameter: $\gamma(\vec{n},m)$. Physical consequences of such a dependence are generally known as "Marangoni effects". The difference between Marangoni effect and curvature-related effects is discussed in 3.5.1.

For an interface in equilibrium, but situated in an inhomogeneous field of $A_{|}$ (i.e. T or C), the capillaity effect results in a varying interface curvature. It is relevant, e.g., for the shape of grain-boundary grooves in a solid/liquid interface in a temperature gradient.

9. Dendritic growth (8.1.1)

The equation of motion provides a straightforward explanation of the experimental fact that dendritic branches are oriented in specific crystallographic directions, even in cases where the orientation dependence of γ and/or R is slight. Once a planar and a periodic cellular interface are unstable, i.e. when also the cellular side interfaces are destabilized ("sidebranching instability"), the orientation dependence itself provides a triggering mechanism which strongly favours the growth in specific directions.

10. Systematic catalogue of crystal growth situations (1.5 and ch.12)

The equation of motion mentioned in conclusion 7 is only a self-consistent equation (from which the evolution of a given initial shape can be calculated) when the argument A is available as a function of the position on the interface for all times. When, however, this A itself is to be computed from another equation, e.g. a diffusion equation, then the interface EOM is just a boundary condition for this diffusion equation, and the coupled set of equations is to be solved. ("Stefan problem").

Any real crystal growth situation in nature or in a technical process, (e.g. growth from solution, melt, Bridgman, CVD etc.) can be characterized mathematically by the proper EOM or set of coupled EOM's, boundary conditions and initial conditions. On the other hand, the type of interface behaviour in such a catalogue has to be characterized by the γ and R functions (or by the parameters occurring in these functions).

11. Analytical functions for $\gamma(\vec{n},A)$ and $R(\vec{n},A)$: representation by nonlinear networks (chs. 9 - 12)

The formal framework of EOM's as discussed in conclusion 7 is useless for practical application until explicit functions for $\gamma(\vec{n})$ and $R(\vec{n},A)$ are available.

We have developed a general mathematical construction principle for these functions which can be characterized by a *nonlinear network*. This principle is highly flexible: it allows for a wide choice of physical models for the surface. Details of a model correspond with a) the topology of the network and b) its discrete elements. In particular, the elements of the network contain

1) vector products $|\vec{n}\cdot\vec{p}_i|$ and/or $|\vec{n}\wedge\vec{f}_i|$, where $\vec{n}$ is the orientation (normal) vector and $\vec{p}_i$, $\vec{f}_i$ are PBC and facet-oriented vectors, respectively;
2) parameters such as $v_{\parallel}$ (ledge velocity - with suffix i for crystallographically different ledges), M (orientation-independent limitation of growth rate by, e.g., diffusion or chemical reaction rate in a boundary layer) and ε (roughening parameter).

Of particular relevance is the availability of the *roughening parameter* ε (one for each crystallographic "family" of facets). It can range between the values 0 and 1, corresponding with the transition from "completely faceted" to "completely roughened", i.e. liquid-like behaviour of the surface. For $\varepsilon = 1$ the facet family disappears completely from any evolving shape (and from the $R(\vec{n},A)$ and $\gamma(\vec{n})$ functions, of course).

The parameters occurring in the γ and R functions can be used quantitatively to characterize interface behaviour, e.g. in dimensionless-number combinations as a distinguishing tool in the systematic catalogue of conclusion 10. These parameters will in general depend on external variables such as the growth affinity and temperature.

The "construction" of analytical R or γ functions by this method does not imply, in any form, a theory of crystal growth. Rather it is a practical method of parameter fitting for such functions. The strength of the method lies in the fact that details of $R(\vec{n})$ functions which are implicit in *all* atomistic models automatically emerge as limit behaviour of the "networks". Examples are the $|\vec{n} \wedge \vec{f}|$ shape of cusps, limitation by an external limiting mechanism and the shape of smoothed minima in the case of partial roughening).

12 "Geometrical Chemistry of Crystal Growth": extension to polynary growth (8.5)

The description of the growth process for mixed crystals, for compound crystals which are dissociated in the parent phase (such as electrolyte solutions) or for crystals growing by a CVD process, implies the transport of $(n+1)$ species under the combined action of $(n+1)$ driving forces (n chemical species plus entropy. Strictly taken "unary systems do not exist" because we have to do with combined transport of heat *and* matter even for a "pure" crystal/melt interface). The consequence of *non-equilibrium interface kinetics* is that the chemical composition of the grown crystal becomes orientation-dependent (which has consequences for mechanical stress in bulk crystals and for doping levels in semiconductor epitaxy).

As compared with the unary case, such an extension amounts to the replacement of unary fluxes and affinities by "species-vector" fluxes and affinities. It can be compared with the transition from a black-and-white to a coloured world. With n components, the possible interactions increase by a factor n^2, but *in principle* the extension can be made.

In section 8.5 we have presented a rough sketch of such an extension. Mathematically and physically, the subject of combined multi-component diffusion and (chemical) reaction mechanisms is covered by multiport circuit theory, a branch of mathematical physics which was well developed during the first half of the century. Later, it was applied to problems in biochemistry in the form of "network thermodynamics", a term coined by its developers OSTER, PERELSON and KATCHALSKY (1973); for further references see 8.5). Although the basic tools have been established on a firm mathematical basis and seem quite appropriate for application to crystal growth situations, they seem somehow to have escaped attention by the crystal growth community so far.

13. The effect of a diffusion boundary layer (8.2)

The case where the growth rate is controlled by the simultaneous action of "surface kinetics" and of diffusion in a thin boundary layer is both of technical importance (LPE) and of theoretical interest, because the diffusion leads obviously to a curvature-dependent growth rate. Moreover, the case can be considered as intermediate between fully surface-controlled kinetics and surface-plus-diffusion kinetics with "far-away" boundary conditions (Stefan-type problem). For this reason, the case has been worked out in some detail in 8.2. The basic approach is to represent the resulting growth rate function by a single $R(\vec{n},A)$ function which describes interfaces of low curvature, and to account for curvature effects at highly curved sections.

Apart from the limitation of the growth rate by the diffusion in the boundary layer, the "mathematical" effects of the boundary layer are *dispersion* (wavelength-dependent $\bar{\mathfrak{R}}$) and *curvature-dependence* of the growth rate, i.e. *nonlinearity* of the EOM. In combination, these effects can lead to soliton solutions, i.e. stabilized nonplanar stationary structures.

14. The detailed shape of moving junctions and their rôle as velocity source (6.2 and 6.4)

As the equilibrium- and the stationary-growth configuration of junctions are not necessarily identical, (see conclusion 2), a stationary moving junction develops a region where some of its interfaces are curved so as to match the equilibrium condition at the junction origin with the stationary-growth condition for the asymptotic interface directions. The size of this region is typically the critical-nucleus size for the affinity involved. The exact shape of an interface near a junction can be solved from its EOM. When this equation is soluble, the behaviour of the junction with respect to this particular interface is called "regular". When it is not soluble, the junction acts as a *velocity source* for the interface. We say that "the interface is nucleated by the junction". The vector velocity of the junction is then imposed by the equilibrium condition, and the interface near the junction is not curved. As a consequence of a sort of "phase rule", a 3-junction nucleates at least one of its interfaces, a 4-junction at least three. To be precise, we have to discern "equilibrium" and "kinetic" nucleation by junctions. The well-known cases of a screw-dislocation D-junction velocity source (spiral growth mechanism) and a twin-plane 3-junction source (TPRE mechanism) should both be considered as cases of kinetic nucleation.

15. Mutual interaction of junctions in a moving inhomogeneous affinity field; periodic structures of junctions (8.4)

In parallel with the analysis of the single-junction shapes mentioned in conclusions 8 and 14, we can also analyse the equation for two junctions within close distance. From the result, we can derive the interaction force. (the situation is comparable to the horizontal interaction between two small wax spheres floating on a water surface, as a result of their interacting surface depressions). For the equilibrium case, the interaction is always attractive (just as for the wax particles). For the moving-interface case, a repulsive component can result from the effect of the bulk-diffusion field. Such a repulsive component can stabilize periodic patterns of D-junction depressions or grain-boundary grooves, as can be observed experimentally in unidirectional-solidification situations.

16. Transients in the approach to equilibrium shape: the RAE concept (8.3)

Ageing and sintering processes provide examples of interface motion induced not by an externally applied affinity but by differences in the capillarity-induced driving force (capillarity shift). The evolution of such processes can adequately be described macroscopically by simple power laws, with constants and exponents which are characteristic for the predominant transport mechanism. On this basis, we defined the concept of the "range of approached equilibrium" (RAE) which is essentially the range within which deviations from the equilibrium shape have been smoothed out to less than 1/e of their original value (it is therefore a functon of time: the general formula is RAE = $(c_n t)^{1/n}$. Applications are discussed in 8.3.

GROUP 3: NONLINEAR ASPECTS - CURVATURE DEPENDENCE OF γ AND R AND THE INSTABILITY OF INTERFACES

Curvature dependence of surface properties.

It is obviously unrealistic to consider the interface as a mathematical surface between continuous media. Intuitively, such a concept should break down for interface curvature exceeding some critical level related with the real inhomogeneity of the "continuous media". Results derived from the equations (without correction terms for curvature effects) are suspect as soon as they involve highly-curved parts or sharp edges and vertices.

Curvature dependence is essential for the solution of certain paradoxes resulting from the purely classical version of the EOM, and more in general for the discussion of nonlinear effects and their consequences for structural differentiation. Therefore, we have given ample room to the discussion of the interrelated subjects of curvature dependence of γ and R, nonlinear partial differential equations of motion, instability of interfaces and periodic pattern formation.

17. Curvature dependence of surface properties expressed as a series expansion in 1/r (2.8)

Curvature dependence of surface properties in general (Ψ) is considered in ch.2, where a series expansion in $1/r$ has been worked out for some relevant cases. We emphasize that "curvature dependence" is not to be associated with one single mechanism or principle, but that several different mechanisms (each leading to some form of curvature dependence) can coexist, each with a different critical curvature. First-order effects in $1/r$ are typically related with asymmetry in the "delocalisation" of the property. The second-order term in $1/r$ contains a contribution proportional to $\mathbf{S}\Psi$.

18. The "fundamental restriction" of the continuum approach (1.3)

The series expansion referred to in 17) obviously becomes suspect when r approaches the lattice-spacing size. The number of atoms involved in a shape detail described by such a radius becomes too small. The "thermodynamic" approach, i.e. the EOM-based description in terms of separable energetic and kinetic terms, is then no longer applicable. Instead, we have to invoke Monte Carlo techniques etc. This is, in short, the "fundamental restriction" as formulated in ch.1.

The FR is relevant for the kinetic behaviour of shape details such as junctions (see concl.13) and stabilized edges and vertices (conclusion 19)

19. Stability of zigzag and pyramid structures; edges and vertices as homogeneous velocity sources (6.3 and 6.4)

For an interface in equilibrium, sawtooth (2-dim.) or pyramid (3-dim.) structures can occur when the $\gamma(\vec{n})$ function has a range where $\mathbf{S}\gamma$ is negative. This zigzag instability is discussed in 3.7. For an interface in motion a similar conclusion cannot decisively be derived from a negative-$\mathbf{S}R$ behaviour alone. We concluded (in 6.3.3, see also conclusion 5) of this chapter) that the $R(\vec{n},A)$ function provides insufficient information in itself, and that the nucleation, the stability and the vector velocity of stable sharp edges and vertices, if any, have to be made available as extra data. These are essentially related with the curvature dependence of the R function (although an explicit expression of this dependence would be of little help in calculating them, because the validity range of the EOM is exceeded at the edges).

Ergo, the "classical" kinematic-wave theory is incomplete for the $\mathbf{S}R < 0$ case (and can even be inadequate in the $\mathbf{S}R > 0$ case, when sharp *protruding* edges are stabilized).

Calculation of the vector velocity, stability and nucleation kinetics of edges and vertices is one of the white areas in the theory which have clearly emerged from the discussion. Such a calculation will necessarily have to be based on an atomistic model of the edge or vertex. Monte-Carlo simulation on such a model may be an alternative.

Although several authors have called attention to the $\mathbf{S}\gamma < 0$ or "zigzag" instability, the status of this prediction seems to be hypothetical up to now by lack of experimental evidence.

20. *Instability types: constitutional supercooling versus instability induced by orientation-dependent γ and R. (6.3; in particular 6.3.5)*

In the "classical" theory of interface instability (based on orientation-independent surface properties) the only mechanism inducing instability of a planar interface is "constitutional supercooling". Conclusion 19 states that the case of orientation-dependent surface properties provides an alternative mechanism which may roughly be characterized as the $\mathbf{S}\gamma < 0$ mechanism.

Obviously, the general case of surface instability requires an analysis in which both aspects are combined (on the basis of the EOM). We have argued (in ch. 4.2 and 6.3) that such an analysis has necessarily to be a nonlinear analysis.

From the generalized point of view, cellular, terrace- and zigzag structures are closely related. Slight orientation dependence of γ and/or R results in cellular structures growing generally in a direction which is partly determined by the crystallographic axes (i.e., not necessarily parallel with the macroscopic growth direction). Strong dependence results either in zigzag structures (when the macroscopic interface is maintained in an off-facet direction by an affinity gradient) or in "terrace" structures ($\simeq$ cellular structure growing almost parallel with a facet).

21. *Interface instability as a branch of soliton physics (ch. 7)*

From a more general standpoint than taken in previous conclusions in consideration of the $\mathbf{S}\gamma <0$ case, we can say that certain nonlinear equations of motion can exhibit soliton-like solutions, i.e. solutions with a stabilized, localized and "particle-like" character. The sharp edges and vertices mentioned in conclusion 5 can be considered as one- and zero-dimensional "edge-like" soliton solutions of the interface EOM, provided the latter adequately accounts for the nonlinear effects.

We have argued that a number of different effects can play this role: e.g. edge stabilisation by an $\mathbf{S}\gamma < 0$ mechanism, curvature dependence of the R function as a result of asymmetric Schwoebel effect, curvature dependence of the R function by diffusion in a boundary layer or in a half-space ("classical" cellular growth induced by constitutional supercooling).

Altogether, interface instability is essentially a nonlinear effect. As a consequence, instability of interfaces with OD γ and R cannot adequately be tackled by the (linear) Mullins-Sekerka type approach (such a linearized analysis can at best provide a correct *stability criterion* for a planar interface with OID γ and R, not predict the ultimate cellular structure and period). This recognition of the mathematical situation leaves us with two aspects of the problem which are still, at least to my knowledge, poorly understood:
a) investigation of nonlinear equations of motion (or sets of coupled EOM's) for an interface, in particular with respect to stability criteria for soliton solutions.
b) As a second step to the calculation of the final interface structure in the case where the planar interface is unstable:
b_1) generation/annihilation kinetics of the solitons; b_2) the multi-soliton equilibrium density and pattern, including an "equation of state" for the soliton continuum. Aspect b) is, effectively, concerned with the "quasi-thermodynamics" of the soliton continuum, and involves the irreversible thermodynamics of the original interface.

The quasi-thermodynamics of the soliton continuum is discussed in 7.2.5.

22. *The family tree of structural differentiation; families of periodic patterns in moving interfaces (7.3)*

As a consequence of the recognition of ledges, edges and other "interfaces" as soliton solutions of nonlinear partial differential equations and of the subsequent possibility of developing equations of motion for the "next-tier field" of the obtained soliton continuum, we have emphasized the concept of a tiered repeatable procedure leading to successive branching in a "family tree of structural differentiation". Each branching corresponds with a phase transition, and successive branchings can be considered to give rise to sub- or superstructures occurring in previous transitions.

In this way, all "structure" occurring in the physical world can be interrelated in a universal family tree which includes, a.o., elementary-particle, cosmological, chemical and even biological branches. Geometrical structures in moving crystal interfaces are only a small part of this universal tree. Nevertheless, the number of transitions is rich: it includes the roughening transition (ledges), terrace and zigzag formation, superterrace formation, cellular growth, periodic grain-boundary and dislocation structures, and surface reconstruction.

Although the family-tree principle is conceptually simple, its quantitative evaluation is quite complicated mathematically. For many of the transitions mentioned, the rigorous formulation can at this moment be considered as an open question which may be a rewarding field for future research. Apart from the direct application to crystal growth physics, the study of morphological phase transitions in interfaces may be instructive as a model system for other branches in the family tree.

23. Applicability of the equation of motion to other interfaces: ferromagnetism and ferroelectricity (ch. 7)

Domain walls in ferromagnetic and ferroelectric media are also characterized by more or less OD surface free energy and mobility. Therefore, all results which are here specified for crystal interfaces, junctions, instabilities and structures can be translated. Examples are

1) Bloch and Néel walls, Bloch lines (as substructures), magnetic-domain equilibrium shapes in thin garnet layers (ribbons, bubbles) and their kinetics. As compared with crystal growth, this field is complicated by the occurrence of the magnetic field and its interaction with the crystal's mechanical stress field by magnetostriction, but the domain-wall equilibrium and kinetics are fully described by their $\gamma(\vec{n})$ and $R(\vec{n},A)$ functions (where A is to be read as the affinity for spin-rotation, i.e. related with the internal magnetic field).

2) Domain walls in ferroelectrics (with the same remarks as for magnetic domain walls).

Note that domain wall kinetics is statistically related with the complex magnetic and dielectric permeability and with the hysteresis loops, as a function of both frequency and field amplitude.

24 "White areas" and their interrelations (chs. 1 through 13)

Below we summarized some areas which might be situated somewhere between "terra incognita" and "promising for future research". We used (rather arbitrarily) three theoretical and three experimental headings - but it is obvious that many cross-relations exist, between the groups as well as between the isolated subjects.

Most of the items were mentioned in other entries of this summary.

Theory of crystal growth

Relations between
- * parameters in assembled $R(\vec{n},A)$ functions (ch. 9-12)
- * parameters in atomistic models
- * Monte Carlo simulation (e.g., on reentrant edges, tilted substrates, $\mathbf{S}\gamma < 0$ orientations)
- * observed experimental data

Application of "network thermodynamics" to (polynary) crystal growth

Computational Physics

Homogeneous-A case:
OD chemical composition
Inhomogeneous-A case:
generalized simulation of growth & etching
(diffusion + OD interface kinetics)$\Rightarrow$dendritic growth, instabilities, terraces)

Mathematics of phase transitions

Theory of NLPDE stability$\Rightarrow$application to crystal EOM
Stationary structures (cellular, eutectics, eutectoids)
Mullins-Sekerka instability theory: extension to OD case with realistic $R(\vec{n},A)$ functions.
Dendritic growth

Experiments focused on theory of CG and surface structure

Sphere growth experiments with emphasis on orientation
mapping ⟹ direct determination of roughening parameters
Precise measurement of roughening parameter as a
function of T, A in vicinity of rougening transition
Verification of $\mathbf{S}\gamma < 0$ instability: is zigzag observable for $\vec{n} \cdot \vec{\nabla} A > 0$?
(better definition of exp. conditions; $\vec{\nabla} T$, $\vec{\nabla} C$)

Experiments focused on technology

Epitaxy and etching on 3D-structured substrates (with masks,
grooves, mesas etc; for IC structure, lasers and waveguides,
quantum devices)
Bulk crystallisation
Single crystals

Experiments focused on interface stability

Terrace growth
Cellular structures in unidirectional solidification (A gradient)
Eutectics, eutectoids
$\mathbf{S}\gamma < 0$ instability hypothesis
Dendritic growth

REFERENCES

ALBERS, W. (1973): Physical Properties of Composite Materials, in *Proceedings Conf. on in situ composites*, Lakeville 1973. (Nat. Acad. of Sciences 1973), **vol.3**, 1-18

ANGENENT, S. and GURTIN, M.E. (1989): Multiphase Thermomechanics with Interfacial Structure; II: Evolution of an Isothermal Interface, *Archive for Rat. Mechanics and Analysis*, **108**, 323-391

ALMGREN, F.J. (1966): *Plateau's Problem, an invitation to varifold geometry*, W.A. Benjamin Inc., 1966

ARBEL, E and CAHN, J.W. (1977): A method for the absolute measurement of anisotropic surface free energies, *Surface Science*, **66**, 14-24

AUDERO, M.A. and BILONI, H (1973): The development of cells during the solidification of dilute Zn-Sn alloys, *J. Cryst. Growth*, **18**, 257-264

AUSLANDER, D.M., OSTER, G.F., PERELSON, A and CLIFFORD, G. (1972): On systems with coupled chemical reaction and diffusion, *J. of Dynamic Systems, Measurement and Control*, Sept.1972, 239-248

BENNEMA, P. and VAN DER EERDEN, J.P. (1987): Crystal graphs, connected nets, roughening transition and the morphology of crystals. Chapter 1 in *Morphology of Crystals, part A* (Ed. I. Sunagawa, Terra Scientific Publishing Company 1987)

BENNEMA, P. and GILMER, G.H. (1973): Kinetics of crystal growth, in "*Crystal Growth, an introduction*", ed. P. Hartman, 263-327. (North-Holland, Amsterdam 1973)

BILONI, H., BOLLING, G.F. and COLE, G.S. (1966): On the origin of the cellular solidification substructure, *Trans. Met. Soc. AIME*, **236**, 930-936

BILONI, H. (1967): On apparent pore formation during cellular solidification, *Trans. Met. Soc. AIME*, **239**, 926-927

BINDER, K. and STAUFFER, D. (1976): Statistical theory of nucleation, condensation and coagulation, *Adv. in Physics*, **25**, 343-396

BIRD, R.B., STEWART, W.W. and LIGHTFOOT, E.N. (1960): *Transport Phenomena* (Wiley, New York 1960)

BISHOP, A.R. and SCHNEIDER, T. (1981): Solitons and condensed matter physics, **vol. 8** in *Solid State Sciences*, Springer Verlag, 1981.

BOLLING, G.F. and TILLER, W.A. (1960): Growth from the melt I: Influence of surface intersections in pure metals, *J.Appl.Phys.*, **31**, 1345-1350

BREMER, E.A. (1990): Effects of surface energy and kinetics on the growth of needle-like dendrites, *J. Cryst. Growth*, **99**, 165-170

BRUSH, L.N. and SEKERKA, R.F. (1989): A numerical study of two-dimensional growth forms in the presence of anisotropic growth kinetics, *J. Cryst. Growth*, **96**, 419-441

BURTON, W.K., CABRERA, N. and FRANK, F.C. (1951): The growth of crystals and the equilibrium structure of their surfaces, *Phil. Trans. Roy. Soc.*, **243**, 299-358

CABRERA, N. and VERMILYEA, D.A. (1958): The growth of crystals from solution, in: *Growth and perfection of crystals* (John Wiley, New York 1958) 393-410

CAHN, J.W. and HILLIARD, J.E. (1958): Free energy of a nonuniform system. I Interfacial free energy, *J. Chem. Phys.*, **28**, 258-267

CAHN, J.W. and HOFFMANN, D.W. (1972): A vector thermodynamics for anisotropic surfaces. Part I: Fundamentals and application to plane surface junctions, *Surface Science*, **31**, 368-388; Part II: Curved and faceted surfaces, *Acta Met.*, **22**, 1205-1214

CAHN, J.W. (1960): Theory of crystal growth and interface motion in crystalline materials, *Acta Met.* **8**, 554-562

CAHN, J.W., HILLIG, W.B., and SEARS, G.W. (1964): The molecular mechanism of solidification, *Acta Met.* **12**, 1421-1439

CAHN, J.W., TAYLOR, J.E. and HANDWERKER, C.A. (1991): in: Sir Charles Frank, OBE, FRS: *An Eightieth Birthday Tribute*, Eds. R.G. Chambers c.s. (Hilger, Bristol 1991; p 88)

CAROLI, B, CAROLI, C and ROULET, B.(1992): Instabilities in planar solidification fronts. Chapter 2 in *"Solids far from equilibrium"*, ed. C. Godrèche, pp 155-296 (Cambridge University Press 1992)

CARPAY, F.M.A. and VAN DEN BOOMGAARD, J. (1971): The relationship between interlamellar period and growth rate in the unidirectionally decomposed eutectoids Co_3Si and Ni-In, *Acta Met.,* **19**, 1279-1286

CARPAY, F.M.A. and CENSE, W.A. (1974): In situ growth of composites from the vitreous state, *J. Crystal Growth,* **24/25**, 551-554.

CHERNOV, A.A. (1961): The spiral growth of crystals, *Sov. Phys. Uspekhi* **4**, 116-148

CHERNOV, A.A. (1962): The kinetics of the growth forms of crystals, *Sov. Physics Crystallography* **7**, 728-730

CHERNOV, A.A. (1963a): Crystal growth forms and their kinetic stability, *Sov. Phys. Crystallography,* **8**, 63 - 67

CHERNOV, A.A. (1963b): Application of the method of characteristics to the theory of the growth forms of crystals, *Sov. Phys. Crystallography,* **8**, 401-405

CHERNOV, A.A. (1979): Stability of planar growth front for anisotropic surface kinetics, in *Growth of Crystals,* vol. 11, Ed. A. Chernov, Consultants Bureau, New York (English translation)

CHERNOV, A.A. , and NISHINAGA, T (1987): Growth shapes and their stability at anisotropic interface kinetics: theoretical aspects for solution growth, Chapter 9 in *Morphology of Crystals, part A* (Ed. I. Sunagawa, Terra Scientific Publishing Company 1987)

CLINE, H.E. (1971): Shape instabilities of eutectic composites at elevated temperatures, *Acta Met.,* **19**, 481-490

CORIELL, S.R., HARDY, S.C. and SEKERKA, R.F. (1971): A nonlinear analysis of experiments on the morphological stability of ice cylinders freezing from aqueous solutions, *J. Cryst. Growth* **11**, 53-67

CORIELL, S.R. and SEKERKA, R.F. (1972): Morphological stability near a grain-boundary groove in a solid-liquid interface during solidification of a pure substance, *J. Cryst. Growth* **19**, 90-104.

CORIELL, S.R. and SEKERKA, R.F. (1976): The effect of the anisotropy of surface tension and interface kinetics on morphological stability, *J. Cryst. Growth,* **34**, 157-163

CURIE, P. (1885): Formation of crystals and capillarity constants of their different faces, *Bull. Soc. Min. France,* **8**, 145 (also in Oeuvres, Gauthier-Villars, Paris 1908, pg 153)

CURRIE, J.F. , FOGEL, M.B. and PALMER, F.L. (1977): Thermodynamics of the sine-Gordon field, *Phys, Rev. A,* **16**, 796-798

CURRIE, J.F., KRUMHANSL, J.A., BISHOP, A.R. and TRULLINGER, S.E. (1980): Statistical mechanics of one-dimensional solitary-wave bearing scalar fields: exact results and ideal-gas phenomenolgy, *Phys. Rev. B,* **22**, 477-496

CROKER, M.N., BARAGAR, D. and SMITH, R.W. (1975): Anomalous eutectic growth, *J. Cryst. Growth,* **30**, 198-212

CROSS, M.C. and HOHENBERG, P.C. (1993): Pattern formation outside of equilibrium, *Rev. Mod. Physics,* **65**, 851-1090

DE GROOT, S.R. and MAZUR, P. (1969): *Non-equilibrium Thermodynamics.* North-Holland Publishin Cy, Amsterdam 1969

DOUGHERTY, A. (1990): Surface tension anisotropy and the dendritic growth of pivalic acid. *J. Cryst. Growth,* **110**, 501-508

EILENBERGER, G (1981): *Solitons, mathematical methods for physicists.* Springer Verlag 1981.

FRANK, F.C. (1951): Capillary equilibria of dislocated crystals, *Acta Cryst.,* **4**, 497-501

FRANK, F.C. (1958): On the kinematic theory of crystal growth and dissolution processes, in: *Growth and perfection of crystals* (John Wiley, New York 1958) 411-419

FRANK, F.C. (1963): The geometrical thermodynamics of surfaces, in *Metal Surfaces,* American Society for Metals (Metals Park, Ohio) pp 1-15

FRIEDEL, J. (1964): *Dislocations* , Pergamon Press, London 1964

GARDENIERS, J.G.E., MOOREN, M.M.W., DE CROON, M.H.J.M and GILING, L.J (1990): The influence of the chlorine-hydrogen ratio in the gas phase on the stability of the {113} faces of silicon in Si-H-Cl CVD, *J. Cryst. Growth,* **102**, 233-244

GARDENIERS, J.G.E, KLEIN DOUWEL, C.H. and GILING, L.J (1991): Reduced pressure silicon CVD on hemispherical substrates, *J. Cryst. Growth,* **108**, 319-334

GIBBS, G.W. (1876): On the equilibrium of heterogeneous substances, *Collected Works,* vol.I, (reprinted) Longmans, Green an Co (New York 1928)

GILING, L.J. and VAN ENCKEVORT, W.P.J (1985): On the influence of surface reconstruction on crystal growth processes, *J. Cryst. Growth,* **161**, 567-583

GILLET, M. (1977): Structure of small metallic particles, *Surface Science,* **67**, 139-157

GLICKSMAN, M.E. and VOLD, C.L. (1969): Determination of absolute solid-liquid interfacial free energies in metals, *Acta Met.*, **17**, 1-11

GLICKSMAN, M.E. and SINGH, N.B. (1989): Effects of crystal-melt interfacial energy anisotropy on dendritic morphology and growth kinetics, *J. Cryst. Growth,* **100**, 277-284

GURTIN, M.E. (1988): Multiphase thermomechanics with interfacial structure; I: Heat conduction and the capillary balance law, *Archive for Rat. Mechanics and Analysis,* **104**, 195-221

GUPTA, N. and SUTHERLAND, B. (1976): Investigation of a class of one-dimensional non-linear fields, *Phys. Rev. A,* **14**, 1790-1801

GUTZOW, I. (1976): Mechanism of crystallisation in undercooled melts and glass-forming systems, in: *Crystal growth and materials,* edited by E. Kaldis and H.J. Scheel, (North-Holland 1977) 379-414

GUYER, R.A. and MILLER, M.D. (1978): The sine-Gordon chain II: nonequilibrium statistical mechanics, *Phys. Rev. A,* **17**, 1774-1791

GUYER, R.A. (1979), The sine-Gordon chain as a model for a two-dimensional interface, *Phys. Rev. A,* **16**, 796-798

HAKEN, H (1977): *Synergetics, an introduction.* Springer Verlag, 1977

HALE, D.K. (1976): The physical properties of composite materials, *J. Mat. Science,* **11**, 2105-2141

HARTMAN, P. and PERDOK, W.G. (1955a): On the relation between structure and morphology I, *Acta Cryst.*, **8**, 49-52

HARTMAN, P. and PERDOK, W.G. (1955b): On the relation between structure and morphology II, *Acta Cryst.*, **8**, 521-524

HARTMAN, P. and PERDOK, W.G. (1955c): On the relation between structure and morphology III, *Acta Cryst.*, **8**, 525-529

HERRING, C. (1950): Effect of change of scale on sintering phenomena, *J. Appl. Phys.*, **21**, 301-303

HERRING, C (1951): Some theorems on the free energy of crystal surfaces, *Phys. Rev.*, **82**, 87-93

HERRING, C (1952): The use of classical macroscopic concepts in surface-energy problems, in: *Structure and properties of solid surfaces*, Ed. R. Gomer and C.S. Smith (Univ. of Chicago Press 1952), 5-81

HILLIG, W.B. (1966): A derivation of classical two-dimensional nucleation kinetics and associated crystal growth laws, *Acta Met.* **14**, 1868-1869

HOFFMANN, D.W. and CAHN, J.W. (1972): A vector thermodynamics for anisotropic surfaces. Part I: Fundamentals and application to plane surface junctions, *Surface Science,* **31**, 368-388; Part II: Curved and faceted surfaces, *Acta Met.*, **22**, 1205-1214

HOLLAN, L. and SCHILLER. C. (1972): Etude de l'anisotropie de la croissance epitaxiale de GaAs en phase vapeur, *J. Cryst. Growth* **13/14**, 319.

HONJO, G., TAKAYANAGI, K., KOHAYASHI, K. and YAGI, K. (1977): Ultra-high vacuum in-situ electron microscopy of growth processes of epitaxial thin films. *J. Crystal Growth* **42**, 98-109

HONJO, G. (1976): Misfit dislocations in epitaxial thin film growth, *Thin Solid Films* **32**, 143 - 150

HUANG, S.C and GLICKSMAN, M.E. (1981): Fundamentals of dendritic solidification, I + II, *Acta Met.* **29**, 701-715 (I) and 717-734 (II)

HULL, D (1965): *Introduction to dislocations.* Pergamon Press, London 1965

HUNT, J.D. and CHILTON, J.P. (1963): An investigation of the lamellar-rod transition in binary eutectics, *J. Inst. of Metals,* **91**, 338-342.

HUNT, J.D. and JACKSON, K.A. (1966): Binary eutectic solidification, *Trans. Met. Soc. AIME*, **236**, 843-852

INFELD, I. and ROWLANDS, G. (1990): *Nonlinear waves, solitons and chaos*, Cambridge University Press 1990

JACKSON, K.A. and HUNT, J.D. (1965): Compounds that freeze as metals, *Acta Met.* **13**, 1212

JACKSON, K.A. and HUNT, J.D. (1966): Lamellar and rod eutectic growth, *Trans. Met. Soc. AIME*, **236**, 1129-1142

JACKSON, K.A., UHLMANN, D.R. and HUNT, J.D. (1967): On the nature of crystal growth from the melt, *J. Cryst. Growth*, **1**. 1-36

JOHNSON, C.A. (1965): Generalisation of the Gibbs-Thomson equation, *Surface Science*, 3, 429-444

KARNOPP, D. and ROSENBERG, R (1968): *Analysis and Simulation of Multiport Systems*, Cambridge, Mass, MIT Press 1968.

KERN, R. (1987): The equilibrium form of a crystal; chapter 2 in *Morphology of Crystals, part A* (Ed. I. Sunagawa, Terra Scientific Publishing Company 1987

LANDMAN, U., c.s. (1986): Faceting at the Si (100) crystal-melt interface: theory and experiment, *Phys. Rev. Letters* **56**, 155-158

LANGER, J.S. and MÜLLER-KRUMBHAAR, H. (1978): Theory of dendritic growth, *Acta Met.*, **26**, 1681-1708

LANGER, J.S. (1980): Instabilities and pattern formation in crystal growth, *Rev. Mod. Phys.*, **52**. 1-28

LARCHE, F. and CAHN, J.W. (1973): A linear theory of thermochemical equilibrium of solids under stess, *Acta Met.*, **21**, 1051-1063

LEAMY, H.J., GILMER, G.H. and JACKSON, K.A. (1975): Statistical Thermodynamics of clean surfaces, in *Surface Physics of Materials, vol. I*, 121-188 (Academic Press, New York 1975)

LI, J.C.M., ORIANI, R.A. and DARKEN, L.S. (1966): The thermodynamics of stressed solids, *Zeitschr. Phys. Chemie* (Neue Folge), **49**, 271-290

LUNDAGER MADSEN, H.E (1992): Affinity and entropy production in electrolyte crystallisation, *J. Cryst. Growth* **118**, 71-77

MORRIS, L.R. and WINEGARD, W.C. (1969): The development of cells during the solidification of a dilute Pb-Sn alloy, *J. Cryst. Growth*, **5**. 371-375

MÜLLER-KRUMBHAAR, H., BURKHARDT, T.W. and KROLL, D.M. (1977): A generalized kinetic equation for crystal growth, *J. Crystal Growth*, **38**, 13-22

MÜLLER-KRUMBHAAR, H. (1987): Theory of dendritic crystal growth, Ch. 9 in *Morphology of Crystals, part B*. Ed. I. Sunagawa, Terra Scientific Publ. Company / D. Reidel Publ. Company, 1987. (pp 615-643)

MULLINS, W.W. (1959): Flattening of a nearly plane solid surface due to capillarity, *J. Appl. Phys.*, **30**, 77-83

MULLINS, W.W. and SEKERKA, R.F.(1963): Morphological stability of a particle growing by diffusion or heat flow, *J. Appl. Phys.*, **34**, 323-329

MULLINS, W.W. and SEKERKA, R.F. (1964): Stability of a planar interface during solidification of a dilute binary alloy, *J. Appl. Phys.*, **35**, 444-451

NAGAMUNE, Y., TSUKAMOTO, S. and ARAKAWA, Y. (1993): Growth process and mechanism of nanometer-scale GaAs dot-structures using MOCVD selective growth, *J. Cryst. Growth*, **126**, 707-717

NASH, G.E. and GLICKSMAN, M.E. (1971), A general method for determining solid-liquid interfacial free energies, *Phil. Mag.*, **24**, 577-592

NASH, G.E. and GLICKSMAN, M.E. (1974), Capillarity-limited steady-state dendritic growth, I+II, *Acta Met.*, **22**, 1283-1299

NASH, G.E. (1976): A self-consistent theory of steady-state lamellar solidification in binary eutectic systems. *NRL Report* 7956, Naval Res. Labs. 1979.

NICOLIS, G and PRIGOGINE, I (1977): *Self-organisation in nonequilibrium systems*, J. Wiley, New York 1977.

NIELSEN, A.E. (1964): Kinetics of Precipitation, *Int. Series of Monographs on Analytical Chemistry* **18**, (Pergamon Press, New York 1964)

NISHIDA, T., SUGIURA, H., NOTOMI, M. and TAMAMURA, T. (1993): InGaAs/InP quantum wires selectively grown by chemical beam epitaxy, *J. Cryst. Growth,* **132**, 91-98

NOZIERES, P (1992): Shape and growth of crystals.; Chapter 1 in *"Solids far from equilibrium"*, ed. C. Godrèche, pp 1-145 (Cambridge University Press 1992)

NOZIERES, P and WOLF, D.E. (1988): Interfacial Properties of Elastically Strained Materials: I Thermodynamics of a planar interface; II Mechanical and melting equilibrium of a curved interface, *Z. Phys. B - Condensed Matter,* **70**, 399-407 and 507-513

OHARA, M and REID, R.C. (1973): *Modeling crystal growth from solution,* Prentice-Hall 1973

ONSAGER, L. and MACHLUP, S. (1953): Fluctuations and irreversible processes, *Phys. Rev.* **91**, 1505-1515

OSTER, G.F., PERELSON, A and KATCHALSKY, A (1973): Network Thermodynamics, *Quarterly Rev. of Biophysics* **6**, 1-134

OSTER, G.F. and PERELSON, A (1974): Chemical Reaction Networks, *IEEE Transactions on circuits and systems* **CAS-21**, 709-721

PARKER, R.L. (1970): Crystal Growth Mechanisms: Energetics, Kinetics, Transport, in: *Solid State Physics* **25**, ed. H. Ehrenreich c.s. (Academic Press, New York 1970), 151-299.

PELCE, P.(1988): *Dynamics of Curved Fronts,* Academic, Boston, 1988

PFEIFFER, L., PAINE, S., GILMER, G.H., VAN SAARLOOS, W. and WEST, K.W (1985): Pattern formation resulting from faxceted growth in zone-melted thin films, *Phys. Rev. Letters* **54**, 1944-1947-

POMEAU, Y and BEN AMAR, M (1992): Dendritic growth and related topics; Chapter 4 in *"Solids far from equilibrium"*, ed. C. Godrèche, pp 365-432 (Cambridge University Press 1992)

RASMUSSEN, H. (1982): Energetics of homogeneous nucleation - approach to a physical spinodal, *J. Cryst. Growth,* **56**, 45-55

ROSENBERGER, F (1979): *Fundamentals of Crystal Growth* (Springer Verlag 1979)

RUKWIED, A., RUFF, A.W. and WILLARD, W.A. (1971): Study of the cellular solidification structure in a continuously cast high purity copper, *Metallurgical Transactions,* **2**, 2105-2114

RUTTER, J.W. and CHALMERS, B. (1953): A prismatic substructure formed during solidification of metals, *Can. J. of Physics,* **31**, 15-39

SAITO, Y., GOLDBECK-WOOD, G. and MüLLER-KRUMBHAAR, H. (1987): Dendritic crystallisation: numerical study of the one-sided model, *Phys. Rev. Letters* 58, 1541-1543

SCHAEFER, R.J., GLICKSMAN, M.E. and AYERS, J.D. (1975): High-confidence measurements of solid-liquid surface energy in a pure material, *Phil. Mag.*, **32**, 725-743

SCHWOEBEL, R.L. and SHIPSEY, E.J. (1966): Step motion on crystal surfaces, *J.A.P.* **37**, 3682-3686

SCHWOEBEL, R.L. (1969): Step motion on crystal surfaces II, *J.A.P.* **40**, 614-618

SCOTT, A.C., CHU, F.Y.F. and MCLAUGHLIN, D.W. (1973): The soliton: a new concept in applied science, *Proceedings IEEE*, **61**, 1443-1481

SCRIVEN, L.E. and STERNLING, C.V. (1964): On cellular convection driven by surface-tension gradients: effects of mean surface tension and surface viscosity, *J. Fluid Mech,* **19**, 321-340

SEKERKA, R.F. (1971) : A Green's function to describe the time-dependent shape of a morphologically unstable solid/liquid interface during solidification. *J. Cryst. Growth* **10**, 239-250

SEKERKA, R.F. (1973): Morphological Stability, in *Crystal Growth, an introduction* , Ed. P. Hartman (North-Holland, 1973) pp 403-443

SETHIAN, J.A. and STRAIN, J. (1992): Crystal Growth and Dendritic Solidification, *J. Comput. Phys.,* **98**, 231-253

SHAW, D. (1979), Morphology analysis in localized crystal growth and dissolution, *J. Cryst. Growth,* **47**, 509-517

SMALL, M.B., BACHEM, K.H. and POTEMSKI, R.M. (1977): An explanation for the phenomenon of meniscus lines on the surfaces of GaAlAs alloys grown by LPE, *J. Cryst. Growth,* **39**, 216-222

SMITH, C.S. (1948): Grains, Phases and Interfaces: an interpretation of microstructure, *Trans. AIME* , **175**, 15-51

SOHNEL, O and GARSIDE, J (1979): The thermodynamic driving force for crystallisation from solution, *J. Cryst. Growth,* **46**, 238-240

TARABAEV, L.P., MASHIKHIN, A.YU. and ESIN, V.O. (1991): Dendritic crystal growth in supercooled melt, *J. Cryst. Growth,* **114**, 603-612

TAYLOR, J.E., CAHN, J.W. and HANDWERKER, C.A. (1992): Geometric Models of Crystal Growth, *Acta Met. Mater.*, **40**, 1443-1474

TAYLOR, J.E. (1992): Mean curvature and weighted mean curvature, *Acta Met. Mater.*, **40**, 1475-1485

TEMKIN, D.E. (1964): Molecular roughness of the crystal-melt boundary, in "*Crystallisation Processes*", Consultants Bureau, New York 1966, 15-23

TEMKIN, D.E. (1972): Analogy between the stability of the crystallisation front and the stability of a metastable phase, *Sov. Phys. Crystallography,* **17**, 982-984

THOM, R (1972): *Stabilité Structurelle et Morphogenèse*, Benjamin, Boston 1972.

TOLKSDORF, W. and WELZ, F. (1978): Crystal growth of magnetic garnets from high-temperature solutions, in *Crystals: Growth, properties and applications* vol. I, pg. 1-52 (Springer Verlag 1978)

VAN DEN BREKEL, C.H.J. and JANSEN, A.K. (1978): Morphological stability analysis in CVD processes I and II, *J. Cryst. Growth,* **43**, 364-377

VAN DER EERDEN, J.P. (1976): Roughening transition in mean-field and pair approximation of Ising models, *Phys. Rev. B,* **13**, 4942-4948

VAN DER HOEK, B., VAN DER EERDEN, J.P. and BENNEMA, P. (1982): The influence of stress on spiral growth, *J. Cryst. Growth,* **58**, 365-380

VAN DER HOEK, B., VAN DER EERDEN, J.P. and BENNEMA, P. (1982): Thermodynamical stability conditions for the occurrence of hollow cores caused by stress of line and planar defects, *J. Cryst. Growth,* **56**, 621-632

VAN LEEUWEN, C (1979): On the driving force for crystallisation: the growth affinity, *J. Cryst. Growth,* **46**, 91-95

VAN SUCHTELEN, J. (1978): Coarsening of eutectic structures during and after unidirectional growth, *J. Crystal Growth,* **43**, 28-46

WILCOX, W.R. (1970): The influence of a temperature gradient on crystal faceting, *J. Cryst. Growth,* **7**, 203-208

WILCOX. W.R. (1976): A generalized treatment of mass transfer in crystal growth, in: *Preparation and Properties of Solid State Materials*, vol. 2, ed. W.R. Wilcox. (Marcel Dekker, New York 1926), p. 129.

WOLF, K.L (1959): *Physik und Chemie der Grenzflächen II*, Springer Verlag 1959

WOLLKIND, D. and SEGEL, L. (1970): A nonlinear stability analysis of the freezing of a dilute binary alloy, *Phil. Trans. Roy. Soc. (London),* **268**, 351-380

WULFF, G. (1901): Zur Frage der Geschwindigkeit des Wachstums und der Auflösung der Kristallflächen. *Z. Krist.,* **34**, pp 449-530

XIAO, R.F., ALEXANDER, J.I.D. and ROSENBERGER, F. (1990): Growth morphology with anisotropic surface kinetics, *J. Cryst. Growth,* **100**, 313-329

XU, Jian-Jun (1990): Global neutral stability of dendritic growth and selection condition of tip stability, *J. Cryst. Growth,* **100**, 481-490

YAGI, K., TAKAYANAGI, K., KOBAYASHI, K., OSAKABE, N., TANISHIRO, Y. and HONJO, G. (1979): Surface study by an UHV electron microscope, *Surface Science,* **86**, 174-181

YOKOYAMA, E. and SEKERKA, R.F. (1992): A numerical study of the combined effect of anisotropic surface tension and interface kinetics on pattern formation during the growth of two-dimensional crystals, *J. Cryst. Growth,* **125**, 389-403

List of symbols, notations and abbreviations

Given the extent of subjects, we could not avoid to use many characters as a symbol for more than one physical quantity, constant or index, in different contexts. In all cases, such multiple use is explained locally, if not obvious from the text and from "convention", such as the use of symbols ε or δ for small fractions; C or K for (local) constants; F, G, H, for (unspecified) functions, etc. In some cases, we have used different fonts to distinguish between symbols.

Loci where symbols are introduced or defined are indicated between brackets.

$\mathcal{A}$	assembled function ⟨10.5⟩
A	affinity ⟨1.5⟩
A_g	growth affinity (as a driving force for crystal growth) ⟨1.5⟩
ΔA_c	shift in growth affinity induced by curvature via capillarity ⟨1.5⟩
$A_{\mid}$	growth affinity at the interface, corresponding with *planar* interface ⟨1.5⟩
A_{total}	Affinity corresponding with boundary conditions imposed at boundaries of the system, not coinciding with the moving interface. ⟨1.5⟩
A_p	asymmetry term ⟨2.8⟩
a	area per PBC in cross-section perpendicular to PBC
a	assembling exponent ⟨10.1⟩
$\bar{\text{a}}$	assembling exponent <0 ⟨10.1⟩
$\underline{\text{a}}$	assembling exponent >0 ⟨10.1⟩
C	concentration
$C_\text{x}(\Delta\text{x})$	contributive function ⟨2.8⟩
$C(\vec{n})$	PBC-counting function ⟨10.1⟩
D	diffusion constant
d	thickness (of epitaxial layer; of boundary layer)
D	"distorting operator" ⟨10.2⟩
$F(\vec{n})$	orientation distribution function of an interface ⟨11.1⟩
ℓ_Ψ	correlation length of physical quantity Ψ ⟨2.8⟩
L	limiting constant for growth rate of specific face ("facet family") ⟨11.4⟩
L^*	extra "close-fitting assembland" ⟨10.4.4⟩
M	overall growth rate maximum imposed by orientation-independent limiting mechanism ⟨10.3⟩
m	general scalar field variable acting on R or γ ("wildcard" parameter for T, C etc, but other than A_g).
P_g	growth pressure ⟨3.1⟩
p	pressure
Δp_m	"Marangoni" pressure shift ⟨3.3⟩
Δp_c	curvature-induced pressure shift ⟨3.3⟩
R	(scalar) growth rate of a surface (measured in perpendicular direction).
$r\ (r_1,r_2)$	curvature radius (two principal radii)
S	"stiffness operator" for scalar quantity depending on only one orientational variable: $\mathsf{S} = (1 + d^2/d\varphi^2)$ ⟨2.4⟩
$\mathfrak{S}$	"spreading rate" = rate of increase of curvature radius along kwt, for one orientational variable ⟨1.6⟩

T	temperature
T_R	temperature corresponding with the roughening transition
$T_i(\vec{n},a)$	"territory" function ⟨10.3.3⟩
V	molar volume (when unspecified, $\equiv V_c$; otherwise V_c, V_p in the crystal and parent phase, resp.)
v	velocity
$v_{//}$	kink velocity parallel with ledge (2D) or or ledge velocity parallel with facet (3D)
$v_{\perp}$	ledge velocity (2D)
$v_{B\&S}$	growth rate due to birth-and-spread process alone ⟨11.1⟩

Greek symbols

α	Jackson's "α factor"
α, α_i	angular variable for angle between normal $\vec{n}$ and PBC vector $\vec{p}$, $\vec{p}_i$ or facet normal $\vec{f}$, $\vec{f}_i$
α,β,γ	suffices to indicate different grains (different in crystallographic orientation, eventually also different phases)
γ	interfacial free energy
δ	1) Dirac delta function (in eq. (2.7))
δ	2) thickness of diffusion boundary layer ⟨2.8 and 8.2⟩
ε	roughening parameter ; range 0 - 1 ⟨10.3⟩
ε'	roughening parameter ; range 0 - ∞ ⟨10.3⟩
ε_{ijkl}	permutation operator ⟨4.1⟩
ϑ, φ	angular variables for polar coordinates
φ	orientation parameter (angle)
η	viscosity
λ, Λ	period of periodic structures; wavelength
μ	chemical potential
ρ_0	"radius" or average radius of critical nucleus
σ	line energy (of ledge or of dislocation)
τ	golden section ratio = 1.618034 . . .
Φ	range of approached equilibrium ⟨8.3⟩
ϕ	angular variable for microscopic scatter in local orientation ⟨11.1..3⟩
χ	surface mobility ⟨eq. (1.7)⟩
ψ	orientation parameter (angle)
Ψ	unspecified physical parameter defined on the interface.

Vectors and tensors

$\vec{d}$	unit vector along dislocation acting as a velocity source
$\vec{e}_1$, $\vec{e}_2$, $\vec{e}_3$	unit vectors in cartesian coordinate frame
$\vec{f}$, $\vec{f}_{ij}$	unit vector in facet-normal direction (eventually with suffix ij: facet defined by PBC's $\vec{p}_i$ and $\vec{p}_j$)
$\vec{F}$	force vector
$\vec{\vec{I}}$	identity tensor
$\vec{j}$	flux density
$\vec{\vec{K}}$	second fundamental tensor of a surface

$\vec{\ell}$	unit vector along dislocation, edge or 3-junction
$\vec{n}$	orientation parameter of a surface = perpendicular unit vector.
$\vec{\vec{O}}$	zero tensor
$\vec{p}$	unit vector in PBC direction
$\vec{\mathsf{P}}$	"plotting" vector operator = "multiplication" with normal vector $\vec{n}$ ⟨2.3⟩
$\vec{\mathsf{Q}}$	"Gibbs-Wulff" vector operator = $\vec{n}$ + d/d$\vec{n}$ ⟨2.3⟩
$\vec{R}$	vector velocity of growth rate function which is only dependent on $\vec{n}$; (= $\vec{\mathsf{Q}}R$)
$\vec{\mathfrak{R}}$	vector velocity, general ⟨1.6⟩
$\vec{\mathfrak{R}}^*$	stabilized vector velocity of a "velocity source" ⟨ 4.5.4 and 4.6⟩
$\vec{\vec{\mathsf{S}}}$	"stiffness" tensor operator = d$\vec{\mathsf{Q}}$/d$\vec{n}$ ⟨2.4⟩
$\vec{\vec{\mathfrak{S}}}$	3-D "spreading rate" = $(\mathrm{d}/\mathrm{d}t)_{\mathrm{kwt}}\ \vec{\vec{K}}^{-1}$ ⟨1.6⟩
$\vec{s}$	curvilinear surface or line parameter
$\vec{\mathsf{T}}$	≡ (d/d$\vec{n}$) ln ⟨13.3.2⟩
$\vec{\xi}$	≡ $\vec{\mathsf{Q}}R(\vec{n})$ in publications by CAHN and HOFFMANN (1972)
$\vec{\varphi}$	normal vector in single-variable $R(\varphi)$ plots; orientation angle φ
$\vec{\nabla}$	gradient
$\vec{\nabla}_s$	interface-tangential component of gradient
$\vec{\nabla}_n$	interface-normal component of gradient
$\vec{Z}$	≡ Σ $\vec{\mathsf{Q}}\gamma_i$ for the three interfaces in a three-junction ⟨4.4⟩
$\vec{\omega}$	tilt rate of moving interface, measured in normal direction

Species vectors and tensors ⟨8.5⟩

In the description of polynary transport, a set of n species-related quantities Q_k can be represented as a "species vector" $\tilde{Q}$. Linear relations between species vectors can be represented as a tensor multiplication, e.g. $\tilde{J} = \tilde{\tilde{L}} \cdot \tilde{X}$. Nonlinear relations between species vectors are more generally represented as "vector maps" $\tilde{J} = \tilde{\tilde{\mathcal{L}}}(\tilde{X})$.

$\tilde{A}$	affinity species vector
$\tilde{C}$	(fractional) concentration, Σ $C_i = 1$
$\tilde{J}$	species flux (vector)
$\tilde{\tilde{L}}$	matrix of constitutive coefficients relating $\tilde{J}$ with $\tilde{X}$
$\tilde{\tilde{\mathcal{L}}}$	"vector map" representing general constitutive relation between $\tilde{J}$ and $\tilde{A}$
$\tilde{\tilde{\ell}}$	matrix relating $\tilde{J}$ with $\tilde{A}$
$\tilde{\tilde{M}}$	matrix of constitutive coefficients relating $\tilde{\mu}$ with $\tilde{C}$
$\tilde{\tilde{\mathcal{M}}}$	"vector map" representing general constitutive relation between $\tilde{\mu}$ and $\tilde{C}$
$\tilde{\mu}$	chemical-potential species vector
$\tilde{P}$	partial pressure species vector
$\tilde{X}$	"thermodynamic force" : $X_k = A_k/T$; $X_{th} = 1/T$

Notation of vector and tensor products

In products involving vectors and/or tensors we use the convention that products written *without* a dot signify an exterior (or ordered) product, whereas a dot signifies a *contracted* product, obtained from the exterior product by equalizing two neighbouring indices belonging to each of the tensors (vectors) respectively, and summing over the resulting dummy index. So, for instance:

$\vec{a}\vec{b}$ is a tensor with $(\vec{a}\vec{b})_{ij} = (\vec{a})_i(\vec{b})_j$,

$\vec{a}\vec{\vec{b}}$ is a tensor with $(\vec{a}\vec{\vec{b}})_{ijk} = (\vec{a})_i(\vec{\vec{b}})_{jk}$,

$\vec{a}\cdot\vec{b}$ $= \Sigma_i\, (\vec{a})_i(\vec{b})_i$,

$\vec{a}\cdot\vec{\vec{b}}$ is a vector with $(\vec{a}\cdot\vec{\vec{b}})_i = \Sigma_k\, (\vec{a})_k(\vec{\vec{b}})_{ki}$,

$\vec{\vec{a}}\cdot\vec{b}$ is a vector with $(\vec{\vec{a}}\cdot\vec{b})_i = \Sigma_k\, (\vec{\vec{a}})_{ik}(\vec{b})_k$,

$\vec{\vec{a}}\cdot\vec{\vec{b}}$ is a tensor with $(\vec{\vec{a}}\cdot\vec{\vec{b}})_{ik} = \Sigma_j\, (\vec{\vec{a}})_{ij}(\vec{\vec{b}})_{jk}$,

$\vec{\vec{a}}:\vec{\vec{b}}$ is a scalar $\Sigma_{i,j}\, (\vec{\vec{a}})_{ij}(\vec{\vec{b}})_{ji}$, ... etc.

The *vector* product of two vectors is written as

$\vec{a}\wedge\vec{b}$ is a vector with $(\vec{a}\wedge\vec{b})_i = (\vec{a})_k\,(\vec{b})_j \;-\; (\vec{a})_j\,(\vec{b})_k$, (cycl.)

Such expressions for vector/tensor products, being independent of the choice of a reference frame, are more compact than an explicit index-summation notation. (Of course, when written out in coordinates, the coordinates for both vector/tensor factors have to be based on the same reference frame!)

Gradients

In equations related with interfaces, we often refer to the *normal* and the *tangential* components of a gradient vector in 3-dim. space. Notation:

$\vec{\nabla}_n$ defined by $\vec{\nabla}_n m = \vec{\nabla} m \cdot \vec{n}$

$\vec{\nabla}_s$ defined by $\vec{\nabla}_s m = \vec{\nabla} m - \vec{\nabla} m \cdot \vec{n}$

d/d$\vec{n}$ *operator:*

A special "gradient" is the tangential gradient on the unit sphere, working on some orientation-dependent parameter $\Psi(\vec{n})$ (which can thus be considered as a "field" defined on this unit sphere). It is written as $d/d\vec{n}$. It is locally an only two-dimensional vector operator, as $\vec{n}$ cannot vary in length: the component of $d\vec{n}$ perpendicular to the surface of the unit sphere is zero as a consequence of the normalisation restriction. So, $d\Psi/d\vec{n}$ is a vector in the tangent plane at orientation $\vec{n}$. Written out in coordinates: when $\vec{e}_2$, $\vec{e}_3$ are the ortho-normal basis set in the tangent plane at $\vec{n}$ $(= \vec{e}_1)$ and φ ,ϑ are the corresponding variables for $\vec{n}$, so that $d\vec{n} = \vec{e}_2\, d\varphi + \vec{e}_3\, d\vartheta$, then $d\Psi(\vec{n})/d\vec{n}$ is a direct vector notation for $\vec{e}_2\, d\Psi/d\varphi + \vec{e}_3\, d\Psi/d\vartheta$.

ABBREVIATIONS:

BC	boundary condition(s)
BCF	Burton-Cabrera-Frank
BL	boundary layer
B&S	birth-and-spread
CE	connective element ⟨ch. 6⟩
CS	constitutional supercooling
CVD	chemical vapour deposition
DE	differential equation
EF	"end function" ⟨ch. 12⟩
EKC	"elementary kaleidoscopic cell" ⟨11.4.5⟩
EOM	equation of motion
FR	"fundamental restriction" ⟨1.3⟩
GT	Gibbs-Thomson
GW	Gibbs-Wulff
KWT	kinematic wave trajectory
LPE	liquid phase epitaxy
MBE	molecular-beam epitaxy
MF	multi-"family" ⟨11.4⟩
MOVPE	metal-organic vapour phase epitaxy
MSC	Mullins-Sekerka-Coriell
NL	nonlinear
NL PDE	nonlinear partial differential equation
OD	orientation-dependent
OID	orientation-independent
PBC	periodic bond chain
PDE	partial differential equation
PE	protruding edge
RE	re-entrant edge
RAE	range of approached equilibrium ⟨8.3⟩
REP	rate of entropy production
RGB	red/green/blue
SBC	"surface bond chain" ⟨10.1⟩
SC	single-cusp ⟨11.4⟩
SF	single-"family" ⟨11.4⟩
SOS	solid-on-solid
TAD	tangential adsorbate drag ⟨8.5.7⟩
TPRE	twin-plane reentrant edge
US	unidirectional solidification
VLS	vapour/liquid/solid
VPE	vapour phase epitaxy
D-j (D-junction)	intersection between dislocation line and interface
3-j (3-junction)	intersection line between three interfaces
4-j (4-junction)	intersection point between four 3-junctions (where four phases or grains meet in a point)

GPSR Compliance

The European Union's (EU) General Product Safety Regulation (GPSR) is a set of rules that requires consumer products to be safe and our obligations to ensure this.

If you have any concerns about our products, you can contact us on ProductSafety@springernature.com

In case Publisher is established outside the EU, the EU authorized representative is:

Springer Nature Customer Service Center GmbH
Europaplatz 3
69115 Heidelberg, Germany

Batch number: 10816038

Printed by Printforce, the Netherlands